Volker Blüm

Vertebrate Reproduction

A Textbook

Translated from the German Edition
by A.C. Whittle

With 109 Figures

Springer-Verlag
Berlin Heidelberg New York Tokyo

Professor Dr. VOLKER BLÜM
Arbeitsgruppe für vergleichende Endokrinologie
Abteilung für Biologie
Ruhr-Universität
Universitätsstr. 150 ND 5/29
4630 Bochum 1, FRG

Translated from the German edition
Volker Blüm, Vergleichende Reproduktionsbiologie der Wirbeltiere
published 1985 by Springer-Verlag Berlin Heidelberg New York Tokyo

ISBN 3-540-16314-X Springer-Verlag Berlin Heidelberg New York Tokyo
ISBN 0-387-16314-X Springer-Verlag New York Heidelberg Berlin Tokyo

Library of Congress Cataloging-in-Publication Data. Blüm, Volker, 1937- . Vertebrate reproduction. Translation of: Vergleichende Reproduktionsbiologie der Wirbeltiere. Bibliography: p. Includes index. 1. Vertebrates–Reproduction. I. Title. QP251.B6613 1986 596'.016 86-1833.

Printed in Germany

Printing and bookbinding: Beltz, Offsetdruck, Hemsbach/Bergstr.
2131/3130-543210

Dedicated to my wife

Contents

Preface

Oral birth in the Australian gastric brooding frog
Rheobatrachus silus

I have deliberately chosen the title picture to the preface, which shows the oral birth of a young gastric brooding frog (*Rheobatrachus silus*), to draw the reader's attention to the preface since I wish to explain why the book was written and illustrated in the way it now appears.

The book has grown out of the material of a lecture course taking 3 to 4 hours per week each semester that I have given for several years at the Ruhr-University Bochum as the theoretical basis for an intensive 4 week practical course on vertebrate reproduction for students of biology.

For various reasons I am a hardened opponent of giving lecture notes to the students. On the other hand I found that my students made great efforts to work through and supplement their own lecture notes by studying the literature. However, there are no suitable textbooks on vertebrate reproduction written in German and the standard work in this field is now several decades old and deals mainly with mammals. The students were therefore often frustrated in their efforts given the limited time they had available. I have therefore decided to present here the most important structural and functional aspects of vertebrate reproduction and include other material to extend the framework of my original lectures.

This textbook is intended primarily for students of biology. Although they should ordinarily know about mitosis, meiosis and protein biosynthesis, such fundamental biological processes are nevertheless covered briefly in the introductory chapter. Since comparative endocrinology is not covered in any greath depth in most German universities I have devoted a relatively large amount of space to this subject. In general I have dealt in full with the major problem areas and other areas which seem to be either important or of interest, although in fitting these into the planned scope of the book I am aware that several deserving topics, population dynamics for example, are mentioned only briefly or not at all.

Though addressed to students of biology, I hope this book will also be of interest to those studying medicine and veterinary medicine, as well as to teachers of the higher grade pupils in secondary schools.

I regard diagrams, schemata and illustrations as essential learning aids and tend to use them to excess, but unfortunately they make the book more costly. I have therefore tried to include as many illustrations as possible without making the book too expensive for students. One way in which I have done this is by using line drawings that I have prepared myself. Some of them began as blackboard sketches made during lectures whereas others are based on material from the literature. In this I have endeavoured to maintain a certain continuity of style, and in many cases have used schemata to make the complicated relationships more easily understandable.

A further concession to reduce production costs is that I have labelled the illustrations with numbers, which admittedly must often be carefully sought for in the figure legend but this may have the benefit that the reader is forced to concentrate more closely upon the illustration and its explanation.

I have intentionally not given literature citations in the text. At the end of the book there is a list of selected literature that covers the material presented in each chapter and goes beyond it to some extent. In each case I have included the most important and relevant monographs and review articles and only in relatively few cases referred to original articles.

I have endeavoured not only to describe structure and function separately, but also to treat them in an integrated manner. This reflects my conviction that biology can only be understood in this way. This is of particular importance with respect to the complex events of reproduction, perhaps the most important aspect of life. Indeed I hope that this book can extend beyond the mere transmission of knowledge to enrich the reader's understanding of him or herself and of the place of human beings in nature.

Dr. A.C. Whittle, who translated the book in 1985, came to our institute from the University of Newcastle upon Tyne and has worked here in Bochum, where he now lives, as a Humboldt stipendiate and scientific collaborator for several years. I think that he understood my initial intentions and has skilfully combined his native language with my way of explaining things, as a comparison with the German edition shows. Moreover, he has eliminated some mistakes in the latter so that, in my opinion, the text has gained in quality through his work.

I wish to thank my family for their patience during the time I was writing the manuscript. My thanks go also to my research assistants, and in particular Dr. R. Schulz who relieved me of many official duties and corrected half of the original German text. Lastly, I am greatly indebted to Mrs. R. Oberstebrink-Scholl who typed the text and rectified the numerous slips of the pen I made while drawing the illustrations.

Bochum, April 1986 Volker Blüm

CHAPTER 1

Introduction

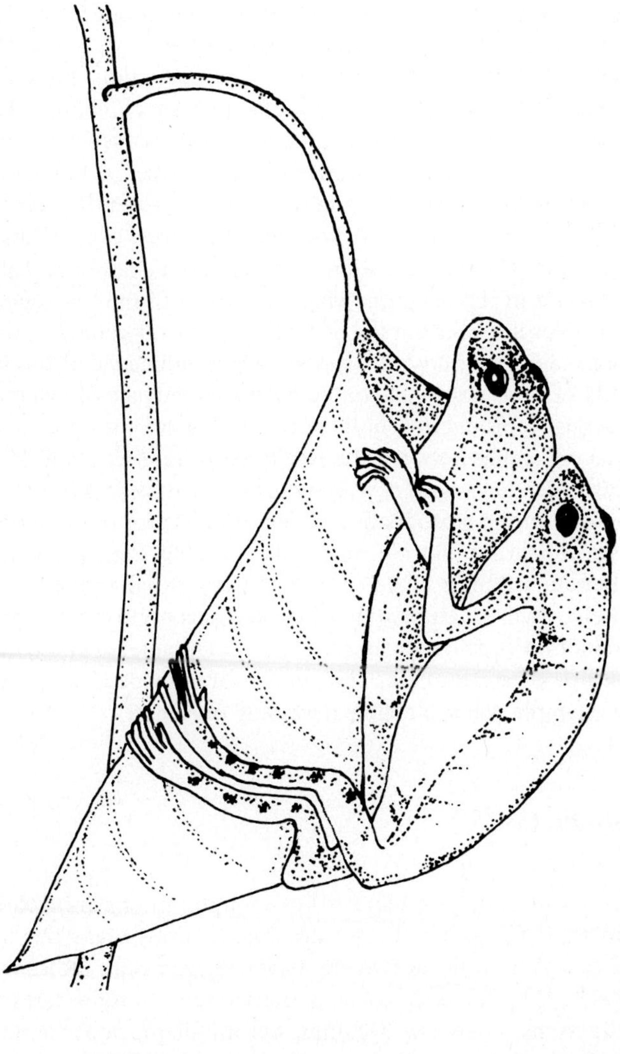

A pair of grasping frogs *(Phyllomedusa hypochondrialis)* spawning on a rolled-up leaf

1.1 Types of Reproduction

A characteristic of all forms of life is their ability to reproduce. The individuals of each species reproduce virtually identical copies of themselves and in doing so pass their hereditary material on to the next generation. While reproduction is biologically meaningless with respect to the individual, it is essential to the preservation of the species over many generations and can, therefore, be regarded as the most important concern of all living things.

Reproduction can occur in two fundamentally different ways: asexual and sexual. Asexual reproduction involves growth of the body tissues and produces new individuals that are genetically identical to those from which they stem. It is, therefore, also called vegetative reproduction. In the case of sexual reproduction, the growth of a new individual is preceded by the fusion of two cells, thereby combining the genetic material from two individuals so that the descendant represents a genetic reformation. This phenomenon, of two individual sets of hereditary material combining to form a genetically unique descendant, is called sexuality. Whereas vegetative reproduction can be accomplished by a single individual, two sexual partners are necessary in the case of sexual reproduction. The few exceptions to this rule which exist probably represent secondary adaptations of the organism which do not fulfil the "biological sense" of sexuality in that it does not result in the new reformation of genetic material that greatly increases the genetic variability of a species. The significance of this last point in the complex events of evolution is illustrated by the abundance of living things on earth. Asexual and sexual reproduction both serve to preserve the species, although there is a fundamental difference between the two types of reproduction. The former merely replaces dying individuals with new, genetically identical individuals, whereas the inclusion of sexuality in the reproductive cycle not only provides replacement individuals but results in offspring that are genetically variable. Sexual and asexual reproduction may alternate within a single species so that sexual and asexual generations occur successively. Such an alternation of generations occurs in a wide range of invertebrates. Reproduction is generally associated with an increase in the number of individuals but this is not always the case. The number of individuals in a species may even decrease as, for example, when a couple have only one child.

1.2 Asexual Reproduction

Asexual reproduction is rare among vertebrates but common among invertebrates. The simplest form of this type of reproduction is division. In protozoans this involves a simple mitotic division which produces two identical daughter cells. Metazoans usually produce new individuals by the body cleaving transversely. In some cases there is a rapid succession of divisions, a new one beginning before the previous division is complete so that a chain of animals is formed: certain Turbellaria and Polychaeta reproduce in this way. In the process of budding the surface of the body becomes swollen to form buds which develop into daughter animals. These may detach themselves or remain attached to the parent individual so as to form a colony. This type of reproduction is restricted to sessile or semi-sessile forms. The formation of a stolon represents a special form of budding and division. A proliferating projection, the stolo prolifer, extends from the body of the animal, and new individuals are formed from this by budding or by division. Another form of vegetative reproduction is the formation of so-called long-lived buds during periods when conditions are unfavourable. The only type of asexual reproduction which also occurs in vertebrates is polyembryony. This involves the disassociation of the embryo into identical multuplets which all derive from a single egg. Polyembryony is not uncommon in armadillos of the genus *Dasypus*, a single zygote giving rise to from 4 to 8 young animals of the same sex. The phenomenon of single-egg multuplets occasionally occurs in other mammals, including man.

1.3 Sexual Reproduction

1.3.1 Basic Genetics

The most essential aspect of sexual reproduction is that it produces new combinations of genetic material within a species. The material carriers of all the hereditable characteristics of an organism are the chromosomes. These are localised within the cell nucleus and their number (n) is constant in every individual of a species. Chromosomes are thread-like structures that can only be observed under certain conditions, which prevail during the division of the nucleus. They are composed of desoxyribonucleic acid (DNA), histones, protein and ribonucleic acid (RNA). Ordinarily the nuclei of the somatic cells of animals possess two sets of chromosomes which are almost identical, one from the mother and one from the father. These cells are diploid (2n). In the course of normal cell division, the two sets of chromosomes are duplicated and then distributed equally to each of the daughter cells. Before sexual reproduction can take place, the double set of chromosomes in the prospective daughter cells must be reduced to a single set, so that after the fusion of the gametes a diploid organism is again produced. The gametes are therefore haploid (1n). During the division of the nucleus the substance of the chromosomes is highly condensed and can, therefore, be visualised. At this stage the chromosomes have a length of between 0.2 and 20 μm. The primary site at which the pair of chromosomes are joined together is the centromere (or kinetochore). During division, the fibres of the spindle apparatus attach to the centromere and hold the long strands of the chromosome together after they have divided along their length at the start of normal cell division. The centromere has a characteristic position along each chromosome.

The carrier of the genetic information in animals is DNA. The DNA molecule possesses a chain of desoxyribose molecules (desoxyribose is a pentose sugar) linked together by phosphate residues. Each desoxyribose molecule bears one of five organic bases which derive from either purine or pyrimidine. The purine bases are adenine and guanine, the pyrimidine bases cytosine, uracil and thymine (Fig. 1A). Uracil very rarely occurs in DNA but is a characteristic component of RNA. The latter, whose sugar component is ribose, plays an important role in the translation of the genetic

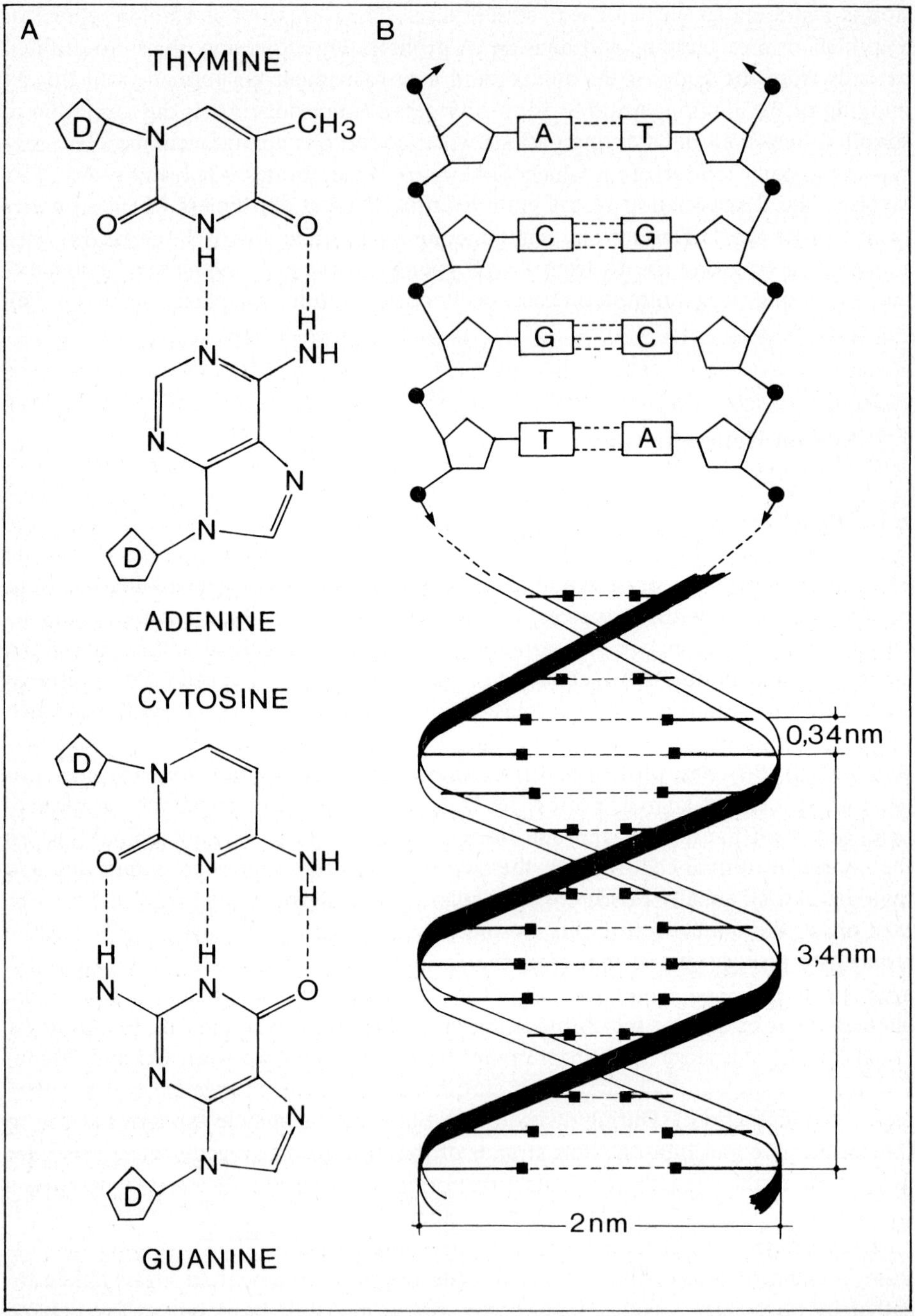

Fig. 1. A The four bases of desoxyribonucleic acid. D = desoxyribose. In RNA thymine is replaced by uracil, which has the same structure as thymine except that it lacks the methyl group. **B** The basic structure of the DNA double helix. The *upper part* shows the nucleotide linkages in DNA. • phosphate; ⬠ desoxyribose; A = adenine; C = cytosine; G = guanine; T = thymine. *Broken lines* indicate hydrogen bonds. The *lower part* shows the dimensions of DNA. ▪ bases

information. In contrast to uracil, the occurrence of thymine is mainly restricted to DNA. The combination of a base and a pentose sugar is called a nucleoside. The nucleosides are coupled together by phosphate residues to form a long chain molecule. The phosphoric acid esters of the nucleosides are referred to as nucleotides, the entire nucleic acid molecule being a polynucleotide. The DNA molecule of every animal is composed of two strands of polynucleotides which are aligned parallel to one another and are joined across their bases. However, not every base can be paired with every other: only thymine or uracil can be paired with adenine, and guanine can only be paired with cytosine. The two linked strands form a double spiral, the double helix, which cannot be separated without the helix being untwisted. In the coiled configuration the bases lie on the inside of the molecule with the complementary pairs being bound to each other by hydrogen bonds. The double helix has a diameter of about 20 Å, each turn being built up of ten nucleotides. The structure of the DNA double helix is shown in Fig. 1B.

As mentioned, DNA is the carrier of the genetic information. The passing on of this system in its entirety is made possible by the ability of DNA to duplicate itself. This process is known as reduplication or replication and it takes place during the time interval between successive cell divisions. Replication probably involves the partial opening of the double helix by some as yet unknown mechanism whereby the complementary bases are separated. New complementary bases, in the form of desoxyribonucleotide triphosphates, then bind to the exposed bases. The new bases are linked together by the enzymatic splitting of pyrophosphate to form a new, complementary strand of DNA. This step-wise biosynthesis is mediated by DNA polymerases and, when complete, there are two identical double helices, each comprised of one old and one new strand of polynucleotide. Thus, every daughter chromosome resulting from the division of the chromosomes contains a complete double helix. The genetic information is "stored" as the sequence of bases in the DNA molecule. The smallest unit of information is a series of three consecutive bases referred to as a triplet or codon, which represents a letter in the genetic alphabet. The bases in DNA are thymine (T), adenine (A), guanine (G) and cytosine (C) and there are therefore $4^3 = 64$ possible triplet combinations. The triplets code for the 20 amino acids from which animal proteins are built. The individual amino acids are coded for by more than one triplet, and up to a maximum of six. It is, therefore, called a degenerate code. Three triplets do not correspond to any amino acid and are referred to as nonsense codons or terminator codons and it is highly probable that they mark the end of a polypeptide. Triplets which code for the same amino acid usually have the same initial pair of bases and differ only in their third base. The first pair of bases can therefore be regarded as the determining elements. Thus the amino acid sequence of body proteins such as enzymes which regulate particular metabolic pathways is laid down in the sequence of triplets in DNA, although, the biosynthesis of proteins takes place outside the cell nucleus on the ribosomes in the cytoplasm. The information stored in DNA must, therefore, be "read" and brought to the sites of protein synthesis. This process of reading off the information is referred to as transcription. Transcription proceeds in essentially the same way as does the replication of DNA, but differs in that only one strand of a DNA molecule, the so-called codogene strand, is restored by the addition of the appropriate complementary bases to the exposed region of the molecule, and also that

Table 1. The genetic code of DNA and mRNA

First nucleotide	Second nucleotide: A	*U*		G	*C*		T	*A*		C	*G*		Third nucleotide	
A	AAA	*UUU*	Phe	AGA	*UCU*	Ser	ATA	*UAU*	Tyr	ACA	*UGU*	Cys	A	*U*
	AAG	*UUC*	Phe	AGG	*UCC*	Ser	ATG	*UAC*	Tyr	ACG	*UGC*	Cys	G	*C*
U	AAT	*UUA*	Leu	AGT	*UCA*	Ser	ATT	*UAA*	Stop	ACT	*UGA*	Stop	T	*A*
	AAC	*UUG*	Leu	AGC	*UCG*	Ser	ATC	*UAG*	Stop	ACC	*UGC*	Try	C	*G*
G	GAA	*CUU*	Leu	GGA	*CCU*	Pro	GTA	*CAU*	His	CCA	*CGU*	Arg	A	*U*
	GAG	*CUC*	Leu	GGG	*CCC*	Pro	GTG	*CAC*	His	GCG	*CGC*	Arg	G	*C*
C	GAT	*CUA*	Leu	CGT	*GCA*	Pro	GTT	*CAA*	Gln	GCT	*CGA*	Arg	T	*A*
	GAC	*CUG*	Leu	GGC	*CCG*	Pro	GTC	*CAG*	Gln	GCC	*CGG*	Arg	C	*G*
T	TAA	*AUU*	Ile	TGA	*ACU*	Thr	TTA	*AAU*	Asn	TCA	*AGU*	Ser	A	*U*
	TAG	*AUC*	Ile	TGG	*ACC*	Thr	TTG	*AAC*	Asn	TCG	*AGC*	Ser	G	*C*
A	TAT	*AUA*	Met	TGT	*ACA*	Thr	TTT	*AAA*	Lys	TCT	*AGA*	Arg	T	*A*
	TAC	*AUG*	Met	TGC	*ACG*	Thr	TTC	*AAG*	Lys	TCC	*AGG*	Arg	C	*G*
C	CAA	*GUU*	Val	CGA	*GCU*	Ala	CTA	*GAU*	Asp	CCA	*GGU*	Gly	A	*U*
	CAG	*GUC*	Val	CGG	*GCC*	Ala	CTG	*GAC*	Glu	CCG	*GGC*	Gly	G	*C*
G	CAT	*GUA*	Val	CGT	*GCA*	Ala	CTT	*GAA*	Glu	CCT	*GGA*	Gly	T	*A*
	CAC	*GUG*	Val	CGC	*GCG*	Als	CTC	*GAG*	Glu	CCC	*GGG*	Gly	C	*G*

Base nucleotides: A = adenine, T = thymine, C = cytosine, G = guanine, U = uracil.
Amino acids coded for by the base triplets: Ala = alanine, Arg = arginine, Asp = aspartic acid, Asn = asparagine, Cys = cysteine, Gln = glutamine, Glu = glutamic acid, Gly = glycine, His = histidine, Ile = isoleucine, Leu = leucine, Lys = lysine, Met = methionine, Phe = phenylalanine, Pro = proline, Ser = serine, Thr = threonine, Try = tryptophan, Tyr = tyrosine, Val = valine

the molecules of nucleoside triphosphate synthesised contain ribose as their sugar component. In addition, the ribonucleic acid synthesised contains uracil instead of thymine. The completed molecule of RNA dissociates from the DNA and respresents a complementary copy of the genetic information determining the sequence of amino acids in a particular polypeptide. Such RNA is called messenger RNA (mRNA) since it leaves the nucleus, migrates into the cytoplasm and delivers the information stored within it to the site of protein synthesis. The mRNA is then transposed into a sequence of amino acids in the process of translation wherein the mRNA first makes contact with a ribosome. Activated amino acids are then individually bound to a molecule of transfer RNA (tRNA), there being more than one specific tRNA for most amino acids. All tRNA molecules are evidently composed of a single strand which carries the base sequence C-C-A on one end and G on the other. One loop of the molecule represents the anticodon, which is the exact complementary triplet to that encoded by the amino acids of mRNA to which the tRNA becomes attached. The appropriate activated amino acid is attached to the C-C-A end. Through the attachment of two tRNA molecules to the mRNA, their respective amino acids are so positioned next to each other that a peptide bond can be formed between them: this is possible because the tRNA molecules all have the same length. The biosynthesis of a polypeptide chain is an energy-dependent process which proceeds in a step-wise fashion by the displacement of the bound tRNAs, together with the mRNA, from one to the other of the two binding sites on the ribosome. The completed polypeptide chain is finally released from the ribosome. The entire genetic code of DNA and mRNA is shown in Table 1.

Mutation, a suddenly occurring change in the inherited character of an organism, is a fundamental aspect of evolution. A change in only one of the triplets in DNA may produce a new sequence of amino acids in a protein and this may, in turn, affect several different aspects of physiology. Each chromosome probably contains only one DNA macromolecule, whose molecular weight may exceed one billion dalton. The double helix spirals around globular molecules of histoprotein to form a superhelix that by further tertiary folding or spiraling forms a super-superhelix. Prior to the division of the nucleus, the chromosomes exhibit yet another, quaternary order of coiling, whereby they become shortened and condensed. DNA is divided up along its length into functional segments called genes which perform different types of function. Some for example, are structural genes, others are regulator genes.

1.3.2 Mitotic and Meiotic Cell Division

The growth of a multicellular organism is due primarily to the increase in cell number resulting from the division of a mother cell into two daughter cells. The daughter cells, in turn, grow into mother cells which divide into daugher cells that grow ... and so on. Every time the nucleus divides the full diploid set of chromosomes is passed on to the nucleus of each daughter cell, the replication of the DNA double helices ensuring that every cell is provided with a complete copy of the genetic information. This type of division is called mitosis. Mitotic cell division is the basic mechanism underlying all types of asexual propagation.

The period between successive cell divisions, which takes up most of the life of the cell, is called the interphase, and the cell characteristically has a high level of metabolic activity during this period. The onset of division is referred to as the prophase. The chromosomes become visible and have by this time already divided along their length to form two identical chromatids joined together at the centromere. A pair of centrioles lying outside the nucleus duplicate themselves during this phase. The centrioles are cylindrical organelles, each consisting of nine sets of microtubules. They later give rise to the microtubules of the spindle apparatus. During the ensuing metaphase the centrioles migrate to diametrically opposing poles of the nucleus and the structure of the membrane enveloping the nucleus breaks down. The chromosomes condense, becoming shorter and thicker, and arrange themselves along the microtubules of the spindle apparatus arising from the centrioles to form a row marking the equatorial plate. The microtubules of the spindle are attached to the centromeres joining the chromatids together. During the following anaphase, the chromatids separate completely from each other and withdraw to opposite halves of the nucleus in the direction of the centrioles. The completion of this process marks the beginning of the telophase during which the cytoplasm divides and two new nuclear membranes are formed, thereby concluding cell division. The entire process can be divided into nuclear division, or karyokinesis, and cell division, or cytokinesis. The individual phases of mitosis are shown in Fig. 2.

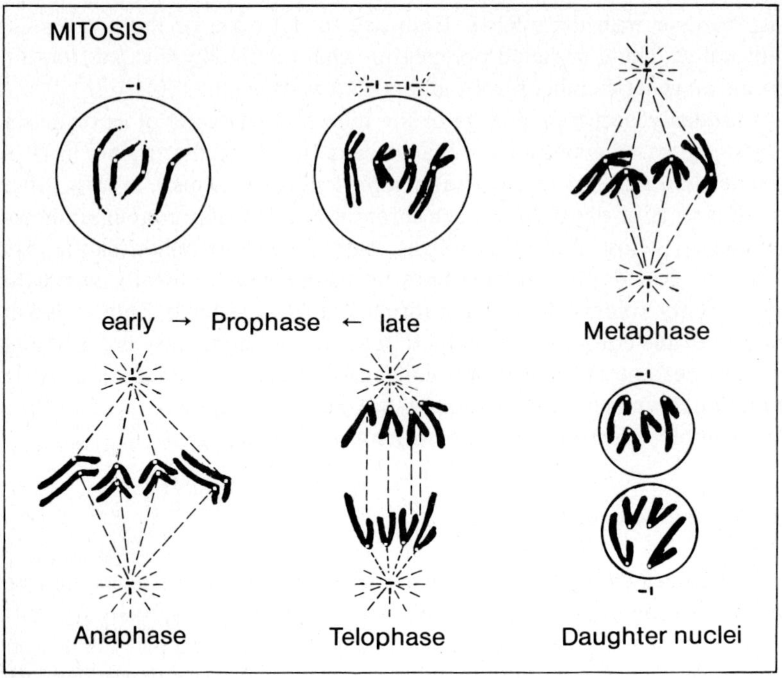

Fig. 2. Mitosis

As mentioned above, a necessary preliminary to sexual reproduction is the reduction in the number of chromosomes in certain cells from two sets (2n) to a single set (1n). This occurs during the course of maturation division, or meiosis. In contrast to mitosis, there are two stages in meiosis at which the cells divide. During the first of these, the reduction division (or first maturation division), the chromosomes are divided among the daughter cells so that each receives a single set. The second, segregation division (or second maturation division) follows the same course as mitosis. From this it follows that four haploid cells are produced from a single diploid cell. The reduction division can come before (pre-reduction types) or after the segregation division (post-reduction types). The individual phases of meiosis are identified as in mitosis. In many cases the prophase of meiosis is greatly prolonged, lasting several years in some cases. It can be divided into five sub-stages. During the leptotene stage the individual chromosomes are recognisable. The crucial feature which distinguishes meiosis from mitosis occurs in the ensuing zygotene stage when the homologous paternal and maternal chromosomes pair together in a process known as synapsis or syndesis. In the pachytene stage which follows the individual chromatids can be seen to be joined together in the form of chromosome tetrads or, more precisely, chromomere tetrads. During the diplotene stage the chromosomes condense and become shorter and thicker. The homologous pairs of chromosomes are referred to as bivalents. The pairs of chromosomes then separate somewhat from each other and, in doing so, one member from each of the two homologous pairs of chromosomes may cross over and stick to another. Importantly, the phenomenon of crossing over is responsible for the ex-

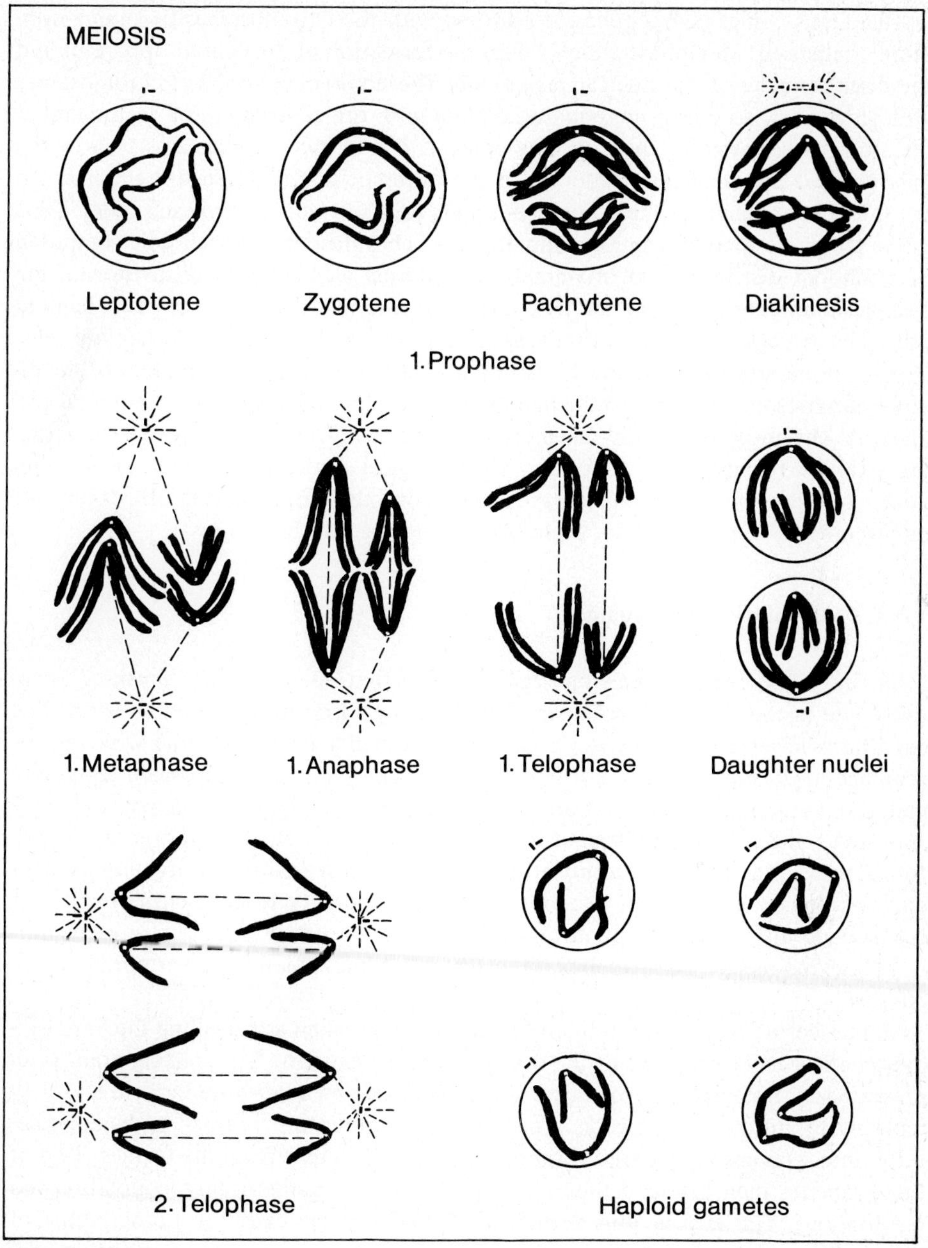

Fig. 3. Meiosis

change of genes between two homologous chromosomes: the chromosomes involved can break at the point at which they cross and by an unknown mechanism join up with the broken-off segment of the other chromomere. This exchange of chromomere segments is called chiasma formation.

Diakinesis brings the prophase to an end with the chromosomes becoming even more condensed. Metaphase follows with the formation of the spindle apparatus and the disappearance of the nuclear membrane. The homologous pairs of chromosomes arrange themselves along the equatorial plate with the centromere of each member directed towards one of the opposing poles. It is a matter of chance as to how the paternal and maternal chromosomes of the different tetrads arrange themselves. In the subsequent anaphase the homologous pairs finally separate from each other. Thus in the telophase there is a single complete set of chromosomes at each pole composed of a random distribution of maternal and paternal elements. After division of the cytoplasm and formation of the new nuclear membrane there are then two haploid cells. The segregation division proceeds as in mitosis and produces four haploid cells. Overall, three important functions can be attributed to meiosis: reduction of the diploid chromosome number to the haploid number, the redistribution of paternal and maternal chromosomes in the daughter cells and, lastly, the exchange of genetic material via the formation of chiasmata. The biological significance of the last two phenomena with respect to evolution goes far beyond the simple halving of a double set of chromosomes. The course of events in meiosis is shown in Fig. 3.

1.3.3 The Evolution of Sexuality

From studies of fossil animals one can conclude that sexuality was "invented" relatively late in the three milliard years that living organisms have existed on earth. The conditions necessary for the development of sexuality probably came with the appearance of organisms possessing a nucleus, the eukaryotes. This happened about one milliard years ago, and the first organisms to exhibit the phenomenon may well have appeared some time thereafter. At that time there was a virtual explosion of new life forms, which was possibly a result of the emergence of sexuality and its consequences on evolutionary events. This does not mean that sexuality is an essential feature of eukaryotes: there are many such organisms which reproduce primarily or secondarily by asexual means. Nevertheless, sexuality is a typical feature of vertebrate reproduction.

A number of features of modern Protozoa can be taken as providing the basis of a hypothetical reconstruction of the development of sexuality. The starting point is represented by hologamy. In this case the individuals of a species are haploid and they multiply by mitotic cell division. At some time or other the "normal" individuals develop into gametes which are morphologically no different from the former. Two of these gametes then fuse and their nuclei melt together (gametogamy, copulation and fertilisation). The zygote thus formed is diploid and can exist for a long period of time without dividing. Eventually the zygote goes through meiosis, resulting in haploid individuals which revert to being normal individuals. Notably, fertilisation by the union of gametes halves their number (Fig. 4A). Merogamy, the second evolutionary stage, is characterised by the fact that the normal individuals and the gametes are morphologically distinct. The normal individuals are diploid and increase in number by mitotic division: the individuals are called agamonts and their multiplication agamogony. Agamonts become gamonts by undergoing meiosis and forming haploid ga-

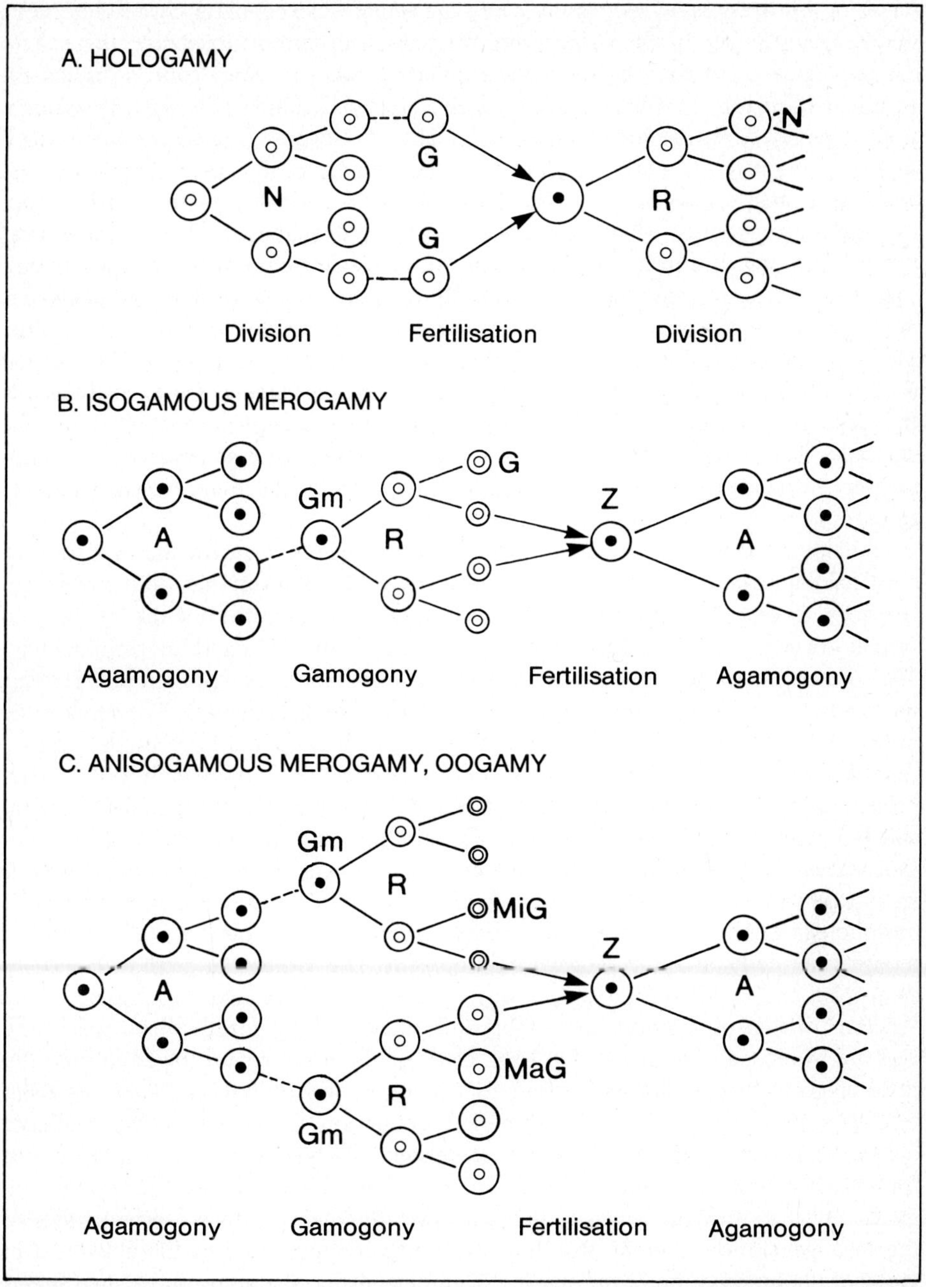

Fig. 4 A–C. The development of sexuality. • 2n; ○ 1n; N = normal individual; G = gamete; Gm = gamont; MiG = microgamete (spermatozoon); MaG = macrogamete (egg); Z = zygote; A = agamont; R = reduction division

metes in a process called gamogony. Only the gametes can copulate. Two subdivisions of merogamy exist. In the more primitive form, isogamous merogamy, there is no morphological distinction between the copulating partners. They form a diploid zygote which, in turn, divides to give normal diploid individuals (Fig. 4B). The formation of gametes from gamonts is generally called gamogony. The second form is anisogamous merogamy and in this case there are two types of gametes, large macrogametes and small microgametes (Fig. 4C). Individuals can therefore arise either from two different gamonts, which corresponds to being dioecious, or from a single type, this being an instance of hermaphroditism. All metazoans reproduce by means of anisogamous merogamy involving a non-motile macrogamete known as the egg and a microgamete, the spermatozoan, which bears a flagellum and is, therefore, motile. This extreme case of anisogamous merogamy is referred to as oogamy. The normal individuals of metazoans are the body cells, whereas the gametes are specialised cells that separate from the other cells to become the primordial germ cells (or primordial sex cells). The development of the embryo and body growth in general represents agamogony, whereas the formation of eggs and sperm within the body represents gamogony.

Individual protozoans are potentially immortal. However, with the evolution of multicellular organisms a heavy price had to be paid – the very individuals of which the body is composed became mortal. According to the germinal line theory, the metazoan organism can be reduced to two components, the soma and the germinal line. The former constitutes the bulk of the body. The somatic cells represent proliferating agamonts which increase their number by mitosis (i.e. agamogony). The soma is the part which dies, its life limited to a time-span which is by and large predetermined for every species. The germinal line is, on the contrary, potentially immortal. It is represented by the gamonts and by the male and female gametes which arise from them and link the successive generations of a species into a continous lineage. The body thus serves as a host tissue for the elements of the germinal line and in the course of its ontogenetic development it grows and differentiates into an organism which is morphologically and physiologically capable of reproducing. The germinal line is often recognisable early in embryological development. The primordial germ cells usually stem from endoderm and in various ways, either active or passive, reach their actual "shelter" within the embryo, the embryonic anlagen of the gonads where sperm and eggs are produced. The primordial germ cells, or the products arising from their mitotic division, remain dormant often for a long period of time until the organsim reaches a state in which it is capable of reproduction. Only then do they undergo meiosis to become gametes. Metazoan organisms only have the ability to reproduce for a limited time and after a certain age is reached the gonads cease to produce gametes. By that time the germinal line has normally been passed on several times to the next generation, thereby fulfilling the biological "purpose" of the individual in terms of the propagation of the species. Seen objectively, the individual is then superfluous and the body usually dies soon thereafter.

1.3.4 Determination of the Sexes

From the above considerations it will be evident that the crux of sexual reproduction is the coming together of the gametes from a male and female individual. In the case of hermaphroditic species, the male and female gametes stem from different individuals. However, hermaphroditism is rare among vertebrates, which are predominantly of separate sexes, male and female, that produce gametes referred to as sperm and eggs respectively. The sexes of many vertebrate species do not differ from one another in their outward appearance, or they only do so temporarily during the reproductive phase. In other cases the males and females are sexually dimorphic and have a different and characteristic appearance throughout life, or from the age of sexual maturity on. The sex ratio in most species of vertebrates is roughly 1:1.

Sex is usually determined genetically and is irreversible. This type of sex determination needs to be clearly distinguished from the differentiation of sexes, which is an ontogenetic process that follows the genetic determination of the sexes and gives expression to, or realises, the genetic information governing sex. All the genes within the chromosomes of an organism which code for the male sex are generally referred to as the A complex and those pertaining to the female sex as the G complex. Hermaphrodites are prescribed by a combined AG complex. This does not mean, however, that males have only the A complex and females only the G complex. Dioecious species, with separate sexes, presumably have a latent bisexual potency, i.e. they are availed of both complexes and either the A or the G complex is actively selected or inhibited. As a result, the development of one or the other sex is normally an irreversible process. The factors operating in this process are still only partly known. They prevent bisexuality by activating or suppressing either the A or the G complex and are known as sex realisers: M realises the male sex, F the female sex. When both are operant sex is determined by their relative strength. A male develops when M is dominant over F and a female when F is stronger than M. If the sex realisers are genes, passed on to an individual via fertilisation, then one speaks of genotypic sex determination. Sometimes external factors can also be effective. In such a case there is phenotypic or modificator determination of sex; it is quite possible for the male and female individuals to have an identical genetic complement.

Genotypic determination of sex is the rule in vertebrates, whereby only the combination of maternal and paternal chromosome sets associated with the fusion of the gametes determines whether the new individual will be male or female. This means that the combination of chromosomes determining sex is first present in a definitive form only in the diploid phase. In this case one speaks of sex determination being diplogenotypic. The gametes themselves are likewise sexually differentiated although this is independent of the chromosome set they happen to receive through meiosis. Thus, the gametes that receive an F realiser are eggs if produced in a female gonad and sperm if produced in a male gonad.

The diplogenotypic determination of sex usually follows the inheritance pattern of a back-cross between a monohybrid bastard and its homozygous recessive parent, the genotype of the offspring corresponding to that of the parents in the ratio 1:1. In the case where the female sex realiser is homozygous it means that FF is present in the chromosomes. The male is, therefore, heterozygous (FM), the M realiser being "stronger" than F: in other words M is dominant and F recessive. The production of gametes in females results only in gametes containing F, whereas in males there is the same number with F as with M. The first example is a case of homogamety, the second of heterogamety. The determination of sex occurs at fertilisation: an F sperm and an F egg result in a female, an M sperm and an F egg form a male. Since the number of F and M sperm is the same, the offspring will theoretically exhibit a sex ratio of 1:1.

Female heterogamety also occurs in vertebrates, the male being homozygous and the female heterozygous. The same fundamental scheme of inheritance applies here as in male heterogamety. The chromosome set of males possesses MM, that of females FM, F being dominant. The time at which the sex of the offspring is decided is different in this case. It is made at the time of the meiosis producing the female sex cells: sex is already determined by F and M respectively in the oocyte. As before, the sex ratio of the offspring is theoretically 1:1.

In most vertebrates the sex-determining chromosome differs from others of the set in its form, behaviour and homologous partners. It is called a heterosoma, others being autosomes. Whenever the former occurs twice in homogametic gametes and once in heterogametic gametes it is referred to as the X chromosome. In the latter case it has either no partner at all (XO type) or a morphologically different partner, the Y chromosome (XY type). In such cases one can regard the determination of sex simply as the inheritance of the heterosome. Accordingly, there are four theoretical possibilities of heterogamety: XX = female XY = male and XX = female XO male for male heterogamety and XY = female XX = male and XO = female XX = male for female heterogamety.

The realiser combination of FF for female and FM for male used so far in connection with male heterogamety is only one of several possible models. The F need not be in the X chromosome nor M in the Y chromosome. Nevertheless, this seems to be the case for some mammals, including man. Aberrations of the heterosomes have been found which show that a male appearance will result when at least one Y chromosome is present (XXY, XXXY, XXXXY) and that female characteristics always result when it is absent (XO, XXX, XXXX, XXXXX). Evidently, the dominant M realiser in this case is located on the Y chromosome.

In contrast to the above, it is of great significance that the genetic sex of vertebrates determines only the direction of development of the undifferentiated gonadal anlage which later becomes the site of gamete production. Other factors, primarily hormones, are involved in the differentiation of the gonads and the development of the characteristic appearance of the sexes and these will be dealt with in detail later (see p. 202). Genotypic determination of the sexes occurs in all classes of vertebrates, whereas phenotypic determination is rare. Male and female heterogamety with XY types exists in teleosts. It was first identified in the guppy (*Lebistes reticulatus*: now *Poecilia reticulata*), in species of *Platypoecilius* and in *Limia nigrofasciata*. It is interesting that in the genus *Platypoecilius*, female heterogamety occurs in *P. maculatis*, as it does in two species of paradise fish (*Macropodus*). No heterosomes have been found in the swordtail (*Xiphophorus helleri*), two species of *Limia* and *Macropodus concolor*. These cases evidence a phenotypic sex determination, although there are indications of some genetic control which works in two directions: a Tm complex "receives" exogenous masculinising factors and ensures that not only females are produced, whereas a Tf complex receives exogenous feminising factors and prevents a population of only males arising. In some *Limia* species a predominance in the number of males or females occurs. This has been explained either as due to an imbalance of M and F factors or as a difference in the sensitivity of the Tm and Tf complexes.

Genotypic sex determination likewise occurs in Amphibia. One group, to which *Siredon mexicanum* (the axolotl), *Bufo bufo* (the European toad) and *Xenopus laevis* (the clawed toad) belong, shows female heterogamety of the XY type. A second group, including *Rana esculenta* (the water frog) and *Rana temporaria* (the grass frog), has male heterogamety of the XY type. In the last-named species one can distinguish differentiated and undifferentiated races. The former have a fixed sex ratio from the beginning, whereas in the latter genetic males first go through a female phase before finally becoming male. A reason for this is thought to be an imbalance between the factors which determine sex.

Male and female heterogamety occur in Reptilia. There are male XO, XX and XXY types as well as female XXXX, XX, XY and XO types The relationships here are not very clear. In many species one can find no heterosomes at all and in some others it is uncertain whether XO or XY types are involved. In birds it is probable that female heterogamety of the XO type predominates.

Mammals exhibit male heterogamety of the XY type exclusively. Sex is evidently determined by the presence or absence of the Y chromosome. Two types are recognised: female XX = MM FF, male XY = MM FO where F is dominant, and female XX = FF and male XY = FM in which case M is dominant. In many species of mammal it is apparent that males are present in excess. The reasons for this are not completely known; it could be that sperm containing the Y chromosome have a greater mobility, although other factors are undoubtedly involved in this phenomenon.

1.3.5 Problems Associated with Vertebrate Reproduction

The majority of vertebrates have separate sexes, male and female. The male gonads, the testes, produce spermatozoa and are usually paired organs. In most species the sperm leave the body via special ducts, the so-called spermiducts. The sperm are flagellated and therefore motile. They "swim" in a fluid produced by the epithelium of the spermiducts or by specialised glands that, together with the sperm, makes up the semen. The female gonads, the ovaries, are likewise usually paired organs. In the vast majority of vertebrates the eggs produced within the ovaries are borne out of the body through the oviducts. In contrast to sperm, the prospective eggs within the ovaries are each surrounded by nurse cells which form a follicle enclosing the egg. In order to leave the ovary the follicle must rupture, thereby setting the egg free: this process is called ovulation.

Reproduction is not essential to the survival of the individual – its significance lies in securing the preservation of the species. Sexual reproduction involves the coming together of the sperm and eggs, the fusion of their nuclei in the process of fertilisation and, finally, every zygote goes through a series of cell divisions and processes of differentiation to form a new individual. The primary requirement of sexual reproduction is, therefore, the coming together of two different gametes. The statistical probability of this happening would be highest in an hermaphroditic organism in which the sperm and eggs mature and become fertile at the same time. However, self-fertilisation within the body of a single individual contradicts the "biological sense" of sexuality as part of adaptive evolution. Accordingly, self-fertilisation occurs rarely in vertebrates and only as an "emergency measure" in those species having a very low population density where the statistical probability of sexual partners meeting is vanishingly small. From this point of view, having two sexes is a disadvantage which must be put up with. The problem of finding a partner is, however, not the only difficulty. When the male and female of a species have met, two further conditions must be fulfilled if reproduction is to be successful: they must recognise each other as members of the same species and the gametes must be mature. The next problem is to guarantee that the gametes come together in conditions which allow fertilisation to take place. The following link in the chain of problems is that of assuring that the developing offspring survives until it is able to exist on its own, so that it too can reproduce itself.

A myriad of solutions to these problems have been created in the course of evolution, both at different levels and with different degrees of complexity, their success being manifest in the modern species of vertebrate. A high population density is the best precondition for ensuring a high probability of finding a partner and many vertebrates use this to their advantage. The formation of swarms or societies is very common. Some species always live together in large populations, others meet together only at certain times for the purpose of reproduction. Within such animal societies, social structures of greater or lesser complexity can form, most of which also have a reproductive significance. Species which lead more solitary lives have developed mechanisms to attract a partner by scent or acoustic signals over even large distances. The chances of finding a partner are double in the extreme case of hermaphroditism, which does not involve self-fertilisation, since all fertile animals of such a species could at one meeting not only mate, but also be mated with, another individual whereas only 50% can be fertilised in species with separate sexes and a sex ratio of 1:1.

Recognition of the sexes is a many-sided problem in which visual, acoustic, olfactory and tactile stimuli play an important role. In sexually dimorphic species, body form and colour can serve as features for identification.

Such animals acquire secondary sex characteristics which are responsible, to a greater or lesser degree, for the distinctive appearance of the sexes and in some cases their distinctive scent. These signals are frequently insufficient to bring the female to reproductive readiness, for example. In such cases a complicated chain of action and reaction in the behaviour of the partners is necessary: this phenomenon is called courtship.

It is common among vertebrates that the maturation of gonads in the two sexes is synchronised. This maturation is effected with the aid of vegetative regulatory processes in which hormones play an important role. Sometimes maturation follows a genetically fixed, endogenous programme which can, however, be modulated by environmental factors such as day length, temperature and the availability of food: the determining factors are often purely exogenous. It is not essential to vertebrate reproduction in general that the gonads mature at the same time and some species avoid this problem in that the sperm are stored in the female ducts or ovaries for different periods of time.

The conditions for the successful meeting of gametes and for fertilisation in water are quite good since it is a relatively stable physical and chemical environment. Most lower vertebrates that live in water therefore release their sexual products directly into the water; i.e. they spawn. This often occurs synchronously in species which form swarms, so that the eggs and sperm of many individuals are present at the same time and mix together in the water, where fertilisation then takes place. This type of synchronous social reproduction appears to be primitive; it is not the rule however. In the vast majority of cases two sexual partners come together and deposit their gametes at the same time or one after the other; i.e. they pair. This usually occurs more than once and with different partners. Pairing is the basic pattern of reproduction in most vertebrates. Here again there are two possibilities. In the more simple case, eggs and sperm are laid in water where external fertilisation then takes place. This phenomenon is called ovuliparity and is characterised by the fact that eggs leave the female's body unfertilised. The second possibility is internal fertilisation, that is the fusion of egg and sperm in the female duct system. This method is essential for all vertebrates which reproduce out of water but also occurs in Pisces. Special accessory copulatory organs are normally developed for the transfer of the semen, the organs being introduced through the female sexual opening and the semen guided into the female via grooves or ducts. Copulatory organs are, however, not essential for internal fertilisation: the majority of birds manage this by simply pressing together their cloacal openings (the cloaca is the common termination of the gastrointestinal system and the urogenital system). Numerous urodeles transfer sperm in a spermatophore, a packet formed in the male's cloaca out of a coagulating secretion and spermatozoa. He lays the packet on the ground and the female then actively picks it up with her cloacal lips.

The introduction of the male's accessory organ into the female's genital tract or the transfer of sperm into the body of the female without the aid of such a structure is called copulation. The phenomenon of internal fertilisation leads on to the next problem in reproduction, that of ensuring the undisturbed development of a new individual. Despite the fact that successful fertilisation is possible within the female's body (a wet environment), other measures are required to ensure that the subsequent development of the embryo can take place in air. These include the formation of primary and secondary egg membranes that physically protect the egg and prevent it from drying up. The primary membranes are made up of follicle and germ cells in the ovary, whereas the secondary membranes are formed by glands in the female duct system. The shells formed by these glands are usually quite firm or are encrusted with minerals that make them hard. In the latter case the penetration of the egg by the sperm must necessarily precede the formation of the shell and this is only possible if fertilisation is internal. An animal which lays a fertilised egg surrounded by a shell is referred to as oviparous. This term is often used to mean the whole phenomenon of egg-laying, irrespective of whether or not the egg has been fertilised, different authors adopting different definitions of the term. The more restrictive definition given above will be used here. Oviparous species of higher vertebrate possess an additional means of protecting the embryo against desiccation which is formed by the embryo itself, namely, the amnion. The amnion develops in one of several ways, the most primitive being that the embryo gives rise to folds of ectoderm which meet dorsally and grow together to form an amniotic sac filled with amniotic fluid. This sac represents a miniature aquarium in which the new individual goes through an aquatic phase in the course of its development.

It is also possible that the fertilised egg never leaves the body of the mother, wherein it undergoes part or all of its development. In the case of ovoviviparity, the eggs, which are usually surrounded by a shell, are simply retained in the ducts of the female reproductive system and may even hatch there, although it is more usual that they are laid beforehand. A characteristic of ovoviviparity is that no close structural or physiological relationship is established between the embryo and the mother such as might, for example, serve to nourish the embryo. Such dependencies are, however, the defining characteristics of viviparity, whereby the provision of the offspring is partly or completely taken over by the mother. This is effected in diverse ways via often highly complex structures. Viviparity occurs not only in mammals but in members of every class of vertebrate, with the exception of Aves, to a sometimes striking degree of perfection. The young are not retained within the genital tract in all cases. In Pisces and Amphibia, for example, there are a range of special adaptations by virtue of which other body cavities are used for this purpose, the simplest case being that in which the young are kept in the mouth. However, such adaptations cannot be regarded as instances of ovoviviparity or of viviparity, which always involves the embryo being retained within a specialised part of the female reproductive tract, the uterus.

The efforts undertaken by parent animals to secure the survival of their brood can all be gathered under the general heading of parental care. Thus ovoviviparity and viviparity and the special adaptations mentioned, whereby the young are kept in cavities of the body which are not part of the genital tract, are all aspects of caring for the young within the body. In contrast, the wide range of behavioural actions which are, in one way or the other, directed towards ensuring the survival of the offspring are aspects of caring for the young outside the body. This includes, for instance, guarding and cleaning the eggs, brooding the eggs, feeding the young etc. Care of the young, in its widest sense, also includes one aspect of the maturation of the female gametes, namely the provision of the egg cells with energy reserves for the development of the embryo. This process, called vitellogenesis or yolk formation, occurs in the ovary and involves the deposition of proteins, lipids and other substances within the female gametes prior to the occurrence of meiosis. Vitellogenesis is a feature of all submammalians. Other reserve materials which stem from special glands in the oviduct can also be laid down round about the egg before the shell is formed.

Several reproductive strategies of differing levels of complexity can be recognised among vertebrates. The simplest of these is the "statistical method", whereby large numbers of eggs and sperm are laid in water and the parents play no further role in their fate. The large number of gametes usually guarantees that a sufficient number of offspring result, since the proportion that do not survive is typically very high. The next level of complexity involves a minimal degree of parental care in that the parents seek out or prepare a favourable or sheltered place in which to deposit their gametes, which are then left to develop on their own. The parents can be said to truely care for their young when one or both guard the eggs or attend to them in some way. This level leads on to the parent animals also attending to the young after they have hatched out by, for example, providing them with food which may be a product of the parent's body. All these activities require a considerable investment of energy on the part of the parents and therefore limit the number of young that can be cared for at any given time. The same applies with respect to the physical and physiological capacity of the mother's body in the case of ovoviviparity and viviparity. There is a clear tendency among vertebrates that the more the parents invest in their young, the smaller is the number of eggs or progeny, which seems to make biological sense. Without doubt, the crucial limiting factor in this is the capacity of the parents to care for their young: the more intensive and time-consuming their attention is, the less time and energy remains for further reproduction. The number of eggs or offspring produced by vertebrates varies from between one and millions. A reduction in the number of offspring is most evident in those vertebrates that reproduce out of water but is especially marked in mammals: many bring only one offspring to maturity (e.g. many primates, ungulates and Chiroptera), although some birds lay only one egg during their breeding season, as does a species of frog (*Sminthillus* sp.) and as do several species of lizards.

The size of the brood or hatch produced by individuals of a species also depends upon environmental influences. It increases with the degree of latitude and, as a rule, populations on islands produce fewer eggs than does the same species on the mainland. Moreover, in many species of vertebrate the number of eggs produced during the breeding season increases up to a certain limit with increasing body weight or age. As mentioned, the size of the brood also depends on the intensity with which the parents care for their young. Thus when a species which takes good care of its young produces more eggs or offspring, the mortality rate usually rises. Female birds that mate with several males, which then take care of the resulting brood, can generally lay more eggs than those which mate with only one male, in which case the female takes care of the brood. This is basically true of mammals as well. Nutrition is an important factor limiting the number of offspring in species which do not care for their young. The availability of food is, in turn, dependent on the general level of fertility and the degree of latitude of the biotope. In the higher latitudes the day-length and temperature increase in the spring, with the result that there is a gradual increase in plant growth and, coupled with this, a multiplication in the number of small animals such as insects that feed on the plants. This means, on the one hand, that foodstuff is available for the young animals and, on the other, that females have sufficient energy resources for the production of eggs. In addition, the longer days provide more time for the animals to search for food and to feed the young. In general, an adequate supply of food appears to be essential if reproduction is to take place. In some species of bird, for example, it has been found that an above-average number of offspring are produced when food is abundant but that reproduction may cease altogether in times of deprivation. Although the quality and quantitiy of food undoubtedly affect reproduction, they probably do not do so directly.

1.3.6 Reproductive Cycles

Many vertebrates reproduce periodically, meaning that there is an alternation between periods of reproductive activity and inactivity. These phases frequently follow changes in environmental factors and this is particularly evident in higher and mid latitudes where the reproductive phases are coupled with the changing of the seasons. The reproductive phases are, in turn, coupled with cyclical changes in the gonads, these being particularly well defined in females. This does not mean that the ovary only goes through one cycle during each breeding season: it may very well go through several cycles. Species which have only one period of reproduction each year are called monocyclic and those which reproduce several times in the course of a year polycyclic. The ovarian cycle involves the ripening of a set of eggs (or of a single egg) in the ovary, their (its) ovulation, and a period of time thereafter until a new set begins to ripen. Around the time of ovulation there is usually a phase of increased sexual activity and readiness to mate in which mammals are said to be in heat or in oestrus. Species having only one oestrus during the reproductive season are monoestrous, whereas those with more than one are polyoestrous. The number of oestrous cycles which occur in succession depends on whether or not fertilisation is successful. If not, the onset of a new cycle brings with it a resurgence of reproductive readiness.

An exact definition of the term breeding season is rather difficult. In ovuliparous and oviparous species which do not care for their young, the term implies only that there are one or more periods of mating, which may involve courtship followed by egg-laying or copulation. In species which care for their young the breeding season is divided in two, the period of mating being followed by a period of parental care. This last can, in turn, be broken down into two subphases: firstly, a period of incubating or of guarding and tending the eggs and secondly, a period of caring for the brood after they have hatched. This scheme also applies, in essence, to ovoviviparous and viviparous species which carry the developing young in the uterus during gestation and continue to care for their young after they are born. The second period of caring for the young is not evident in every case. Gestation can last a considerable length of time, up to several years, so that a long period of sexual inactivity may intervene between mating and the phase of parental care. In some vertebrate species the sperm is stored in the female's body and the eggs only mature to ovulation and are finally fertilised long after the end of the mating season. Between copulation and egg-laying or the beginning of gestation there is therefore a period of complete reproductive inactivity. Although copulation may lead to successful fertilisation, in some mammals the development of the embryo is arrested at a very early stage and resumes only after what is in some cases a lengthy interval.

The linking of the breeding season to particular seasons of the year is of great importance in polar regions and mid latitudes. The offspring only have a good chance of survival when they are born in a favourable season that, for example, offers them sufficient food. Thus a temporal factor comes into play which depends on how long the young take to develop and in what manner they develop. This point is particularly well illustrated by mammals, which have markedly different gestation times. Most species which have a short gestation begin mating in spring, whereas others, such as sheep and deer, mate in autumn and the young come into the world the following year. In species which are monocyclic and ovuliparous such as species of salmonid and anuran, the gonads of the adult animals mature during summer, when there is an abundance of food, so that the gametes are already fully developed in winter before the beginning of the reproductive season the following spring. In such cases the period of gonadal maturation therefore occurs much earlier than the period of mating.

The gonads of the two sexes of species living in equatorial or polar latitudes mature more or less synchronously. Differences in the time the eggs take to ripen also occur as a consequence of ageing in monocyclic species which live for more than a year: older females come into reproductive readiness earlier than younger ones.

From the above considerations it is clear that favourable climatic and nutritional conditions for the developing and growing young are important factors determining the timing of reproduction. This is evidenced by the common phenomenon that most species which live in a particular environment give birth at the same time. Other factors, however, are also involved, one of them being that it seems important that the conditions are favourable for the reproducing adult members of a species. For example, that the humidity is appropriate in the case of amphibians.

The significance of the role played by favourable conditions also appears to be demonstrated by a variety of domestic and laboratory animal species. These animals reproduce in cycles, one following the other irrespective of the time of year. Correspondingly, the reproduction of vertebrate species living in equatorial regions takes place throughout the year in accordance with the individual needs of the species. Tropical regions do not, however, have a uniform and unchanging climate: there may be great differences in the temperature by day and by night; there are cold regions in the highlands, rainy seasons, and periods of drought etc. On the other hand, there is a gradual transition from the tropics to the zones of temperate climate on either side of the equator, the tropics giving way to the subtropics and these in turn to the temperate latitudes. There is, however, a distinct difference between the regions near the equator and those distant from it: the rhythm of day and night is more clear-cut and invariant in the former. Theoretically, the length of time the sun shines every day is virtually constant, the only possible cause of variation being the persistence of cloud cover. The seasons of the year are therefore less clearly defined in regions close to the equator. There are, nevertheless, considerable differences in the timing of the breeding season of tropical species. These are, as mentioned, primarily related to the individual needs of each species, the dry season being optimal for some, the rainy season for others. It is a characteristic of tropical and subtropical species in general that the breeding seasons of different species occur at widely different times of the year.

The question arises as to which factors trigger the onset of the reproductive period of species living in the different climatic zones. In temperate latitudes the changing relationship between daylight and darkness throughout the course of the year is a conspicuous and determining factor for all living things. The days become longer following the winter solstice, whereas after the summer solstice the days become shorter and the nights longer. In many vertebrate species it has been shown that these changes in photoperiod stimulate the maturation of the gonads and thus the beginning of reproduction. This phenomenon has been studied intensively in birds and less so in mammals and poikilotherms. The inactive winter gonads of many species of bird increase in weight dramatically under the influence of the lengthening day. This activation can be produced at any time of the year by experimentally lengthening the day. The gonads of mammals also respond to light. Thus, when female laboratory rats are kept in continuous light, the weight of their ovaries increases. Continuous darkness produces the opposite effect, although it has also been found to disrupt the reproductive system of female rats and may even lead to sterility. The reproductive cycle of the ferret (*Mustela putorius domesticus*) also depends on the length of daylight (photoperiod). In this case light per se does not appear to be the determining factor, it is the change-over from light to darkness. However, an increasing level of light is not always the trigger: there are apparently "long-day" and "short-day" animals. The former include many species of bird, the ferret, horse and raccoon which all begin to reproduce as the days lengthen in spring. The latter include goats, deer and sheep which breed in autumn when the days become shorter. The photoperiod appears to be the most important factor determining the onset of the breeding season in many vertebrates. However, this does not mean that this principle applies universally within a class of vertebrates. Among mammals, for example, the reproduction of rabbits and guinea-pigs does not depend on light: in the former case this may be a consequence of domestication, whereas in the latter case it could relate to the animals equatorial origin.

The basic principle behind the physiological link between light and the stimulation of the gonads appears to be common to many species. The reception site is evidently the appropriate sense organ, the eye. The information is conveyed by the optic nerve to the central nervous system where it causes the excitation of certain hormone-producing nerve cells which release their product into the bloodstream. This hormone then stimulates part of the pituitary gland, the adenohypophysis, to produce or secrete hormones which, again via the bloodstream, reach the gonads and promote the production of gametes and sex hormones. Such an "optic–hypothalamic–hypophysial–gonadal axis" exists in numerous vertebrates. Blinding such animals usually eliminates the stimulatory effect of light on the gonads. The domestic duck (*Anas platyrhynchos*) is exceptional in that this operation does not disrupt the normal course of the reproductive cycle. One suspects in this case that there is direct photostimulation of certain brain areas through the somewhat transparent skull.

Light does not seem to play such an important role in vertebrates that live in regions close to the equator. For example, when deer are transferred to a temperate region they reproduce throughout the year, as in their original homeland, irrespective of the harsher winter. From this and other examples one can conclude that genetically determined endogenous programmes, perhaps modulated by exogenous factors, can also be effective in the regulation of the reproductive cycle.

A second important factor influencing reproduction in vertebrates is temperature. This is, understandably, not of great importance for homoiotherms, although such animals often do not reproduce at extremes of temperature, but is of considerable importance in poikilotherms and especially reptiles, where a rise in temperature is the principle stimulus for the start of the breeding season. The effects of temperature may occur on several levels. For example, it is thought that temperature can stimulate the central nervous system directly and, via hormone-producing neurons of the hypothalamus, activate the adenohypophysis: alternatively the latter could be activated directly. It has been found that raising the temperature increases the affinity of gonadal tissues for the adenohypophyseal hormones. Moreover, changes in the hormone profile of blood can be attributed to a shift in the temperature-dependent rate of hormone breakdown. On the other hand, the metabolic rate of most poikilotherms increases or decreases respectively with a rise or fall in temperature, so that a low temperature could also represent a limiting factor with regard to reproduction.

An interesting aspect of reproduction has been revealed by the results of investigations on the vole (*Microtus montanus*). The shoots of young plants contain the substance 6-methoxybenzoazolinone. When voles are fed this substance, they start performing the activities associated with reproduction. It is possible that this or a similar substance serves as a chemical trigger for reproduction in other species as well.

The third factor of importance with respect to the initiation of reproduction is the availability of water and food, the one being closely connected with the other. Two aspects of water need to be considered. Firstly, it can serve as a temporary environment in which animals reproduce and, secondly, it supports the growth of plants and thus directly or indirectly affects the availability of food. The first aspect is of primary importance with respect to the reproduction of amphibians, for which the formation or filling up of spawning pools is essential. Accordingly, the beginning of the breeding season is coupled to periods of rainfall in many species of Amphibia. Some species of anuran living in regions close to the equator, where the humidity is high, reproduce on land, and at all times of the year. Contrasting with these are species which spawn in water and whose reproduction is correlated with the fall of rain. This phenomenon is exemplified by the North American spade food toads of the genus *Scaphiopus*. These animals live in desert regions and prairies and spend dry periods underground in a state of aestivation. After rain falls they emerge to reproduce in the pools and puddles. The tadpoles hatch out after only 2 days and metamorphose about 10 days later, in a race against time as the pools dry up. It has been suggested that if they "lose the race" their remains facilitate the more rapid development of the next generation the following season. A similar though not strictly comparable phenomenon occurs in Teleostei and concerns so-called season fish which live in Africa and South America. The latter are species of the genus *Aphyosemion*, whose eggs survive in the ground after the pool in which they lived has dried up. In this case the development of the embryo is delayed until after it has rained. The young animals then hatch out, grow very quickly, reproduce and finally die when the pool dries up. Hatching of the eggs laid by tropical urodeles is usually correlated with falls of rain. In such animals, one of the parents winds itself around the hatch of eggs and protects it, sometimes for months, from being eaten or from drying out in the dry season: the young then hatch at the beginning of the rainy season (e.g. species of the genera *Bolitoglossa, Parvimolge* and *Pseudoeurycea*).

Indirect influences of rain on reproduction via its effects on the production of food are extensive and many-sided. Examples of such phenomena can be found in reptiles, birds and mammals. For example, oviparous species of reptile lay their eggs at the beginning of the rainy season, and species of bird living in the Australian desert begin reproducing immediately after rain has fallen. Among mammals a positive correlation has been found between rainfall and fertility in the black buffalo (*Syncerus caffer*). In general it seems that rain is of considerably greater importance as a factor influencing reproduction in tropical regions than it is in temperate climes. Rain is undoubtedly of more crucial importance than food for the higher vertebrates, being a contingent factor enabling animals to reach an optimal condition for reproduction and ensuring that the offspring have an adequate supply of food. This does not mean that reproduction is contingent on the animal being in optimal physiological condition. A range of species reproduce directly after a prolonged period of aestivation or hibernation, such as the spade foot toad mentioned above or the Dipnoi. The same applies to vertebrates which undertake migrations in association with reproduction, as do for example the European eel (*Anguilla anguilla*), *Oncorhynchus* and *Salmo* species, and the Petromyzontia.

In most cases, several of the factors influencing reproduction work together and it is often very difficult to distinguish which are the determining factors and which have only a modifying or modulating effect. Moreover, the exogenous factors described may also act in co-operation with endogenous programmes, and yet other stimulatory or inhibitory influences could be operating which are hard to define. Not surprisingly therefore, it is common for animals in captivity not to reproduce, perhaps for the lack of social contact with their own kind. Indeed it is known that the sight of a displaying male can stimulate ovarian activity in the female.

A synchrony of reproduction with the phases of the moon, or the changing tides they cause, occurs in teleosts, the most spectacular example of which is the American silverfish or grunion (*Leurethes tenuis*). This fish lets itself be thrown onto the wet beach by a wave of the spring tide, 1 or 2 days after full moon or new moon. The females dig themselves into the sand where they lay their eggs, which are then immediately fertilised by the males. The animals then return again to the sea with the next outgoing wave. The next spring tide follows in 14 to 15 days, during which time the eggs develop in the wet sand. They eventually hatch and the tide washes them into the sea. The related species *Hubsiella sardina* apparently reproduces in a similar manner.

A New Zealand trout of the genus *Galaxias* spawns when the spring tides flood the marshland, as does the killifish *Fundulus heteroclitus*. As in the above cases, the brood develop until the next spring tide comes and takes them into the water.

Not all vertebrates reproduce each year and there are evidently different reasons why this is so. One is the duration of the gestation period, or of parental care, and another is the exhaustion of energy reserves within the female body. Many large mammals have a very long gestation. Fin whales of the genus *Balaenoptera* have a pregnancy lasting 12 months and suckle their young thereafter. They apparently mate only every second year, although suckling mothers which were already pregnant again have been caught. Bears (Ursidae) also reproduce every 2 years, as do some big cats (Felidae) and the Indian elephant (*Elephas maximus*), but this phenomenon is not limited to mammals. Thus the Alpine salamander (*Salamandra atra*) has a gestation period lasting 3 to 4 years and birds with slowly developing young reproduce every 2 years (e.g. the condor *Sarcorhamphus gryphus* and the albatros *Diomedea exulans*). The 2 year rhythm of reproduction observed in certain reptiles is apparently due to the necessity of females to "restock" their bodies with energy reserves: such rhythms occur in Viperidae, the loggerhead turtle (*Caretta caretta*), the Gila monster (*Heloderma suspectum*) and have also been observed in tropical species of urodele.

1.3.7 Types of Mating

As has been explained in Section 1.2.5, the sexes of a species must recognise each other before actually mating and they do this in different ways with colours, scents, signalling behaviour etc. The probability of two sexual partners initially meeting each other is increased greatly by the aggregation of individuals of a species and this applies in principle to a wide variety of vertebrates. Thus many species of fish form schools which, with the onset of the breeding season, either become communal schools for spawning or break up temporarily due to the formation of pairs. The males of migratory vertebrate species may reach their breeding ground before the females, whose arrival they then await by signalling their presence in different ways: this applies to species of Anura, passerine birds, and seals (Pinnipedia). Some species form aggregations and then mate in autumn, such as insectivorous bats (Chiroptera), or in winter, such as many species of duck (Anatidae). A further possibility is that individuals of species which hibernate during the winter assemble long before the breeding season. In spring the animals all wake up and begin reproducing. In vertebrate species which neither migrate nor hibernate the chances of finding a partner are often increased by the formation of groups within the population.

There are basically two types of mating in vertebrates, polygamy and monogamy. The first is characterised by one individual mating with more than one partner of the opposite sex. When a male mates with several females it is referred to as polygyny and when a female mates with several males it is polyandry. Monogamy refers to the mating of an individual with only one sexual partner and is often characterised by a strong pair bonding between the individuals which may endure a lifetime. Strictly speaking monogamy also applies when an individual only has one sexual partner per breeding season, but a different partner each season. This represents a boundary case between monogamy and polygamy. Polygamy contributes to the variability of a species in that one individual mixes its genetic material with that of several others in the course of a breeding season, whereas with monogamy each partner contributes equally to the genetic constitution of the offspring.

Each type of mating is represented more or less equally in fishes. The range extends from the no doubt most primitive kind of polygamy, that of communal spawning as exhibited by herring (Clupeidae), to the life-long pair bond as represented by some monogamous tropical cichlids. Polygamy is the predominant type of mating in amphibians. This often involves a purely random mating between individuals in the breeding ponds, although very complicated mating displays can also play a role in pairing of Urodela in particular. Such displays usually do not exclude the possibility of mating with more than one individual. The situation with respect of Reptilia is less clear. Some species of lizard are definitely polygamous, whereas crocodiles are evidently monogamous, as are a large number of birds. In the case of at least some birds, however, studies show that their common bond to the nest seems to be of more importance than the bond between the two individual birds. Some birds are polygamous, most of these being polyandrous.

In mammals, the burden of parental care is shifted to be borne almost completely by the female. In those species which have young that require parental care, the female can take over the nutrition of the young with her mammary glands. Accordingly, the females often have a strong bond to the nest. Monogamy is rare in mammals. In the different types of polygamy, the male has a diversity of roles. Many species of mammal form polygamous family groups; e.g. Pinnipedia, Artiodactyla, and Ungulata. A territory is often held and defended by a dominat male, the territory in some way offering the females favourable conditions which attract them to the territory. Upon entering the breeding ground the females come under the control of the male and constitute his harem. Such is the case with sea lions (Otariidae), the bull defending a crag which is apparently very attractive to the female as a place for giving birth. By defending his females against intruding rivals, the male of such a system at the same time defends the young and thereby contributes indirectly to the parental care of the young. The formation of a harem is also a permanent feature of sociological systems in Islamic cultures. Promiscuity also belongs to the overall phenomenon of polygamy and occurs in many of the smaller species of mammal. In this case every fertile male mates with every female and vice versa. Polyandry does not occur in mammals with the exception – although certainly not directly comparable – of man: the asiatic Sherpa women marry more than one man.

The reason for the variety of mating types in vertebrates lies in the fact that it depends on many factors such as population density, the availability of appropriate mating and breeding grounds, and the type of parental care. A monogamous male that does not participate in the parental care of the young becomes superfluous after mating as far as reproduction is concerned, unless he contributes indirectly by defending his territory in which there are also females and offspring. If his territory should also offer reserves of food or other favourable conditions which are more scarce elsewhere in the environment, other females could be attracted to it. Since after mating the male is not "busy" with the first female, it is possible that he will turn his attention to and mate with these other females. It is possible that polygyny could have originated in this way. However, when the male engages in parental care he has less energy and time to attend to other females. One can accordingly explain why polygyny does not arise in many species of bird with male parental care whose nests lie close to one another and which have the opportunity so to speak, since the males are "busy" taking care of the young, as they do in the breeding colonies of gulls (Laridae) and cormorants (Phalacrocoracidae). In general polygynous species usually have a sex ratio of about 1:1 and are characterised by the fact that a large number of sexually inactive males are present in the population. The dominant male is commonly the only male to copulate with females in condition to mate, but this is not always the case: subordinate males also mate, particularly when several females are in oestrus at the same time. In some species of gallinaceous bird the fertile males all congregate at a display ground and establish a hierarchy among themselves. The females attracted by the males are usually mated by the dominant male.

A precondition for the development of polyandry seems to be that parental care of the young is taken over entirely by the males. This occurs in the Teleostei, a single species of Anura, and birds. In the latter it is quite possible that the male tends or broods the eggs even though the female does not mate for a second time. The males of some species of pheasant (Phasianinae) brood the first lay of eggs, whereas the second is brooded by the female. Polyandry does not necessarily occur in such cases.

1.3.8 Parthenogenesis

Reproduction from unfertilised eggs is called parthenogenesis or virgin birth and is a secondary "emergency measure" since it does not make use of the biological advantage of sexuality. It occurs rarely in vertebrates and primarily in so-called hybrids, the offspring of a mating between individuals of different but closely related species. The eggs of such animals do not go through meiosis and are therefore diploid. The offspring are all female and genetically identical to the mother. Examples of such all female species are known in the Teleostei, Amphibia and Reptilia.

In Teleostei, parthenogenesis occurs only in a few hybrid species of the family *Poeciliidae.* An example is *Poecilia formosa*, a cross between *Poecilia sphenops* and *Poecilia latipinna*, which illustrates a variation of parthenogenesis, namely gynogenesis. The females mate with a male of one of the parent species, although the sperm penetrates but does not fertilise the egg. Cleavage of the egg is nevertheless induced, giving rise to daughters which are genetically identical to the mother. In exceptional cases a true fertilisation takes place and the resulting offspring are triploid. Parthenogenetic species also exist in the genus *Poeciliopsis*, the hybrids being closely related species. These species illustrate another variation referred to as "imitative" sexuality or hybridogenesis. The females mate with and are fertilised by males from one of the parent species. The purely female offspring are triploid and have the characteristics of both parents. However, the paternal chromosome material is "lost" during development of the eggs so that the now diploid eggs only contain genetic material from the hybrid mother. Parthenogenesis is considerably more common in amphibians and reptiles than it is in the teleosts. Parthenogenetic species of the urodele genus *Ambystoma* form all-female populations which may be diploid or triploid. In both cases, spermatophores from males of closely related species can be successfully transferred to the females. The sperm penetrates the egg without fertilising it but nevertheless induces cleavage: i.e. this is a further example of gynogenesis.

Most of the parthenogenetic species which occur in six families of lizards appear to be hybrids, as their chromosome sets have been found to differ from the diploid form. Mating between these species and males of a related species occasionally leads to fertilisation, which results in sterile triploid females. The gecko *Leptodactylus lugubris* is diploid and as yet no males have been found. Some individuals of this species and of another parthenogenetic species *Lepidohyma flavimaculatum* have chromosome sets which look the same, so these species are apparently not hybrids. An Australian species of gecko of the genus *Gehyra* and the agame *Leiolepis triploida* exist only as triploid females. On the other hand, diploid, triploid and tetraploid females occur in members of the genus *Cnemidophorus*. It is known that tetraploid offspring result from a mating between triploid females and normal males of closely related species. The chameleon species *Brooksia spectrum affinis* is also only female and one therefore suspects it to be parthenogenetic. The same applies to the burrowing blind snake *Typhlops braminus*.

Natural parthenogenesis does not occur in birds and mammals, although parthenogenetic offspring have been bred experimentally in the domestic chicken (*Gallus domesticus*) and the rabbit (*Oryctolagus cuniculus*).

Parthenogenesis is perhaps so successful because it effectively doubles the fertility of species reproducing in this way since all individuals of a female-only species can raise young, not just half of the population, as in species with separate sexes. However, the fundamental shortcoming of parthenogenesis remains the very lack of sexuality characterising this form of reproduction and of the variability that sexuality engenders. Theoretically, parthenogenesis appears to be an evolutionary cul-de-sac lacking sufficient possibilities for further adaptation. This problem can be resolved by regarding hybridisation as the primary reason for parthenogenesis, based on the fact that hybridisation has arisen not once but time and again in different species sharing the same biotope. Thus the parents, being fully adapted to their common environment but of different species, pass on their genetic material to hybrid parthenogenetic females that possess a new genetic make-up. These new individuals, together with the original populations, represent the only possible basis for continued adaptation.

1.3.9 Hermaphroditism

The great majority of vertebrates are dioecious, that is they have separate sexes, whereas a minority are monoecious, every such individual possessing testes as well as ovaries in their body; they are hermaphrodites. When the male and female gonads of such an animal mature and produce gametes at the same time it is referred to as simultaneous hermaphroditism, but when one precedes the other it is called consecutive hermaphroditism. Two possibilities therefore exist in the latter case; protandry, when the male phase comes first, and protogyny, or proterogyny, when the female phase has precedence.

The group of vertebrates exhibiting by far the greatest proportion of hermaphroditic species is the Teleostei, which in any case appear to have a certain lability with regard to sex. Hermaphroditism has been studied experimentally in Cyprinodontidae where heterogametic males can be turned into functional females by treatment with female sex hormone and young females become functional males under the influence of male sex hormone. Simultaneous hermaphroditism is rare in environments supporting only a very low population density and in which there is only a minimal chance of two sexual partners meeting. This chance is increased by 100% in species in which there is no self-fertilisation if the ovaries and testes mature at the same time. In fact, a variety of deep sea fish are simultaneous hermaphrodites, including, for example, species of Aulepiformes. These latter have separate testes and ovaries which are functional at the same time. A comparable situation is found in benthic species of the families Chlorophalmidae, Ipnopidae, Evermannelidae, Alepisauridae and Paraleninidae. It is not certain whether any of these species is self-fertilising.

The best-known example of a consecutive hermaphrodite among the teleosts is the wrasse *Labroides dimidiatus*, a marine cleaner fish. Each dominant male possesses a harem of several females. When the male is removed, the harem is either taken over by another male or the dominant female of the harem changes itself into a "new" male within 2 to 4 days. In other species of Labridae and parrot fish (Scaridae), not all genetic males develop directly into functional males, some first go through a female phase before changing sex. The hermaphroditic Cyprinodontid *Rivulus marmoratus* has been particularly well studied. This species has a hermaphroditic gonad whose testicular part becomes larger with age until egg laying eventually stops. Male secondary sex characteristics then develop and a functional male is formed. This sex change is accomplished at a time of short day length, and the amount of time required is apparently dependent on temperature and on the genetic constitution of each individual. A further example of protogyny occurs in the East Asian swamp eel *Monopterus albus*. Animals at an age of about 2 years first develop ovaries and become females but by the end of their third year, the animals change into males with functional testes.

The plasticity of the gonads is limited to the early stages of life in amphibians and reptiles: naturally occurring hermaphrodites are rare. In these animals, low temperatures generally seem to suppress the development of the testes. An example of consecutive hermaphroditism in amphibia has already been mentioned while dealing with the genetic determination of sex: Northern and Alpine races of *Rana temporaria* all begin to develop ovaries shortly after metamorphosis and only later do the genetic males develop testes. Such protogyny is also shown by certain populations of *Rana catesbeiana* which live on Formosa. All the animals are initially female after metamorphosis, then 6 months later roughly half transform themselves into males. Sex change in amphibians has been induced experimentally by treating the larvae with sex hormones or to changes in temperature. Temperatures from 25° to 30°C generally promote the development of the testes whereas at 10°C undifferentiated gonads develop into ovaries. Basically the same effects have been obtained using high and low incubation temperatures in lizards and chelonians.

Naturally occurring hermaphrodites are unknown in Reptilia, nor do they occur in Aves and Mammalia.

1.3.10 Storage of Sperm

In numerous vertebrates internal fertilisation results in the fusion of the egg and sperm relatively soon after copulation, but in a wide range of species the sperm are stored. In such cases copulation takes place before the eggs ripen, sometimes even years before. The sperm are usually kept in folds, pits or pockets of the oviduct called receptacula seminis, but in other cases in lacunae, or spermatheca, in the tissue of the ovary. Mechanisms for the nutrition of the sperm involving the formation of suitable secretions appear to have been developed in some species. The spermatozoa are usually immobile while in these storage structures.

In Elasmobranchii, sperm are stored in glandular extensions of the oviducts, the glandulae nidamentariae, or shell glands, and there they penetrate the egg shortly before the egg shell is formed. Sperm are stored in the ovaries of teleosts, including the Poeciliidae, which give birth to live young (e.g. the guppy *Lebistes reticulatus*). In this species the sperm are held in spermatheca of the unpaired ovary and at certain intervals penetrate and fertilise the ripe eggs: the sperm are stored for about 8 months.

Only one case of sperm storage is known among Anura and that is in the North American bell toad *Ascaphus truei*. Fertilisation is internal in this species and the sperm survive for several months in the genital tract. Urodeles are in almost every case fertilised internally by spermatophores and in two species, *Notophthalmus viridescens* and *Desmognathus fuscus*, the sperm are stored.

In vertebrates, sperm storage occurs most frequently in the reptiles, being found in chelonians, lizards and snakes. Sperm remain for about a year in the female genital tract of marine turtles. These animals mate in water, at which time, however, shelled eggs are already present in the body which were fertilised by sperm from the previous copulation that took place 1 year earlier: i.e. the sperm are stored until the next reproductive period. Sperm storage also occurs in pond turtles of the family Emydidae. In some species of snake the sperm are preserved in the oviduct for up to 6 years. The phenomenon is particularly common in the families Gekkonidae, Eublephyridae, Chamaeleonidae, Iguanidae and Agamidae. In general reptiles store sperm in folds, crypts or pockets of the oviduct, mostly in the cranial region prior to the segment which forms the egg shell, but in some cases free in the lumen of the oviduct. After a single copulation, the Madagascan gecko *Phelsuma dubia* can lay several hatches, each of two eggs, at intervals during the breeding season, every egg of which hatches to give a young animal. In this case the spermatozoa are stored in the lumina of the extended oviductal glands.

The longest time sperm have been observed to survive in the female genital tract of a bird is roughly 6 weeks, this being in the turkey (*Meleagris gallopavo*). Sperm still capable of fertilisation can remain 3 weeks in the oviduct of the common pheasant *Phaisanus colchicus*, but only about 1 week in the ring dove *Streptopelia risoria*.

Among mammals the long-term storage of sperm only occurs in insectivorous bats (Microchiroptera). Mating takes place between late summer and autumn and the sperm are stored for about 5 months during the period of hibernation and then the ripe eggs are finally ovulated and fertilised. Some of these species of bat overwinter in separate male and female colonies and in the latter it has been observed that ovulation and fertilisation tend to occur simultaneously in different individuals. Other than this, the longest survival time for sperm in the female genital tract is that recorded in the European hare *Lepus europaeus*: approximately 30 days. In dogs, living sperm can be found 1 week but not longer than 2 weeks after copulation.

With respect to the phenomenon of sperm storage in general, it is still not clear how sperm are immobilised in the female genital tract and activated again when needed.

1.3.11 Delayed Implantation

The mammalian egg is not provided with yolk and after fertilisation develops differently from the eggs of other vertebrates. It forms at first a blastocyst which anchors itself in the wall of the uterus, in other words it implants. Following this, a complicated structure is formed to support the transport of substances both to and from the developing embryo, the placenta, which is composed of embryonic as well as maternal tissues. In a range of mammals, the birth of the young is delayed because the development of the blastocyst is interrupted or "frozen". This phenomenon is generally known as diapause and includes the postponement of implantation until development is renewed. Such delayed implantation occurs in species which live in particularly constraining conditions with respect to their reproduction and survival.

Male and female migratory seals (Phocidae) and sea lions (Otariidae) live separately outside of the breeding season. This means that females already pregnant with young from the previous breeding season must given birth and suckle their young relatively soon after arriving at the breeding ground, but at the same time also find a new mate. This problem of timing is solved with the aid of delayed implantation, and has been particularly thoroughly studied in the fur seal *Calorhinus ursinus*. Females of this species give birth half a day or one day after reaching the breeding ground. Within the next week a postpartum oestrus follows and ovulation culminating in copulation with the male partner. A blastocyst is formed but its development is quickly arrested, thereby delaying implantation. Implantation only occurs when the newly born young stop suckling, that is when milk production in the mother's mammary glands ceases. However, there appears to be no causal relationship between deferment of implantation and suckling of the young, since females which copulate for the first time and therefore have no young also show delayed implantation. Two things are achieved by this kind of reproduction. Firstly, the newly born young can be fed and at the same time another conceived within the relatively short period of reproduction. Secondly, the delay in development means that at the end of gestation the young will be born at the beginning of the next breeding season. Species which care for their young for a short length of time only begin mating when free of their young. Again in this case, the blastocyst "sleeps" for a shorter or longer period of time until it finally implants and develops further.

Delayed implantation is also a component of reproduction in many species of marten which live in temperate or cold climates. As a rule these animals mate in summer and give birth to their young the following spring. The ermine (*Mustela erminea*) mates in spring but implantation does not occur until winter or even the following spring. The fisher marten (*Martes pennanti*) has a similar mode of reproduction. The spotted skunk (*Spilogale putorius*) shows a delay of implantation of about 200 days. In this species it has been found that the "awakening" of the blastocyst and implantation are evidently under the influence of the hormone progesterone produced within the ovaries by the corpus luteum. In most species of marten which have delayed implantation, the further development of the blastocyst occurs at times of increasing day length, but with shortening day length an a minority of species.

Delayed implantation also occurs in the brown bear (*Ursus arctos*), deer *Capreolus capreolus* and nine-banded armadillo (*Dasypus novemcinctus*), as well as in some species of bat (Microchiroptera). The latter cases deviate from the pattern described so far. In one tropical species of bat the development of the embryo is not arrested but merely slowed down. In another, the blastocyst implants immediately but then develops only slowly. A slightly modified type of delayed implantation occurs in a range of marsupials and these special cases will be dealt with in detail later in another connection (see p. 190).

1.3.12 Sexual Dimorphism

The key criterion for deciding the sex of an animal is the sex of the gonads, and these structures therefore represent sensu stricto the primary sex characteristics. In general, however, this term also refers to the respective reproductive ducts, the gonoducts. Other than this anatomical difference the sexes of a number of vertebrate species do not differ in their physical form. The majority of species nevertheless exhibit additional features specific to each sex which, taken together, give each sex a different characteristic appearance, manifest as sexual dimorphism. Such features of the sexes are called secondary sex characteristics. In the simplest case males and females differ only in their body size. Thus in lower vertebrates the female is frequently bigger than the male. In mammals the male is often more solidly built than the female. There is, in addition, a wide spectrum of differences in body form, ranging from a simple shift in the proportions of the body to the formation of accessory structures which are specific to each sex, including body colouring, extensions and appendages to the body, hair, feathers and structures for the transfer of sperm or the laying of eggs, respectively the copulatory organs and ovipositors. The secondary sex characteristics also include organs which are involved in parental care, for example, the mammary glands of mammals. When the sexes differ only in their colouring, one speaks of sexual dichromatism. All the above features are of course genetically predetermined, although whether or not they are in fact manifested depends in many cases on the influence of sex hormones. The secondary sex characteristics frequently develop in an ontogenetic sequence which is correlated with the process of sexual maturation and can be irreversible or reversible. Some characteristics develop and persist throughout life, whereas others are formed anew every breeding season and disappear again thereafter. The secondary sex characteristics play an important role in reproduction: they serve as a foundation for the recognition of the sexes and for the necessary synchronisation between the sexual displays of the partners, they make successful copulation possible in many cases, and in the context of parental care contribute to the survival of the offspring.

Outside the breeding season there is no difference between the sexes of Petromyzontia. When the season begins, a swelling forms in the region around the anus and the common opening of the excretory and reproductive system of females. In males the ventral fin becomes larger and the urogenital papilla enlarges. Males from the class Chondrichthyes have permanent copulatory organs which derive from parts of the pelvic fins, the so-called mixopterygium or pterygopodium. There is a greater variety and wider distribution of secondary sex characteristics in Osteichthyes, occurring predominantly in males. At the onset of the breeding season the males commonly change colour, becoming more colourful and conspicuous: it is their so-called wedding dress or display coloration. The fins are often larger and more colourful than those of the female and function as signalling organs in the mating display. A particularly spectacular example of this is the sword-like process of the tail fin of the male Poeciliid *Xiphophorus helleri*, which gives this species the name of sword tail. Males of species which fertilise internally have a copulatory organ, the gonopodium, which derives from the anal fin. Male salmonids grow a hook on their upper or lower jaw at the time of spawning and, in salmon of the genus *Oncorhynchus* especially, this is accompanied by changes in body form (Fig. 5A). In Cyprinidae and Catostomidae the males develop horny spawning papillae over the body, mostly the head, which disappear again after the spawning period: their function is, however, unknown. In some species of Blenniinae and Salariinae a flap of skin grows on the head of the males during the spawning period. An extreme and spectacular example of sexual dimorphism occurs in the deep sea species *Photocorynus spiniceps* and *Endriolychnus schmidti*.

Fig. 5A–E. Examples of sexually dimorphic vertebrates. **A** The pink salmon (*Oncorhynchus gorbuscha*) (after Lalgler et al.). **B** Parasitic dwarf males on a female of the species *Endriolychnus schmidti* (after Buchner). **C** Dimorphism of the forelegs in the frog *Leptodactylus ocellatus* (after Kingsley Noble). **D** Dimorphic "hair styles" of the lion (*Panthera leo*). **E** Body form and body hair of the human sexes

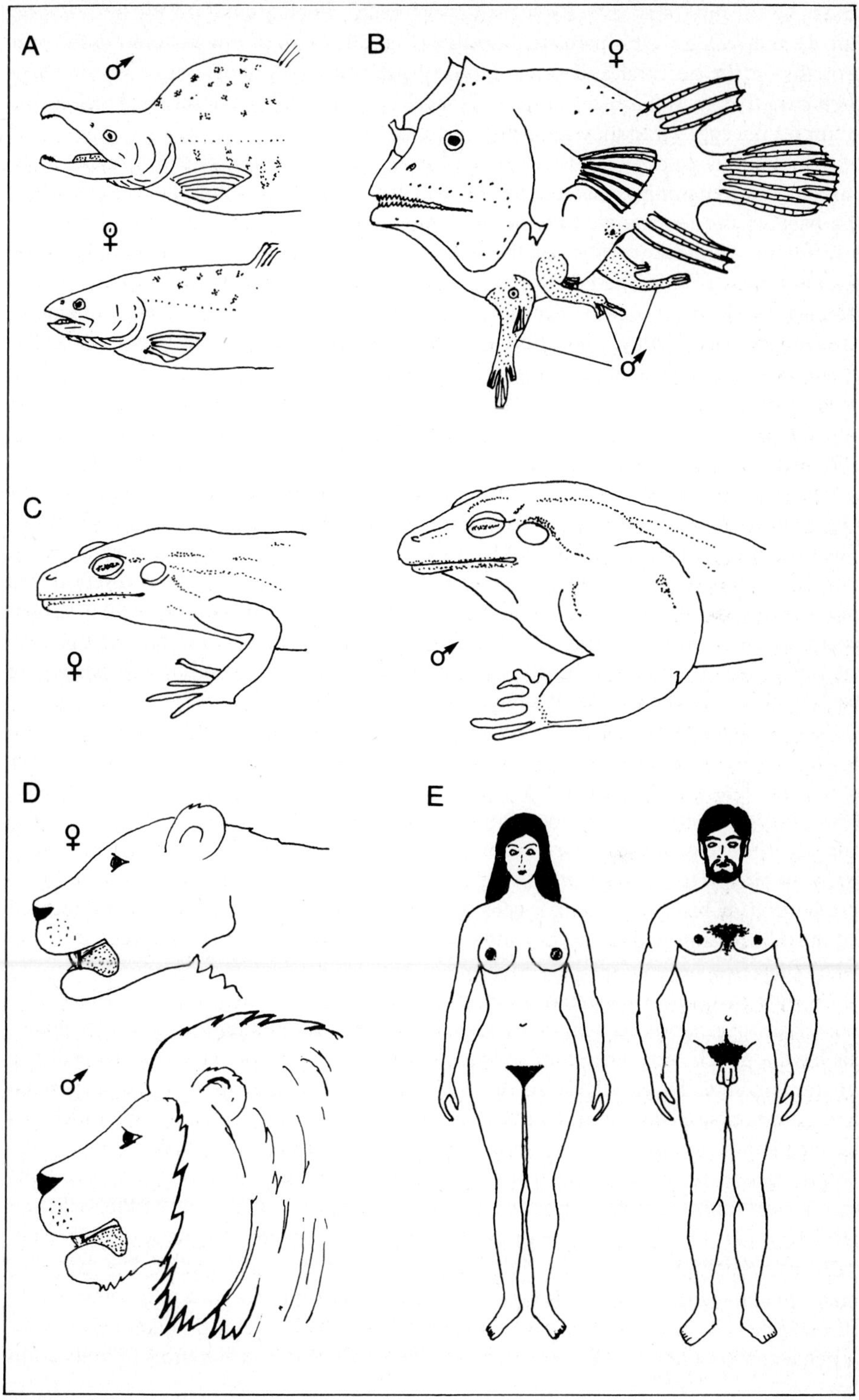

Fig. 5A–E

The males of these species are so-called dwarf males which grow onto the body of the female and which have a parasitic life style (Fig. 5B). Also of interest are the filiform growths on the pectoral and pelvic limbs of male South American lungfish (*Lepidosiren paradoxus*). It has been suggested that these growths secrete oxygen which is taken up by the eggs while they are being cared for.

Conspicuous differences between the sexes are less common in Amphibia. Male anurans are frequently smaller than the females. Prior to spawning the male rides on the back of the female, clasping her with his "arms". The arms of the male often have a different structure to those of the female, the musculature and humerus being of stronger build (Fig. 5C). Male Ranidae have in addition so-called nuptual pads on the thumbs which make it easier for him to hold onto the female. Sex differences among urodeles include display coloration and an enlarged crest on the back and tail, as for example in the crested newt (*Triturus cristatus cristatus*).

In many species of reptile the sexes appear virtually the same. Some saurians have appendages on the head, back and throat (see title picture to Chap. 5). Differences in coloration and the presence of gland-like structures also occur. The tail of male chelonians is often somewhat longer than the female's. There is nevertheless a marked difference between the sexes, albeit one that is only evident at the time of copulation. Thus the males of many species of lizard and snake have a copulatory organ, the so-called hemipenes, which normally remains hidden in the cloaca. Females also possess a rudimentary hemipenes. The copulatory organ of chelonians and crocodiles is very probably a homologue to that of mammals. This "penis" lies on the floor of the cloaca and prior to copulation becomes filled with blood, thus swelling and extending backwards out of the cloacal opening.

Sexual dimorphism in birds relates primarily to the structure and colour of the feathers. The males of many species have a characteristic colouring of the feathers and beak. The display plumage of male birds of paradise (Paradiseidae) and peacock (*Pavo cristatus*) are particularly conspicuous. In the domestic chicken (*Gallus domesticus*), the size of the comb is a secondary sex characteristic. Throat appendages also occur in birds. The production of sounds plays a particularly significant role in the reproduction of birds. Bird song is generated by the syrinx, a specialised region located at the bifurcation of the trachea into the bronchi. This structure develops in males when they first become sexually mature.

A wide spectrum of secondary sex characteristics exists in mammals. The female sex of all species possess mammary glands, or mammae. In addition, female marsupials have a pouch, or marsupium, in which the young are kept. The external genitalia of the female differ greatly from those of the male, which in apes are much enlarged and conspicuously coloured. In males there is a wide variety of structures which can be regarded as weapons, for example the antlers of Cervidae, the tusks of the walrus (*Odobaenus rosmarus rosmarus*) and the bull elephant, and the horns of Artiodactyla (ungulates) which are often more well developed in males. The mane of the male lion (*Panthera leo*, Fig. 5B) is an example of hair which is specific to one sex. Male mammals commonly have a better developed musculature than females. The testes of many mammalian species no longer lie in the coelom but descend to lie temporarily or permanently in the scrotal sacs hanging from the body. All male mammals possess a penis as a copulatory organ. In man there are conspicuous differences in body form between the sexes. The female is characterised by wide hips and mammae whereas males usually possess broad shoulders and a more powerful musculature. Further differences occur in the pattern of facial and body hair, the male being distinctly more hirsute, although there is considerable racial diversity (Fig. 5E).

1.3.13 Reproduction and Migration

Many vertebrates undertake migrations that are related directly or indirectly to reproduction. These changes of habitat can occur once, or every year. In many cases of vertebrate migration the foremost causes are to be found in the living conditions in regions distant from the equator. Many animals exploit the favourable conditions of temperate climes in spring and summer, resulting from the increasing growth of plants and the consequent proliferation of animals on which they can feed, to reproduce at this time. In autumn they normally move on to regions nearer the equator. Aside from such climatic influences, it is also possible that a particular biotope is favourable for egg laying and the development of the young but does not represent an adequate biotope for the adult animals, so that after reaching a certain age the young animals leave the biotope where they were born and only return to reproduce themselves when they reach sexual maturity. The phenomenon of animal migration is highly complex, involving a profusion of ecological, metabolic, physiological, osmoregulatory, and sensory problems and it would go beyond the scope of this book to cover all of these points in detail.

The term migration means, from the short description given above, that the animal leaves the place where it was born and returns again some time later. This homing ability implies that the animal has somehow imprinted on its place of birth and that it can find its way back by means of orientation and navigational mechanisms. When that place is the breeding ground of a species one speaks of a spawning or breeding migration. These terms describe only one part of the migratory cycle, which as a rule includes reproduction in one of its phases. Thus vertebrate migration can be separated into two phases, the first being food migration which is not directed towards reproduction. Rather, the individual seeks out a region favourable to itself in which to to spend the winter or where food is available.

The second phase is directly or indirectly orientated towards reproduction and involves migration to the spawning or mating ground or to the place where the young are born or looked after by the parents. The special migrations to over-wintering sites can be included among the latter.

Species belonging to the suborder Petromyzontia (Cyclostomata) have a migratory phase in their life-cycle. After metamorphosis and before sexual maturation the river lamprey *Lampetra fluviatilis* migrates down river to coastal regions where they live in brackish or sea water. They remain there for several years, although it has also been reported that they may migrate back up river in the same year. The gonads mature while the animals remain in sea water and there is evidently a coincident change in the animals' tolerance to salinity such that they have to return to freshwater. The spawning migration begins in summer and may continue until the following March. The lampreys do not feed at all on their way up river, and their gut atrophies. They spawn in the upper courses of streams and subsequently die. Immediately after the larvae hatch they begin the first phase of their migration to the sea, being carried passively by the current to still waters where they bury themselves in the slimey river bed and remain for several years. The marine lamprey *Petromyzon marinus* similarly spawns in freshwater. In North America this species has invaded the great lakes but also changed its life style since, after making its spawning migration up river, it does not return to the sea but swims into lakes. The lamprey is a diadromous species which migrates from freshwater to sea water and then back again. On the other hand, the spawning migration is anadromous since it goes up river from the sea. Species which migrate from freshwater into the sea where they reproduce are catadromous.

Some species of Teleostei undertake extremely long migrations. The most spectacular is that of the European eel *Anguilla anguilla*. Its spawning ground is the Sargasso Sea where mating takes place between March and April, possibly at a depth of about 400 m. The fish die after spawning. The developing eggs are pelagic and when the larvae hatch they grow into "willow leaf" larvae (*Leptocephalus*). These drift eastward, mostly passively with the Gulf Stream, reaching the coasts of Spain and Ireland some time in October of their third year of life by which time they have a length of about 7 cm. The willow leaf larvae then go through a first metamorphosis, losing their leaf-like form and becoming thin elvers, their originally long, thin teeth being replaced by conical teeth. The elvers reach the mouths of German rivers by April or May and begin to climb up river primarily at night, at which stage the animals have a dark pigmentation and are called climbing eels, which subsequently develop into yellow eels. Some time after their fifth year the male eels stop growing and remain in the lower courses of the river, whereas the females migrate further up river and continue growing, becoming silver eels with an age of between 9 and 15 years. The females then stop feeding and, together with the males, migrate in late summer or autumn down river back into the sea. It takes them about 18 months to cover the more than 6000 km to the Sargasso Sea. The American eel, *A. rostrata*, also spawns in the sea but more to the southwest. Since the larvae have a considerably shorter distance over which to migrate to the American rivers, the larval stage in this case lasts only about 1 year. The spawning grounds of *A. japonica, A. dieffenbachi*, and *A. australis* in the Pacific Ocean are unknown and they never climb the rivers of the American west coast.

Many species of salmonids are also migratory. The Atlantic salmon *Salmo salar* spawns in the clear upper reaches of rivers. At first the young have eight to ten dark transverse bands over their body. They then change into so-called smolts which migrate downstream where maturation of the gonads begins. The fish then make the return migration into the rivers but no longer feed. During this journey the males develop their secondary sex characteristic (see p. 37), becoming hook salmon. After spawning in the upper courses of rivers, the salmon return once more to the sea. The spawning migration of this species is an annual event. The migration of the Pacific salmon of the genus *Oncorhynchus* is similar but, as a rule, they only undertake one spawning migration during their life, reproduction being followed by death. These species cover a considerable distance during their migration, about 4000 km in the case of *Oncorhynchus tschawytscha*. Migratory forms of the European trout (*Salmo trutta*) swim up river to spawn in the upper courses of rivers.

Certain forms of the three-spined stickleback (*Gasterosteus aculeatus*) make a spawning migration from brackish water up into rivers (i.e. they are anadromous). Spawning migrations are also undertaken in the sea by herring (*Clupea harengus*), cod (*Gadus morrhua*) and mackerel (*Scomber scombrus*).

Many amphibians make seasonal or yearly spawning migrations, commonly returning again to reproduce in the waters where they were born (e.g. anuran and *Triturus* species indigenous to middle Europe). However, no great distances are covered in these cases. Mass migration occurs in the North American species of salamander *Notophthalmus* (= *Diemyctilus*) *viridescens*. After metamorphosis this species leaves the water as an immature red eft that spends 2 to 3 years on the land. With the onset of sexual maturity the animals go through a second metamorphosis involving changes in the structure of the skin, the formation of a keel on the tail, and a change in body colour to green. This green eft then undertakes a spawning migration back into the water. The tendency of this animal – and probably of amphibia in general – to seek out water (water drive) is triggered by the hormone prolactin.

Among reptiles, phenomena comparable to those described above are known only for turtles (Chelonia). The green turtle (*Chelonia mydas*) is pelagic and reproduces every 2 to 3 years. Populations living off the Brazilian coast undertake a reproductive migration of some 1200 km to Ascension Island where they lay their eggs on land. Exactly where this species lives outside of the egg-laying period is not known.

The best known example of animal migrations are those of birds. The delaying of reproduction until regions with favourable feeding conditions are reached is of foremost importance in these cases. Only a few species of bird (resident birds) winter in temperate latitudes, most (migratory birds) move more or less directly to regions nearer the equator before the less favourable season begins. As a broad generalisation one can say that migratory birds usually breed in the polar regions during the season when conditions there are favourable and, before the harsher season sets in, move back to more equatorial regions to winter. Different populations of the same species sometimes take different migration routes and also winter in different regions. For example, the West European populations of the white stork (*Ciconia ciconia*) go over the Mediterranean or the Straits of Gibraltar to West Africa, while the East European populations take a route over Asia Minor and Egypt to East and South Africa. Contrary to this generalisation, some species of bird take extreme migratory routes going from one pole to the other, an example being the sea swallow (*Sterna paradisea*), which breeds on the coasts of Eurasia and North America north of the polar circle. It migrates over the Atlantic and Indian Oceans as far as the Weddell Sea south of the Antarctic polar circle. On the other hand, Wilson's tern (*Oceanites oceanicus*) nests on islands between Cape Horn and the Antarctic as well as on Mauritius and the Kerguels in the Indian Ocean. After reproducing, this species migrates to the Northern Hemisphere reaching the latitude of Labrador and the British coast.

A number of mammals are migratory, among them the Pinnipedia as has already been mentioned in dealing with delayed implantation. A further example is the Northern fur seal (*Calorhinus alascana*). This species reproduces on the Pribilof Islands, among others, to the north of the Aleutians. In autumn the animals leave their breeding ground, the males remaining in the North whereas the females and young migrate south to the coast of North Mexico to return every year to their breeding ground.

Many whales make extensive migrations every year. For example, the baleen whales (Mystacoceti) follow their major source of food (krill) to the polar regions. In autumn they migrate south to their breeding grounds with some species taking no food for months on end. These animals live off their enormous amounts of reserves, primarily fat, which also suffice for the production of large amounts of milk after the birth of the young. Each year these animals migrate a distance of over 7000 km. A further example is the humpback whale (*Megaptera nodosa*), which migrates from the Antarctic Ocean where the young are born to equatorial waters.

The migration of the North American caribu (reindeer: *Rangifer tarandus*) is primarily directed towards food. They go from lowland pastures in summer to the mountains in winter where storms frequently clear the snow from the ground. On the other hand, they also make nomadic migrations over great distances. European reindeer live in forests during the summer and move into the mountains in winter.

Similarly, a number of species of bat (Microchiroptera) usually migrate annually from, and then return to, a specific breeding cave. The North American grey bat (*Myotis grisescens*), for example, gives birth in warm caves in the south-east of the U.S.A. then migrates to colder caves, which may lie up to 400 km away, to winter.

CHAPTER 2

Comparative Anatomy of the Urogenital System

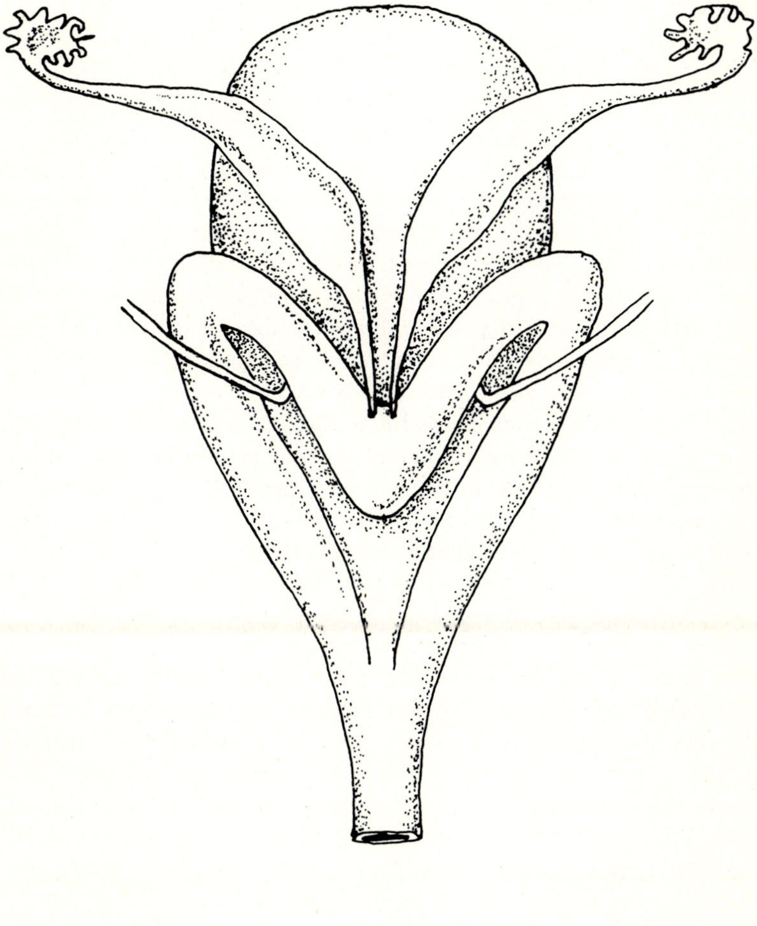

Genital system of a female opossum *(Didelphis virginiana)*

2.1 The Gonads of Vertebrates

The gonads of all vertebrates have two functions. In the first place they produce the gametes (eggs and sperm) but they also function as glands for the production of hormones which have an important role in the regulation of reproduction. This last aspect will be dealt with in Chapter 5. The existence of separate sexes is the rule in vertebrates, as has already been mentioned. The gonads of the male are the testes, those of the female the ovaries. Basically they are paired structures, deviations from this rule always being secondary. The Acrania have segmentally arranged gonads all their life, as do the Elasmobranchii and Amphibia at the embryonic stage, but in these latter they rapidly become fused together into a single organ. All other vertebrates have gonads that arise as compact organs.

Strictly speaking, the testes lie outside the coelomic cavity and only project or hang into it, each being invested in a peritoneal duplicature (or omentum) called the mesorchium. The mesorchia attach the testes to the dorsal wall of the coelom, except in mammals in which these organs are secondarily displaced (see p. 62). The form of the gonads varies greatly. The primitive form is that they extend through the whole length of the peritoneal cavity, but there is a later tendency for them to become more compact round or ovoid structures, a tendency that is evident even in fishes (Holocephala). Considerable differences in the size of the gonads are also commonly seen within an individual: these always accompany sexual maturation either in the course of ontogenesis or during the reproductive cycle. The Cyclostomata represent a secondary deviation from the rule that the embryonic anlage of the testes is paired. In some Elasmobranchii, the caudal part of the testes fuse together. More or less well-defined asymmetries of the testes also occur in the latter.

Since testes are suspended in the mesorchium, they are necessarily overlain by coelomic epithelium, which is erroneously identified as the germinal epithelium in some old textbooks. In mammals, whose testes descend, this covering is double due to the secondary repositioning of the testes (see p. 62). Beneath the coelomic epithelium there is a layer of connective tissue called the tunica albuginea. The testes themselves contain two principle cell types: cells of the germinal line derived from the primitive germ cells, which become the male gametes in the process of spermatogenesis, and somatic cells, which form the remaining structures of the testes.

The basic construction of the testes is relatively uniform and varies little within the vertebrates. Gametogenesis takes place in spherical or tubular compartments, respectively the ampullae and tubuli (seminiferous tubules) of the testes. These are invested by a basal membrane and a more or less well-developed connective tissue and contain primarily cells of the germinal line at different stages of spermatogenesis. The Elasmobranchii and Urodela have ampulla testes, whereas all other vertebrates have tubular testes. The Sertoli cells are the only somatic cells within the ampullae or tubuli. They play an important role in spermatogenesis and in some vertebrates are capable of producing hormones. The tubuli or ampullae are separated by interstitial tissue composed of connective tissue fibres, fibroblasts, hormone-producing Leydig cells and, allegedly, boundary cells in the case of lower vertebrates (see p. 173). It is also provided with blood vessels and nerves. The structure of the vertebrate testes is illustrated in Fig. 6A.

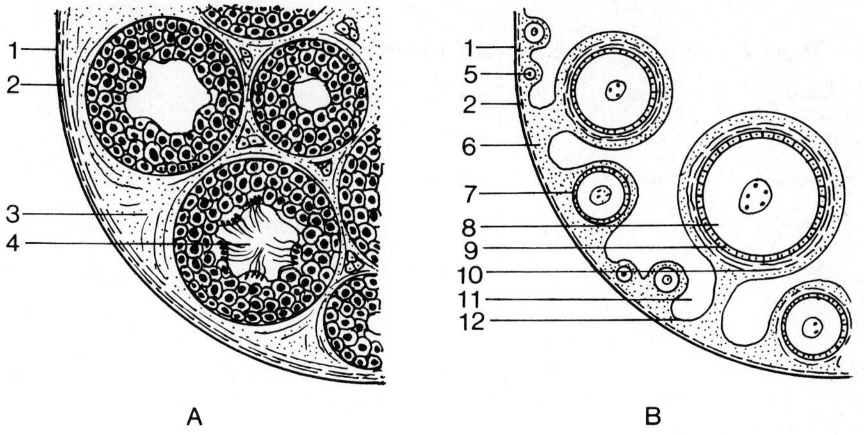

Fig. 6A,B. Diagram of the structure of the gonads. **A** Testis. **B** Ovary (of an amphibian). 1 = peritoneal epithelium; 2 = tunica albuginea; 3 = somatic tissues of the testis including connective tissue, nerves, blood vessels, boundary cells and Leydig cells; 4 = seminiferous tubule consisting of Sertoli cells and cells at different stages of spermatogenesis; 5 = early follicle; 6 = ovarian stroma including connective tissue, nerves and blood vessels; 7 = growing follicle; 8 = mature follicle; 9 = granulosa; 10 = theca folliculi (connective tissue); 11 = ovarian cavity; 12 = internal epithelium

The ampullae or tubuli of the testes merge together into a system of collecting ducts which leave the testes as ductuli efferentes. The ducts leading the sperm out of the body derive from and represent a part of the excretory system, although this is not always the case in fishes where the seminal ducts can have a different origin. This system of ducts will be dealt with in detail below (see p. 56). Different arrangements of the collecting ducts within the testes occur and the most important of these will be described. In the simplest case the tubuli of the testes join a longitudinal canal running along the margin of the testes from which the ductuli efferentes lead out of the testes (e.g. *Acipenser*: Fig. 7A). When this canal lies at the centre of the testis with the tubuli arranged more or less radially around it, it is referred to as the central testicular canal. Such an arrangement is found, for example, in fishes, in many of which the central canal extends out of the testis forming a secondary spermiduct which derives not from the excretory system but from the testis itself (Fig. 7C). A variant of this arrangement is the epitesticular type in which the canal for the sperm lies wholly or partly on the surface of the testis. The tubuli are connected to this canal singly or in groups (as in *Salmo gairdneri*; Fig. 7D). In the case of a rete testis, the ampullae or tubuli lead into a network of collecting canals situated marginally or centrally within the testes. Running from this network are the ductuli efferentes which pass out of the testis (e.g. Elasmobranchii, Aves, Mammalia; Fig. 7B). A different arrangement is found in some birds where the tubuli feed into a cavity or testicular antrum on the border of the testis which is drained by the ductuli efferentes (Fig. 7E).

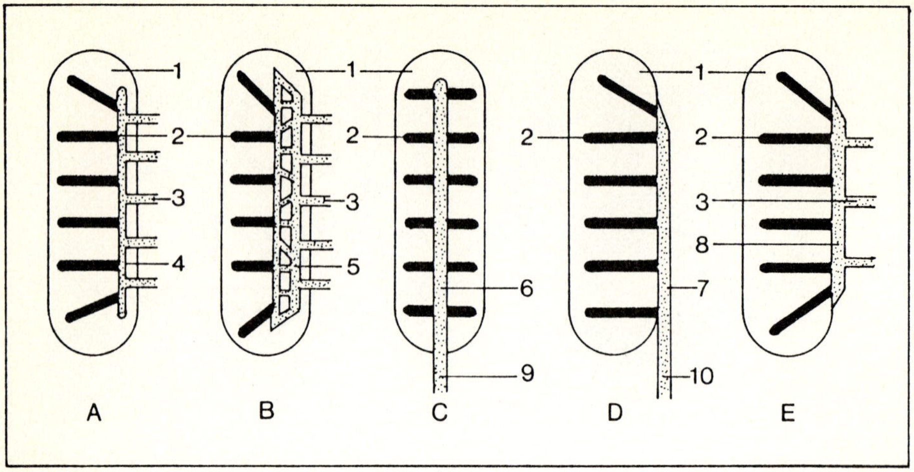

Fig. 7A–E. Diagram of the principle types of testis. **A** Testis with a longitudinal canal. **B** Testis with a rete testis. **C** Testis with a central canal and a secondary spermiduct. **D** Testis of the epitesticular type with a secondary spermiduct. **E** Testis with an antrum testis.
1 = testis; 2 = ampulla or tubule of the testis; 3 = ductulus efferens; 4 = marginal longitudinal canal; 5 = rete testis; 6 = central testicular canal; 7 = epitesticular duct; 8 = antrum testis; 9 = continuation of the central canal; 10 = continuation of the epitesticular duct (this duct and the latter canal function as secondary spermiducts)

The ovaries, like the testes, hang from the dorsal wall of the peritoneal cavity in duplicatures called mesovaria. The size and shape of the ovaries varies significantly according to the number of eggs and the amount of yolk they contain. As with the testes, the ovaries tend to have a more restricted distribution than in the primitive condition, where they extend along the whole length of the peritoneal cavity. The outer investment of the ovaries corresponds to that of the testes with the peritoneal epithelium overlying the tunica albuginea, which in this case is only poorly developed. Other than the female germinal cells in their different stages of development, the ovary also contains somatic cells which form, among other things, the characteristic follicles of the vertebrate ovary. Each follicle consists chiefly of a single cell layer, the membrane granulosa (or simply granulosa) surrounding the developing egg. This layer becomes multilayered during the course of the maturation of the egg, although not in fishes and amphibians (with the exception of Elasmobranchii). The granulosa cells of Monotremata and other vertebrates which do not suckle their young are involved in the transfer of yolk into the oocyte in some way that is as yet not fully understood. These cells are also a probable source of hormone. Since the eggs of viviparous mammals contain practically no yolk, they are very small compared to those of submammalians and only fill the follicle during the early stages of their development. As development proceeds a cavity forms in the follicle (antrum folliculi), which is filled with a liquid (liquor folliculi) secreted by the granulosa cells. The unripe egg (oocyte) is attached to the follicle wall by a single, or occasionally more than one, stalk composed of granulosa cells (see p. 174). At this stage the follicle is called a Graafian follicle. In submammalians the granulosa becomes enveloped by a layer of connective tissue (theca folliculi) as the follicle matures. In mammals this casing is com-

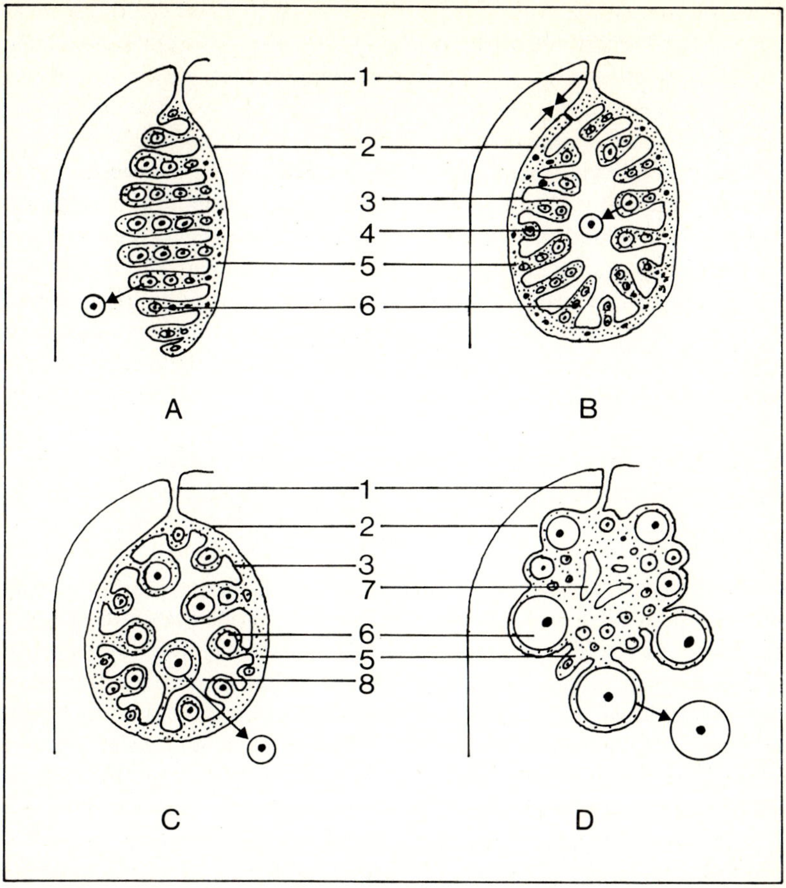

Fig. 8A–D. Diagram of the principle types of ovary. **A** Lamellar type (Teleostei). **B** Lamellar type with entovarian duct (Teleostei). **C** Ovary with an ovarian cavity (Amphibia). **D** Compact ovary (Amniota). *Arrow* indicates the direction of ovulation. *Opposing arrows* indicate where the lateral fold meets the body of the ovary.
1 = mesovarium; 2 = peritoneal epithelium; 3 = internal epithelium; 4 = entovarian duct; 5 = ovarian stroma; 6 = follicle containing an oocyte; 7 = lymph lacuna; 8 = entovarian cavity

posed of an outer layer of connective tissue (theca externa) and an inner (theca interna) which is cellular and separated from the granulosa by a basal membrane. The cells of the theca interna produce the female sex hormone (see p. 183). The connective tissue surrounding the follicles, the stroma ovarii, is usually poorly developed in fish and amphibians, but is increasingly evident in Amniota to a degree corresponding to the number of follicles (see Fig. 6B).

The ovaries of all vertebrates have the same basic structure. Variations occur primarily in the formation of the ovarian ducts and cavities. In the simplest case the ovary is a long, undivided organ containing follicles embedded in the ovarian stroma. Segmented ovaries having the form of lamellae occur in primitive teleosts (lamellar type; Fig. 8A). Ovarian cavities are formed by the lamellar aspect of the ovary rolling up to contact the dorsal or dorsolateral wall of the peritoneal cavity. These cavities serve as secondary oviducts that extend out of the ovary and open into the environment. In such cases one speaks of entovarian and parovarian oviducts (Fig. 8B). The follicles rupture when ripe and the oocytes leave the ovary, this being the process of ovulation. In the last-mentioned cases the oocytes are released into the entovarian or parovarian oviducts respectively, via which they leave the body in the process of being laid. Bony fish also possess this type of ovary. The ovaries of amphibians contain a cavity, lined with an epithelium, into which the follicles project. This type of ovary may be further divided by transverse folds to form ovarian pouches. In this case the eggs are not ovulated into the ovarian cavity but directly into the body cavity where they are taken up by the funnel-shaped opening of one of the oviducts, via which they leave the body (Fig. 8C). In Amniota the ovarian cavity is lost and the ovarian stroma is better developed and pervaded with more or less voluminous lymph spaces (Fig. 8D), although these are absent from the typical ovary of mammals.

As stated above, the ovaries are laid down as paired structures and usually develop as such. In some Elasmobranchii and most birds, however, there is a complete reduction of the ovary on one side. Asymmetrical ovaries are a common feature of vertebrates in general.

The ovaries of vertebrates contain additional structures which derive from the follicles, namely the corpora lutea and corpora atretica. The corpus luteum is of great functional significance as a source of hormone in mammals (see p. 193): it is formed from the follicle as a result of the ovulation of the ovum. The granulosa cells change into so-called granulosa luteal cells and together with cells of the theca interna completely fill the antrum of the follicle. Such corpora lutea postovulatoria also occur in other vertebrates. After its functional phase, the corpus luteum becomes a corpus albicans, consisting mainly of connective tissue. Corpora atretica are follicles from which the egg is not ovulated and are commonly found in submammalians and have a different appearance according to the stage of maturity of the follicle. They often resemble the corpus luteum morphologically, although they have never been unequivocally shown to be a source of hormone. They are therefore also called corpora lutea praeovulatoria. Such atretic follicles also occur in mammals.

2.2 Development of the Gonads

The gonads of all vertebrates arise during the course of embryonic development from the lateral plate mesoderm in the cranial part of the body. The anlage of the gonad on each side of the body lies on the dorsal wall of the body cavity just lateral to the mesentery. Initially these anlagen are simple epithelial ridges in the wall of the coelom which are referred to as the genital ridges (Fig. 9A). As development proceeds they become multilayered and project to a greater or lesser extent into the body cavity. There is a close spatial relationship between the genital ridges and the anlagen of the excretory system (see p. 53), although in the two sexes the anlagen of the gonads and excretory organs are originally separate. The epithelium of the genital ridges is composed of two different types of cell. The bulk of the cell mass is made up of cells re-

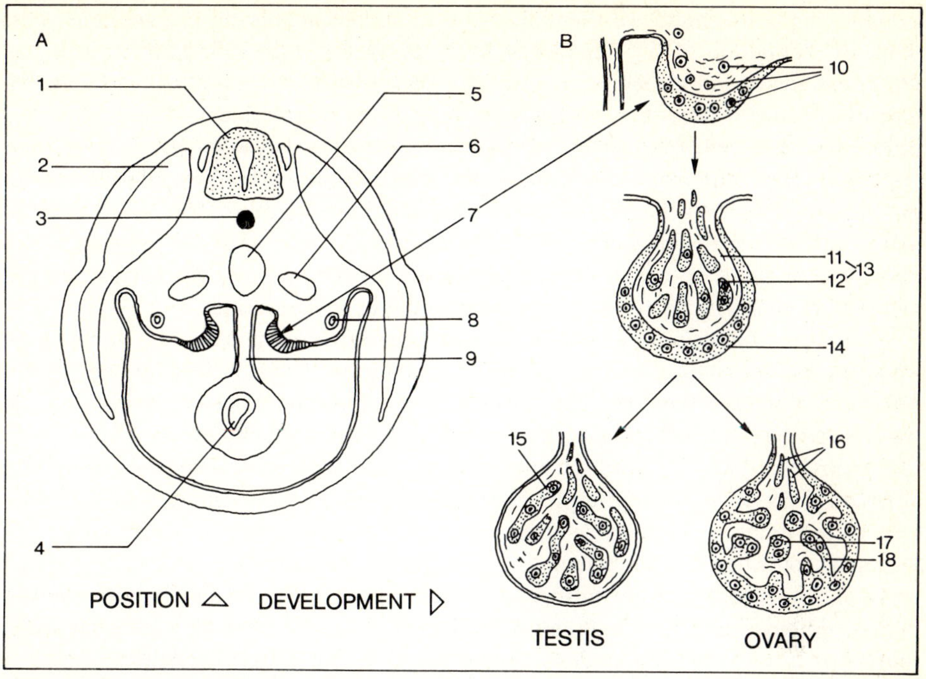

Fig. 9A,B. Position of the genital ridges in the embryo (**A**) and the development of the testes and ovaries (**B**). (After Balinsky).
1 = neural tube; 2 = myotome; 3 = chorda dorsalis; 4 = gut; 5 = aorta; 6 = posterior cardinal vein; 7 = genital ridge; 8 = pronephric duct; 9 = dorsal mesentery; 10 = primordial germ cells; 11 = mesenchyme; 12 = primary sex cords; 13 = medulla; 14 = cortex; 15 = anlage of the tubule; 16 = remains of the primary sex cords; 17 = egg clusters; 18 = secondary sex cords

sembling those of the peritoneal epithelium, with a pronounced cuboidal shape in the early stages of development. The second type of cell has a completely different appearance and is considerably less numerous. They are also markedly larger than the mesodermal cells, have a larger cell nucleus and their cytoplasm has different staining characteristics. Their form is spherical and they do not form a true epithelium but lie scattered within the epithelium of the genital ridges. They are the stem cells of the gametes and are therefore called primordial germ cells. The germinal ridge epithelium only gives rise to somatic elements of the gonad.

The primordial germ cells have an origin different to that of other cells of the genital ridge. At first they are usually to be found in other parts of the embryo, from whence they migrate secondarily into the genital ridges. For example, the primordial germ cells of amphibians can first be identified in mesenchyme above the dorsal mesentery and outside the genital ridges, from where they migrate into the germinal ridge epithelium. The position of the primordial germ cells outside the genital ridges is itself probably secondary, that is, they migrate there having arisen somewhere else. The primordial germ cells of anurans contain unusual cytoplasmic inclusions which can be traced as far back as the egg prior to cleavage. The inclusions, called germinal plasma, lie in a sub-cortical layer at the yolky pole of the egg and after cleavage can be found in cells lying in the middle of the yolk-rich endoderm, the presumptive germinal cells. After the formation of the nervous system (neurulation), the cells come to lie in the dorsal endoderm and subsequently migrate into the mesenchyme overlying the dorsal mesentery, from where they probably move into the genital ridges. It seems that the primordial germ cells of urodeles have a different origin. They apparently do not stem from endoderm but from lateral plate mesoderm, from where they migrate via the dorsal mesentery into the genital ridges. One can interpret this difference broadly to mean that the primordial germ cells do not derive from any particular germ layer but have their origin at an earlier stage of development. The somatic elements of the gonads are, in contrast, clearly of mesodermal origin in all vertebrates.

Cells with the staining characteristics of primordial germ cells can similarly be found in the extraembryonic endoderm at a very early stage of development in sauropsids. In chickens these cells lie in front of the embryonic head. If they are selectively destroyed experimentally, the genital ridges subsequently contain no primordial germ cells. Early in development the primordial germ cells of the mouse contain large amounts of alkaline phosphatase, which is almost totally lacking in the other cells of the embryo. These cells migrate out of the endoderm into the dorsal mesentery and from there into the genital ridges. During this migration the cells are apparently able to proliferate by mitosis. Primordial germ cells can also be identified in humans among elements of endoderm close to the so-called allantois stalk from where they move into the bordering mesenchyme and then into the genital ridges.

The question remains as to how the primordial germ cells accomplish their migration within the embryo. In vertebrates these remarkable translocations are made in two ways. In amphibians and mammals the primordial germ cells move by active amoeboid locomotion—they "creep" through the tissue. Those of sauropsids are apparently also carried passively in the bloodstream. Thus they first leave the endoderm and cross the intervening space to the mesoderm by amoeboid locomotion, then actively penetrate into blood vessels of the area vasculosa. Primordial germ cells can then be found in the embryonic blood. The majority evidently find their way into the genital ridges, but a minority end up in other parts of the body, where they probably degenerate.

The genital ridges are the final destination of the primordial germ cells, although these cells undergo further displacement in the course of the differentiation of the gonads. This process begins by the genital ridge epithelium bulging out into the peritoneal cavity forming a space behind it (i.e. dorsal) which is filled with mesenchyme. As the gonads develop, this tissue is rapidly replaced by cells which aggregate together to form compact primary sex cords. Two versions of how these cords originate can be found in the relevant textbooks. The most prevalent view is that the cords represent ingrowths of the genital ridge epithelium from which they become detached. According to the second view, the primary sex cords stem not from the genital ridge epithelium but are formed from cells which migrate from the anlage of the embryonic kidney. Whichever is the case, the anlagen of the gonads differentiate into two parts, a cortex and a medulla. The former is composed of the original epithelium of the genital ridge and primordial germ cells which migrate into it. The medulla is composed of the primary sex cords and scattered remnants of mesenchyme (Fig. 9B). This represents the indifferent stage in the development of the gonads.

The subsequent development of the gonads proceeds in fundamentally different ways in females and males. The formation of testes involves the migration of the primordial germ cells out of the cortex and into the primary sex cords of the medulla where they form the ampullae or tubuli of the testes: the primordial germ cells themselves later become spermatozoa in the course of the maturation of the gonads. In mammals it has been found that the Y chromosome carries a gene which codes for a histocompatibility antigen (HY antigen). This antigen so changes the nature of the primordial germ cell surface that the cells can penetrate into the medulla and induce it to grow considerably. The Sertoli cells of the seminiferous tubules derive from the tissue of the primary sex cords. The hormone producing Leydig cells of the interstitial tissue between the tubuli arise from the mesenchyme, as does the intertubular connective tissue. Adjacent to the seminiferous tubules of mammals there is usually a rete testis composed of small canals which, in the dorsal part of the gonad, probably arise from the same tissue as the seminiferous tubules. Thus the medulla is the functional part of the testis whereas the cortex is reduced to a thin skin covering the organ (Fig. 9B).

The relationships are exactly the reverse in females. During the differentiation of an ovary, the primary sex cords are degraded, thereby reducing the extent of the medulla until all that remains is a loose mesenchyme provided with blood vessels. At the same time the cortex becomes thicker: the primordial germ cells remain in this part of the gonad. As development continues, cords and groups of cells from the cortex grow towards the centre of the ovary anlage. Among them are the primoridal germ cells which later become eggs during the maturation of the ovary (see p. 95). The ingrowing cords and groups of cortical cells containing the primordial germ cells are referred to as the secondary sex cords. The granulosa cells of the follicle derive from the secondary sex cords, whereas the stroma and theca folliculi are derived from the mesenchyme.

The two courses of development described above apply in pinciple to the Elasmobranchii and all tetrapods. However, certain modifications to this basic pattern exist which must be mentioned. During the development of the amphibian ovary the medulla degenerates to such an extent that the fully mature gonad is merely a hollow sack into which project structures bearing the follicles (see p. 47). Eggs are released from this after ovulation by the rupturing of the outer ovary wall. In birds only the left ovary is normally fully developed. Even during the early stages of development, many more primordial germ cells migrate into the left genital ridge than into the right. Later, the typical structures of the ovarian cortex develop only in the left ovary whereas the right fails to differentiate. More exactly, the primary sex cords remain intact but the primordial germ cells present probably perish during the course of development. If the left genital ridge is removed operatively, the right can nevertheless develop into a testis. A comparable phenomenon occurs in male toads (Bufonidae) whereby the cranial portion of the genital ridge does not differentiate but remains in an "embryonic" condition: it is known as Bidder's organ. Experimental removal of the testes results in this element becoming an ovary. In this case the sex reversal occurs in the opposite direction to that found in birds. One can deduce from this that the primordial germ cells are undetermined with respect to their sex. It seems to be a simple question of whether the primordial germ cells remain in the cortex or end up in the medulla: in the first case they become eggs, in the second sperm. The situation in the teleosts is somewhat different from the general pattern in vertebrates. The gonads of teleosts derive solely from the peritoneal epithelium underlying the genital ridge which contains the primordial germ cells: a medulla is not formed. This condition is interpreted as being primitive and it is possible that this atypical type of gonadal development is one reason for the comparatively common occurrence of intersexuality in bony fish. The unpaired gonads present in Cyclostomata arise either by the suppression of the development of one anlage (Myxinoidea) or the fusion of both anlage (Petromyzontia).

The above-mentioned discrepancy about the origin of the primary sex cords in the accounts given in embryology textbooks is also evident in the original literature, which provides the following picture. The primary sex cords of amphibians arise from cells of the embryonic kidney anlage, whereas in amniotes they arise from the kidney anlage in some cases but the gonad anlage in others. The emerging picture from more recent evidence, however, argues strongly for a change of approach and may also throw light upon the source of earlier confusion. Thus the results of electron microscopical investigations show clearly that the development of the cortex and the medulla, as described above, does not take place successively. Rather, it appears that both parts differentiate from the same mesoderm: the superficial genital ridge epithelium is formed from an epithelial component, and the primary sex cords (i.e. the medulla) are compounded of epithelial cells and elements of the mesenchyme. This pattern applies to the Sauropsida and mammals. Until now the view that the medulla of amphibians is formed from cells of the embryonic anlage of the primitive kidney has not been refuted or been seriously put in question.

In this connection it must be mentioned that the idea of there being male and female compartments (medulla and cortex respectively) in the embryonic gonads was developed in accordance with the view that the fate of the primordial germ cells is determined by their location: they become spermatozoa if they reach the medulla and develop into eggs if they remain in the cortex. More recently this concept has been found in need of modification. Interactions between the inner epithelium, that is the primary sex cords, and the mesenchyme probably take place in the presumptive testis and these induce the development of Leydig cells in the mesenchyme. These cells then produce a hormone which influences the course of further development. This influence does not come into play if the indifferent gonads develop into an ovary, in which case the interaction between the epithelium and mesenchyme (stroma) occurs only after the ingrowth of the secondary sex cords. This latter process is probably mediated by the primordial germ cells.

The genital ridge is divided into three parts: the pars progonalis lying cranially, a medial pars gonalis and a caudal pars epigonalis. The gonads themselves always arise from the pars gonalis, the other two regions remaining undeveloped in most vertebrates. In some elasmobranchs, however, the caudal region develops into the so-called epigonal organ composed primarily of connective tissue and having an unknown function. The pars progonalis and pars epigonalis form the fat body of the gonads in Urodela and Gymnophiona. The pars epigonalis of anurans remains rudimentary, the fat body being formed exclusively from the pars progonalis. In Bufonidae the cranial portion of the pars gonalis also differentiates into Bidder's organ.

2.3 Structure and Development of the Kidneys: the Uro-genital Connection

It has already been mentioned that there is a close anatomical relationship between the genital ridges and the anlagen of the excretory system at a very early stage in the ontogenesis of all vertebrates. In most vertebrates this is of considerable functional significance in males: the excretory system contributes the final elements serving the discharge of sperm by the fusion of the seminal ducts arising from the testes with the kidney ducts. In other words a uro-genital connection is established which can lead to whole sections of the kidney functioning not as an excretory organ but only in the service of reproduction. To clarify this phenomenon it is necessary to introduce the essential facts concerning the ontogeny and phylogeny of the vertebrate excretory system. The typical excretory organ of vertebrates are the kidneys. They arise as paired structures in the course of embryonic development in close association with the genital ridges in the dorsal wall of the coelom. The morphological and functional unit of the kidney is the nephron. It is composed of a kidney duct for the excretion of urine and, in the primitive condition, a ciliated funnel which is open to the coelom via the nephrostome. Even early in the evolution of the vertebrates this ciliated funnel was replaced by a closed Bowman's capsule which is always intimately associated with a ball of tangled blood vessels, the glomerulus. The capsule and the glomerulus form a kidney corpuscle (corpusculum renis; Malpighian corpuscle). Nephrons which have a glomerulus are called glomerular, at least those which open into the coelom via a nephrostome, since as a rule the glomerulus is contained in a protrusion of the coelomic wall opposite the nephrostome that projects into the peritoneal cavity. In this last-mentioned case the glomerulus is referred to as being external: an internal glomerulus is represented by a corpusculum renis. Kidney corpuscles are absent in some marine bony fish, in which case the nephrons are said to be aglomerular. This condition is probably a secondary variant of the basic type.

The kidneys of modern vertebrates probably derive from a primitive type occurring in forms which are now extinct. The nephrons of this primitive type were probably situated lateral to the dorsal mesentery and were laid down segmentally throughout the length of the body cavity. Such a holonephros does not occur in its primitive form in any extant species, although similar forms can be observed during the embryonic development of Myxinoidea. The course of ontogenetic development in amniotes goes through the three successive generations of kidney which have arisen during vertebrate evolution. The first generation is the differentiation of the nephrons of the cranial part of the kidney anlage into a pronephros. The kidney tubules all lead into a common pronephric duct, or primary urinary duct, that extends caudally. A pronephros is formed in all vertebrates but persists throughout life as a functional kidney only in the Myxinoidea and some bony fish. It is not composed of many nephrons. In Petromyzontia and numerous bony fish it becomes modified in the course of ontogenesis into a lymphoid organ, but is rudimentary in all other vertebrates. The next two generations of kidney are commonly referred to together by the term opisthonephros and are the mesonephros, or primitive kidney, and the metanephros. The mesonephros is situated just caudally of the pronephros and is the functional excretory organ of most fish and of amphibians. Its basic structure corresponds to that of the pronephros, each nephron having a Bowman's capsule though open nephrostomes are also present.

The nephrons develop during embryogenesis from nephrotomes lying between the mesoderm of the somites and that of the lateral plate on each side of the body. The nephrotomes situated most cranially give rise to tubules which fuse into a pronephric duct that extends caudally. Tubules similarly arise from the cranial aspect of the nephrotomes lying behind those mentioned above and merge with the pronephric duct. The pronephros subsequently degenerates and only the caudal section of the pronephric duct persists as a metanephric duct (or archinephric duct) and is called the wolffian duct. The segmental character of the kidney becomes increasingly obscured, and secondary and tertiary nephrons are formed.

It is the mesonephric generation which plays a role in the transportation of the semen in many male vertebrates. In the simplest case, the ductuli efferentes of the testes make contact, directly or indirectly, with the nephrons of the primitive kidney so that the former can function to convey either urine or semen. In this case the wolffian duct serves as a urospermiduct. The next level of complexity is where the mesonephros differentiates into a cranial part serving exclusively the transport of the semen (pars sexualis) and a caudal part which is excretory (pars renalis). The former becomes the epididymis. In Elasmobranchii the pars sexualis is, in addition, differentiated into the epididymis and Leydig's gland, in which case the wolffian duct has its original function as a urospermiduct, although in some elasmobranchs it serves only as a seminal duct. In such cases the primitive kidney canals which serve as conduits for sperm are called the ductuli epididymis, that part of the primary kidney duct which runs through the epididymis is called the ductus epididymis, and that part outside the kidney the ductus deferens. Secondary urinary ducts are formed from the pars renalis of the mesonephrons in such cases and are sometimes falsely called ureters.

The third and final generation of this developmental series, the metanephros, is that in which the entire primitive kidney becomes an appendix to the testes with the metanephros taking over the task of excretion. This is the case in all amniotes. The metanephros differs from the mesonephros in that it arises from two different types of tissue. The part which includes the nephron develops from so-called metanephrogenic tissue, while the urinary part represents a derivative of the caudal section of the wolffian duct from which a diverticulum extends into the metanephrogenic tissue. Thus, a truly secondary kidney duct, the ureter, is formed outside the metanephros, the proximal end of which leads into the collecting ducts in the metanephros. The metanephros itself functions solely as an excretory organ and has no direct relationship to the reproductive system and therefore will not be described in any further detail.

The pattern described above also applies to amniotes of the female sex. In these cases the mesonephros is undeveloped, its rudiments being the epoophoron and associated Gartner's duct, the appendices vesiculosae or morgagnic hydatids, and the paroophoron. Functionless rudiments of the mesonephros are also found in the male, namely the paradidymis and appendices testis.

As mentioned, there is a tendency in males for the mesonephros to become increasingly more involved with the transportation of semen as phylogenetic devel-

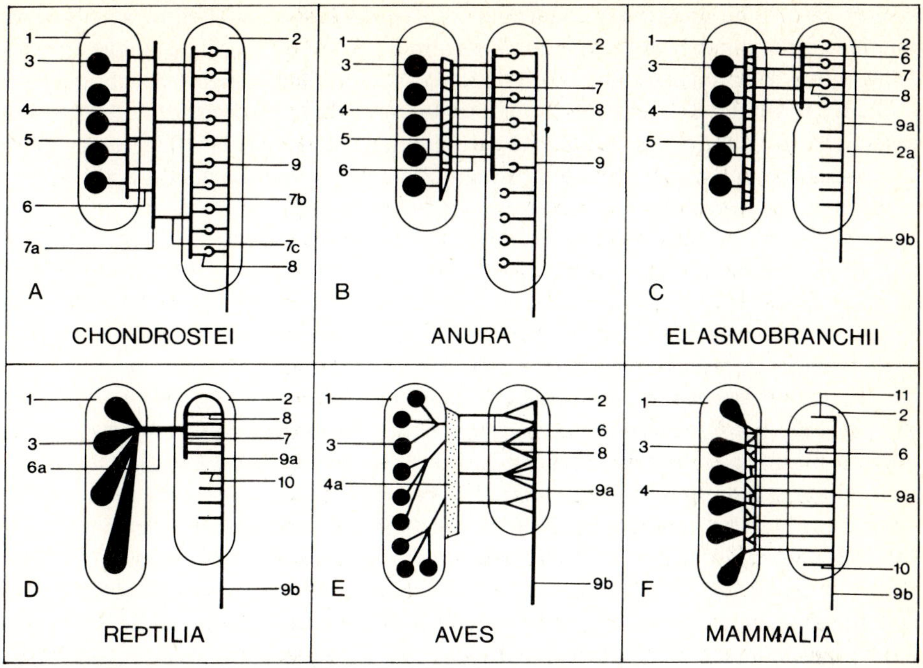

Fig. 10. Diagram of the different types of uro-genital connection.
1 = testis; 2 = epididymis; 2a = Leydig's gland; 3 = seminiferous tubule or ampulla; 4 = longitudinal canal or its equivalent (e.g. rete testis); 4a = antrum testis; 5 = connecting duct from the tubules or ampullae (ductulus rectus); 6 = ductulus efferens; 6a = ductus efferens; 7 = lateral kidney canal; 7a = outer lateral kidney canal; 7b = inner lateral kidney canal; 7c = connecting duct between the inner and outer lateral kidney canals; 8 = ductulus epididymidis; 9 = wolffian duct; 9a = ductus epididymidis; 9b = ductus deferens; 10 = ductulus aberrans; 11 = hydatid

opment proceeds towards the Amniota. Different types of uro-genital connections arise in the course of this development and the most important of these need to be described. It has already been established that the fusion between the testes' own ducts and those of the nephron may be direct or indirect. In the case of the indirect uro-genital connection the link is the so-called lateral kidney canal. In the primitive case this canal extends through the whole length of the primitive kidney and is fed by the ductuli efferentes of the testes. Side branches of this canal make contact with the nephrons and together these form the ductuli epididymis. Two such lateral kidney canals occur in Chondrostei: the ductuli efferentes lead into an outer lateral kidney canal which is confluent with an inner lateral kidney canal via connecting ducts from which the ductuli epididymidis arise (Fig. 10A). The next developmental stage is exhibited by anurans. The ductuli efferentes lead directly into the lateral kidney canal, which is restricted to the cranial part of the mesonephros (Fig. 10B), although in this order other also other construction principles are present which will be dealt with in detail later. The lateral kidney canal of elasmobranchs is limited to the epididymis region of the mesonephros. In addition, Leydig's gland represents a non-excretory portion of the primitive kidney (Fig. 10C). In some species there is no lateral kidney canal. The uro-genital connection of Reptilia is characterised by a marked reduction in the number of ductuli efferentes: in extreme cases a single ductus efferens feeds into a short lateral kidney canal (Fig. 10D). Birds and all higher vertebrates have a direct uro-genital connection. The lateral kidney canal is no longer in evidence and several ductuli epididymis merge into each ductulus efferens (Fig. 10E). In mammals the ductuli epididymis are probably totally reduced so that the ductus epididymidis connects directly with the ductuli efferentes (Fig. 10F).

2.4 Organisation of the Male Urogenital System

The tendency described above for the primary kidney ducts to be also used for the discharge of sperm is evident from early on in the evolution of the vertebrates. Among modern forms only the Cyclostomata and the Teleostei have no uro-genital connection. The situation in the Cyclostomata is very probably primitive and can be considered as the starting point for the following account of development. In Petromyzontia the testes are laid down as paired structures which fuse in the larval stage to form an unpaired organ that extends throughout the length of the body cavity. Only the right gonad is fully developed in Myxinoidea and it differentiates into a testis (or ovary) late in development. When mature the tubules of the testes burst, releasing sperm into the body cavity from where they pass via genital funnels out of the body and into the surrounding water. The latter structures are extensions of the coelom and are also called abdominal funnels. They open into the terminal, unpaired segment of the primary urinary duct to form a sinus urogenitalis whose external orifice is borne on an urogenital papilla (Fig. 11A).

The male urogenital system of teleosts of the family Salmonidae is very similar to that of Cyclostomata but should be interpreted as a special case. The paired testes extend throughout the entire length of the body cavity and form their own epitesticular ducts. These ducts converge into a short, unpaired end section which opens into an adjacent, medially situated coelomic funnel, the genital funnel or Lickteig's funnel. This latter opens to the exterior in a genital pore between the anus and the excretory pore (Fig. 11B).

A uro-genital connection is present in Elasmobranchii and Chondrostei. In the latter the testes are drawn out and contain a central testicular canal which lies somewhat peripherally. Ductuli efferentes arise along the whole length of this canal and connect with an external lateral kidney canal. Connecting ducts join the latter with an internal lateral kidney canal in the mesonephros, whose extensions make contact with the primitive kidney canals and so become ductuli epididymidis. Thus the entire length of the wolffian ducts serve as urospermiducts which open to the outside via an unpaired terminal section (Fig. 11C). In Elasmobranchii the external lateral kidney canal is missing and the mesonephros undergoes extensive differentiation and specialisation. The urogenital system of this group of animals will be dealt with later. The types of uro-genital connection found in Dipnoi can, theoretically, be regarded as antecedent to those in teleosts. Thus in the South American lungfish *Lepidosiren paradoxus*, a central testicular canal extends through the elongate testis and projects beyond it cau-

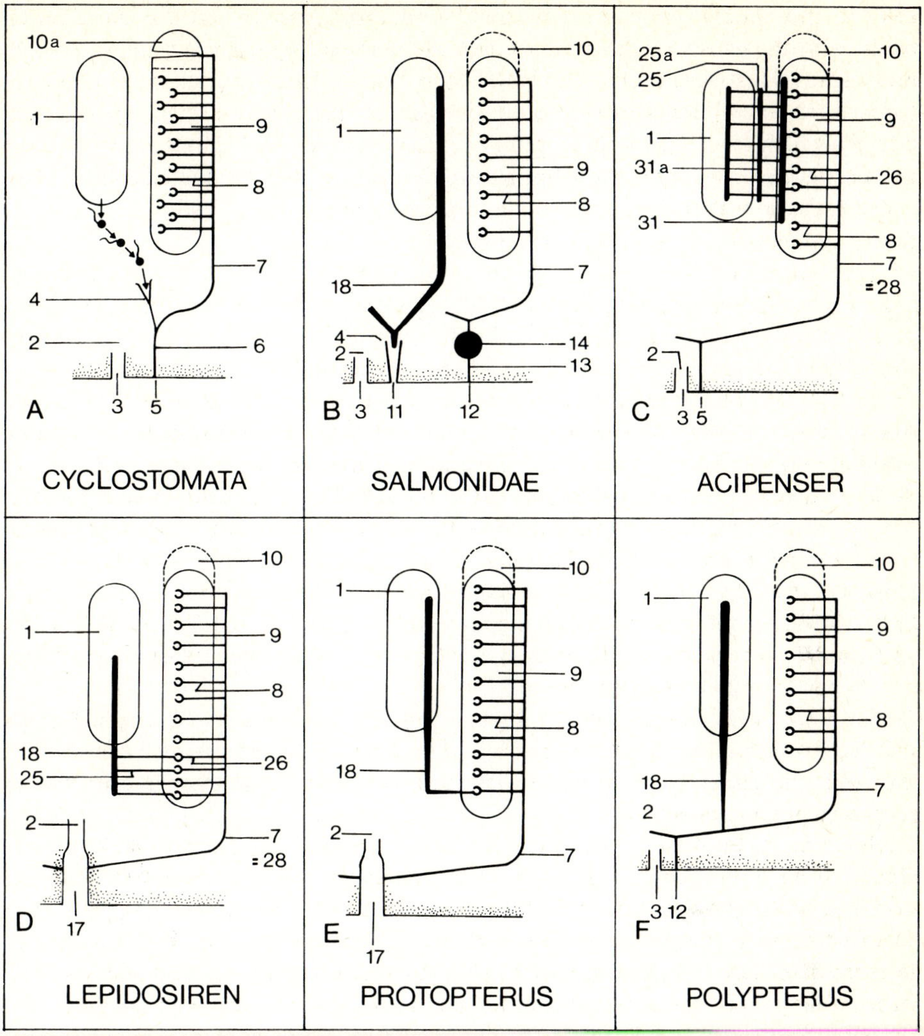

Fig. 11

Figs. 11–13. Organisation of the male urogenital system.
1 = testis; 2 = rectum; 3 = anus; 4 = genital funnel; 5 = porus urogenitalis; 6 = sinus (canalis) urogenitalis; 7 = primitive kidney duct, primary urinary duct, or the wolffian duct; 8 = primitive kidney tubule; 9 = mesonephros; 9a = pars sexualis of the mesonephros; 9b = Leydig's gland; 9c = pars renalis of the mesonephros; 10 = rudimentary pronephros; 10a = pronephros; 11 = genital pore; 12 = excretory pore; 13 = urethra; 14 = vesica urinaria; (15 and 16 are not used here) 17 = cloaca; 18 = secondary spermiduct or ductus spermaticus (but not a wolffian duct); 19 = secondary urinary duct; 20 = ureter; 21 = metanephros; (22 to 24 are not used here) 25 = ductulus efferens; 25a = cross-connecting duct; 26 = ductulus epididymidis; 27 = ductus epididymidis; 28 = urospermiduct; 29 = ductus deferens; 30 = ampulla ductus deferentis; 31 = lateral kidney canal; 31a = outer lateral kidney canal; 32 = vena renalis reheventis; 33 = ductus efferens; 34 = paradidymis; 35 = hydatid; 36 = glandula vesicularis; 37 = glandula prostatica

dally to end blindly. A small number of ductuli efferentes connect this end piece directly with the canals of the mesonephros. A lateral kidney canal is completely lacking (Fig. 11D). Basically the same situation is found in the African lungfish *Protopterus annectens*, however the uro-genital connection is restricted to a single ductulus efferens which leads into a single primitive kidney canal (Fig. 11E). In both cases an extratesticular collecting duct is formed which serves as a spermiduct. The next stage in the reduction of the uro-genital connection is represented in Holostei. A connection is completely lost in bichirs of the genus *Polypterus* and the extratesticular duct, representing a secondary spermiduct, joins the terminal segment of the wolffian duct (Fig. 11F), although the extratesticular duct is not a simple tube but a network of canals. The uro-genital connection of other Holostei, in contrast, extends over the whole length of the testes and an internal kidney canal is lacking.

This developmental series ends with the Teleostei. The terminal sections of the secondary spermiducts unite forming an unpaired section which usually has a separate opening to the exterior between the anus and the excretory pore (Fig. 12A).

A uro-genital connection is present in all but a few species of the three amphibian orders. In Gymnophiona (Apoda) the testis is composed of up to ten lobes which have a common central testicular canal. Ductuli efferentes arise from between the lobes and connect with a single lateral kidney canal which has branches that anastomose with the nephrons of the mesonephron to form ductuli epididymidis (Fig. 12B). The caudal part of the wolffian duct serves as a urospermiduct. The two wolffian ducts open separately into the cloaca.

The uro-genital connection in Anura is typically restricted to the cranial part of the mesonephros. The ductuli efferentes which originate from the ovoid testes feed into a lateral kidney canal whose extensions form ductuli epididymidis with the nephrons of the mesonephrons. The wolffian duct functions as a urospermiduct and its caudal end section may be temporarily widened into an ampulla prior to its opening separately into the cloaca (Fig. 12C). A developmental series can be recognised within the Anura which similarly ends with the elimination of the uro-genital connection. The first stage in this series is represented by the genus *Bombinator*. The ductuli efferentes connect to a lateral kidney canal from which transverse canals make contact with cranial canals of the mesonephros that finally lead into the wolffian duct. This situation represents, in full, the basic anuran pattern, although there is here and additional connection between the cranial end of the lateral kidney canal and the corresponding part of the primary urinary duct which provides an additional, direct path (the ductus efferens) to the outside (Fig. 12D). The next developmental stage, represented by the genus *Discoglossus*, sees the loss of the ductuli efferentes and the lateral kidney canal together with its connection to the tubules of the mesonephros. A direct connection is formed between the testes and the wolffian duct. The latter also drains the cranial part of the mesonephros, then separates off to finally end in the wall of the cloaca via a urinary bladder. The caudal nephrons of the mesonephros come together as a secondary urinary duct which similarly extends to the urinary bladder (Fig. 12E). The final stage is reached by members of the genus *Alytes*. The ductus efferens unites with the wolffian duct which projects to the urinary bladder without connections being made to the nephrons. The drainage of urine is taken over by a secondary urinary duct which serves all the nephrons of the mesonephros. The primary urinary duct therefore becomes simply a spermiduct (Fig. 12F).

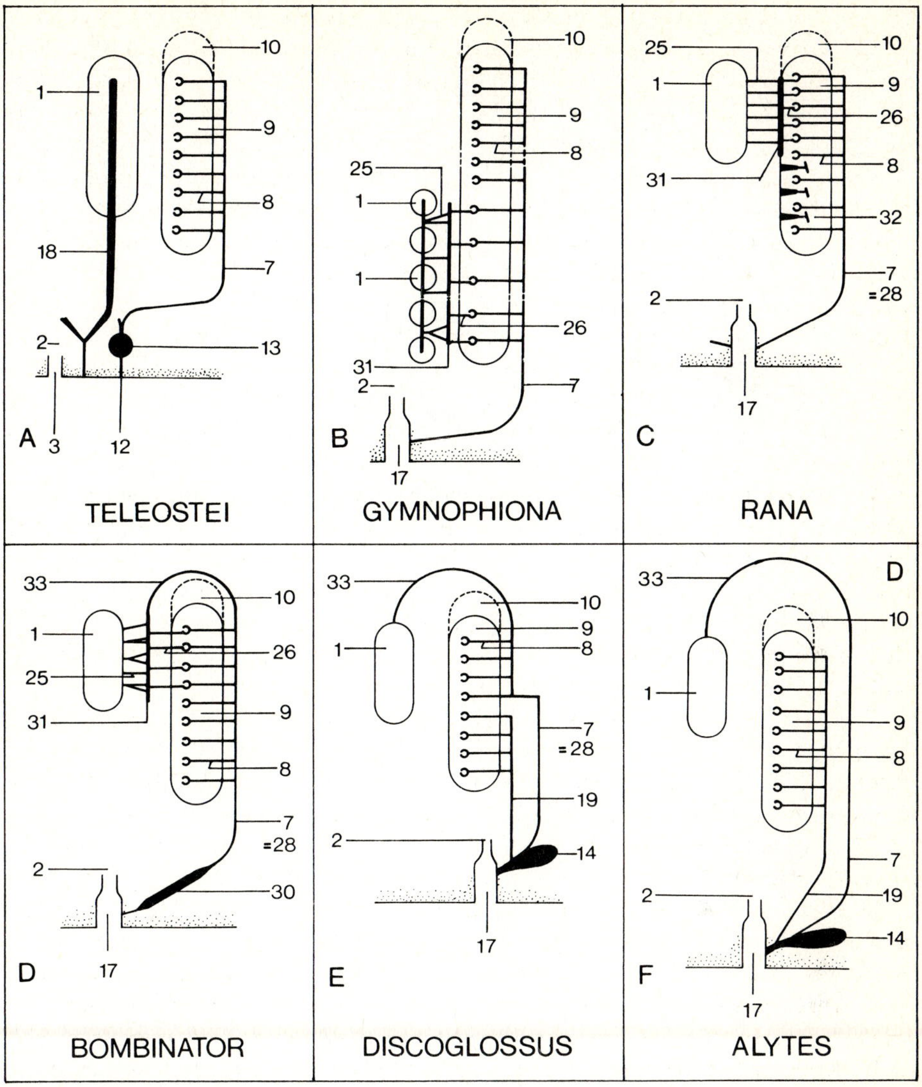

Fig. 12 (Legend see p. 57)

The mesonephros of Urodela has undergone a differentiation into a cranial pars sexualis and a caudal pars renalis. The testis usually has up to three or four lobes and gives rise to ductuli efferentes which feed into a lateral kidney canal. Transverse connections join this canal to the nephrons in the pars sexualis of the mesonephros. Although this part of the kidney is normally only poorly developed, it is still capable of some excretory activity in all cases. The wolffian duct only serves the nephrons of the pars sexualis and therefore is not a sperm duct but strictly speaking a urospermiduct. It is nevertheless sometimes referred to as a ductus deferens since it functions primarily as a sperm duct. The caudal end of this duct may be temporarily slightly enlarged

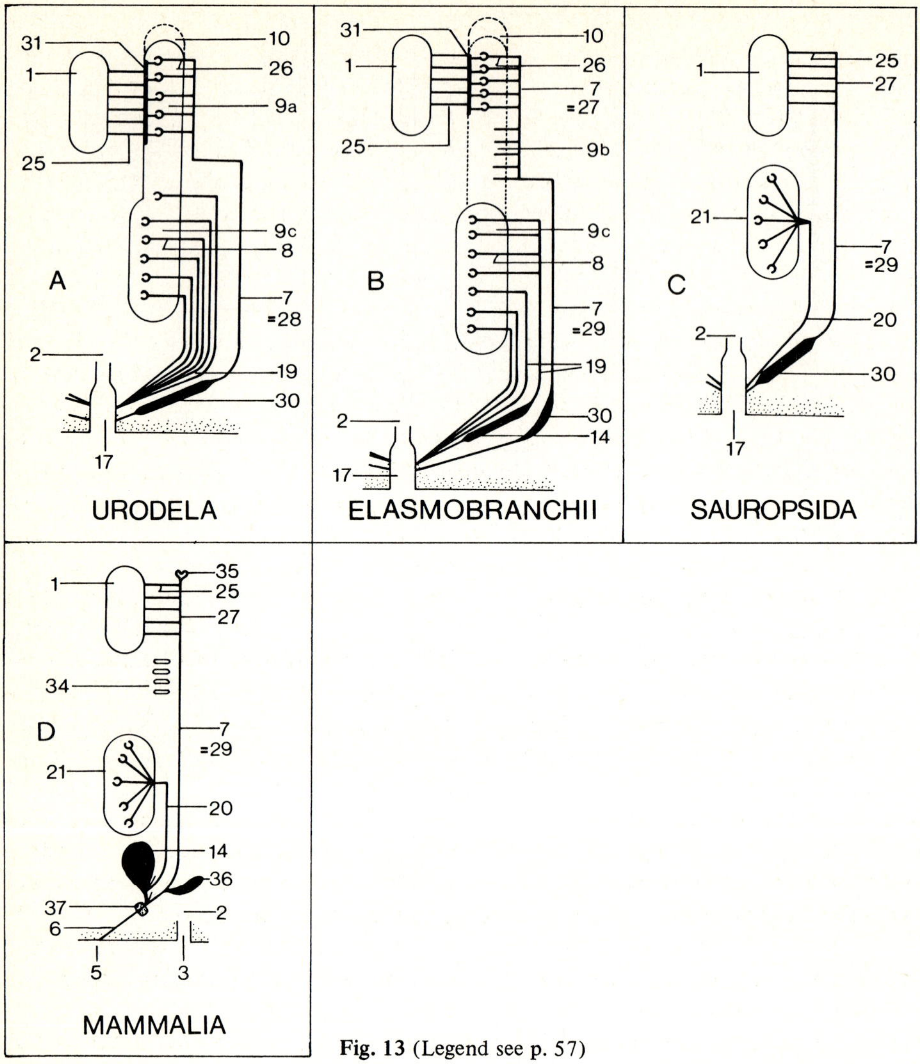

Fig. 13 (Legend see p. 57)

into an ampulla (ampulla ductus deferentis) prior to its opening into the cloaca. The drainage of urine from the caudal part of the kidney is taken over by secondary urinary ducts whose number varies from species to species. They each project to the cloaca without joining the wolffian duct (Fig. 13A). The interelationships existing in the Elasmobranchii are either comparable to those in urodeles or go beyond them in terms of their position in the developmental series. The mesonephros is normally more clearly differentiated into a pars sexualis and a pars renalis in Elasmobranchii, the former being exclusively involved with the transport or storage of sperm. The ductuli efferentes coming from the testis lead into a short lateral kidney canal which gives rise to transverse extensions that join the mesonephric tubules to the ductuli epididymidis. The cranial section of the wolffian duct is highly coiled and together with the ductuli epididymidis forms the epididymis. This part of the primary urinary duct is now called the ductus epididymis. It extends from the epididymis as a non-excretory part of the mesonephros called Leydig's gland composed of modified nephrons whose secretion supposedly activates the sperm. The wolffian duct then separates from the rest of the mesonephros, becoming a true ductus deferens, and extends to the cloaca, where it forms an ampulla ductus deferentis. The pars renalis of the primitive kidney is drained by secondary urinary ducts. The nephrons in the cranial part of the pars renalis often run in a common duct which expands into an urinary bladder (vesica urinaria) before opening into the cloaca. In addition, several secondary urinary ducts arise from the caudal section of the pars renalis and extend to the cloaca (Fig. 13B).

The entire mesonephros loses its excretory function in amniotes and in all cases the wolffian duct becomes a seminal duct. The excretory functions are taken over by the metanephros which is drained of urine by the ureter. Thus the reproductive and excretory systems are completely separate, if one disregards the fact that they end in a short common duct. The testes of Sauropsida (Reptilia and Aves) are egg-shaped. In reptiles the ductuli efferentes, which may be reduced to a single ductus efferens, leave the testis and join a short lateral kidney canal in the pars sexualis, the whole of which has now become an epididymidis. The ductuli epididymidis lead on from the epididymidis to a knotted section of the wolffian duct, the ductus epididymidis which leaves the epididymis as the ductus deferens and extends to the cloaca and before ending usually forms an ampulla ductus deferentis. The lateral kidney canal is lost in Aves, so that the ductuli efferentes lead on directly to the ductuli epididymidis, but other than this the configuration of ducts corresponds to that in Reptilia (Fig. 13C). The positions of the opening of the ductus deferentes and ureters in the cloaca are highly variable in Sauropsida. The cloaca of Chelonia is partly divided into an upper and lower chamber by duplicatures and the ductus deferentes and ureters open into the latter which is therefore referred to as the sinus urogenitalis. The cloaca of crocodiles and birds is divided by transverse folds in its wall into three sections, respectively the coprodaeum, urodaeum and proctodaeum. The urinary and seminal ducts of birds open into the urodaeum, whereas in crocodiles they open into the urodaeum and proctodaeum respectively. The cloaca of lizards is undivided and the urinary and seminal ducts usually end in a common duct which opens into the cloaca.

The Monotremata represent a link between reptiles and didelphic and placental mammals. The urinary and seminal ducts lead into a sinus urogenitalis, the end section of which is differentiated into a urinary canal that opens directly into the cloaca and a seminal canal that opens via a penis (see Fig. 20: p. 72).

In the remaining mammals the openings of the gastrointestinal and urogenital systems are separated by the formation of a perineum. A direct uro-genital connection is established between the testes and the wolffian duct. The ductuli efferentes run directly to the ductus epididymidis, which is highly convoluted. The epididymis is divided into three parts. The initial segment of the ductus epididymidis forms the caput epididymidis whereas the middle segment, the corpus epididymidis, runs alongside the testis. The last segment is highly coiled and forms the cauda epididymidis from which the ductus deferens arises. The end of the ductus deferens is enlarged into an ampulla ductus deferentis. The excretion of urine occurs via the ureters, which open into the vesica urinaria whose outlet duct is called the urethra. The end sections of

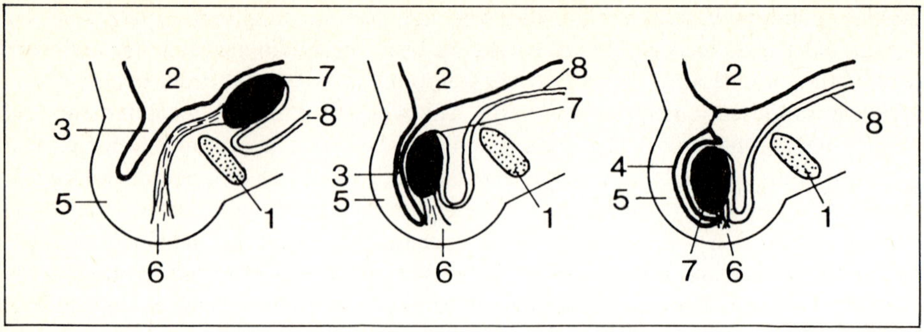

Fig. 14. The descensus testiculorum of mammals. (After Turner).
1 = os pubis; 2 = body cavity; 3 = processus vaginalis; 4 = tunica vaginalis; 5 = scrotum; 6 = gubernaculum testis; 7 = testis and epididymis; 8 = ductus deferens

the urethra and ductus deferens unite to form the canalis urogenitalis (male urethra) which runs through to the end of the penis where it opens to the exterior (see p. 73). Mammals characteristically have a number of accessory glands to the excretory ducts. The glandula vesicularis and glandula coagulans derive from the ductus deferens and are situated close to where the latter merges into the canalis urogenitalis. The glandula prostatica and glandulae cowperi (bulbourethral glands) arise from the canalis urogenitalis, the former lying at the juncture of the ductus deferens and the canalis urogenitalis, the latter being situated more caudally. Other, smaller urethral glands also occur within the wall of the canalis urogenitalis (Fig. 13D).

The position of the testes can vary in different mammals. In testicond forms the testes lie dorsally within the body cavity throughout life but in other mammals there is a more or less pronounced descensus testiculorum whereby the testes come to lie, temporarily or permanently, in a scrotal sack hanging from the body. This also applies to the epididymis and the originally cranial section of the ductus deferens. The scrotum develops during the course of ontogenesis from the so-called sex swellings and can be regarded as an originally paired bulge in the abdominal wall (the cremaster sacks) which is covered by skin. The testes themselves are anchored to the caudal end of the scrotum (at the conus inguinalis) by the gubernaculum testis which runs alongside the inguinal canal. An extension of the body cavity, the processus vaginalis, projects into and lines the inguinal canal and the testes and epididymis are drawn into this cavity as they descend into the scrotum. The descent of the testes is in part caused by the fact that the gubernaculum retains its original length whereas the embryo continues to grow in size. The processus vaginalis is closed in some cases, thereby forming a sinus vaginalis. Since the testes lie outside the body cavity in the scrotum, each becomes partly enveloped by two layers of peritoneal epithelium, the periorchium and epiorchium. The inguinal canal is closed in most animals in which the descensus testiculorum is permanent but remains open in those in which it is only temporary. The process of the descensus testiculorum is shown in Fig. 14. The physiological aspects of this process are obscure but the shifting of the testes to a cooler place was presumably of selective advantage (see p. 93). Notably there is also a descensus ovariorum in females, although it is much less prominent than the descent of the testes and, at most, only changes the position of the ovaries in the region of the cranial wall of the pelvis.

2.5 Organisation of the Female Urogenital System

A uro-genital connection is lacking in the female sex of all vertebrates. On the other hand, there is a tendency for special oviducts to be formed. Typically these ducts are the müllerian ducts and they have two different origins within the different classes of vertebrate. In Elasmobranchii and Amphibia they represent duplicates of the primary urinary duct, but in all other cases they are "new" structures formed from a fold of peritoneum lying lateral to the genital ridges which closes upon itself from its cranial to its caudal end. A rudimentary müllerian duct also occurs in males as a uterus masculinus or utriculus prostaticus.

In Cyclostomata there are no ducts in the female for the transport of oocytes, their anatomy in this respect corresponding to that of males (see p. 57). The ovulated eggs fall into the body cavity and thence into an abdominal funnel which functions as a genital funnel. Together with the end section of the primary urinary duct, this funnel forms a sinus urogenitalis which opens to the outside at the summit of a urogenital papilla (Fig. 15A). A similar situation occurs as a special case in Teleostei in the family Salmonidae. The ovaries are paired and extend throughout the whole length of the body cavity and, as above, the ovulated eggs fall freely into this cavity. They are then taken up by an unpaired genital funnel of coelomic origin, Lickteig's funnel, which finally opens to the outside as a genital pore situated between the anus and the excretory pore (Fig. 15B).

A müllerian duct first appears in Acipenseriformes. As in the above case, the eggs are released into the body cavity from the elongated ovaries. They leave the body via a short oviduct which feeds into the end of the primary urinary duct forming a sinus urogenitalis (Fig. 15C). It must nevertheless be noted that it is not certain whether the oviduct of Acipenseriformes is in fact a homologue of the müllerian duct. It is interpreted as such in most of the relevant textbooks, and accordingly one can regard it as the most primitive stage exhibited by any modern species.

A somewhat more advanced stage of development can be found in Holostei of the genus *Amia*, in which case the müllerian ducts are considerably longer. They open into a pair of urinary bladders which are the enlarged sections of the primary urinary duct (Fig. 15D).

A complete separation of the oviduct from the urinary ducts occurs in the Dipnoi. In *Protopterus* the funnel-shaped opening of the müllerian duct, the ostium abdominale, lies against the cranial portion of the ovary. The two oviducts extend through the body cavity and unite into an unpaired terminal section shortly before entering the cloaca (Fig. 15E). A comparable anatomical arrangement is found in the genus *Neoceratodus*, except that the oviducts are not of uniform structure but are enlarged shortly before they unite into glandular sections which provide the eggs with a gelatinous shell (Fig. 15F).

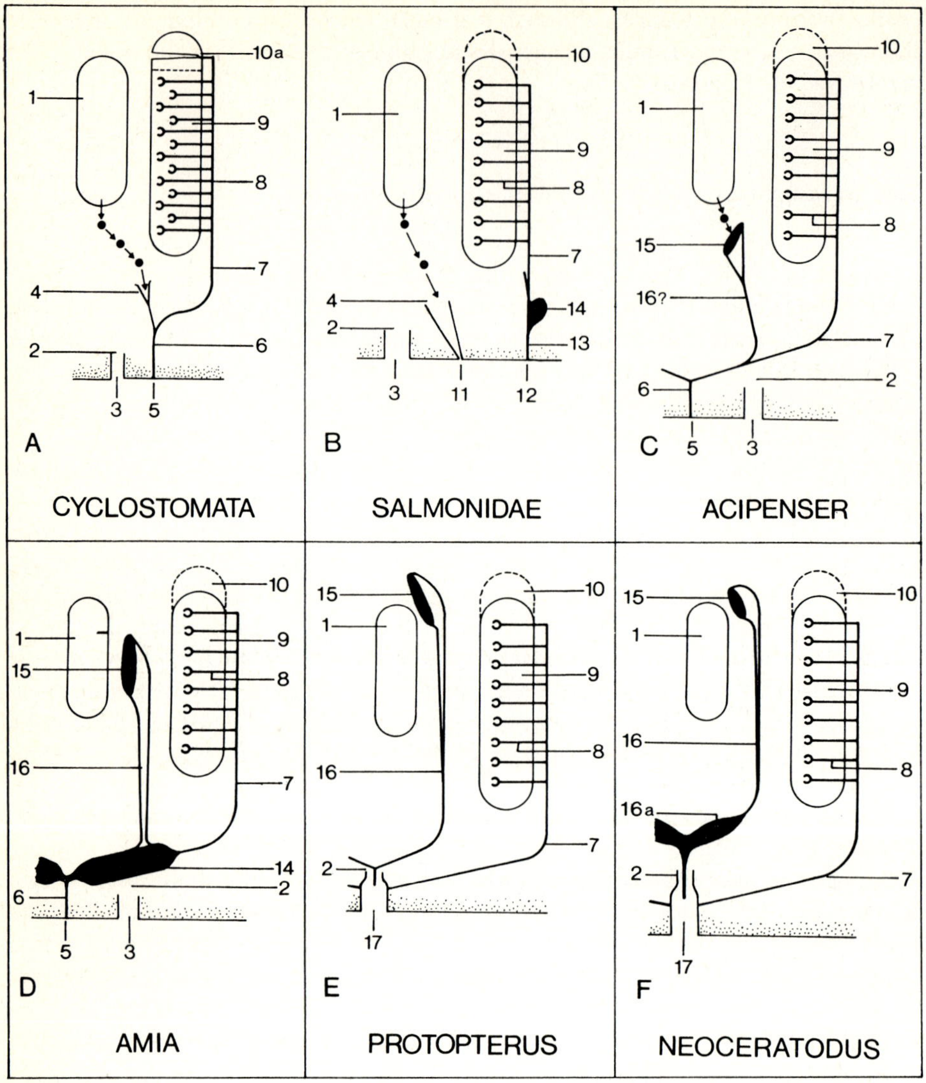

Fig. 15

Figs. 15 and 16. Organisation of the female urogenital system.
1 = ovary; 2 = rectum; 3 = anus; 4 = genital funnel; 5 = porus urogenitalis; 6 = sinus (canalis) urogenitalis; 7 = primitive kidney duct or primary urinary duct; 8 = primitive kidney tubules; 9 = mesonephros; 10 = rudimentary pronephros; 10a = pronephros; 11 = genital pore; 12 = excretory pore; 13 = urethra; 14 = vesica urinaria; 15 = ostium abdominale; 16 = müllerian duct or oviduct; 16a = glandular expansion; 16b = infundibulum; 16c = tuba uterina; 16d = glandula nidamentaria; 16e = uterus; 16f = vagina; 17 = cloaca; 18 = secondary oviduct (but not müllerian duct); 19 = secondary urinary duct; 20 = ureter; 21 = metanephros; 22 = epoophoron; 23 = paroophoron; 24 = Gartner's duct

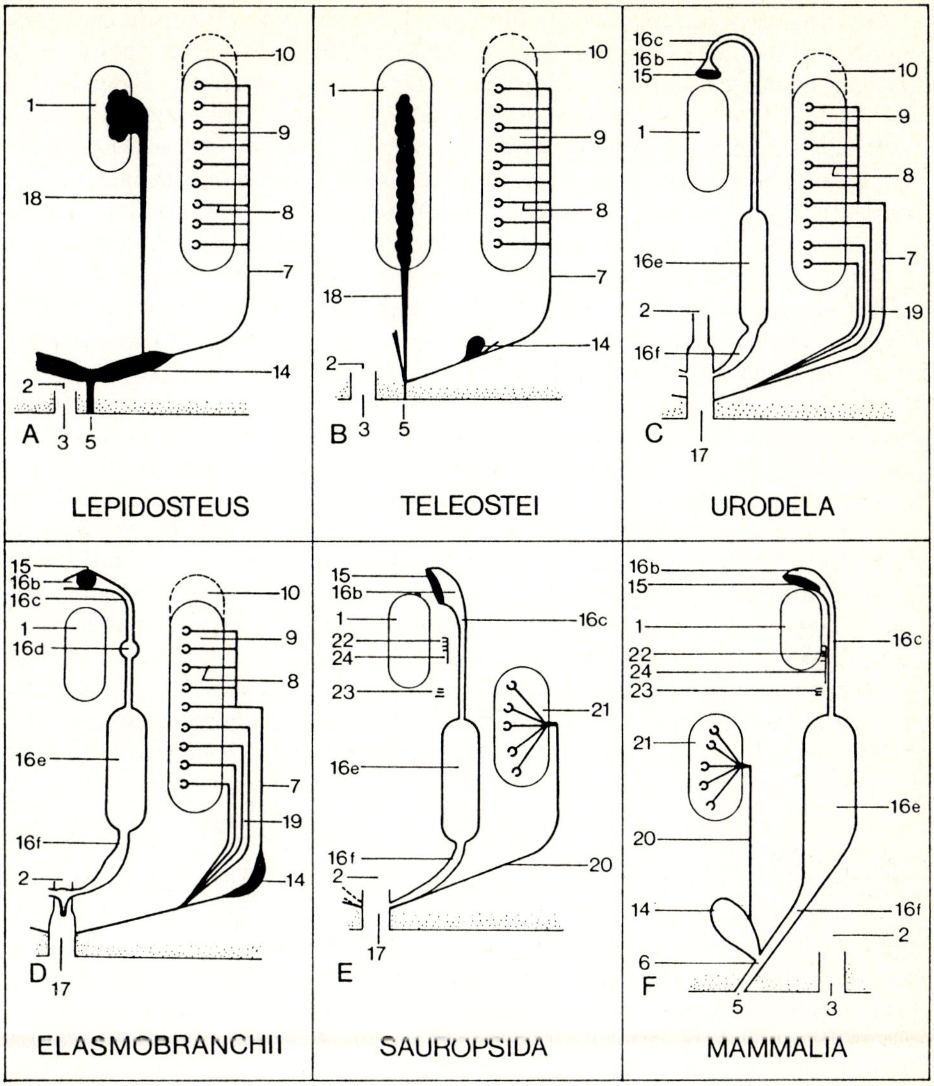

Fig. 16

Teleostei have no müllerian duct, instead they usually possess a secondary oviduct that represents an extension of the entovarian cavity (see p. 47). A similar structure occurs in Holostei of the order Lepidosteiformes: in this case the ovary forms its own duct from an extension of the entovarian cavity forming a thin-walled tube that opens into the terminal section of the primary urinary duct, which is enlarged into a urinary bladder (Fig. 16A). The ovaries of teleosts with entovarian or parovarian oviducts are mostly elongated and the free part of the ducts are therefore only short. The latter commonly form a sinus urogenitalis together with the terminal portions of the primary urinary duct, although this is often simply a very short section at the tip of a urogenital papilla (Fig. 16B).

All amphibians possess müllerian ducts which become greatly modified during the reproductive period. The ostium abdominale, the opening of the funnel-shaped infundibulum, is situated cranial to the ovaries, which are either elongated (Gymnophiona and Urodela) or more compact (Anura). Joined to the infundibulum there is a convoluted tuba uterina whose wall is densely packed with tubular glands which encapsulate the eggs in a gelatinous secondary egg membrane. The end of the müllerian duct is normally expanded into a uterus which stores the eggs or forms them into clusters or strings. The uterus is where embryonic development takes place in ovoviparous or viviparous species (see p. 22). It opens into the cloaca via a vagina (Fig. 16C).

The modifications of the müllerian ducts in Elasmobranchii are even more pronounced than those described above. In these animals the ovaries are laid down as paired structures but they seldom undergo an identical development (as they do in Notidanidae and Torpedinae for example): as a rule the development of the left ovary is suppressed. In a large number of cases the openings of the two müllerian ducts are fused together so that there is a single ostium abdominale situated cranially within the body cavity and ventral to the oesophagus. The tubae following on from the common infundibulum have glandular extensions along their length, the glandulae nidamentariae (shell glands), which provide the eggs with additional foodstuffs and a horny shell as the eggs pass through. Each tuba uterina opens into a uterus whose epithelial lining may be differentiated into specialised structures in viviparous forms (see p. 241). The uteri open into the cloaca via vaginae (Fig. 16D). Despite the lack of a uro-genital connection in females, secondary urinary ducts are also present in this sex in Urodela and Elasmobranchii. Whereas the cranial part of the mesonephros is still drained by a primary urinary duct, the nephrons of the caudal part are connected to "false ureters" which are usually less numerous than they are in males.

The female urogenital system of Amniota represents a step forward in development from the stage just described. In reptiles both ovaries are developed and the müllerian ducts are divided into an infundibulum, tuba uterina, uterus and vagina. In oviparous species the uterus produces the egg shell, whereas in ovoviparous species it serves to contain the egg or embryo within the body. In birds only the left ovary is fully developed as a rule, the right ovary remains undeveloped in what may be interpreted as the rudimentary state of a testis anlage (see p. 57). The various differentiations of the müllerian duct are particularly distinct at the time of reproduction. They are firstly the infundibulum with its ostium abdominale, followed by a tuba uterina, which is also referred to as the pars albuminifera since it is in this region that the yolky egg is enveloped in egg white. At the junction to the uterus there is an isthmus in which the shell membrane is formed. The hard and, in some cases, pigmented outer egg shell is produced in the uterus, which leads to a short vagina that opens into the urodaeum of the cloaca. As only the left ovary is developed, only the left müllerian duct is fully developed in most cases, the right being present as a rudiment (Fig. 16E).

The anatomy of the urogenital system of Monotremata is comparable to that of reptiles. Other mammals characteristically possess a well-developed uterus but no cloaca. The müllerian ducts begin in an ostium abdominale which is often bordered by fringe-like growths (fimbriae) of the infundibulum. The wall of the infundibulum envelops the ovary to a greater or lesser extent forming an ovarian pouch (bursa ovarii) which greatly increases the chances of an egg finding its way into the tuba uterina.

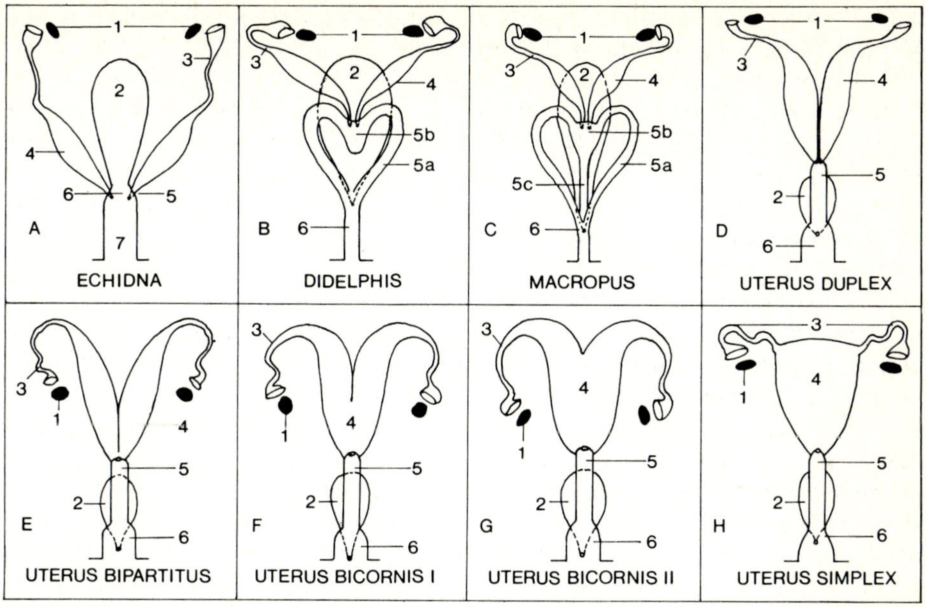

Fig. 17A–H. The different types of uterus and vagina in mammals.
1 = ovary; 2 = vesica urinaria; 3 = tuba uterina; 4 = uterus; 5 = vagina; 5a = vagina lateralis; 5b = sinus vaginalis; 5c = pseudovagina, birth canal; 6 = sinus urogenitalis; 7 = cloaca

The tuba uterina leads on to the uterus which is either a paired or an unpaired structure. One or two vaginas link the uterus or uteri to the outside (Fig. 16F).

The structure of the uterus and vagina differ markedly in the different orders of Mammalia. In Monotremata both uteri open separately via short vaginas into a sinus urogenitalis which in turn opens into the cloaca (Fig. 17A). In marsupials, the cranial sections of two vaginae laterales merge into a medial sinus vaginalis into which the two uteri open separately. The vaginas, together with the urethra of the bladder, form an unpaired sinus urogenitalis whose opening to the outside is separate from that of the intestine (Fig. 17B). In macropodid marsupials there is, in addition, a birth canal, or pseudovagina, extending caudally from the middle of the sinus vaginalis which opens into the sinus urogenitalis and through which the embryo is born (Fig. 17C). This canal may exist either temporarily or permanently.

An unpaired vagina is a characteristic of Eutheria. It derives from varying proportions of the müllerian duct (forming the cranial part) and the sinus urogenitalis (forming the caudal part) in the different orders of animals. One can distinguish four main types of uterus. A uterus duplex is composed of two separate uteri which have separate openings into the vagina and this configuration exists in Rodentia and elephants for example (Fig. 17D). Two separate uteri are also present in the case of the uterus bipartitus but they have a common opening into the vagina. Such an uterus occurs in the Carnivora, Cetacea and Suidae (Fig. 17E). The uterus bicornis possesses an unpaired caudal section and a cranial section with paired horns, the relative proportion of the two sections being different in the different animals in which it exists, namely Insectivora, Perissodactyla and some Ruminantia and Prosimiae (Fig. 17F,G). Finally, the uterus simplex as found in Simiae and Hominidae is completely unpaired (Fig. 17H).

2.6 Copulatory Organs

The primitive method of fertilisation in vertebrates is the shedding of eggs and sperm into water where the gametes meet more or less by chance and fuse into a zygote. This "act of chance" depends on there being a large number of gametes. The presence of a hard egg shell usually requires that fertilisation be internal, that is the sperm must be within the female's body before the egg shell is formed. Some bony fish in which fertilisation is external are exceptional in that the eggs have a hard shell in which there is a small opening, the micropyle, to admit the spermatozoa. Fertilisation must be internal when embryonic development takes place partly or completely in the uterus. Internal fertilisation in vertebrates is achieved in three different ways. In the simplest case the cloacae are everted to some extent then pressed together and the sperm transferred into the female's cloaca. This is the case in the majority of birds. The second method involves the formation of a spermatophore in the cloaca and this occurs in a range of Urodela. An exact mould of the cloaca is produced by a coagulating secretion onto which the sperm are deposited. The male sets the spermatophore down on a firm piece of ground and the female picks it up with her cloaca. The third method is by the means of specialised copulatory organs developed by the males. These organs are inserted into the sexual opening or cloaca of the female, whereby the sperm are transferred directly into the female's genital tract. Copulatory organs can have one of two different origins: specially modified fins serve as a pterygopodium or gonopodium in fishes whereas in all other vertebrates the copulatory organ derives from the cloaca. Gonopodia occur in Cyprinodontidae which give birth to live young, the anal fin (or analis) being modified into an intromittent organ which is prolongated and bears hooks and spines and moreover can be flexed by the action of muscles. A joint allows it to be folded forwards. The semen flows through two lateral grooves which run along the length of the organ. The gonopodium of Anablepidae is similarly a modified analis but has a canal running along its length. It is not inserted into the female, however, but serves merely to anchor the male while he deposits a spermatophore in front of the female's sexual opening. Therefore, strictly speaking, it is not a copulatory organ. The same applies to the clasp organ of Phallostetheidea, the priapum, which is composed of skeletal elements from the ventral fin, ribs and pectoral girdle.

The Elasmobranchii and Holocephala also possess a copulatory organ, the pterygopodium or mixopterygium, an appendage of the medial aspect of the ventral fins. A groove runs through it which opens cranially as the apopyle and caudally as the hypopyle, the former lying in front of the cloacal opening. A muscular sack, the sipho, is situated beneath the skin of the ventral body wall to either side and in front of the pterygopodium. It extends caudally in a sipho tube. Although paired, only one sipho tube is used. The sipho tube of sharks opens close to the apopyle whereas that of rays opens further caudally. In the former case it is filled with sea water and in the latter

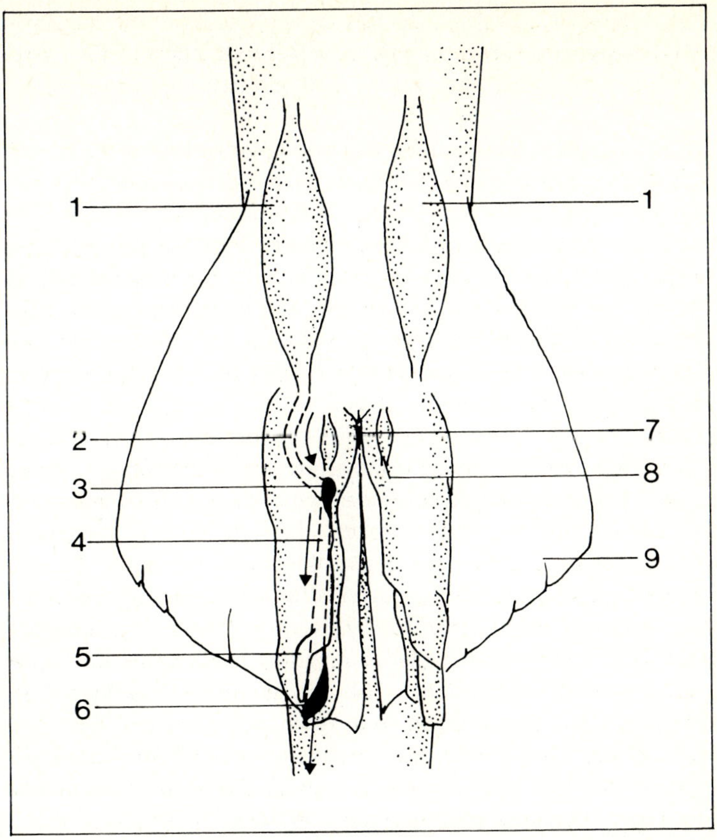

Fig. 18. The pterygopodium of an elasmobranch. (After Leigh-Sharpe).
1 = sipho; 2 = sipho tube; 3 = apopyle; 4 = grooved pterygopodium; 5 = rhipidion; 6 = hypopyle; 7 = orifice of the cloaca; 8 = parasipho; 9 = pelvic fin. *Arrows* indicate the direction in which the contents of the sipho flow when released

with a glandular secretion. The pterygopodium and the sipho function together as a copulatory organ.

During copulation one pterygopodium is inserted into the female's cloaca whereby the apopyle is pressed firmly against the male's cloaca so that semen can be transferred via the apopyle into the grooves of the pterygopodium. The muscles of the sipho contract, expelling its contents into the apopyle and propelling the semen in the grooves through the hypopyle into the female's cloaca or vagina. Whether these details apply in all cases is not known but the description given nevertheless provides a model of how this type of copulatory organ functions. There is often another smaller appendage next to the apopyle, the rhipidion, which can be splayed out and whose function is still a matter of speculation. Figure 18 shows the overall structure and arrangement of the pterygopodia and siphos.

Among Amphibians, copulatory organs exist only in one species of anuran (*Ascaphus truei*) and in the Gymnophiona. These organs are either extensions of the cloaca or its eversible end portion, one muscle serving to evert it and two others to retract it back into the cloaca.

The copulatory organs of reptiles are, without exception, derivatives of the cloaca. Squamata possess a hemipenis, a paired organ formed from a pair of blindly ending sacks in the caudal wall of the cloaca which are everted during copulation. The inner surface of the sacks, which bears papillae or spines, therefore becomes the outer surface of the hemipenis. A seminal groove runs from the base to the rounded ends of the hemipenis and guides the semen from the opening of the wolffian duct in the male's cloaca into the cloaca of the female. Only one limb of the hemipenis is used during copulation. The organ consists of a cavernous tissue and two large retractor muscles.

Chelonia and Crocodilia have an unpaired "penis". It is formed out of two longitudinal ridges forming a groove in the ventral floor of the cloaca. Each ridge is composed of a coelomic canal and an erectile tissue, the corpus cavernosum. During erection the corpora cavernosa fill with blood and become distended making the groove into a tube that guides the semen from the opening of the ductus deferentes situated at the base of the penis into the female's cloaca. An additional supporting tissue, the corpus fibrosum, lies ventral to the groove and extends up to the tip, or glans penis, of the copulatory organ. A retractor muscle assists in pulling the penis back into the cloaca after the corpora cavernosa have deflated. This description of the penis of Chelonia also applies with only slight modification to the penis of Crocodilia, which has the same basic structure. In the latter animals the corpus fibrosum is more prominent but the corpora cavernosa are apparently lacking. Figure 19 shows the position and general structure of the penis in members of the genus *Testudo*.

A few birds are exceptional in also having penises with a structure which basically corresponds to that described for *Testudo* and *Crocodilus*. Thus the corpora cavernosa form a seminal groove above a corpus fibrosum. An asymmetry in the former structure leads to the penis of ducks, for example, having a spiral form. Some species have acquired an eversible sack at the base of the penis which greatly extends the erection. A penis occurs in the following genera and families: *Struthio, Apteryx, Dromaeus, Rhea, Casuarius,* Anatidae, Penelopidae and Tinamidae.

The penis of mammals is also structured like that of Chelonia. It stems not from the cloaca directly but from a sex protuberance or phallus which results from the partitioning of the cloaca and the formation of the perineum. It is nevertheless a derivative of the ventral part of the cloaca, the sinus urogenitalis. This sinus always extends through the penis as the male urethra and is also called the canalis urogenitalis (see p. 62). However, this does not apply to the Monotremata where the penis is provided only with a seminal tube.

The penis of Monotremata arises from the ventral wall of the cloaca, as in reptiles. A poorly developed corpus cavernosum envelops a seminal tube that branches off from the sinus urogenitalis and itself branches into several tubules at the end of the penis (the glans penis). The corpus cavernosum of Monotremata is called the corpus spongiosum. The glans penis lies in a preputial pouch when the penis is in the retracted state. Beneath the corpus spongiosum lies a well-developed corpus fibrosum. In

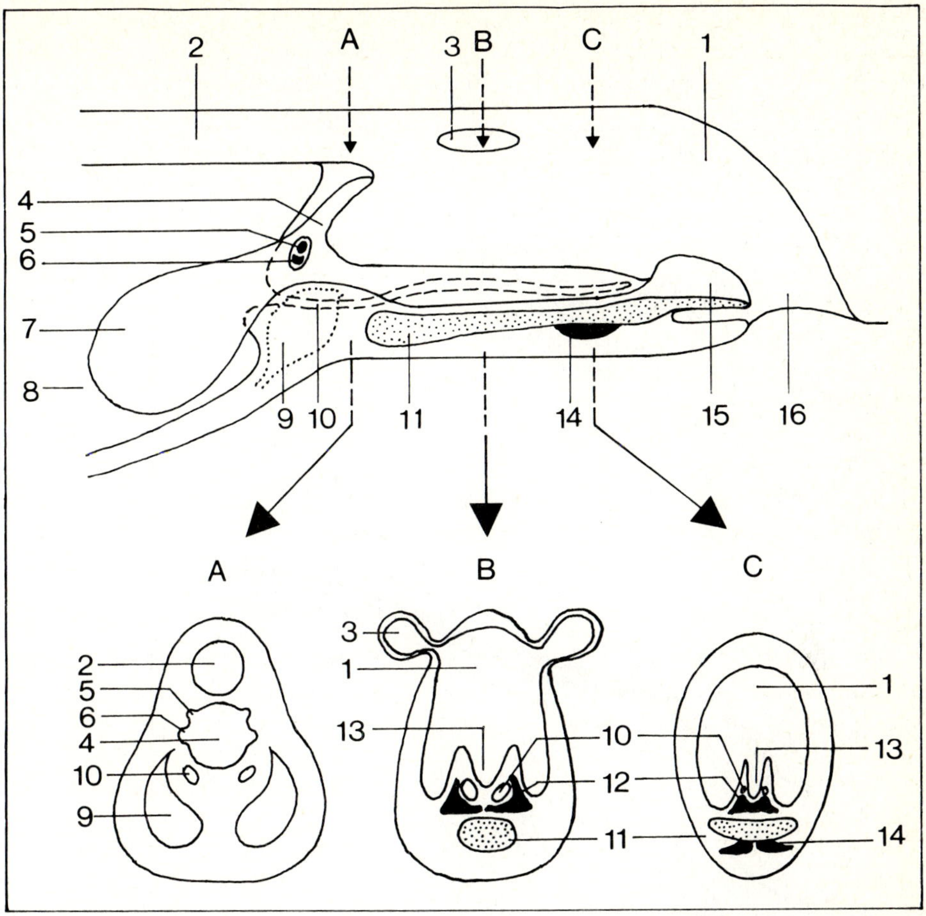

Fig. 19A–C. Diagram showing the position and structure of the penis of *Testudo*. (After Wibault-Isebree Moens).
1 = cloaca; 2 = rectum; 3 = cloacal vesicle; 4 = sinus urogenitalis; 5 = orifice of the ureter; 6 = orifice of the wolffian duct; 7 = vesica urinaria; 8 = coelom; 9 = bulbus urethralis; 10 = coelomic canal; 11 = corpus fibrosum; 12 = corpus cavernosum; 13 = seminal groove; 14 = musculus retractor penis; 15 = glans penis; 16 = orifice of the cloaca

monotremes the urine and semen are expelled separately, as mentioned previously. Urine reaches the cloaca via a special urinary orifice of the sinus urogenitalis, whereas semen passes through the seminal tube. Erection of the penis projects it caudally out of the cloaca thereby blocking the urinary orifice. The skin of the preputial pouch serves to extend the length which the erect penis can reach (Fig. 20).

The penises of all other mammals have a basically uniform structure, although there are a multiplicity of variations. Primarily, the tube carrying the urine and semen, the canalis urogenitalis, is enveloped by an unpaired corpus spongiosum urogenitale, which corresponds to the corpus cavernosum urogenitale. Above or below this is a corpus cavernosum penis, corresponding to the corpus fibrosum, which is more or

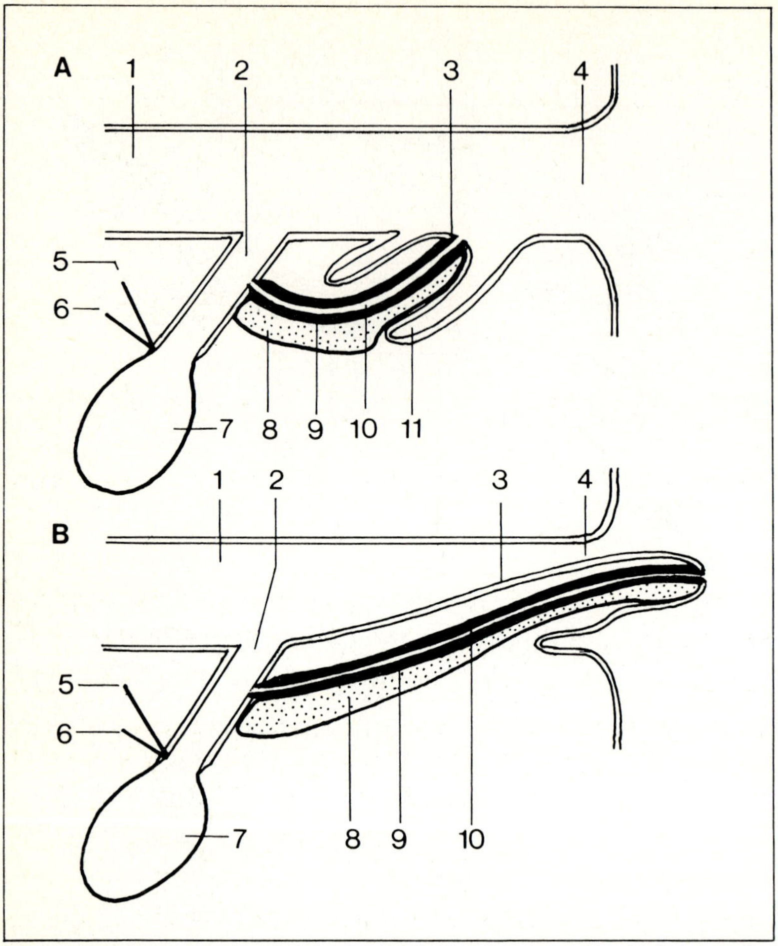

Fig. 20A,B. Diagram showing the position and structure of the penis of Monotremata. (After Boas). **A** Retracted. **B** Errect.
1 = cloaca; 2 = urinary orifice; 3 = penis; 4 = cloacal orifice; 5 = ductus deferens; 6 = ureter; 7 = vesica urinaria; 8 = corpus fibrosum; 9 = corpus cavernosum = corpus spongiosum; 10 = seminal tube; 11 = preputial pouch

less partitioned by a septum. The terms corpus spongiosum and corpus fibrosum as used with respect to the penis of Monotremata will continue to be used here in the interests of maintaining a uniform terminology, although this deviates from accepted usage. The corpus spongiosum also usually fills the glans penis in mammals, however this body of tissue is laid down separately and later joined to that in the shaft of the penis. The relative proportion of the corpus spongiosum and corpus fibrosum varies from species to species. The glans penis and sometimes part of the shaft of the penis is covered by a preputium that extends the length attained by the errect penis, as in Monotremata. The glans penis of many Marsupialia is bifurcated so as to fit the female's vaginae laterales, a phenomenon which is foreshadowed in the Monotremata.

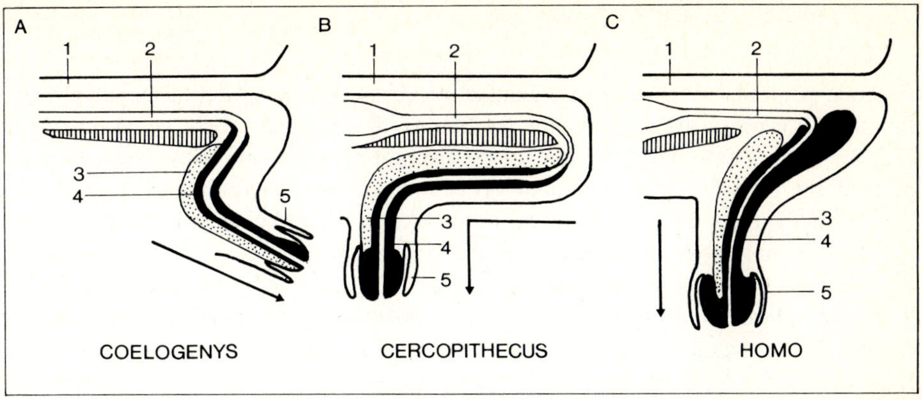

Fig. 21A–C. Schematic representation of the stages in the evolution of the penis pendulus (**A**, **B** after Boas; **C** after Broek). **A** Initial stage; **B** Penis appositus; **C** Penis pendulus.
1 = rectum; 2 = canalis urogenitalis; 3 = corpus cavernosum = corpus fibrosum; 4 = corpus spongiosum urogenitale = corpus cavernosum urogenitale; 5 = preputium

Intermediary forms with respect to the organisation of the urinary and seminal ducts in Monotremata and higher mammals can be found in the genera *Perameles* and *Didelphis*. In the former a separate urinary canal opens into the cloaca, whereas in the latter it passes up into the preputial pouch. A common duct for urine and semen first occurs in the Macropodidae. In certain orders the septum and tunica albuginea of the corpus fibrosum contain an ossified element, the os penis (penis bone), which reinforces the penis (namely primates other than man, Rodentia, Carnivora, Chiroptera and some Insectivora).

The corpus spongiosum serves primarily to strengthen the wall of the canalis urogenitalis and to keep this duct open. Erection of the penis is brought about by blood filling the cavities of the sponge-like corpus fibrosum due to an increase in the inflow of arterial blood and a decrease in venous return. At first sight it may seem remarkable that the corpus fibrosum lies dorsal to the corpus spongiosum whereas in monotremes it is found ventral to this structure. The reason for this is that the direction of the penis has altered from the original orientation as in Monotremata. In Marsupialia as well as in some Fissipedia and Rodentia the penis similarly points backwards but is bent into an "S" shape (Fig. 21A). In most mammals that walk on four legs and in those which live in water, the aspect of the preputial pouch which was originally ventral has grown together with the abdominal wall so that the penis comes to be directed cranially. Due to this reorientation the corpus fibrosum now lies on the dorsal side of the penis. This type of penis is called a penis appositus (Fig. 21B). The connection between the penis and the abdominal wall is lost again in animals with a more or less upright posture: the penis succumbs to the influence of gravity and falls down again, pointing in the direction of the ground (Fig. 21C). Such a penis pendulus occurs in Chiroptera and many primates.

The glandulae praeputiales open out into the preputial pouch. They produce the smegma which keeps the glans penis moist and supple. Various modifications of these glands occur (e.g. the musk glands).

2.7 External Genitalia of Females

External genitalia are rare in the female sex of anamniotes. Genital papillae which function as ovipositors occur in the female sex of bony fish, there being no significant differences between these and the corresponding organs in males. The genital papilla of the bitterling (*Rhodeus amarus*) becomes much longer during the spawning period and with it the female injects her eggs into the gill chamber of a freshwater muscle. The hemipenes of squamatous reptiles are also laid down in females early in development but they do not develop further. The same applies to the penis of Chelonia and Crocodilia. In mammals the female possesses a homologue of the penis, the clitoris. This develops from the anlage of the phallus and represents a miniature penis lying directly in front of the opening of the urethra in most cases. Only rarely does the clitoris contain the urethra (e.g. in some Insectivora and Talpidae), although in a range of primates the urethra begins as a groove along the underside of the clitoris. In the former case the clitoris contains all the erectile elements of the penis whereas in all other cases the corpus spongiosum is lacking, as is the erectile body in the single case of *Ateles.* In some of the species where the male has an os penis the female also has a small os clitoridis. The sinus urogenitalis forms the vestibulum vaginae in the female sex, the opening to the adjacent vagina being constricted by a flap-like hymen in humans and the horse. The penetration of the penis during the first copulation causes the hymen to stretch or tear. The opening of the vestibulum vaginae is bordered by the labia vulvae and the preputium of the clitoris. In primates a pair of labia majora lie lateral to the labia vulvae (= labia minora). These arise from the sex swellings and are homologous to the scrotum of males. In some species of ape the labia, clitoris and the surrounding area become swollen and conspicuously coloured during the course of the reproductive cycle.

CHAPTER 3

The Formation of Gametes

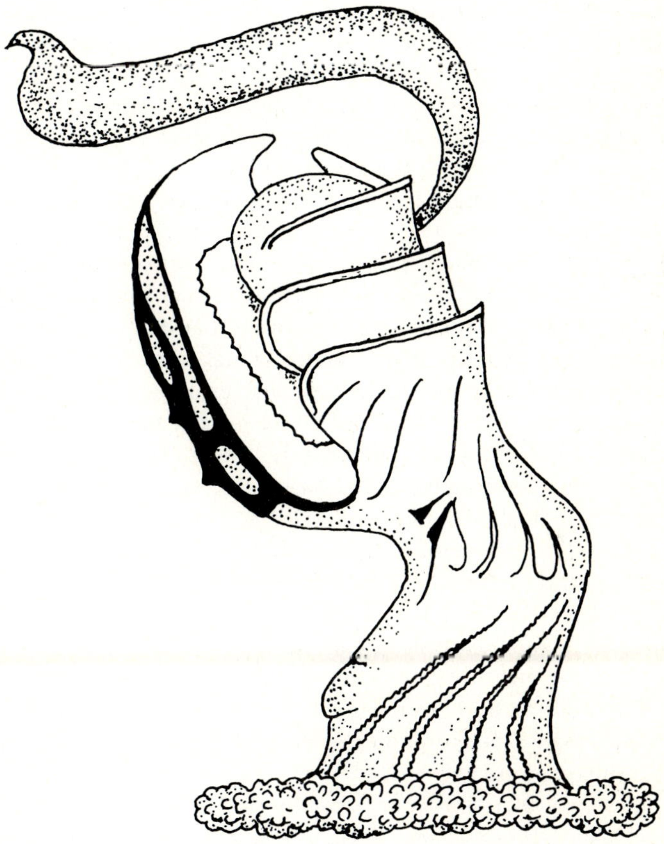

Spermatophore of the palmate newt *(Triturus helveticus)*

3.1 Sexual Maturity

Vertebrates are normally not able to reproduce at the time when they hatch out of the egg or are born. Rather, they go through a phase of growth first by the end of which the gonads have become functional and secondary sex characteristics have developed. The entire process, from the beginning of the maturation of the gonads to the animal becoming fully capable of reproducing, is called pubertas, or puberty. This term refers not only to the formation of functional gonads and thus the attainment of sexual maturity but the whole change over from the growth phase to the reproudctive phase of life. For domesticated animals one can divide puberty into a period of sexual maturity and a time of breeding readiness. For example, horses become sexually mature at an age of between 15 and 18 months but reproduction first takes place when they have reached the stage of breeding readiness at an age of between 36 and 60 months. In many cases the transition from the juvenile to the adult state is not complete when the gonads become functional because the animal continues to develop until it is fully able to reproduce. Species with a short life-span have a very short puberty as a rule and by the time the gonads are functional the body has also matured in all other respects necessary for reproduction. Puberty takes a much longer time in species with a long life-span.

Different species of vertebrate reach sexual maturity at very different times. It has been suggested that the point in time at which sexual maturity is reached falls within the first 10% of the entire life-span, or alternatively that the life-span of a species corresponds to about ten times the age at which sexual maturity is reached. However, this crude rule of thumb is only of limited use. For example, the females of many species mature earlier than the males but nevertheless reach a greater age. For other species the relationship is simply not applicable. In general it seems that to within certain limits the age at which reproduction first takes place depends on the growth rate, on whether or not the animal hibernates, on social factors, which are influenced by population density, and in some birds it depends directly on the availability of food within the population and, especially in poikilothermic vertebrates, upon the ambient temperature.

Sexual maturity comes relatively late in the life cycle of Petromyzontia, beginning between the fifth and eighth year of life in the case of *Petromyzon marinus*. The river lamprey indigenous to Europe (*Lampetra planeri*) lives 3 to 4 years as an ammocoete larva then rapidly metamorphoses and within the same year the mature animals reproduce and subsequently die.

The time at which teleosts become sexually mature varies greatly. Thus male surf perch (*Cymatogaster aggregata*) may already be mature at the time of birth. The closely related Tule Lake perch (*Heterocarpus traski*) mates about three months after birth, the sperm being stored in the female for 6 or 7 months until the following spring when fertilisation occurs. Sexual maturity is similarly reached quickly by Season fish which live for only a very short time; within about 1 month of birth in the case of the tournament perch *Nothobranchius guentheri*. A range of species are able to reproduce by the age of twelve months or less, for example the guppy *Lebistes reticulatus* and the river silverfish *Labidesthes sicculus*. Many teleosts which have a life-span of from 4 to

9 years become sexually mature in their second year, as for example the spiny mackerel (*Trachrus* sp.). The river perch *Perca fluviatilis* and the rainbow trout *Salmo gairdneri* reach the end of puberty after 2 to 3 years, the minnow *Aspius aspius* only after 4 to 7 years. In the pike *Esox lucius* it has been found that, within certain limits, age is much less important than achieving a particular body size (approximately 38 to 46 cm long) which can occur between the first and fifth year of life. The European eel *Anguilla anguilla* is a decidedly late developer, becoming sexually mature between its tenth and fourteenth year of life. Red perch of the genus *Sebastes* belong to the Scorpaenidae which give birth to live young and first reproduce at about the same age as the eel.

Sexual maturity frequently appears to be linked to a certain body weight in amphibians. There are in addition indications that there is generally a phase of rapid growth preceding the initial maturation of the gonads followed by a period of slow growth which can continue throughout life in some species. The age at which sexual maturity is reached can also vary within a species from population to population, just as it can between males and females. Most amphibians are not able to reproduce before the end of their first year of life. An exception to this is the Gulf toad *Bufo valliceps* whose testes may be functional at an age of 10 months. There are a range of anuran species which become sexually mature after 1 or 2 years, for example the clawed toad *Xenopus laevis* and the leopard frog *Rana pipiens*. Others require 2 to 3 years, such as the bronze frog *Rana clamitans*, or even 3 to 4 years as in the case of the bullfrog *Rana catesbeiana*. The European toad *Bufo bufo* and the grass frog *Rana temporaria* first reproduce when they are between 4- and 5-years-old. Urodela become sexually mature after a similar length of time; for example the tiger salamander *Ambystoma tigrinum* after 1 year, the alpine newt *Triturus alpestris* after 2 or 3 years, the alpine salamander *Salamandra atra* after 2 to 4 years, and the three-toed newt *Amphiuma tridactylum* after 4 or 5 years. Nothing is known of the Gymnophiona in this respect.

The attainment of a certain body size also appears to be more important for the onset of puberty in reptiles than reaching a particular age. The Madagascan chameleon *Chamaeleo lateralis* is already mature at an age of about 3 months. Species which are small in size are often able to reproduce relatively early, for example, after a year or less in the case of the American anole (*Anolis carolinensis*). The European lizards *Lacerta agilis* (sand lizard) and *L. vivipara* (common lizard) become sexually mature when they are 2-years-old and sometimes younger. A range of reptilian species from temperate climes similarly become sexually mature after 2 years, by which time they have usually reached about two thirds of their final size. The grass snake (*Natrix natrix*) and adder (*Vipera berus*) first reproduce at an age of from 3 to 5 years. There are species of Chelonia which only reach puberty at a much greater age, for example, the South Chinese box turtle (*Cuora trifasciata*) at an age of between 8 and 11 years and the alligator snapping turtle (*Macrochelys temminckii*) at an age of between 11 and 13 years. The Mississippi alligator (*Alligator mississipiensis*) is able to reproduce when it has a body length of about 1.8 m, which it reaches at an age of from 6 to 10 years. The tuatara (*Sphenodon punctatus*) requires some 20 years to become sexually mature.

In birds, the amount of time before the onset of the reproductive phase of life varies between a few months and several years. Captive quail (*Coturnix coturnix*) have been observed to become sexually mature only 1 or 2 months after hatching. Chickens have functional gonads when 4- to 7-months-old but reproduce only at an age of 9 to 12 months at the earliest. Species of small birds commonly become sexually mature after 1 or 2 years, 1 year in the case of the dunlin (*Calidris alpina*) and 1 to 2 years for the starling (*Sturnus vulgaris*).

The first opportunity for migratory passerines to become sexually mature falls at a time during their first year of life when the days get shorter. Since the maturation of the gonads in such species is controlled by the photoperiod, and specifically by a lengthening of the day (see p. 25), the onset of sexual maturation is delayed for 6 to 7 months. Therefore, many species which are small in size start breeding only in the following year whereas larger birds begin even later, at the end of their second year of life. This delay in breeding is probably influenced by other factors in addition to light, such as the limited availability of nesting sites or the presence of a dominant older male of the same species. A comparable phenomenon occurs in gulls and other sea birds, which take between 2 and 10 years to reach sexual maturity but frequently breed only some years later. What causes this delay is still unknown, although one important factor may be that there is insufficient food for the whole population. Clearly, in such a situation, only a fully grown individual would be able to provide sufficient food for its brood. The result is a very low rate of reproduction that only species which live for a long time can withstand. It has been shown that in populations in which the annual mortality rate for adult animals is less than 5%, enough individuals live for 50 years or more that a very low rate of reproduction suffices to maintain the species. The gray goose (*Anser anser*) is another species of bird which first reproduces at an age of 2 or 3 years whereas the common cormorant (*Phalacrocorax carbo*) and the pink pelican (*Pelicanus onocrotalus*) take 3 to 4 years to become sexually mature. The Californian condor (*Gymnogyps californianus*) only begins to reproduce at an age of about 12 years.

The onset of puberty in mammals usually occurs at an age when body growth is slowed. Voles (*Microtus* sp.) become sexually mature extremely quickly. Individuals born in spring reproduce when they are only 1-month-old, however those born in late summer first mate the following spring when about 6-months-old. The house mouse (*Mus musculus*) can reproduce about 6 weeks after being born. In the laboratory variety of this species, social factors also appear to play a role: females become mature quicker in a bisexual atmosphere which includes the scent of mature males than in isolation. Several species of larger mammal also sometimes have functional gonads before the end of their first year of life, for example, at an age of 8 to 14 months in cattle (*Bos taurus*), 6 to 9 months in the pig (*Sus scrofa domesticus*), and 7 to 12 months in the sheep (*Ovis aries*). The beaver (*Castor canadensis*) becomes mature after 2 years, the Rhesus monkey (*Macaca mulatta*) after 3 years, and the jumping or mule deer (*Odocoileus hemionus*) after 4 years. The sea lion (*Otaria jubata*) first reproduces at an age of 6 years. A distinct acceleration of sexual maturation can be observed in the Pinnipedia which have been culled by man. Thus, the harp seal (*Pagophilus groenlandicus*) has a natural life-span of about 40 years and mating first takes place at an age of about 6 years. In populations culled by man, the animals currently grow faster and become sexually mature 1 year earlier than in past years. Similarly, elephant seals (*Mirounga leonina*) first become pregnant between their fourth and seventh year of life in undisturbed populations but in their third year on average in culled populations. A comparable situation has been found in whales. Up until 1930, females of balaenopteran species became sexually mature in about their tenth year. On the other hand, between 1930 and 1960 the first pregnancy occurred after only 5 or 6 years. It was found, however, that the average size of the female's body had not changed. It appears, therefore, that keeping the population density low by culling had the effect of increasing the amount of food available to the survivors so that they grew more quickly. This example provides a further, clear indication that there is a correlation between the age at which sexual maturity is reached and body weight. This probably also applies to man. The mean body weight at which women reach sexual maturity has been found to be 48 kg, and this value is largely independent of age and height. Since body weight depends on how much we eat, the age at which puberty begins has steadily decreased during the last century in the more developed countries. In 1840 in Norway, girls had their first menstrual period (menarche) at an age of about 17, whereas in 1970 the average age was about 13 years. Almost identical results are available for other northern European countries and the U.S.A. In the years from 1945 to 1970 alone the value for Sweden fell from 14 to 13 years. It must be remembered, however, that the menarche does not necessarily indicate that sexual maturity has been reached. It demonstrates that the reproductive tract is functional, but need not be coupled with an ovulation in every case: it merely provides a reference point for the onset of sexual maturity.

From the above considerations it follows that vertebrates reach sexual maturity as fully differentiated, adult animals. There are nevertheless several exceptions to this rule. Some species of Petromyzontia and Amphibia go through a prolonged larval stage before metamorphosing, whereas others never even reach this stage of development. This phenomenon is generally called neoteny. A variety of neotenous amphibian species belonging exclusively to the Urodela reproduce facultatively or obligatorily in the larval stage. An example of facultative neoteny and of larval reproduction is provided by the North American tiger salamander (*Ambystoma tigrinum*). Different races and varieties of this species which live all over the U.S.A. undergo normal metamorphosis and subsequently reproduce. In the east of the country, however, the species is neotenous and reproduction takes place during the larval stage. The reason for this is that there is an extreme deficiency of iodine in the waters of the biotope and this results in the synthesis of insufficient thyroid hormone, the hormone which promotes metamorphosis. Low water temperatures also inhibit the activity of the thyroid glands and lead to neoteny. Another well-known example of a neotenous species which reproduces as a larva is the axolotl (*Siredon mexicanum*). Other examples of the same phenomenon occur in water newts of the genus *Eurycea* (*E. tynerensis, E. neotenes*) as well as gill bearing Urodela including the genera *Necturus* (mud puppies), *Siren,* and *Pseudobranchus* (sirens), *Cryptobranchus* (mud devil), and the European cave salamander, or olm (*Proteus anguinus*).

3.2 Production of Gametes

In all vertebrates the formation of gametes, generally called gametogenesis, takes place in the gonads, the development of functional spermatozoa being referred to as spermatogenesis, and that of functional eggs as oogenesis. In the majority of metazoans, gametogenesis proceeds according to a uniform plan which has four phases. It starts with cells, the so-called gonia, produced by the mitotic division of the primordial germ cells that migrate into the embryonic gonad. In males one speaks of these as spermatogonia, in females as oogonia. Initially these cells go through a phase of mitotic proliferation whereby the cytoplasm is not fully restored after each division so that the cells become progressively smaller. The spermatogonia divide more often than oogonia as a rule. During the next phase of growth there is no further division but the cytoplasm increases in volume. As they enter this phase the cells are respectively called first order spermatocytes (primary spermatocytes) and first order oocytes (primary oocytes). In the former there is only a small increase in the volume of the cytoplasm, by a factor of about two. This phenomenon is considerably more pronounced in first order oocytes: for example, in chickens there is an approximately 200-fold increase in volume during the last growth phase 6 to 14 days before the ovulation of the egg. During this phase of development the eggs of all submammalians are provided with a greater or lesser amount of reserve material called yolk or vitellus. The process of yolk deposition is generally referred to as vitellogenesis. The third phase of gametogenesis includes the kernel of the whole process, the reduction of the chromosome set from diploid to haploid. This occurs in the course of meiosis, which is accomplished in two steps, respectively the first and second maturation divisions, which have already been described (see p. 12). The cells which result from the first maturation division of the first order spermatocytes are called second order spermatocytes (secondary spermatocytes, or prespermatids). These then become spermatids after the second maturation division. The two maturation divisions are equal so that four spermatids arise from each first order spermatocyte. In exactly the same way, first order oocytes give rise to second order oocytes from which eggs are formed. In contrast to spermatogenesis, the divisions are unequal in the case of oogenesis, the first meiotic division giving rise to a large and a small second order oocyte, the smaller being referred to as the primary polar body, or polocyte. As a result of the second maturation division the large second order oocyte is again split unequally, whereas the primary polar body divides into equal parts. In this way only a single functional egg arises from a first order oocyte: the three polar bodies, although theoretically representing haploid gametes, are not viable. The division of the primary polar body may not even take place, in which case the egg is accompanied by only two polar bodies. The last phase of gametogenesis is the conversion of potential gametes into functional gametes and is only pronounced in males. It embraces the dramatic transformation of the spermatids into flagellated, and therefore motile spermatozoa which all vertebrates possess. The entire process is called spermiohistogenesis, or more correctly, spermioteleosis. A synonym commonly used in the literature written in English is the term spermiogenesis.

3.2.1 Spermatogenesis

Two types of spermatogenesis occur in vertebrates, differing with respect to the compartment in which the process described above take place. The first type is cystic spermatogenesis and it occurs in anamniotes which in most cases require a large number of sperm for external fertilisation. Thus the cells derived from an early spermatogonium, representing the different stages of spermatogenesis, are contained within a cyst built of somatic cells which are probably homologous to the Sertoli cells. Every cyst, therefore, contains a clone of cells which normally develop synchronously. Two variants of this type occur. In one case the cysts containing the germ cells are distributed throughout the tubules or ampullae of the testes (the unlimited type), in the other the cysts are situated in particular regions, usually at the terminal end of the tubules (limited type). The cysts constitute the primary compartment of spermatogenesis, the secondary compartment being represented by the small or large number of cysts filling the tubule or ampulla, and lastly, the whole testis forms the tertiary compartment. The sperm are released out of the cysts when spermiohistogenesis has been completed.

Acystic spermatogenesis, in contrast to the above, takes place within the seminiferous tubules, which in this case represent the primary compartment. This type of spermatogenesis is typical of amniotes. The lining of the tubule is made up of the different developmental stages of the germinal line cells together with the somatic Sertoli cells. Thus spermatogenesis takes place in close contact with the Sertoli cells, which serve as mediators between the cells of the germinal line and external influences such as hormones (see p. 185). The resulting sperm are released into the lumen of the tubules. Acystic spermatogenesis is therefore one degree less complex than the cystic variant. It either proceeds continuously or in spurts.

In vertebrates the course of spermatogenesis in time differs according, for example, to whether or not reproduction is seasonal or whether the species is a poikilotherm or a homoiotherm. The majority of poikilothermic vertebrates which reproduce seasonally have so-called postnuptial spermatogenesis. This means that spermatogenesis is begun during the summer, after mating or spawning has taken place in spring, and provides the gametes that will be used during reproduction the following year. There are three main variations on this theme: (a) the whole process of spermatogenesis elapses before winter; (b) spermatogenesis begins in summer, procedes more slowly during the winter, and is completed next spring; (c) the spermatogenesis begun in summer is interrupted during the winter and begins again in spring. Postnuptial spermatogenesis occurs in many fish, amphibians and reptiles of the temperate zones.

The second, prenuptial type of spermatogenesis is characteristic of homoiothermic vertebrates. In this case, after one reproductive period has ended little or no spermatogenesis takes place in the testis until shortly before the next reproductive period in spring, when it begins again in full. Both of the types described are variants of discontinuous, or cyclic spermatogenesis which is characteristic of vertebrate species living in temperate latitudes whose reproduction is cyclical. Contrasting with this is continuous spermatogenesis, which is particularly common among tropical species but also occurs in highly domesticated species of the temperate climatic zones. Curiously the water frog *Rana esculenta* shows both types of spermatogenesis: its sperm production is discontinuous in temperate latitudes but continuous in the sub-tropics.

The general scheme of spermatogenesis described in the introduction to this chapter provides a life history of a spermatozoan and a terminology for referring to the different stages of its genesis but does not elucidate in full the true complexity of the whole process. It disregards one important aspect, that a spermatozoan never develops alone or in isolation. Rather, the different steps of division and differentiation take place in a syncytium of clonal cells which arise because the division of the cells is incomplete. Thus, the individual cells are joined by cytoplasmic bridges which are only broken, and the spermatozoa set free after spermiohistogenesis. As has already been mentioned, spermatogenesis only takes place in co-operation with accessory somatic cells: these are the Sertoli cells in the case of amniotes and, in anamniotes, their presumptive homologues the cyst cells, or more exactly the Sertoli-cyst cells.

The primordial sex cell represents the starting point for the formation of gametes. Such cells are initially located outside the gonads, are not sexually determined, and can divide mitotically. They migrate into the rudimentary gonad and become sexually determined according to whether the gonad develops into a testis or an ovary. There they proliferate by mitotic division and undergo a certain degree of differentiation into so-called gonocytes, the whole process whereby they are formed in the testis being prespermatogenesis. This is a very clearly defined process in mammals but not in other classes of vertebrate. Upon reaching a certain degree of differentiation these cells become early primary spermatogonia. At this stage the process of spermatogenesis, whereby these mother cells give rise to a clone of spermatozoa, really begins. The primary spermatogonia enter a phase of mitotic proliferation producing daughter cells called secondary spermatogonia which, as has already been mentioned, are joined together by cytoplasmic bridges. The division of these cells takes place in the space within a cyst (a spermatocyst) or in close association with a Sertoli cell. This phase of proliferation is also called spermatocytogenesis. The mitotic division of the spermatogonia is evidently a precondition for their going on into meiosis: they must undergo at least one division. The actual number of divisions is specific to each species, vertebrates going through between three and fourteen. It is possible that every division in the formation of the gametes involves differentiation, which is certainly the case with respect to the secondary spermatogonia. The nuclei of these cells apparently change in some way with every division, gradually approaching the state necessary for the onset of meiosis. The last division of the spermatogonia results in first order spermatocytes, the cytoplasm of which then grows in volume to a greater or lesser degree, and for this reason these cells have also been referred to as auxocytes. The first order spermatocytes then enter meiosis, the kernel of spermatogenesis. The haploid daughter cells arising from the first maturation division are second order spermatocytes, or prespermatids. After the second maturation division these become spermatids, i.e. gametes which are not yet functional. The number of spermatids arising from a single primary spermatogonium depends on the number of times the spermatogonia divide. Thus, counting the primary spermatogonia as the first generation, the number of generations produced is always one more than the number of divisions. For example, four divisions give rise to five generations producing 16 first order spermatocytes which go on to give 32 second order spermatocytes, and 64 spermatids. Up until the spermatid stage, the mitotic and meiotic divisions of a cell clone are incomplete, as mentioned, so that all these cells are connected to one another by cytoplasmic bridges. It is prob-

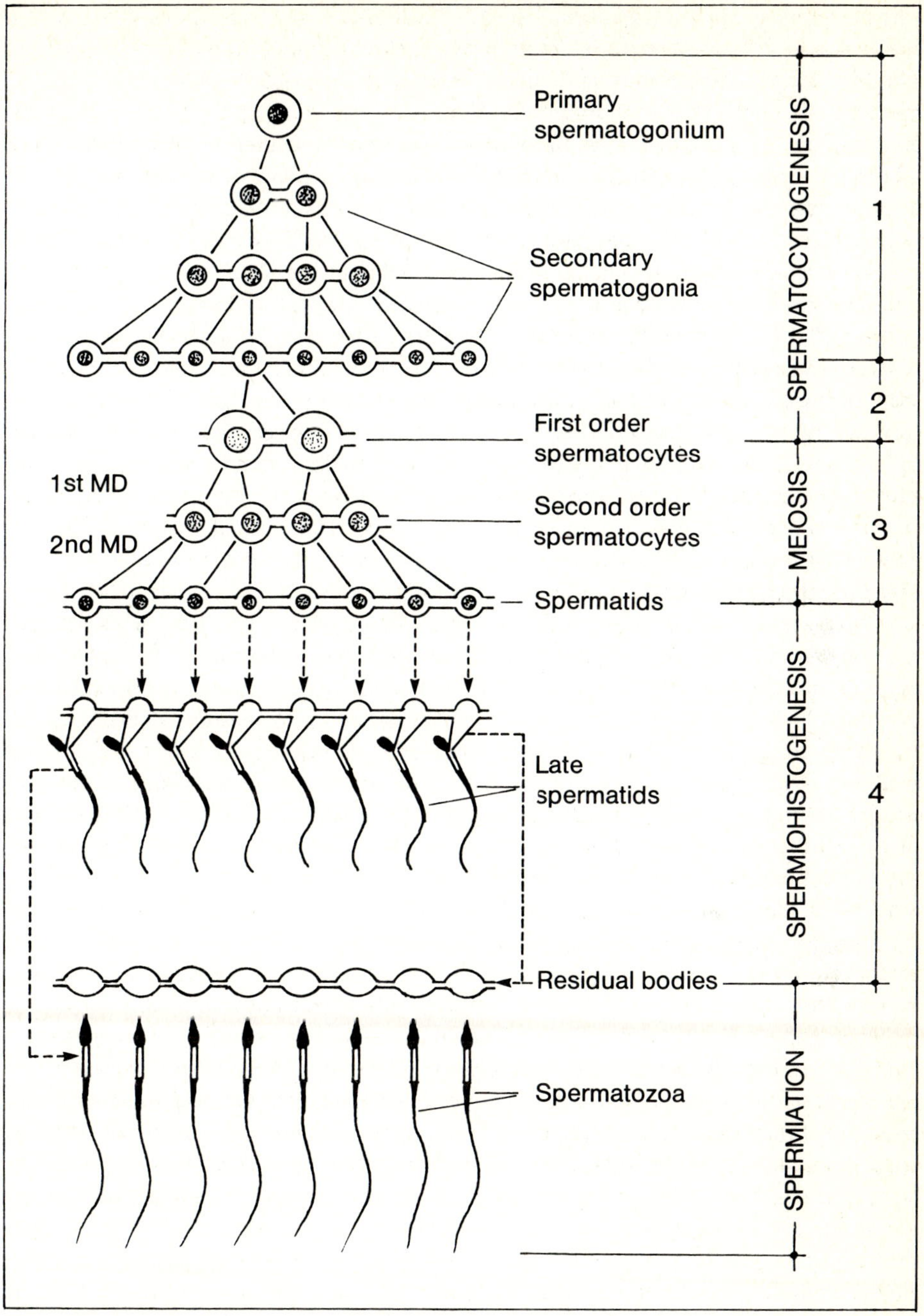

Fig. 22. Schematic representation of spermatogenesis in vertebrates.
1 = proliferative phase; 2 = growth phase; 3 = phase of meiosis; 4 = phase of differentiation; MD = maturation division

able that this syncytial organisation greatly facilitates the synchronous division of a clone, which is what happens in the majority of cases throughout the animal kingdom. The final stage of spermatogenesis is the transformation of the spermatids into spermatozoa, a process known as spermiohistogenesis. This complex process involves fundamental changes in the morphology of the spermatids leading to the formation of flagellated spermatozoa which, together with the other members of their clone, are released into the lumen of the tubule or ampulla, or into the seminal ducts (the process of spermiation). The subject of spermiohistogenesis will be returned to later. The course of spermatogenesis is illustrated schematically in Fig. 22.

From the above description of spermatogenesis it would appear that the differentiation of the cell clone has a certain degree of autonomy. In reality, however, the cells depend greatly on the somatic tissue of the individual, which exercises a regulatory influence via, for example, hormones. The mediator between the soma and the germinal line is represented by the accessory or supporting cells. These are the Sertoli cells in amniotes and the cyst cells in anamniotes, which are very probably homologues of the former. As mentioned, spermatogenesis takes place in primary compartments composed of cells and, at least partly, of extracellular substances. In general within the animal kingdom, the compartments have a diameter of between 30 and 300 μm, this being exceeded only by the seminiferous tubules of marsupials which have a diameter of 360–520 μm. The primary compartments of the vertebrates are either spermatocysts (Pisces and Amphibia) or tubules (Amniota). The spermatocysts of Teleostei are in many cases organised so that the Sertoli-cyst cells surround and directly border a cell clone. In a variety of species a single such cell takes part in the formation of two or more spermatocysts and, therefore, is in direct contact to a corresponding number of cell clones (Fig. 23A). After the cysts rupture in the process of spermiation the Sertoli-cyst cells can transform themselves into epithelial cells of the reproductive ducts or possibly the seminiferous tubules. The spermatocysts of amphibians are made up of flat, indistinctive cells which enclose the cell clone for most of the duration of spermatogenesis. In many species, however, the spermatocysts rupture at an early stage of spermatogenesis and the tubule becomes the primary compartment, as also occurs in the Teleostei. Thus the evolutionary change-over from cystic to acystic spermatogenesis is already evident in teleosts and amphibians.

The Sertoli cells of mammals have been particularly well studied. They are highly differentiated cells which also constitute the elements of a dynamic skeleton supporting the developing cells of the germinal line. The Sertoli cells are held together by intercellular junctions and their branched and deformed shape forms compartments that accommodate the spermatogonia, spermatocytes and spermatids. The spermatogonia lie against the wall of the tubule and are covered on their lumenal side by the spermatocytes and spermatids: spermiohistogenesis takes place at the border of the lumen. Experiments using lanthanum clearly demonstrate that the Sertoli cells form two subcompartments, a basal compartment containing the spermatogonia and early first order spermatocytes and an adluminal compartment containing spermatocytes and spermatids, the two being separated from each other by cell junctions (Fig. 23B). However, the degree of isolation of the individual cell clones is much less in acystic spermatogenesis than in the cystic type. Moreover, the Sertoli cells are capable of phagocytosis, a characteristic that enables them to take up the cytoplasmic remnants pinch-

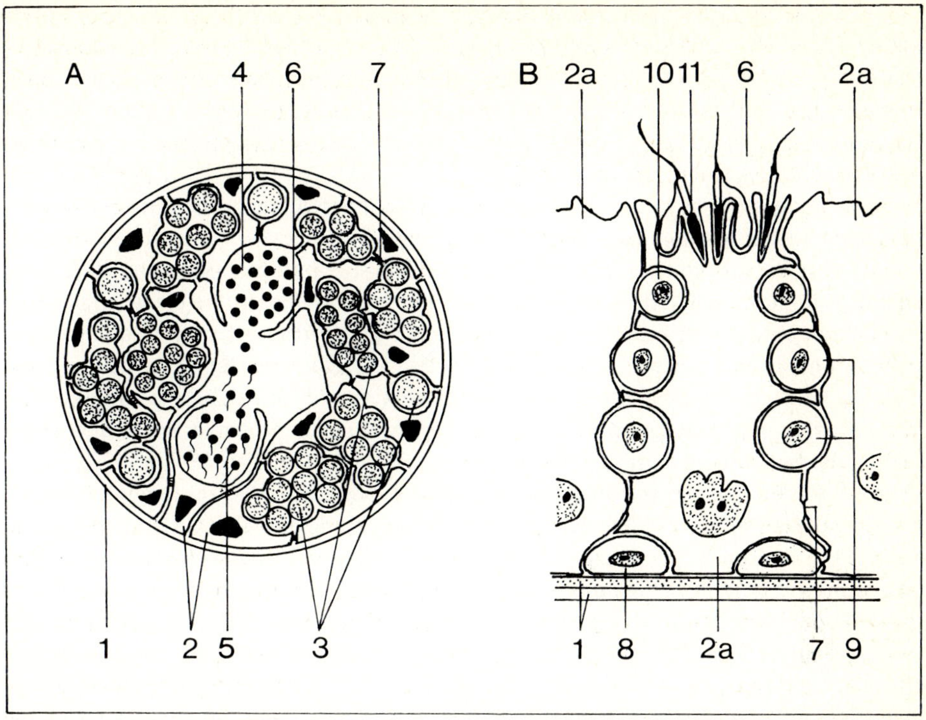

Fig. 23A,B. Spatial relationships between cells of the germinal line and Sertoli cells (cyst cells). **A** Cross-section of a seminiferous tubule of a teleost. (After Billard et al. modified considerably). **B** Longitudinal section of a seminiferous tubule of a mammal. (After Dym and Fawcett).
1 = tubule wall; 2 = Sertoli cyst cell; 2a = Sertoli cell; 3 = cells of the germinal line at different stages of development; 4 = spermatids; 5 = spermatozoon; 6 = tubule lumen; 7 = cell junction; 8 = spermatogonium; 9 = spermatocytes; 10 = early spermatid; 11 = late spermatid

ed off in the course of spermiohistogenesis. The Sertoli cells probably also play an important role in the passing on of hormonal "instructions" (see p. 185) and are evidently able to function as endocrine cells.

The basic pattern of spermatogenesis throughout the course of evolution has remained almost unchanged and is the same within the vertebrates. Variations between the different classes and species specific variants relate primarily to the kinetics of the process, the number of mitoses in spermatocytogenesis and therefore the number of cells in a clone, the duration of the individual stages, and finally the number of cells that die during the course of the whole process. Over and above these there is the difference between the amniotes and anamniotes in the compartment where spermatogenesis takes place.

There has been very little study of spermatogenesis in Cyclostomata. The ammocoete larvae of *Petromyzon* species generally go through an initial female phase. In their first and second year of life the primordial germ cells proliferate to give rise to the protogonia which are enclosed within a cyst or cellular envelope. When two or more such cells are thus enclosed they are referred to as deuterogonia: they correspond to the secondary spermatogonia and develop synchronously. At the start of the third year of life the oocytes begin to degenerate, a process which may last until metamorphosis at the end of the sixth year of life. During this time the spermatogonia remain inactive, until metamorphosis or thereafter, when they divide into spermatocytes which then enter meiosis.

Spermatogenesis in Elasmobranchii takes place in ampullae in the testes, each spermatocyst probably being formed by only a single Sertoli cyst cell. In the lesser spotted dogfish (*Scyliorhinus caniculus*), a total of 64 spermatozoa arise from each early spermatogonium, and remain clustered together in a bundle. In the electric ray (*Torpedo marmorata*) and other species of ray, five generations of spermatogonia have been counted, and thirteen in the spiny dogfish (*Squalus acanthias*). In this last species the cells of the cyst apparently go through two different phases: an exocrine phase during spermatocytogenesis which gives way to a probable endocrine phase during the first prophase of meiosis.

A continuum of intermediary forms of testis exists in Teleostei ranging between two extremes represented by the Percomorphi type and the Microcyprini type. The Percomorphi type is the most prevalent and in this case the seminiferous tubules merge individually into the central canal of the testis. The second type occurs in several Microcyprini and in this case there is a network of anastomosing tubules which merge with the spermiduct. The two types differ with respect to the "behaviour" of the spermatocysts: in the first they remain where they are whereas in the second type the cysts migrate during the course of spermatogenesis from the closed end of the tubule to the spermiduct. In the first type the cysts open and release the spermatozoa into the lumen of the tubule whereas in the second type, when the cysts reach the end of the network of tubules their cells merge with and form part of the wall of the spermiduct, thereby setting the spermatozoa free. In teleosts the division of the spermatogonia usually occurs synchronously, but in the surf perch (*Cymatogaster aggregata*) they are asynchronous and the cysts are probably first formed at the spermatocyte stage. With regard to the teleosts, the present view is that the spermatogonia derive from cells of the germinal line which either lie outside the tubules at the periphery of the testis as quiescent stem cells, or persist within the wall of the tubule as residual germ cells. The number of spermatogonial generations determined so far differs from species to species from five to fourteen. The earlier view, that the spermatogonia stem from a skein of connective tissue laying outside and ventral to the testes and migrate into this organ by amoeboid locomotion, has not been confirmed. Outside the reproductive period the cyst cells still line the tubule, with one or two spermatogonia lying between them. At the beginning of spermatogenesis the cyst cells grow and multiply in number forming the cysts around the cell clones. At a later developmental stage they form numerous desmosomes which, in the case of the guppy (*Lebistes reticulatus*), make up about half of the entire area of cell contact. Moreover, in this species they form cell processes up to 20 μm long which project into the cyst. The cyst cells develop a high level of phagocytotic activity during the course of spermiohistogenesis. Throughout the whole process of spermatogenesis an average of about 35% of the cells degenerate. The rate of production of spermatozoa in the guppy is approx. 150×10^6 germ cells g^{-1} (wet weight of testis) day^{-1}.

Spermatogenesis in amphibians has been most intensively studied in those from temperate latitudes, in which it is seasonal. The predominant type of spermatogenesis is the postnuptial variety. In general the course of spermatogenesis is similar to that in Teleostei, differing only in the relatively early rupturing of the spermatocysts, which then become a part of the tubule wall. In anurans this occurs when the spermatids become flagellated. Thus twelve of the cyst cells aggregate together at the periphery of the cyst and grow larger, each cell coming to harbour in a cup-like depression a bundle of from 60–150 spermatozoa whose tails are directed into the cyst lumen. At this stage the cyst cells are called supporting cells and vacuoles form in their cytoplasms via which fluid is apparently released into the lumen of the tubule. The primary spermatogonia for the next cycle are sometimes already present surrounded by cyst cells. So-called multiple testes occur in urodeles, the testes being divided into discrete lobes. This results from a special form of spermatogenesis which passes as a spermatogenetic wave over the length of the testis so that in each lobe of the testis the early stages of spermatogenesis lie cranially, the later stages caudally. The speed with which the spermatogenetic wave passes over the testis differs from species to species: the length of the wave generally increases somewhat every year. Spermiation always takes place from the caudal part of the lobes and these sites then lie dormant, for up to 3 years in some species, thereby redefining the caudal border of each lobe, shifting it cranially (the cranial border being defined by the caudal border of the lobe lying cranial to it). The number of testicular lobes can increase during life up to a limit which differs for each species. The dormant stem cells in the "sterile" sections of the testis become active again after a certain refractory period. In the cave salamander (*Proteus anguinus*), every spermatogonium of the undivided testis is enclosed by 3 to 4 cyst cells and the wave of spermatogonial division begins cranially and passes over the whole length of the testis. In the spermatocyte stage there are 68 to 124 cells present in each cyst which arise from four to six cells. Cysts at the spermatid stage contain 238 to 484 germ cells, indicating that the number of divisions undergone by the clones must differ (i.e. six and seven respectively). About 7% of the cells degenerate. Seven spermatogonial generations have been found in members of the genera *Ambystoma* and *Triturus*. The spermatogonia of amphibia are characteristically relatively large.

Spermatogenesis in amniotes follows a uniform pattern. As already mentioned, this process takes place in different layers of the tubule wall, the younger stages lying peripherally and the later stages more towards the lumen. Spermatogenesis has been thoroughly investigated in man, the rat, and the mouse but correspondingly detailed studies of reptiles and birds are scarce. Two such studies relate to the slowworm (*Anguis fragilis*) and the domestic duck (*Anas platyrhynchos*), respectively. In general the immature or inactive (out of season) testis of mammals contains gonocysts or primary spermatogonia as well as Sertoli cells. When a synchronised colony of spermatogonia reaches the spermatocyte stage the cells migrate at specific intervals towards the lumen of the tubule. The Sertoli cells lying radially around the lumen develop villus-like processes which enclose the spermatocytes and spermatids so that spermiohistogenesis takes place at the apical pole of the Sertoli cells (Fig. 23B). When the first generation of spermatozoa are released into the lumen of the tubule the next moves in from the periphery to take its place, thereby completing one cycle of spermatogenesis. In some species successively advanced stages of development occur along the long axis of the tubule, corresponding to the spermatogenetic wave which occurs in urodeles. In mammals the germ cells which are not participating in the current spermatogenetic cycle probably lie separate from each other against the wall of the tubule: they go through mitosis at regular intervals which, however, bear no demonstrable relationship to spermatogenesis. In the rat and mouse it has been observed that these divisions become incomplete at some point so that pairs or groups of cells arise which nevertheless still undergo cell division at irregular intervals. These latter divisions subsequently become synchronous and lead on to spermatogonial divisions. The transition from stem cells to spermatogonia is often not clearly discernable. A wide range of mammals have four to six generations of spermatogonia but similar data do not exist for reptiles and birds. At every stage of spermatogenesis in mammals, some of the cells degenerate.

The transformation of the spermatids into motile spermatozoa in the course of spermiohistogenesis involves remarkable changes in morphology. It is appropriate here to detail the structure of the fully differentiated germ cell before tackling the complicated processes of spermiohistogenesis, considering first the general structure of the vertebrate spermatozoon.

The spermatozoon represents the most autonomous cell of a metazoan organism. They set themselves free of the cell colony from which they arise and usually leave the body. They "lead their own lives" for a certain length of time until they reach their goal, namely fusion with an egg, or perish. For part of this time spermatozoa may be stored within the female organism for up to several months in a dormant state within specialised storage structures, as mentioned in Chapter 1 (see p. 33). The autonomous phase ends with the fusion of the spermatozoon with the nucleus of the egg. Two types of spermatozoan occur within the Vertebrata, a primitive type and a progressive or modified type. The former is common among invertebrates which have external fertilisation. Among vertebrates this type occurs, with only a single modification, in Teleostei. The primitive spermatozoon consists of a head and a tail piece. The head contains a spherical or ovoid nucleus surrounded by a thin border of cytoplasm. Behind the head there is a small cytoplasmatic region containing mitochondria. At the tip of the nucleus is a cap or vesicle called the acrosome, although this is missing in the case of the primitive spermatozoa of most teleosts. The tail consists of a flagellum with a 9 + 2 pattern of microtubules, as in protozoans, which arise from a centriole located in the cytoplasmic region (Fig. 24D). The progressive type of sper-

Fig. 24A–D. The structure of mammalian spermatozoa. **A–C** The progressive type of spermatozoon of man. (After Holstein and Rosen-Runge). **A** Longitudinal section. **B** Cross-sections at the levels indicated by the *arrows* in **A**. **C** Highly schematic surface view of the spermatozoon (rotated 90° around its long axis with respect of **A**). The *middle piece* is drawn as if it were transparent so as to show the spiral of mitochondria. **D** The primitive type of spermatozoon, lacking an acrosome, which occurs in teleosts.

1 = head; 2 = acrosome; 3 = neck region; 4 = middle piece; 5 = main piece of the tail; 6 = end piece of the tail; 7 = plasmalemma; 8 = nuclear vacuole; 9 = nucleus; 10 = proximal centriole; 11 = outer filaments; 12 = cytoplasm; 13 = mitochondria; 14 = microtubules in a 2 x 9 + 2 arrangement; 15 = annulus; 16 = fibrous sheath; 17 = end piece of the flagellum; 18 = flagellum

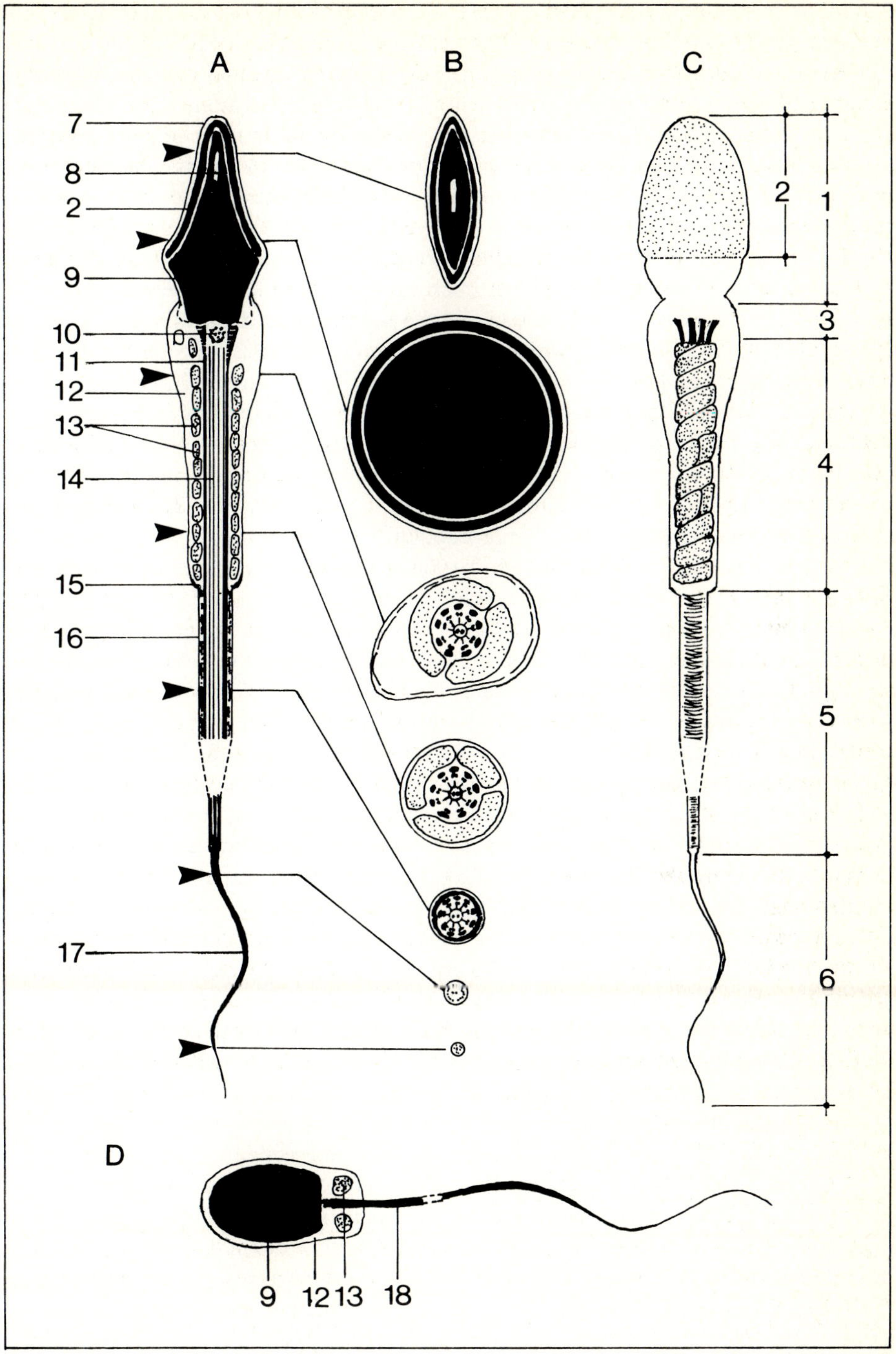

Fig. 24A–D

matozoon possesses additional structures which can be illustrated by taking the human spermatozoon as an example. Here too there is a division into a head and tail. The head is shaped like the tip of a spear when seen in longitudinal section and the front portion is somewhat flattened (Fig. 24A–C). The acrosome caps about two thirds of the nucleus and together these are covered by a thin layer of cytoplasm. Behind the head is the middle piece which corresponds to the cytoplasmic region of the primitive spermatozoon. The neck region of the middle piece contains two centrioles: the proximal centriole is passed on to the zygote where it forms the spindle apparatus during cleavage whereas the flagellum arises from the distal centriole. The latter also possesses a 9 + 2 arrangement of microtubules, to which, however, a non-contractile outer filament is added to each of the nine pairs of microtubules. The mitochondria in the middle piece, which has rather more cytoplasm nearer the nucleus, spiral round the flagellum forming a mitochondrial spiral. The middle piece ends in an annulus. The tail has a main and an end piece, the former being composed of the continuation of the flagellum and its enclosing fibrous sheath. The outer filaments of the flagellum extend along a considerable length of this piece of the tail. The end piece of the tail is composed of the unsheathed flagellum, which lacks the outer filaments. The 9 + 2 arrangement of microtubules is gradually lost towards the end of the tail. The head of the human spermatozoon is 3–5 μm long, the middle piece measures about 6 μm and the tail 40–50 μm.

The transformation of a spermatid into a spermatozoon in the course of spermiohistogenesis occurs in basically the same way within different vertebrates. In the early stages the spherical spermatid contains a nucleus, mitochondria and the two centrioles. Further development first sees the formation of a Golgi apparatus in the vicinity of the nucleus, where this organelle subsequently forms an acrosomal vesicle enclosing an acrosomal granule. At the same time the centrioles migrate to a position at the periphery of the cell roughly opposite the acrosomal vesicle where one of them gives rise to a small flagellum. This centriole subsequently moves towards the nucleus, thereby prolonging the flagellum. On the other side of the nucleus is the acrosomal vesicle, which lies against the nucleus. There is thus a morphological and physiological polarisation of the cell. At this stage the mitochondria have already become aggregated together at the flagellated pole of the cell. The acrosomal vesicle then comes to lie against the nucleus like a cap whereas the remains of the Golgi apparatus move off towards the flagellated pole of the cell. In the next phase the nucleus loses some of its fluid content and condenses, becoming elongate in shape. Microtubules arrange themselves in parallel arrays around the nucleus forming the so-called manchette. The bulk of the cytoplasm shifts gradually to the flagellated cell pole giving the spermatid a more or less spindle-shaped form. This change is accompanied by a grouping of the mitochondria around the flagellum. During the course of further development the cytoplasm continues to shift from around the nucleus to the site of the prospective middle piece until the nucleus and the acrosomal vesicle capping it are covered only by a thin layer of cytoplasm. The mitochondria move to take up a position around the flagellum, that later becomes the axial filament of the middle piece. In the last phase of spermiohistogenesis the acrosome takes on its final form and the head of the spermatozoon is formed. The mitochondria become arranged in a spiral around the axial filament, which is covered only by an extremely thin layer of cytoplasm as is

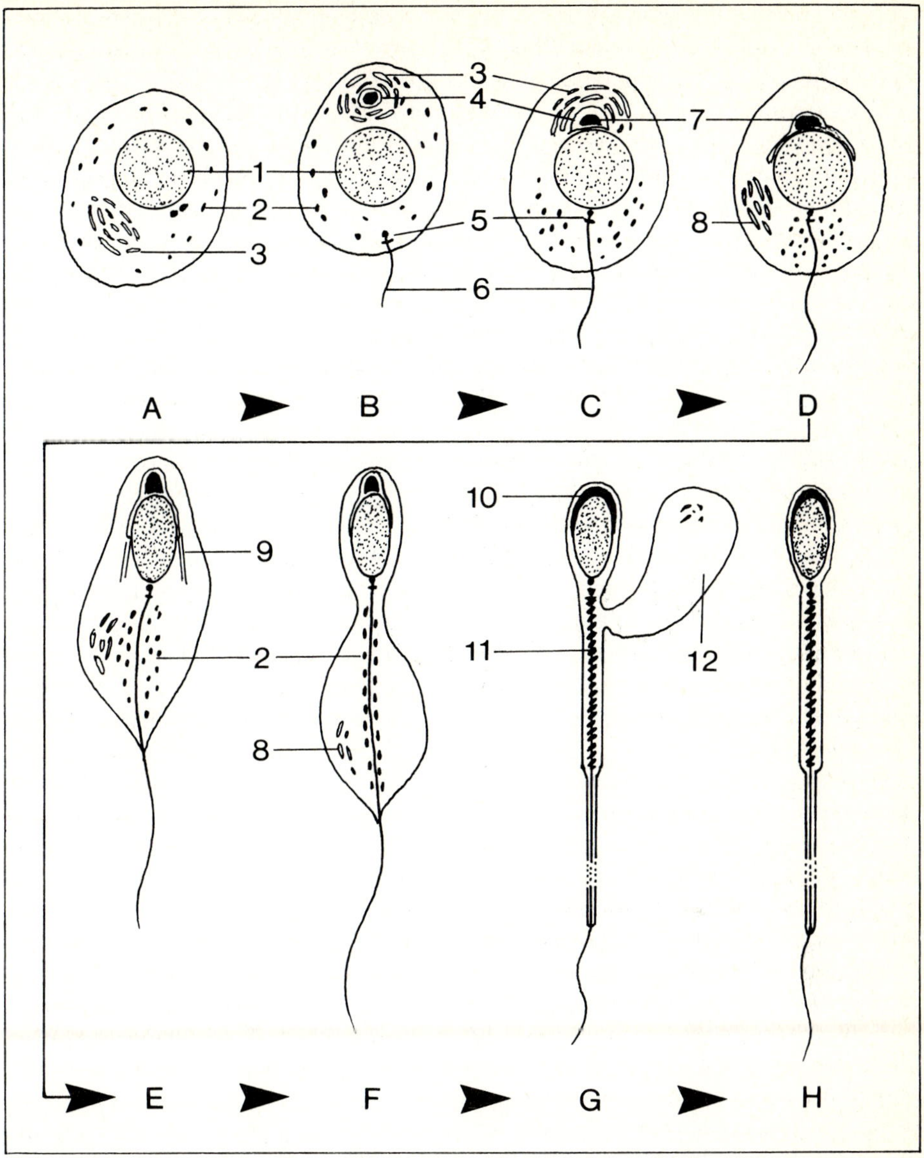

Fig. 25A–H. Highly simplified diagram showing various stages of spermatohistogenesis (**A–H**) in mammals.
1 = cell nucleus; 2 = mitochondria; 3 = Golgi apparatus; 4 acrosome vesicle; 5 = centrioles; 6 = flagellum; 7 = acrosome granule; 8 = remains of the Golgi; 9 = manchette; 10 = acrosome; 11 = mitochondrial spiral; 12 = residual bodies

the main piece of the tail. The remaining cytoplasm forms a residual body at the beginning of the middle piece which is pinched off before spermiation. The course of spermiohistogenesis described above applies in principle to all progressive or modified types of spermatozoa. With respect to the primitive spermatozoa of Teleostei, neither an acrosome nor the mitochondrial spiral is formed in the majority of cases. A summary of the course of spermiohistogenesis is shown in Fig. 25.

The individual organelles of the spermatozoon have quite distinct functions. It has been found that in anurans und urodeles the acrosome contains a material with a carbohydrate moiety in addition to an enzyme (lysin) which dissolves the gelatenous shell and plasma membrane of the egg thus allowing the spermatozoon to penetrate the egg. The predominant acrosomal enzyme in mammals is hyaluronidase, which dissolves the intercellular substance between the granulosa cells of the corona radiata that surrounds the oocyte. The same enzyme has been demonstrated in birds. The acrosome vesicle of mammals also contains acid phosphatase, β-glucoronidase, arylamidase, arylsulphatase, β-N-glucosaminidase, phosphorylase A, unspecific esterases and β-aspartyl-N-glucosamine-aminohydrolase. Lysin has characteristics similar to trypsin, i.e. it is a protease. Its existence has been demonstrated in chickens, sheep, cattle, rabbits and the boar. The corresponding enzyme in humans is called acrosin. The acrosome, therefore, represents a specialised lysosome which is essential for the dissolution of the egg membranes. In various species of Petromyzontia, Elasmobranchii, Amphibia (excluding Ranidae), Reptilia and Aves other than passerines, a rod or thread-like structure composed of a fibrous material has been found between the acrosome and the nucleus of the sperm which forms a canal that partly or completely penetrates through the nucleus. This subacrosomal material, also called the perforatorium, shows no trace of enzymatic activity. However, it plays an important role in the so-called acrosome reaction which will be dealt with fully in a later section (see p. 118).

Fibrous subacrosomal material or a comparable structure is completely absent in Mammalia. Instead, the space between the innermost membrane of the acrosome and the nuclear membrane is filled with an amorphous material that has sometimes falsely been referred to as a perforatorium.

The function of the nucleus of the spermatozoon revolves around its role in sexual reproduction: it contains the haploid genome which is carried over to the egg cell. Finally, the mitochondria lying behind the nucleus deliver the energy needed for the movement of the axial filaments by means of which the spermatozoon is propelled.

Both the primitive and progressive, or modified, types of spermatozoa exist in three variant forms respectively having a single flagellum, a pair of flagella and no flagellum at all. Primitive and modified spermatozoa with one flagellum occur in Cyclostomata. The spermatozoa of Petromyzontia are of the progressive type with an elongate nucleus about 50 μm long, only the first third of which contains chromatin. The nucleus is tubular with a long subacrosomal filament running through it that extends into the middle piece and which is thrust out in the presence of eggs. This structure is referred to as the head bristle in the older literature. The Elasmobranchii have modified spermatozoa with one or two flagella. In those species studied to date, the head piece is shaped like a worm and bears an acrosome. The middle piece is much shorter than the nucleus and surrounds the axial filaments of the tail which arise from the distal centriole. In *Raja clavata*, the thornback, the nucleus is penetrated by an intranuclear filament which probably represents subacrosomal material. In addition, a centriole lying in front of the acrosome apparently gives rise to a stereocilium. An extranuclear flagellum is wound around the nucleus of the spermatozoa of the spotted dogfish (*Scyliorhinus caniculus*). The spermatozoa of the crossopterygian *Latimeria chalumnae* and of *Raja clavata* are similar in having an undulating membrane on the tail. Primitive spermatozoa which lack an acrosome occur in teleosts. These may bear one or two flagella, as, for example, do those of the toadfish (*Opsanus tau*). In the guppy (*Lebistes reticulatus*) the spermatozoa have a long middle piece but the mitochondria within it are not arranged in any recognisable way. The spermatozoa of the European eel (*Anguilla anguilla*) presumably possess an acrosome and subacrosomal material within the nucleus. The only known example of spermatozoa lacking a flagellum also occur in the Teleostei, in mormyrids of the families Gymnarchidae and Mormyridae. The male gametes are cells which are not quite spherical and lack a flagellum or a structure resembling an acrosome. However, they contain an organelle, the so-called spongy chamber, whose function is unknown. These electric fish presumably have external fertilisation although the mechanism of gamete fusion is unknown. The spermatozoa of amphibians exhibit a multiplicity of forms. In Urodela the head piece and its acrosome are typically elongate and drawn out to a needle-like point. The middle piece is relatively short and gives rise to two flagella, one centrally and another marginally which is connected to the former by an undulating membrane. The spermatozoa of Anura have a highly variable form. Ranidae, for example, have spermatozoa with long, pointed heads and a short middle piece. On the other hand the spermatozoon of the red-bellied toad (*Bombinator igneus*) has an elongated, spindle-shaped head from the front of which arise two flagella joined together by an undulating membrane, as in urodeles. Thus in this case the head of the spermatozoon and the locomotory apparatus do not lie one behind the other but next to each other. Two forms of spermatozoa exist in Sauropsida, a reptilian type and a passeriform type. The former is characteristic of reptiles and all birds with the exception of passerines and is made up of a prolongated, tear-shaped nucleus possessing an intranuclear subacrosome beneath a relatively small acrosome, a spiral of mitochondria in the middle piece, and a single flagellum in the tail. In pigeons (genus *Columba*) the mitochondrial spiral encloses an appreciable length of the flagellum. The passeriform type of spermatozoon is characterised by a pointed nucleus, which appears to be wound into a spiral, and an acrosome, the nucleus being covered by an undulating membrane which gives it the appearance of a drill. A central and marginal flagellum connected by a membrane arise from the middle piece of the spermatozoon.

All mammals possess the modified type of spermatozoon with a single flagellum, as has already been described with respect to human spermatozoa. Some variation exists in the form of the acrosome. In the ejaculate of humans one can find spermatozoa with a range of shapes that do not correspond to the normal shape as described. These represent abnormal forms which evidently result from "faulty" spermiohistogenesis. It has been suggested that this is a consequence of being clothed, in that the temperature in the scrotum is raised slightly thereby counteracting any advantage gained by the descensus testiculorum (see p. 62). The number of abnormal forms present is characteristic of each individual and this provides the basis of a spermiogram which, in a healthy man, changes little over the years. Such abnormal spermatozoa can be induced in the ram by experimentally warming the scrotum to a temperature of 40°C. Humans produce 3 ml of semen in 24 h on average, every millilitre containing 60–120 million spermatozoa. The spermatozoa swim in a mixture of secretions from the epididymis, glandula vesicularis and glandula prostatica which is weakly alkaline, having a pH value of 7.2. This alkalinity is essential since the spermatozoa become immobile in an acid milieu. The secretion of the glandula vesicularis includes fructose as a source of energy for the spermatozoa. The secretion of the epididymis similarly plays an important part in promoting the motility of the spermatozoa since it is responsible for the formation of disulphide bridges between the proteins of the flagellar microtubuli that make movement of the flagellum possible.

The last example illustrates the point that spermiation, which concludes the process of spermiohistogenesis, does not necessarily result in spermatozoa that are fully functional. Indeed, in many animals the spermatozoa cannot even penetrate the egg before certain functional changes have taken place. This final maturation is promoted by substances from the egg itself or from the female genital tract and, in mammals, is known as capacitation. As has already been mentioned, in the case of mammals the spermatozoon must penetrate the corona radiata and zona pellucida of the oocyte, which it does with the aid of the enzymes of the acrosome. Their ability to do this takes 3 to 20 h to develop depending on the species and the particular conditions. It is possible that this development is associated with certain changes that take place in the structure of the acrosome and the membrane of the spermatozoon, which will be considered later in the context of the acrosome reaction. The term capacitation is used only with respect to mammals and is so defined that it also includes processes which establish the conditions necessary for the acrosome reaction to take place. Such processes are thought to involve some form of "demasking", probably by the removal of a protective layer from the spermatozoan membrane. The finding that the spermatozoa of the leopard frog (*Rana pipiens*) must first penetrate through the gelatinous capsule of the egg before one of them is able to fertilise the egg can also be interpreted in this light. The process of capacitation is a slow one. In contrast, the spermatozoa of species which spawn in water do not appear to require any great length of time before they are capable of fertilising the egg. It is not certain whether capacitation occurs in all vertebrates that have internal fertilisation.

In a range of lower vertebrates the spermatozoa are not released freely into the female or the environment but are released in the form of aggregations or encapsulated in packets, i.e. in spermatozeugmata and spermatophores respectively. The basking shark (*Cetorhinus maximus*), which may grow up to 14 m long, forms ball-shaped spermatophores about 3 cm in diameter within pouches of the ampulla ductus deferentis, which contains several litres of semen. Thus each aliquot of spermatozoa is enclosed within several concentric, gelatinous shells. The spermatophores are suspended in the fluid that is transferred into the genital tract of the female during internal fertilisation. A comparable phenomenon occurs in the elephant chimera (*Callorhynchus antarcticus*) although in this case the spermatophores are formed in the ductus epididymidis as bundles of spermatozoa enclosed in a gelatinous material and a membrane.

Among the Teleostei, spermatozeugmata occur in a range of Poeciliidae, each representing the cell clone of a single spermatocyst, including the sword tail (*Xiphophorus helleri*), *Poecilia melanogaster* and *Poecilia latipinna*. True spermatophores also occur in cyprinodontid teleosts. For example, in the genus *Horaichthys*, the spermatozoa lie in the thick end of a club-shaped capsule whose thin end, apparently empty of spermatozoa, bears a frill of fibrils. The male positions the spermatophore on the female, close to her genital opening, and after a few minutes a swelling forms between the fibrils that eventually bursts, setting free the spermatozoa. The spermatophores of Percomorphi from such families as the Aphyonidae and Brotulidae are formed in the testis. More detailed observations have been made on the live-bearing surf perch (*Cymatogaster aggregata*) and here also the contents of the thin-walled spermatophore represent those of a single spermatocyst.

Urodeles, with the exception of the Cryptobranchidae, Sirenidae and Hynobiidae, also produce a spermatophore during mating which is laid by the male then picked up by the female. In these animals the spermatophore is generally made up of a cap containing the spermatozoa and a stem piece. As they pass down the wolffian duct into the cloaca the spermatozoa are encapsulated in a secretion produced by the pelvic glands, and an enveloping membrane as well in some species. The stem is composed exclusively of secretory material from the cloaca and can be simple in shape or, as in some species, be an exact moulding of the cloacal chamber with a complex form (see title picture to this chapter). Spermatophores between different species can vary considerably. The mechanism by which the spermatophore is dissolved within the female's cloaca is not clear: mucolytic or proteolytic enzymes are suspected to be involved.

3.2.2 The Formation and Maturation of Eggs

The process whereby eggs are formed in the ovary is commonly referred to as oogenesis, though this term is not entirely correct since strictly speaking it refers only to the formation of oocytes from oogonia or the growth phase of the oocytes. The formation of eggs follows the general scheme given at the beginning of this chapter. Thus oogonia produced by the division of the primordial sex cells go through a phase of mitotic proliferation. These mitoses divide the oogonia incompletely, as is the case in spermatocytogenesis, so that a cell clone develops which is held together by bridges of cytoplasm. There follows a growth phase during which the oogonia become first order oocytes. In many cases the oocytes are arrested in the prophase of meiosis and continue growing and differentiating by storing in their cytoplasm the reserves necessary for the development of the embryo. This process of vitellogenesis occurs in all vertebrates, with the exception of the Marsurpialia and Eutheria. The build up of reserve materials in the oocyte is mediated by cells which form the follicle of the egg (see p. 46) and, therefore, the formation of the follicle (folliculogenesis) is a prerequisite for the occurrence of vitellogenesis. The maturation of the oocyte can take a considerable length of time (up to several years). The meiosis following on from the growth phase is not completed within the ovary. Thus in vertebrates of lower evolutionary standing than mammals, excepting the dog and fox, the second order oocyte is ovulated from the ovary, that is after the pinching off of the primary polar body and usually at the transition to the second metaphase. Ovulation in the exceptions named occurs before the first maturation division. The second maturation division is normally completed only after the egg has been penetrated by a spermatozoon. Following the phase of meiosis there is no further differentiation of the egg. The formation of the egg is shown schematically in Fig. 26.

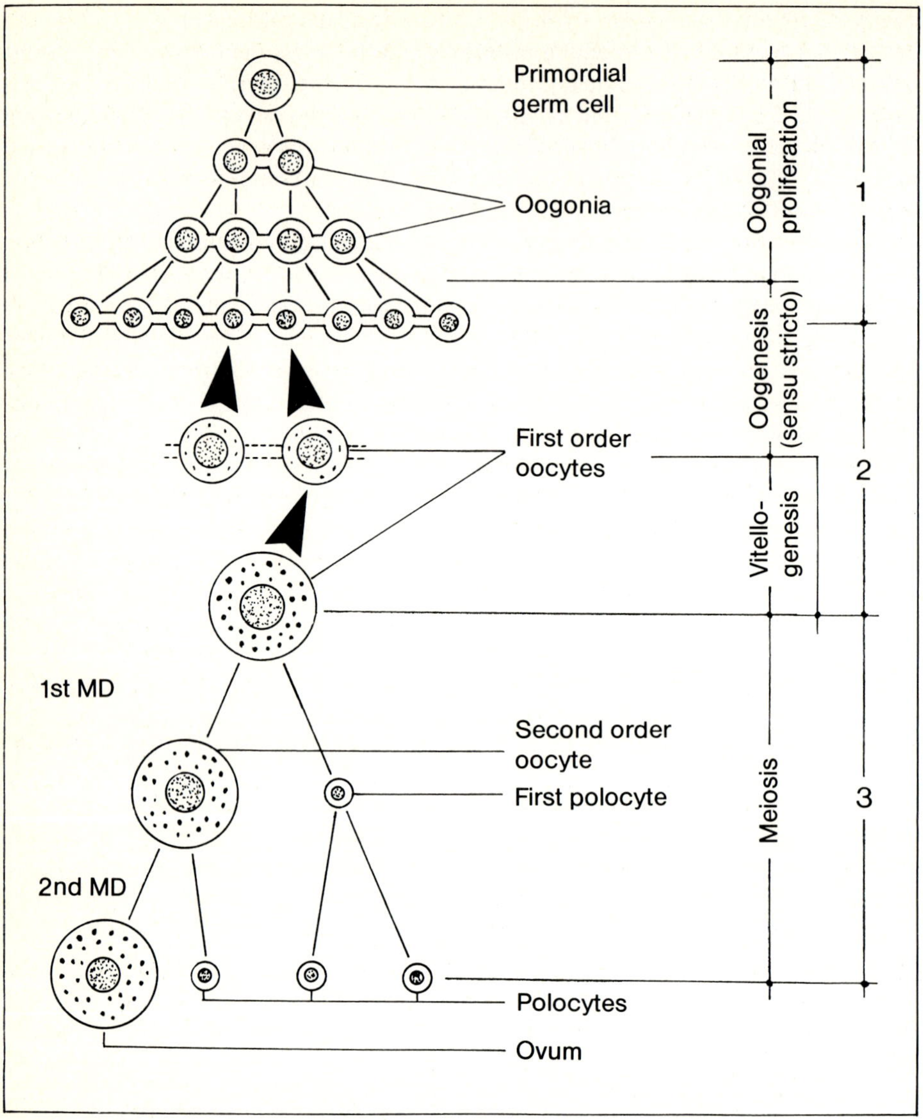

Fig. 26. Schematic representation of egg formation in vertebrates.
1 = proliferative phase; 2 = growth phase; 3 = phase of meiosis; MD = maturation division

It is probable that the cytoplasmic bridges connecting the oogonia together play an important role in the synchronisation of cell division and may also serve to sustain proliferation of the cells within the clone. On the other hand, it is also possible that the bridges are in some way responsible for the death of sister oocytes in those cases where the oocytes degenerate. One can recognise two types of oogonial proliferation in vertebrates according to whether oogonial division is restricted to the embryonic or larval phase of life or occurs during the entire reproductively active phase of adult life. The former is the case in Petromyzontia, a few Teleostei, probably all Elasmobranchii and Aves, and the vast majority of Mammalia. The second group includes most Teleostei, the Amphibia and the Reptilia (with one possible exception that belongs to the first type). A number of hypotheses have been put forward concerning the possible evolution of these two types of proliferation. Equally unresolved are the mechanisms which control the number of oogonial divisions, although in the fish, amphibians and reptiles belonging to the second type there are indications that gonadotrophic hormones of the hypophysis and hormones from the gonads could be involved. On the other hand, hormones appear to have no such influence in Cyclostomata.

As mentioned above, the first order oocytes arising at the beginning of oogenesis are arrested in the prophase of the first meiotic division. In the ensuing growth phase the oocyte undergoes an increase in size whose magnitude depends, among other things, on the amount of stored yolk. For example, the young first order oocytes of the leopard frog (*Rana pipiens*) have a diameter of 50 μm but a diameter of about 1500 μm when mature, corresponding to an increase in size by a factor of 27,000. The diameter of mouse oocytes increases from 20 μm to only about 70 μm, that is a size increase by a factor of approximately 43. The variation in the duration of the growth phase is equally great. Thus, to remain with the examples mentioned, oogonial proliferation in the mouse takes place during the development of the embryo and the oocytes reach their final size 16 days after birth. The same process in the leopard frog takes 3 years, the growth of the oocytes beginning shortly after metamorphosis. This does not mean however that this frog can reproduce only every 3 years since every year a new pool of oocytes is produced which develops separately so that a single ovary may contain up to three generations of oocytes, each at a different stage of development. The growth rate of each generation of oocytes is not linear but increases sharply in the third year. Oocytes can be of immense proportions, witness the egg "yolk" of large birds and reptiles. The largest vertebrate eggs are those of certain elasmobranchs.

During their growth phase the first order oocytes progress steadily through the meiotic prophase but are arrested in their development before the end of the prophase. The nucleus increases in size, primarily by the production of large amounts of karyolymph: the enlarged nucleus is also known as the germinal vesicle. At the same time the chromosomes increase markedly in size, however, without there being a corresponding increase in their content of DNA. The chromosomes of certain species from every class of submammalian vertebrate have been found to have a characteristic appearance during the diplotene stage of meiosis and are referred to as lampbrush chromosomes (Fig. 27). The conspicuous side-loops which indentify this form are uncoiled regions of the chromatids on which the synthesis of mRNA takes place. Lampbrush chromosomes have been observed predominantly in large oocytes and are perhaps a general feature of submammalians, which reflects the high metabolic activity of the growing oocyte. They commonly occur in the eggs of amphibians.

The oocytes of fish and amphibians are notably different from those of all other vertebrates in that the former possess many more nucleoli in the nucleus, these being mostly located at the periphery of the nucleus. This is apparently a consequence of the large amount of RNA the growing oocytes requires. In the clawed toad (*Xenopus laevis*) it has been found that the gene for two of the main ribosome components is

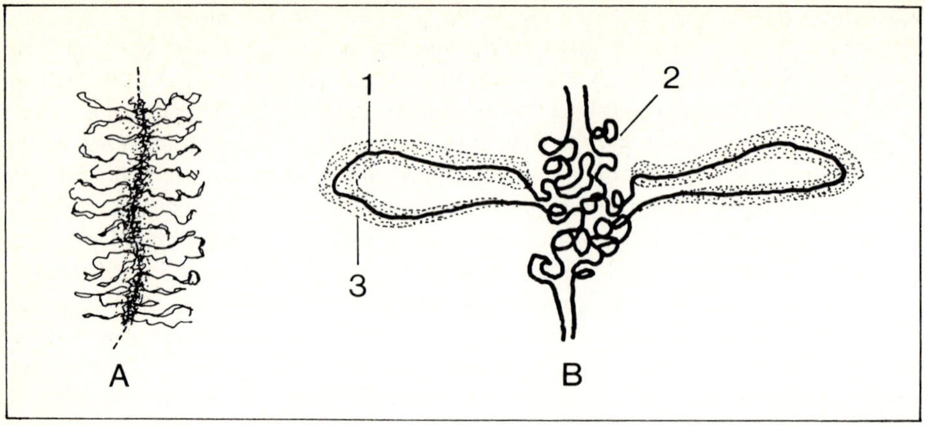

Fig. 27A,B. A lampbrush chromosome. **A** Light microscopical appearance of a small section of the chromosome (stylised). **B** Detailed structure (stylised). Both chromatids have convoluted regions (2) and also form loops (1) laden with RNA and ribonucleoproteins (3) produced by the genes

represented 450 times in the haploid chromosome and that for a third component over 20,000 times. This so-called amplification of the genes takes place without concomitant mitosis, but, rather, is linked to a multiplication of the nucleolar organiser section of the chromosome resulting in the formation of numerous nucleoli. However, only the two main genes are amplified in this way, the third being in any case present in large numbers on various other chromosomes. It is probable that this mechanism also exists in other species which have supernumerary nucleoli in the oocyte nucleus.

The process of the first order oocyte into the prophase of meiosis occurs at different times in members of the different classes of vertebrate depending on the type of oogonial proliferation. In the Petromyzontia, birds and the mammals it occurs during the larval stage or during the development of the embryo. In most Teleostei, Amphibia and Reptilia it follows oogonial proliferation, occurring during the reproductively active period of life and often seasonally after each period of breeding. The first order oocytes of many submammalians aggregate together as cell nests and in a few species the cells of the nest develop synchronously up to the early prophase of meiosis. Cytoplasmic bridges can still be found between the cells in various species and it is possible that synchrony is lost when these bridges are broken. The follicle is usually formed at this stage. The follicle cells probably stem from the coelomic epithelium and migrate, often as pregranulosa cells, into the cell nests where they surround individual oocytes. A close contact exists between the follicle cells and the oocytes in, for example, amphibians, reptiles and birds where contacts resembling desmosomes have been found between the adjacent cell membranes of the follicle cells and oocyte (the oolemma). In addition, both types of cell form microvilli which interdigitate with each other. The formation of the follicle is closely associated with vitellogenesis and the formation of the egg membranes, which are to be dealt with later.

Numerous cells perish during the phase of oogonial proliferation as well as during the phase of oogonial growth, although the rate of degeneration varies from species to species. Among the Cyclostomata the formation of female gametes has been studied only in Petromyzontia. It has been found that the proliferation of oogonia begins when the larvae reach a particular body length, which is different for each species. The number of oogonia in the various species investigated varies between 5000 to 10,000 in the brook lamprey (*Lampetra planeri*) and 150,000 to 300,000 in the marine lamprey (*Petromyzon marinus*). The cells either separate or remain in pairs after each oogonial division so that the cell nests are eventually formed, several nests arising from a single oogonium. How many oogonial divisions take place is not absolutely certain. In *Lampetra planeri* there are 9 or 10 and in *Petromyzon marinus*, 13 or 14. The proportion of degenerate oogonia varies according to the species. No more oogonial divisions occur after metamorphosis, the first order oocytes being arrested in the leptotene stage of the meiotic prophase and it is at this stage that lampbrush chromosomes can be found. It has been estimated that about 80% of the cells which enter the prophase of meiosis degenerate. Vitellogenesis takes place after metamorphosis, with the single exception of the neotenous species *Lampetra zazandrei* which already possesses mature oocytes at the larval stage. In Elasmobranchii there are probably no longer any more oogonia present in adult animals, oogenesis having ended before the attainment of sexual maturity. But since only a few species have been studied in this respect, this course of events cannot be taken to be generally applicable. Oogonial mitoses very likely take place throughout life in most teleosts although there are a few species in which they are limited to either the embryonic, larval, or juvenile phase of life. In the hake (*Merluccius merluccius*) for example, oogenesis is ended by the time the animal first becomes sexually mature. In those species in which oogenesis ends before the larval stage is reached there are two basic patterns of oogonial proliferation. In polycyclic species from the tropics, several successive phases of oogonial proliferation occur throughout the course of a year, or the reproductive period if this is longer. This process can be continuous or proceed in more or less discrete waves. Oogonial proliferation in the surf perch (*Cymatogaster aggregata*), which gives birth to live young, only begins when the ovary contains no more embryos.

In Teleostei, the nucleus of the oocyte becomes enlarged when the oocyte enters the meiotic prophase, and at the same time there is a marked increase in the number of nucleoli. Mitochondria aggregate close to the nucleus to form the so-called yolk nucleus which migrates to the periphery of the cytoplasm in the course of subsequent development. Lampbrush chromosomes are also present and there is apparently an increased synthesis of RNA prior to vitellogenesis. In general among the Teleostei, oogenesis probably begins relatively soon after the phase of oogonial proliferation. The formation of the follicle takes place during the growth phase of the oocyte so that by the diplotene stage the follicle cells usually bear well-developed microvilli. Whether or not desmosomes are established between the follicle cells and the oocyte is not clear: none could be found in the sword tail (*Xiphophorus helleri*).

In some species of amphibian it has been found that there are intercellular bridges between the oogonia. New nests of oogonia are produced every year in species which have only one period of reproduction per year and this commonly occurs after spawning. Oogenesis is initiated in the larval stage and continues on after metamorphosis. In the clawed toad (*Xenopus laevis*) it has been found that the first order oocytes come together in cell nests and go through development synchronously, and it is presumed that intercellular bridges still exist in this case. Numerous nucleoli are formed in the nucleus of the oocytes, their number varying from species to species. The nucleus enlarges considerably during the diplotene stage of the meiotic prophase and appears as as germinal vesicle. A yolk nucleus is present in some species but not in others: it is evidently not directly involved in vitellogenesis but probably represents a speciel mechanism for the transfer of information from the nucleus to the peripheral cytoplasm. Many germ cells perish during the oocyte stage, as do those of fish, although this seldom occurs prior to vitellogenesis. Oogenesis in amphibians often takes a long time from when it begins to when the ripe oocytes are finally ovulated, taking up to three years in the case of *Rana pipiens*. Following ovulation most of the new oogonia enter the prophase of meiosis. The formation of the follicle begins after the separation of the oocytes in the diplotene stage. Direct contacts between the follicle cells and the oocytes have been observed in addition to structures which resemble desmosomes.

Among reptiles, intercellular bridges between the oogonia have been detected in one species of lizard (*Lacerta sicula*). The oogonia of animals in this class are generally relatively large and are present in all adult animals, oogonial proliferation in part having taken place during the development of the embryo. They lie together in so-called germinal beds which occur singly or in greater numbers and in which can also be found oogonia and oocytes at various stages of development. Investigations into oogonial proliferation in reptiles are scarce but it is known that in one species of chelonian oogonial proliferation begins after ovulation. In some species the oogonia have already entered the prophase of meiosis in the embryonic stage of life, examples being the tuatara (*Sphenodon punctatus*) and the common lizard (*Lacerta vivipara*). The former species is evidently exceptional within Reptilia. In *Lacerta sicula* it has been found that the follicle first begins to form early during the diplotene stage. In this process the nucleus of oocyte becomes larger and may fragment. Multiple nucleoli situated at the periphery of the nucleus are not present in this case. Lampbrush chromosomes occur in the oocytes or reptiles. The yolk nucleus is a spherical mass composed of protein, RNA and closely associated mitochondria. It is probable that this structure is involved in the synthesis of yolk, which is rich in protein, although no yolk particles are formed within it. The process of oogenesis varies in different species according to the type of reproductive cycle. In some species of snake a new cohort of oocytes is formed after every ovulation. In general it seems that the oocytes of reptilians migrate into the ovarian stroma shortly before the formation of the follicle and only then do the prospective follicle cells proliferate. When the oocyte is surrounded by a layer of flat cell, the whole structure is referred to as a primordial follicle. The degeneration of oocytes (follicular atresia) can occur at various stages of development although previtellogenic atresia is as rare in this as in other classes of vertebrates already dealt with. Observations on birds relate primarily to the chicken (*Gallus domesticus*). In this case the oogonia are larger than the somatic cells but smaller than the primordial sex cells, and lie in nests at the periphery of the ovary. Oogonial mitoses do not occur synchronously over the entire ovary and probably no longer take place after hatching, although whether one can generalise this to all birds is not certain. As a rule oogenesis has already begun during the development of the embryo and is complete by the time of hatching. However, exceptions to this rule probably exist. The yolk nucleus is initially formed from the Golgi complex in close proximity to the nucleus. The Golgi contains material rich in RNA and is surrounded by mitochondria. About five days after hatching the yolk nucleus begins to gradually break up and the mitochondria disassociate themselves, becoming dispersed throughout the cytoplasm until finally, after about 3 weeks, they form a layer at the periphery of the oocyte, by which time the yolk nucleus has completely disappeared. In the chicken, the intercellular bridges between the primary oocytes still exist at the prophase of meiosis. After hatching, that is during the diplotene stage, the follicle cells become flattened and overlap each other to some extent thereby forming the follicle around the oocyte. Degeneration of the germ cells can occur in either the oogonial or oocyte stage.

In most mammals the proliferation of oogonia and oogenesis take place in the foetus but occur neonatally in some others. In the mouse, for example, 95% of the germ cells present in the foetal gonad are oogonia on the thirteenth day of gestation, the remaining 5% being oocytes. One day later there are only 43% oogonia present and on the seventeenth day there are none at all. In contrast, the ovary of the rabbit at birth contains only oogonia, which then develop into oocytes within the following ten days. A small group of primates of the suborder Prosimiae are exceptional in that oogonial proliferation still occurs in the adults, a phenomenon which is regarded as an embryological curiosity. Very large numbers of cells often degenerate during the course of oogenesis: the foetal human ovary in the fifth month of gestation contains approx. 6 to 7 million oocytes but only two million at birth, half of which are already degenerate. The number of germ cells in the ovary of the foetal pig (*Sus scrofa domesticus*) increases between the twentieth and fiftieth day of gestation from about 5000 to 1.1 million but subsequently decreases again sharply to about 500,000 at birth. In the rat there are approximately 70,000 oogonia present in the foetus on the seventeenth day of gestation but only 12,000 2 days after birth. The reason for this marked degeneration of germ cells in unknown. In general it has been found that the majority of oocytes which degenerate do so in the follicle, i.e. by true atresia. The degeneration of a single oogonium is often accompanied by the degeneration of all the cells of the nest. Three discrete waves of degeneration occur in humans, respectively involving oogonia in the course of mitotic proliferation, oocytes in the pachytene stage of the meiotic prophase, and oocytes in the diplotene stage. A possible fourth wave may occur immediately after the beginning of meiosis. The degenerate germ cells are phagocytosed by granulosa cells. In the mouse they are also expelled from the ovary into the body cavity. Overall, the actual number of oocytes varies greatly between the different species of mammal and variations also occur within species: for example, in humans the number of oogonia lies between 350,000 and 1.25 million.

The formation of first order oocytes capable of going through meiosis can be divided into two phases: a so-called previtellogenetic phase extending from the last oogonial division up to the formation of the follicle, and a vitellogenetic phase in which the reserve materials required for the provision of the embryo are stored. The oocyte grows during both of these phases, although much less quickly in the first than in the second, which is therefore called the period of rapid growth. The processes of oogenesis described so far end with the formation of the follicle. In many vertebrates one can find desmosomes between the oocyte and the granulosa cells of the follicle epithelium during the early stages of follicle development. As development continues the

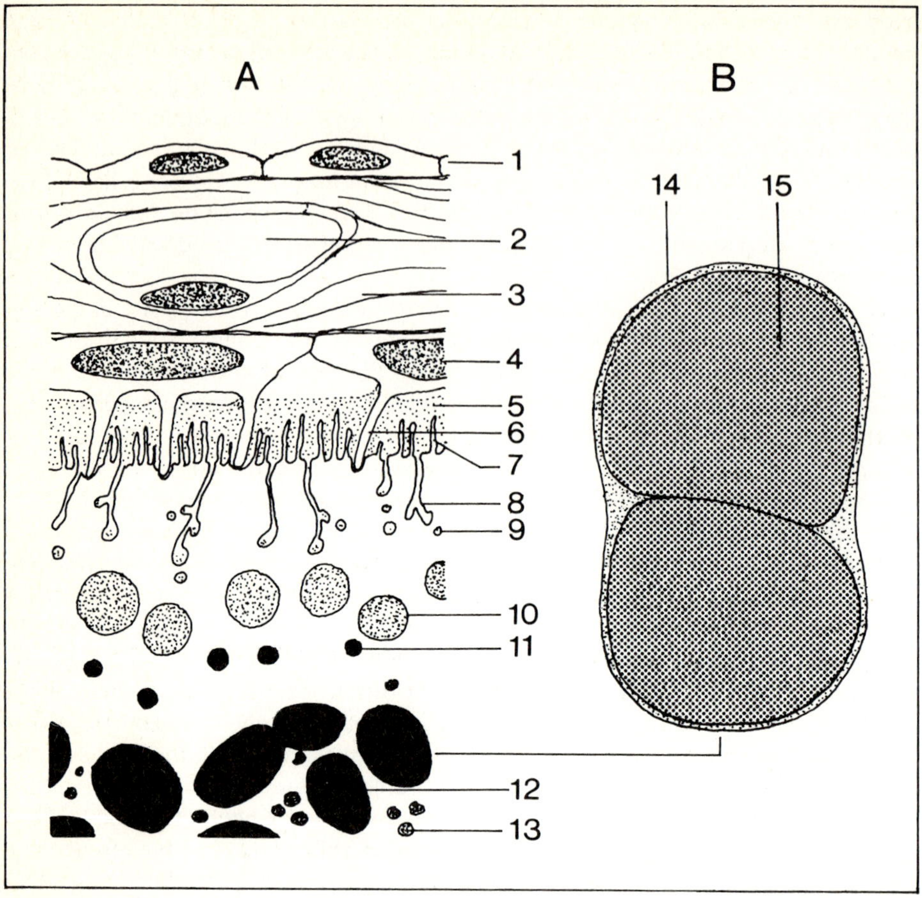

Fig. 28A,B. Diagram of the morphological relationship between the oocyte and follicle epithelium in an amphibian ovary (**A**). The structure of an amphibian yolk platelet is shown in **B**.
1 = ovarian epithelium; 2 = capillary; 3 = connective tissue fibres; 4 = granulosa cells; 5 = egg membrane; 6 = microvillus of a granulosa cell (macrovillus); 7 = microvillus of the oocyte; 8 = crypt; 9 = pinocytotic vesicle; 10 = cortical granule; 11 = pigment granule; 12 = yolk platelet; 13 = lipochondrium; 14 = superficial layer; 15 = main body

area of contact is gradually reduced and a space forms between the oocyte and the follicle epithelium. They do not become completely separated however. The desmosomes of the granulosa cells apparently become plastic and give rise to large projections, also called macrovilli because they look like large microvilli, that extend between numerous microvilli which arise from the surface of the oocyte at the same time (Fig. 28A). Individual microvilli are too small to be seen with the light microscope but collectively they appear as a radially striated zone around the oocyte, the so-called zona radiata. The microvilli greatly enlarge the surface area of the oocyte (by a factor of about 35 in the frog) and subsequently play an essential role in the exchange of substances between the oocyte and the surrounding medium and the uptake of materials for storage within the oocyte.

Within the animal kingdom there are basically two ways in which the nutrients within oocytes are formed. The first, referred to as autosynthesis (also as solitary yolk formation), is where the oocyte produces its own nutrients, but this is of minor importance in vertebrates. The second, heterosynthesis, involves the formation of yolk by somatic elements of the organism (so-called alimentary yolk formation), for example, by special nutrient cells. In vertebrates the most important constituents of yolk are synthesised in the liver in a form which can be transported in the blood, via which they reach the follicle. Here they are absorbed into the oocyte by pinocytosis and converted into yolk. The uptake process can be demonstrated in members of all the vertebrate classes using the electron microscope which shows that there are crypts at the base of the microvilli from which pinocytotic vesicles pinch off (see Fig. 28A). At a later stage of development the space between the granulosa cells and the oocyte is filled with a fibrous material which becomes one of the egg membranes: this will be dealt with later.

Vitellogenesis is a more complicated process which includes the formation of all the different components of the yolk, whose chemical nature is typically very heterogeneous. The terms "yolk" and "vitellus" should therefore be thought of more as morphological terms which make no reference to the chemical nature or composition of the reserve materials within the oocyte. Thus they are used in a collective sense to denote the so-called yolk platelets of Cyclostomata, Elasmobranchii, Chondrostei, Holostei, Dipnoi, Amphibia and Reptilia or the yolk spherules in the oocytes of Teleostei and Aves. These structures are composed primarily of proteins, phospholipids and small amounts of neutral lipids. There are also reserves of other material within the oocyte, such as lipids and glycogen. In amphibians, lipids are stored in the so-called lipochondria (see Fig. 28A) which consist of a lipid core enclosed in a protein shell. Oil droplets occur in the cytoplasm of teleost oocytes, their abundance often being characteristic of the different teleost families. Glycogen is stored in the form of cytoplasmic granules. Thus about 45% of the total dry weight of an amphibian oocyte is yolk protein, 25% is lipid, about 8% is glycogen, and only about 20% represents the cytoplasm itself.

The yolk content of vertebrate eggs often varies considerably between the members of different classes and also within each class. The eggs can be classified according to the amount and distribution of yolk, so that eggs with large amounts of yolk are called macrolecithal, or polylecithal, those with moderate amounts mesolecithal, and those with very little microlecithal. When the yolk is distributed evenly throughout the cytoplasm the egg is referred to as isolecithal. If there is a gradient in the amount of yolk from one pole to the other, the egg is telolecithal, the yolk-rich pole becoming the vegetal pole and that with little or no yolk the animal pole. Naturally enough, this polarisation is more pronounced in macrolecithal than in microlecithal eggs. The Elasmobranchii, Teleostei, Reptilia and Aves have macrolecithal eggs which are telolecithal whereas the Cyclostomata, most Amphibia and the Osteichthyes (excluding the Teleostei) have mesolecithal eggs which are similarly telolecithal. The eggs of mammals contain very little yolk and are, therefore, microlecithal.

The physiological aspects of vitellogenesis have been particularly well studied in amphibians, and especially in the clawed toad *Xenopus laevis*, which can therefore serve as a model for other vertebrates. As stated above, in vertebrates most of the yolk is not produced by the oocyte itself but by the liver, and is, therefore, heterosynthetic. The substance produced in the liver is vitellogenin, a lipoglycophosphoprotein of high molecular weight. Its synthesis is stimulated by the female sex hormone oestrogen and takes place only during the reproductive period. Vitellogenin is transported as a serum protein in the blood to the ovary where it is selectively taken up by the oocyte by pinocytosis. In *Xenopus* this process if apparently dependent, directly or indirectly, on hypophyseal gonadotrophin whereas oestrogen has no influence.

The vitellogenin of *Xenopus* has a molecular weight of approximately 460,000 Dalton and contains 12% lipid, about 1% covalently bound carbohydrate, and 1.3% phosphate which is primarily bound to serine. Its green colour is probably due to the presence of biliverdin which is not covalently bound. Under certain conditions vitellogenin degrades into two subunits, each of about 200,000 Dalton. In serum every phosphate group bound to the vitellogenin protein has an associated calcium atom. The capillaries of the theca transport vitellogenin into the follicle, the granulosa cells being permeable to vitellogenin, unlike the cells of the inner epithelium of the ovary. Vitellogenin is taken up by the oocyte some fifty times quicker than other serum proteins then rapidly converted into two different proteins, lipovitellin and phosvitin. These are then recombined to form the crystalline structure of the yolk platelets, which are flattened, oval-shaped bodies. They are composed of one or two main bodies and a superficial layer, the former exhibiting a characteristic hexagonal substructure in the electron microscope whereas the latter is composed of an amorphous matrix with filaments embedded in it (Fig. 28B). An enclosing membrane can sometimes be seen around small yolk platelets. Phosvitin and lipovitellin are stored within the main bodies, the former being water soluble in its partially phosphorylated form but insoluble in its fully phosphorylated state. The partially phosphorylated form of phosvitin in the yolk platelets is phosphorylated by a protein kinase, for which ATP is required. This mechanism does not apply to the water soluble lipovitellin. Two molecules of phosvitin combine with one of lipovitellin, which thereby becomes insoluble in water, like phosvitin. Seven of these molecular complexes come together to form the hexagonal crystalline structure recognisable in electron micrographs as the subunit of the main bodies. The phosphorylation of phosvitin is a reversible process and it is conceivable that this represents the means by which yolk proteins are mobilised when required by the developing embryo.

Lipovitellin has a molecular weight of approximately 200,000 Dalton and contains about 20% lipid. Denaturation produces two peptides with molecular weights of 31,000 and 120,000 Dalton respectively, only the smaller of which is phosphorylated. The molecular weight of phosvitin lies between 35,000 and 40,000 Dalton. Over half of its amino acids are serine, most of which are esterified with phosphate. When the pinocytotic vesicles containing vitellogenin reach the yolk platelets the vitellogenin is converted into lipovitellin and phosvitin by a specific proteolytic enzyme. In *Xenopus* it has been found that 99% of the protein in the yolk platelets is heterosynthetic.

The scheme of vitellogenesis in *Xenopus* appears to hold good for other amphibians as well. Thus it has been found that phosvitin and lipovitellin are stored only in the main bodies of the yolk platelets. The oocytes can evidently take up other substances of high molecular weight in addition to vitellogenin but these are incorporated into the peripheral layer of the yolk platelets. In the tadpoles of Ranidae it has been shown that yolk crystals are formed in the mitochondria of the oocyte. Moreover, observations on the North American bullfrog (*Rana catesbeiana*) show that these crystals are expelled from the mitochondria and form hexagonal crystalloids in the cytoplasm. This phenomenon appears to be peculiar to the Ranidae. Some old textbooks refer to the mitochondrial yolk crystals as precursors of the yolk platelets, but this is evidently not the case. The mitochondrial yolk crystals constitute only a relatively low percentage of the total number of yolk crystals in an oocyte and they also have a somewhat different structure to the main bodies. Their function is unknown.

The oocytes of Cyclostomata, Elasmobranchii, Chondrostei, Holostei and Dipnoi have, as far as is known, yolk platelets resembling those of amphibians. In one species of shark (*Scyliorhinus stellaris*) a substance resembling lipovitellin with a molecular weight of about 560,000 Dalton has been isolated, as has a phosvitin with a molecular weight of 40,000 Dalton from a species of cyclostome (*Ichthyomyzon unicuspis*).

Vitellogenesis in teleosts follows the same basic course as in amphibians, though there are many specific differences. For example, in most of the teleosts which have been investigated the yolk proteins are atypical: the lipovitellins are heterogeneous in their composition and contain only small amounts of protein bound phosphate. Furthermore, they do not have the same solubility as those of other vertebrates. The phosvitins are similarly heterogeneous, have a low molecular weight, and a variable phosphate content. Some of the components of the yolk are completely different from phosvitin and lipovitellin, such as the beta component of the yolk of salmonids for example which contains neither lipid nor protein bound phosphate. The vitellogenins of teleosts are presumably split at several positions, of the same or of a different type, by a proteolytic enzyme. Moreover, teleosts do not have typical platelets but yolk spherules – fluid filled vesicles containing yolk. This is presumably a consequence of the high solubility of the yolk proteins. Furthermore, a certain amount of autosynthesis of yolk occurs in addition to its heterosynthesis. Thus, in the early oocyte there is a well-developed rough endoplasmatic reticulum which produces a substance that passes through the smooth endoplasmatic reticulum to become a second type of yolk inclusion. It has been found that, before the appearance of lipovitellin-like substances in the oocyte, a glycoprotein is apparently produced which is not present in the liver or serum. Finally, the oocytes of many Teleostei, primarily marine species, are further distinguished by the fact that they undergo a second phase of enlargement after the completion of vitellogenesis. However, this process probably has nothing to do with vitellogenesis but rather represents an adaptation connected with the production of pelagic eggs.

Few data are available with respect to vitellogenesis in Reptilia. Proteins resembling vitellogenin have been found in the serum of a snake, a lizard and a chelonian. That of the snake *Thamnophis elegans* has an unusually high lipid content of about 43%. The synthesis of vitellogenin in the liver of reptiles is apparently stimulated by oestrogen as it is in other vertebrates: hypophyseal growth hormone is also necessary, at least in the case of the Carolina anolis (*Anolis carolinensis*). Pinocytosis has also been observed to occur in the oocytes of this species. Aside from these data, phosvitin has been isolated from the eggs of the snapping turtle (*Chelydra serpentina*) and Nile crocodile (*Crocodylus niloticus*).

Most of the studies relating to Aves have, understandably, been carried out on the chicken (*Gallus domesticus*). Although contradictory and often confusing results are to be found in the older literature, a much clearer picture has emerged in more recent times which, however, reveals a more complex situation than in lower vertebrates. The vitellogenin of chickens has a molecular weight of 480,000 Dalton and consists of two peptides, each with a molecular weight of 240,000 Dalton. Protein-bound phosphorus makes up about 1% of its content: a value for lipid is not available. The synthesis of vitellogenin in the liver depends on oestrogen, as does that of a second yolk substance, low density lipoprotein (LDL), which has an extremely high lipid content of 82 to 88%. The latter is associated with an apoprotein with a molecular weight of about 275,000 Dalton which is evidently not phosphorylated. The serum of laying chickens contains 2.0 to 2.5 g of vitellogenin and about 2.0 g of LDL 100 ml^{-1}. As in other cases, the uptake of yolk precursors into the oocytes probably occurs by pinocytosis. The yolk consists of a gelatinous substance and granules. The former contains a low density fraction (LDF), which undoubtedly corresponds to the LDL, and soluble proteins called livetins. In the granules there are equal proportions of two lipovitellins (α and β), phosvitin and also LDF, which is probably a contaminant from the gelatinous fraction. The native lipovitellin dimers each have a molecular weight of 400,000 Dalton and are chemically similar to one another, although they differ with respect to certain physical properties such as their electrophoretic mobility and dissociation characteristics. Phosvitin has two components with molecular weights of approximately 36,000 and 40,000 Dalton, respectively. The LDF is composed of about 89% lipid and probably makes up the bulk of the yolk spheroids in chickens. Native LDF in aqueous solution has a very high molecular weight of between 5×10^6 and 17×10^6 Dalton. Phosvitin, LDL and LDF have also been found in some other species of birds.

No investigations have been made on vitellogenesis in Monotremata. Among placental mammals pinocytotic activity has been observed in the oocytes of the guinea-pig and mouse oocytes have been found to take up serum proteins. It has been suggested that the gene for vitellogenin was lost during the evolutionary step from the Prototheria to the Theria.

It will be recalled that in telolecithal eggs there is a more or less prominent gradient in the distribution of yolk, resulting in a polarisation of the egg into a vegetal pole rich in yolk and an animal pole which contains the nucleus but little yolk. This polarisation develops gradually during the course of vitellogenesis but the accumulation of yolk evidently does not cause the polarisation since isolecithal eggs are also differentiated into an animal and vegetable pole. What actually does cause this polarisation is not yet known.

The eggs of numerous species and in particular the eggs of amphibians contain pigment granules in addition to yolk platelets or yolk spheroids, lipochondria, glycogen granules and lipid droplets. The number of pigment granules increases steadily from when the number yolk platelets begins to be produced. The majority are located in the peripheral cytoplasm of the oocyte, others are located more centrally. The pigment granules are, however, not evenly distributed but are much more numerous in the animal half of the oocyte than in the vegetal half. One can see a sharp boundary between the two hemispheres with the naked eye although there is in fact a thin marginal zone of transition. Within the animal half of the oocyte there is a ring of pigment, one side of which is thicker and lies closer to the cell surface than the other:

this side later becomes the dorsal aspect of the embryo. In some species the distribution of pigment granules within the cytoplasm has been found to follow the distribution of RNA. The physiological significance of this coincidence is unclear: it does not appear to be essential for the subsequent development of the embryo since the eggs of a range of amphibian species are unpigmented and nevertheless develop in a similar way to pigmented eggs.

The cytoplasm of the ovum, or oocyte, is not homogeneous but divided into a gelatinous cortex of high viscosity and the central cytoplasm consisting of a cytoplasmic sol in which organelles and particles are suspended and moved about by cytoplasmic streaming. The oocytes of frogs, many fish and a few mammals (rabbits and humans for example) contain so-called cortical granules within their cortex. These are spherical, membrane-bound inclusions which arise from the Golgi bodies and have a diameter of 1–2 μm (Fig. 28A). They contain acid mucopolysaccharides and play a part in the process of fertilisation, as will be described later.

The transformation of a first order oocyte into an egg occurs during meiosis, which can only be begun by mature oocytes. The onset of meiosis is presaged by the dissolution of the membrane surrounding the swollen nucleus (the germinal vesicle), which usually happens shortly before the oocyte is ovulated. The karyolymph thus mixes with the cytoplasm and the chromosomes, which have become condensed, always migrate to the animal pole of the cell. There the chromosomes form homologous pairs that subsequently arrange themselves along the equatorial plate and give rise to a spindle apparatus. Prior to division, which is extremely unequal, the spindle apparatus forms an achromatic figure parallel to the surface of the cell which then rotates so as to lie perpendicular to the surface while the homologous pairs of chromosomes arrange themselves along the equatorial plate. The chromosome pairs then separate from one another and a bulge containing one of the haploid sets of chromosomes forms at the animal pole of the cell. This subsequently pinches off to give rise to a small primary polar body and a large second order oocyte. The second maturation division proceeds in the same way and results in a single large egg and a secondary polar body. The primary polar body may divide at the same time, though it frequently does not. In most vertebrates it is the second order oocyte which is ovulated, at which stage it is fully capable of being fertilised. The oocyte is usually released when the spindle reforms during the second metaphase of meiosis, except in the case of the axolotl (*Siredon mexicanum*), where this takes place during the second anaphase, and the dog and fox in which case first order oocytes with an intact germinal vesicle are released from the ovary. However, in these last two species it is probable that the spermatozoa can only penetrate the egg when it has reached the metaphase of the first maturation division.

To conclude this chapter it remains only to describe the egg membranes. In most cases one can distinguish two types of membrane: primary egg membranes, which are formed in the ovary, and secondary egg membranes, which arise in the reproductive ducts. Some textbooks propose a different classification according to which primary egg membranes are produced only by the oocytes, secondary egg membranes are produced by the follicle cells, and tertiary egg membranes by the oviducts. The former classification will be adapted here since in vertebrates one of the egg membranes is usually formed by both the oocyte and the granulosa cells. In dealing with vitellogenesis it was stated that the follicle cells and the oocyte form microvilli which interdigi-

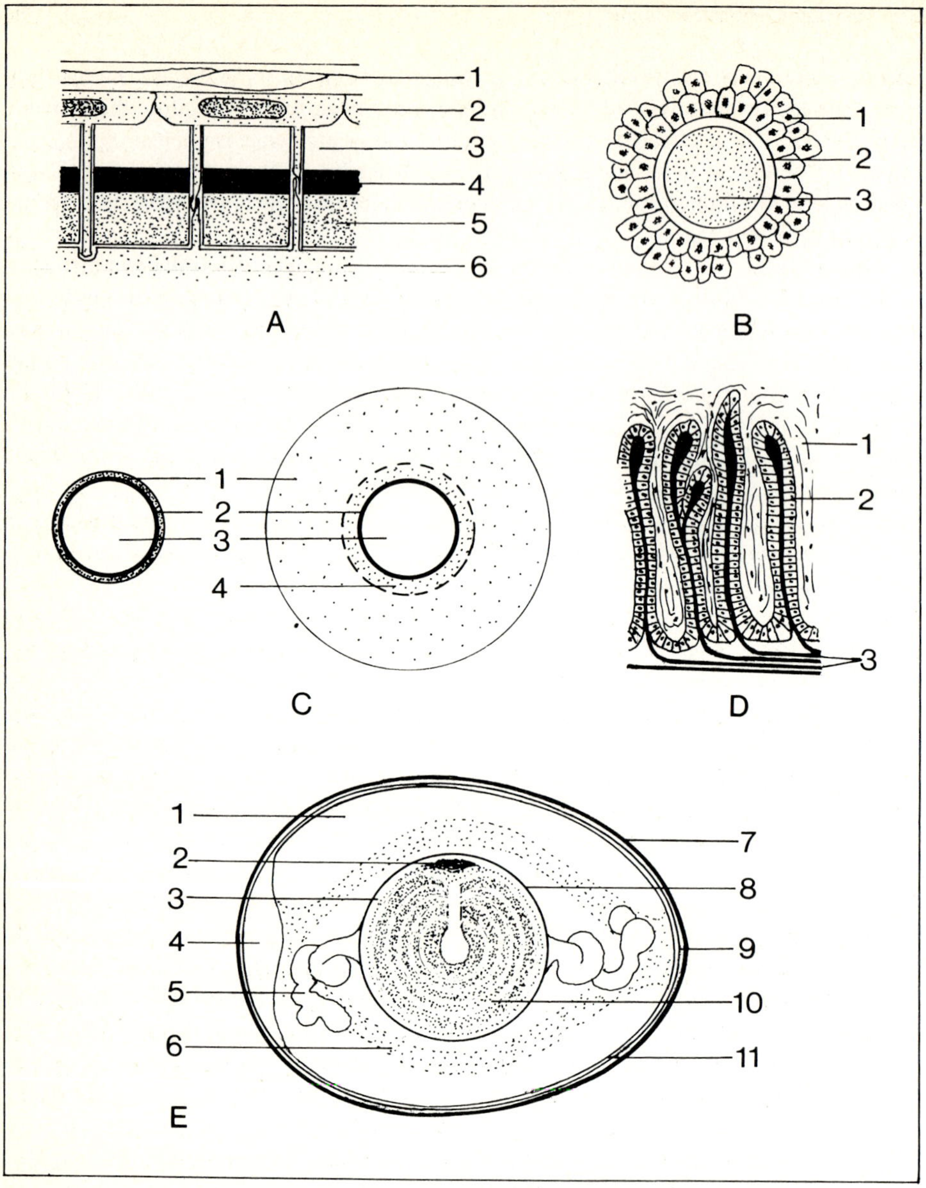

Fig. 29A–E. Egg membranes.

A Chorion of a teleost. (After Götting). 1 = theca; 2 = granulosa; 3 = zona pellucida (sensu stricto); 4 = cortex radiatus externus; 5 = cortex radiatus internus; 6 = oocyte.

B Ovulated oocyte of a mammal. 1 = corona radiata composed of adhering granulosa cells; 2 = zona pellucida; 3 = oocyte.

C Anuran egg in the oviduct (*left*) and in water (*right*). (After Balinsky). 1 = gelatinous capsule (swollen with water); 2 = vitelline membrane; 3 = oocyte; 4 = inner gelatinous capsule.

D Histological section through the wall of the glandula nidamentaria of an elasmobranch. (After Borcea, much changed). 1 = connective tissue; 2 = shell gland; 3 = layers of shell material.

E Schematic diagram of a section through a chicken's egg. (After Lillie). 1 = egg white; 2 = cytoplasm of the egg; 3 = "white" yolk; 4 = air space; 5 = chalaza; 6 = dense region of the egg white; 7 = calcerous shell; 8 = vitelline membrane with a fibrous covering; 9 = outer shell membrane; 10 = yellow yolk; 11 = inner shell membrane tate with each other and which can be recognised under the light microscope as the zona radiata. At a later stage the spaces between the microvilli are filled with a fibrous material which becomes the egg membrane, the microvilli frequently being "withdrawn" during the final stages of oocyte development. The egg membrane of fishes is called the chorion, that of amphibians, reptiles and birds the vitelline membrane, and that of mammals the zona pellucida. Thus the zona radiata is not an egg membrane but merely the microvilli of the oocyte and granulosa cells as they appear in the light microscope. More recently it has been proposed that the egg membrane of all vertebrates should be referred to as a zona pellucida. Both the oocyte and the follicle cells are evidently involved in the formation of the egg membrane and it is assumed the latter form the peripheral layer of mucopolysaccharides, i.e. the zona pellucida sensu stricto. The underlying cortex radiatus probably stems from the oocyte and consists of an electron dense substance which is homogeneous in amphibians but divided into two layers in many teleosts (Fig. 29A). When ovulated the mammalian egg is enclosed in a further envelope made up of granulosa cells, namely the corona radiata mentioned above (Fig. 29B). According to the definition adopted here, this is a primary egg membrane.

The jelly surrounding the eggs of amphibians is a secondary egg membrane deposited around the oocyte in the oviduct. After the eggs have been spawned in water the membrane swells considerably (Fig. 29C). Complex egg membranes are often formed in the oviduct by the glandulae nidamentariae, the so-called shell glands of oviparous Elasmobranchii. Ovoviviparous and viviparous elasmobranchs also produce egg membranes which are less complex in structure but, on the other hand, contain several eggs. After being fertilised the egg is enclosed in mucus which contains albumin then lamellae are deposited over this to form an often robust shell (Fig. 29D). It was once thought that the outer shells contain keratin but more recent evidence indicates that collagen is the main component. By far the most complex egg membranes are undoubtedly those of birds. The oocyte leaves the ovary surrounded by a fibrous, coarsely textured primary egg membrane, the vitelline membrane, and enters the infundibular part of the oviduct. A second, finely textured fibrous layer is then laid down over the first so that the yolk membrane of the egg is compounded of primary and secondary egg membranes. The egg, which has normally been fertilised by this stage, is then surrounded by another egg membrane, the egg white. The major component of this membrane is water (85%), the rest being a mixture of proteins, though primarily albumin. The egg is held in the centre of this fluid egg membrane by the chalaza. Thereafter two layers of shell membrane composed of keratin fibres are laid down. These two layers are so arranged that they contact each other over most of their surface, except at the blunt end of the egg where they are separated by an air sac. The outer shell of bird's eggs is mainly composed of calcium carbonate and may be pigmented. It contains numerous pores filled with a collagenous substance. In the case of the chicken's egg the pores are about 0.04 to 0.05 mm in diameter and some 7000 of them are distributed over the entire surface of the shell. All five of the secondary egg membranes are formed during the passage of the egg through the oviduct, which takes about a day in the chicken. The formation of the egg white in the upper part of the oviduct requires about 3 h whereas the egg remains for about 20 h in the uterus during which time the rest of the egg white and the other egg membranes are formed (Fig. 29E).

CHAPTER 4

Fertilisation and Early Development

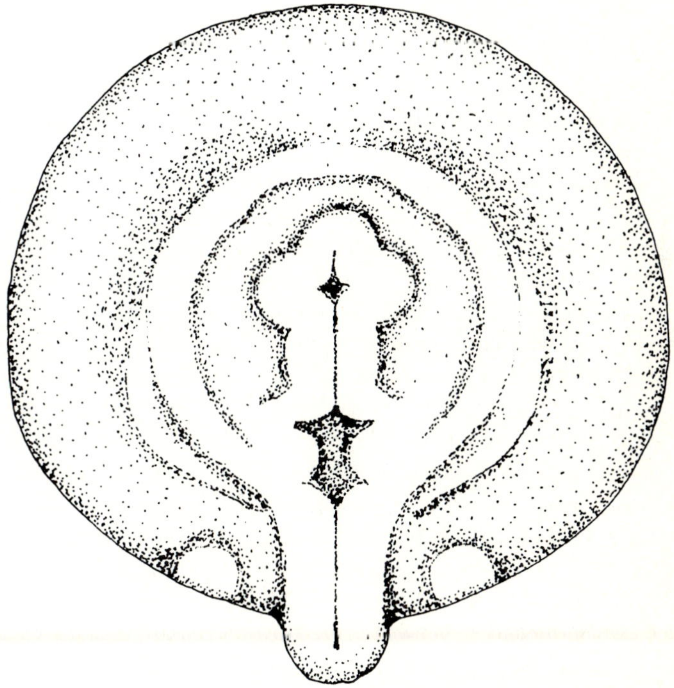

Embryo of the sturgeon *(Acipenser sturio)* at the neurula stage

The steps in the course of reproduction following the formation and maturation of the male and female gametes are the coming together of a spermatozoon and an oocyte and their fusion, the union of the maternal and paternal chromosomes, and the development of a new individual which then differentiates and grows to eventually reproduce itself. Strictly speaking, only then is a generation in the reproductive cycle complete. Only the initial steps in this course of events up to and including early embryonic development will be dealt with in this chapter, the latter being presented as a comparative account. It would go well beyond the intended scope of this book to present a more detailed and extensive account, for which the reader is directed to the several excellent textbooks that exist on embryology and which are listed in the bibliography to this chapter.

4.1 Mechanisms of Bringing the Gametes Together

The Cyclostomata, most Osteichthyes and many amphibians spawn their gametes directly into water. Two problems are associated with this: on the one hand, the danger that water currents carry the eggs and sperm away from each other, effectively diluting them, and on the other hand, the germ cells are suddenly exposed to a medium which differs considerably from the body fluids in its chemical and physical properties, and this is of particular significance in freshwater where osmotic imbalance can have damaging effects. The first problem is ameliorated in three ways, the most primitive of which may well be that already mentioned in Chapter 1, namely collective spawning whereby many individuals of each sex congregate together, and large numbers of eggs and sperm are laid at the same time, assuring a high probability of the different gametes meeting each other. This type of spawning occurs in herring (Clupeidae). The second means is by bringing the genital openings or cloacae as close to each other as possible while the gametes are released. This mechanism occurs in Cyclostomata and many teleosts and amphibians. For example, numerous anurans spawn in pairs with the male riding on top of the female and gripping her with his arms. The eggs are extruded past the male's cloaca, which deposits the sperm on the eggs in the manner of a conveyor belt operation. The third means involves the eggs being laid in a confined space such as a spawning hole or nest where they can then be easily fertilised by the male. Species which adopt this means include the three-spined stickleback *Gasterosteus aculeatus* and the Mediterranean wrasse *Crenilabrus ocellatus* which build nests of plant material in which they deposit their gametes together with one or more females, one after the other (see p. 307). Many species of substrate spawners also belong in this third category. It includes, for example, species of cichlid in which the female lays clusters of eggs on a previously cleaned leaf or stone. Each cluster of eggs is fertilised by the male positioning his genital papilla directly over the eggs and at the same time releasing the sperm. The bitterling (*Rhodeus amarus*) represents a highly specialised case. During the spawning period the female develops an ovipositor that is several centimetres long which she extends into the exhalent siphon of a freshwater mussel (of the genus *Unio* or *Anodonta*) to lay her eggs in the gill chamber. The male then releases his semen over the mussel, which sucks it in to the eggs via the inhalent siphon. The young develop within the gill chamber of the mussel. A few other teleosts lay their eggs in either a sponge, an ascidian, or in the gill chamber of a crab. The release of gametes into another animal is called phoretic spawning.

The second problem associated with releasing the gametes into water, namely avoidance of the injurious effects of this milieu, is solved by enclosing the oocyte in egg membranes (see p. 107) which, in the Cyclostomata and especially in the teleosts, are firm and tough and may further harden when in water. The spermatozoa have no such defense mechanism and are therefore subject to the osmotic effects of their hyperosmotic or hypoosmotic surroundings. In freshwater in particular, this results in the spermatozoa having an extremely short lifetime, in the order of a few minutes. Among teleosts for example, it has been shown that the spermatozoa of the medaka *Oryzias latipes* are no longer capable of fertilising the eggs after remaining for 6 min in freshwater. In isosomotic Ringer solution however they survive for several hours. The mechansims which serve to defend the spermatozoa from these effects and ensure that they can successfully penetrate the eggs will be discused later. The swollen, gelatinous capsule of amphibian eggs can similarly be regarded as a defense mechanism which buffers the egg from the surrounding milieu.

The problems dealt with so far are obviated if fertilisation is internal (sensu stricto), that is, the sperm are transferred from the male directly into the oviducts of the female or are actively taken up by the female when released in the form of a spermatophore. In contrast, two adaptations are known to occur in cichlid teleosts which can be interpreted as a kind of internal fertilisation in that fertilisation takes place not in the oviduct but in the mouth. Thus the female of the mouth-breeding cichlid *Haplochromis wingatii* takes the unfertilised egg into her mouth after it has been laid. She also snaps at the male's so-called egg spots, coloured spots which are secondary sex characteristics on the anal fin that look like the eggs. This action induces the male to release his sperm, which thereby pass into the female's mouth where they fertilise the egg. A comparable phenomenon occurs in *Tilapia macrochir* where the female suckles at the genital papilla of the male so that the sperm are released into her mouth where she holds her eggs.

True internal fertilisation occurs in the elasmobranchs, some teleosts and Gymnophiona, a single anuran species, most urodeles, reptiles, birds and mammals. Typically it involves the use of copulatory organs (see Chapt. 2) except in the case of certain Urodela which employ a spermatophore and in most birds, in which case the semen is transferred by the male pressing his cloaca against that of the female. To recapitulate briefly, elasmobranchs and teleosts possess modified fins (pterygopodia and gonopodia) that function as copulatory organs whereas the frog *Ascaphus truei* and Gymnophiona have an eversible cloaca. Sauropsida and Mammalia have respectively a hemipenes and a penis. In each case these organs are inserted into the genital opening or cloaca of the female so that the semen is released deep within the reproductive tract of the female. Very few investigations have dealt with the physiology of copulation in submammalians. In general, the hemipenes of lizards and snakes must be everted prior to copulation. Similarly, the erectile tissue in the true penises of Chelonia, Crocodilia, Aves and Mammalia fills with blood thereby enlarging the organ and making it firm. The mechanism of erection in mammals is described in Chapter 2 (see p. 73). After the penis has been inserted the semen is ejaculated into the genital tract of the female. In mammals this usually occurs following a series of thrusting movements which serve to excite receptors in the glans penis which leads to ejaculation of the semen. The reflex arc involved in this process is dealt with in Chapter 6 (see p. 212). The ejaculate of mammals is not a homogeneous fluid but a composite of several components which are added together one after the other. The composition of human semen has been described in Chapter 3 (see p. 94).

The introduction of the semen into the female genital tract is called insemination and in mammals one can distinguish five different types of insemination according to the amount of semen transferred and the part of the tract in which it is deposited. The first type is an uterine insemination with a dilatation of the transient part between vagina and uterus, the cervix uteri, and involves a very large volume of semen: the horse serves as an example of this type. The second type is basically the same as the first but the semen is ejaculated into the vagina (e.g. the dog and pig). In the third type a vaginal plug is formed after which the vagina contracts considerably which leads to uterine insemination (e.g. the mouse and the rat). The fourth type involves vaginal insemination and is followed directly by the formation of a vaginal plug or, in other cases, the semen coagulates (e.g. rabbits and humans). In the fifth and final type, insemination takes place in the vagina and involves a small volume of semen which contains a high "concentration" of spermatozoa (e.g. cattle).

As mentioned in Chapter 1 (see p. 5), there are generally a number of conditions which must be fulfilled, at least in part, for fertilisation to be successful. Thus the gametes must become mature at the same time, or if this is not possible then the semen must be able to be stored for some length of time. There must in addition be a close synchronisation between ovulation and spawning or copulation: if this does not occur then it can be compensated for by copulating as often as possible during the reproductive period. Furthermore, in the case of internal fertilisation, the fluid medium within the female genital tract must be such as to ensure the survival of the spermatozoa for the longest possible time. Some species are induced ovulators, copulation being the stimulus which triggers ovulation. In other cases the female is usually only ready to copulate when she is in oestrus, during which time ovulation occurs. Fertilisation is more a matter of chance in species which have no distinct oestrus phase and which must therefore copulate frequently, as in man, for example.

So far as is known fertilisation takes place in the upper part of the oviduct or müllerian duct in Elasmobranchii, some ovoviviparous and viviparous Teleostei, Sauropsida and most Mammalia. In teleosts which have follicular gestation (see p. 247) and in a few mammals fertilisation occurs in the follicle within the ovary, prior to ovulation. This means that the spermatozoa must travel a considerable distance in order to encounter the oocytes. However, little is known concerning the length of time the sperm require to make this journey in submammalian species, only that in chickens and turkeys fertilisation takes place less than 15 min after insemination. The sperm of the sheep, rat, mouse and dog require a similar length of time. In the cow the path between the vagina and the upper part of the tuba uterina is covered in about 2–4 min whereas the corresponding journey in rabbits requires 2–4 h. It might be thought that the spermatozoa actively swim to the site of fertilisation but by comparing the speed with which they move with the sometimes remarkably short time it takes them to cover the necessary distance and considering also that their motion is usually randomly directed, it is evident that this is impossible and that one must postulate the existence of mechanisms for the transport of spermatozoa in the female genital tract. In fact, the movements made by the spermatozoa play only a minor role in all vertebrates which have internal fertilisation, the spermatozoa being transported primarily by contractions of the smooth muscles of the müllerian ducts. To what extent this also applies to the secondary oviducts of ovoviviparous and viviparous teleosts is not clear. Nevertheless, it is possible that the random movements of the spermatozoa become directed by the flow of fluid outwards towards the sexual opening which is created by ciliated epithelial cells in the upper parts of the müllerian duct. Thus the spermatozoa swim against this current, exhibiting a positive rheotactic reaction. This undoubtedly applies to mammals and possibly also to members of the other vertebrate classes. On the other hand, in the müllerian ducts of certain birds and reptiles it has been found that between the ciliated cells which create the outward current there is a narrow band of cilia which produce an opposing inward current that carries the spermatozoa up to the ovulated oocyte in the upper part of the tuba uterina.

Two mechanisms have been proposed to account for sperm transport in mammals, both of which rely on contractions occurring in the female genital tract. Thus the semen of humans and of sheep contains a relatively high concentration of prostaglandins (see p. 185) which could potentially affect the smooth muscles of the female tract, as could hormones from the neurohypophysis of the female. However, to what extent these substances, acting either singly or together, are of physiological relevance in other species is still largely unknown. With respect to submammalian species there is a lack of concrete evidence. Very high concentrations of the neurotransmitter serotonin have been found in the sipho fluid of the spiny dogfish (*Squalus ancanthias*) and in vitro both this substance and sipho fluid evoke strong, short-lived contractions of the uterine muscles which are followed by weaker, spasmodic contractions that continue for several hours.

Insemination in mammals involves very large numbers of spermatozoa being deposited in the vagina or uterus, from tens of millions up to hundreds of millions. Compared to this immense number, the number of spermatozoa which actually reach the site of fertilisation is decidedly small, a few hundred in small animals and up to several thousand in large animals. In the rat, for example, between 12 and 43 spermatozoa have been found to reach the site of fertilisation, the corresponding figure for the rabbit being between 500 and 1000. The main obstacles for the spermatozoa are probably the cervix uteri and the openings of the tubae uterina into the uterus, especially the latter since they frequently have a complex structure.

The transport of semen within the upper parts of the müllerian ducts represents only one side of the story of how internal fertilisation results in the gametes coming together since the oocyte must usually also be transported out of the ovary to the site of fertilisation. This begins with the ovulation of the egg and the first and most important problem is then to make sure that the oocyte actually enters the oviduct. In the majority of vertebrates the müllerian duct opens at the ostium abdominale directly into the body cavity and one would imagine that the oocytes frequently fall into this cavity. In fact, this rarely happens since the infundibulum bearing the ostium abdominale frequently lies close to or around the ovary. In some species, such as the dog and cat and species of rodent, the infundibulum forms a complete, or nearly complete, capsule around the ovary so that the entry of the oocyte into the oviduct is a virtual certainty. In other species of mammals, such as man and the rabbit, the infundibulum is not attached to the ovary. In such cases the border of the infundibulum is formed into finger-like fimbria which brush the oocyte into the ostium abdominale by a swaying motion which the infundibulum makes over the surface of the ovary at the time of ovulation. The oocyte is then conveyed deeper into the infundibulum by the current produced by the ciliated epithelium covering the fimbria and also by the contraction of muscles, which are regulated by the ovarian hormones oestrogen and progesterone (see p. 183). Despite the existence of such transport mechanisms the morphology of the system is such that in some cases it seems impossible or at least very difficult to imagine how the ostium abdominale is able to collect up the oocytes. For example, the giant-sized ovaries of female frogs release very large numbers of eggs that are taken up by the ostia abdominales, which are tiny compared to the eggs. Even more striking is the case of elasmobranchs whose müllerian ducts share a common ostium abdominale (see p. 66) which has a diameter that is usually smaller than that of the eggs. The elasticity of this opening is reportedly not sufficient to allow the oocytes to pass through without them deforming: they must actually be squeezed through. The biological "success" of this mechanism nevertheless attests to its effectiveness, although it remains unexplained in physiological terms.

4.2 Fertilisation

The step following the coming together of the gametes is fertilisation. In general, this term refers to the fusion of the spermatozoon with the oocyte but this is not quite correct according to the strict sense of the definition given in Chapter 1 (see p. 14). Strictly speaking, fertilisation occurs upon the fusion of the nuclei of the two gametes (karyogamy) whereby the diploid complement of chromosomes is restored, this being preceded by the fusion of the cells (cytogamy). But even this is not an exact description of the process. According to the first, more general definition, fertilisation includes a range of processes that occur serially or in parallel which begin when the spermatozoon and oocyte are in close proximity and when they come into actual physical contact, processes in which the two gametes act as co-operating partners.

The first problem is the physical meeting of the spermatozoa and oocytes. As already mentioned, this is primarily a question of statistics. If large numbers of gametes are present then there is a good chance of them making contact. On the other hand, it was stated that in mammals, for example, only relatively few apermatozoa may reach the site of fertilisation so that numerically the chances of the gametes meeting seem to be much less favourable. There nevertheless exists an enormous difference in the sizes of the spermatozoa and the oocytes. Eggs are giants among cells, especially if they contain large amounts of yolk as, for example, do those of elasmobranchs and Sauropsida. But in comparison to the size of the spermatozoa, even the relatively small mammalian egg represents a "target" large enough to increase the chances of it being hit.

The male gametes of algae, mosses and ferns are chemically attracted to the female gamete, raising the question of whether or not such attractants also exist in animals. Despite much speculation and the attempts of a number of experimental studies to demonstrate just this point, only a few cases are known in which the spermatozoa are evidently attracted by a chemical. One such relates to a hydrozoan species of the genus *Campanularia* in which one of the substances eminating from the eggs has this effect. Only one example of such a mechanism relates to a vertebrate and this involves the eggs of teleosts. The chorion surrounding the egg (see p. 109) is a relatively tough egg membrane which cannot be penetrated by the spermatozoa except through a small opening called the micropyle. In herring it has been found that when the spermatozoa approach the micropyle they apparently become activated and respond by swimming straight towards it and finally pass through. An isolated piece of chorion which includes the micropyle has the same effect whereas other portions of the chorion do not. The micropyle still exerts its attractant effect after it has been washed thoroughly in Ringer solution for half an hour. It therefore seems probable that specific spermatozoon attractants are indeed present. However, in both the examples mentioned the sperm do not appear to be attracted to the oocyte itself but to an opening through which it can be reached since in *Campanularia* the spermatozoa congregate at an opening in the gonotheca, a chitinous, cup-shaped structure that contains the eggs.

The spermatozoa frequently stick to the egg upon contact. This so-called agglutination reaction is stronger in some animals than in others but may nevertheless be a general feature of all animal spermatozoa. One can demonstrate agglutination in a number of species simply by suspending spermatozoa in water that has contained eggs. Attempts have been made to treat this phenomenon theoretically in terms analogous to an antigen-antibody reaction, although this has not been carried out on vertebrates but primarily on certain echinoderms, molluscs and annelids. Accordingly, a substance called fertilisin is present in the gelatinous layer surrounding the eggs of echinoderms. It is probably a glycoprotein with a molecular weight of around 300,000 Dalton. The amino acid and carbohydrate composition of fertilisin varies between the different species which have been investigated so that it is more appropriate to speak of fertilisins. These substances have more than one binding site which is capable of specifically binding another class of substances, the antifertilisins. These latter are acid polypeptides with a molecular weight of about 10,000 Dalton which are localised in the spermatozoa. The fertilisin of one species reacts most strongly with its homologous antifertilisin, although cross-reactions between species also occur. Like fertilisin, each antifertilisin has several binding sites so that the fertilisin-antifertilisin reaction involves the formation of a network of fertilisin molecules and bound spermatozoa. Antifertilisin is thought to be localised in the membrane of the spermatozoon whereas fertilisin is thought to diffuse out of the membrane(s) encapsulating the egg. However, the results relating to this phenomenon are to some extent contradictory and therefore the mechanism described here should be considered only as a provisional model of agglutination whose applicability to vertebrates in general remains an open question.

Agglutination serves to anchor the spermatozoa to the egg membrane, which must be penetrated before fertilisation can take place, at least in those cases where there is no micropyle as there is in teleosts. Penetration is facilitated by the acrosome reaction, which has already been mentioned briefly (see p. 92). This reaction, like agglutination, has been intensively studied in invertebrates, the only detailed observations on vertebrates having been made on mammals. As mentioned in Chapter 3, the spermatozoa of submammalians differ from those of mammals in that only the former possess a perforatorium, i.e. fibrous material beneath the acrosome (see p. 92). Making the fairly safe assumption that the spermatozoa of submammalians which bear an acrosome react in a similar way to the spermatozoa of invertebrates which possess a perforatorium, there is nevertheless a distinct difference between these types of acrosome reaction and that which occurs in mammals. Of the many species of invertebrate studied in this respect, one from the genus *Saccoglossus* will be taken here as an example since these animals belong to a group, the hemichordates, which are phylogenetically close to the vertebrates. The spermatozoa of *Saccoglossus* correspond to the primitive type described in Chapter 3 (see p. 88). The acrosome lies in front of the nucleus and consists of an acrosomal vesicle, containing an acrosomal granule and finely granular material, surrounded by the acrosomal membrane (Fig. 30A). The acrosome bursts upon contact with the egg, the plasma membrane of the spermatozoon fusing with that at the front of the acrosome. The acrosomal granule thus comes into contact with and gradually dissolves into the egg membrane, probably as a result of the action of lysins set free from the acrosome (see p. 88). At the same time the membrane at the rear of the acrosome overlying the subacrosomal material forms a swelling which grows into a tubular structure (the acrosomal filament) that penetrates the egg membrane and approaches the plasma membrane of the oocyte (Fig. 30B–D) with which it finally fuses. At this moment the spermatozoon and the oocyte, although still appearing to be two cells, become a single cell, and with this cytogamy begins. There follows a mixing of the content of the two cells, that is the assimilation of the spermatozoon into the oocyte, whereby a pseudopodium arises from the oocyte and forms a fertilisation cone which envelops the nucleus of the spermatozoon as it sinks into the egg (Fig. 30E,F). The entire spermatozoon (nucleus, mitochondria and axial filament) is finally taken into the former ooplasm. Although this process is reminiscent of phagocytosis, it differs in one significant respect, namely that the cell membrane of the oocyte and spermatozoon and that at the rear of the acrosome fuse together to form a mosaic membrane (Fig. 30G,H). It is important to note that the initial contact between the spermatozoon and the oocyte "wakens" the oocyte from a dormant state. After the oocyte has assimilated the spermatozoon it goes through a series of reactions which are different in different species of vertebrate, as will be dealt with later.

The process of fertilisation described above, at least during its initial stages, proceeds differently in those mammals whose spermatozoa do not possess a perforatorium. As mentioned (see p. 94), mammalian spermatozoa are incapable of effecting fertilisation directly after they have completed spermiohistogenesis and must first mature further within the male reproductive tract as well as go through the process of capacitation, which principally affects the acrosome, within the female genital tract. The latter process takes a different length of time in different species. For example, capacitation takes 5 h in rabbits, 3 h in rats, one and a half hours in sheep and about 7 h in man. In each case this is longer than the time required to transport the spermatozoa to the site of fertilisation, where they usually arrive before the oocyte. It is probable that capacitation is a demasking process (see p. 94) in the course of which

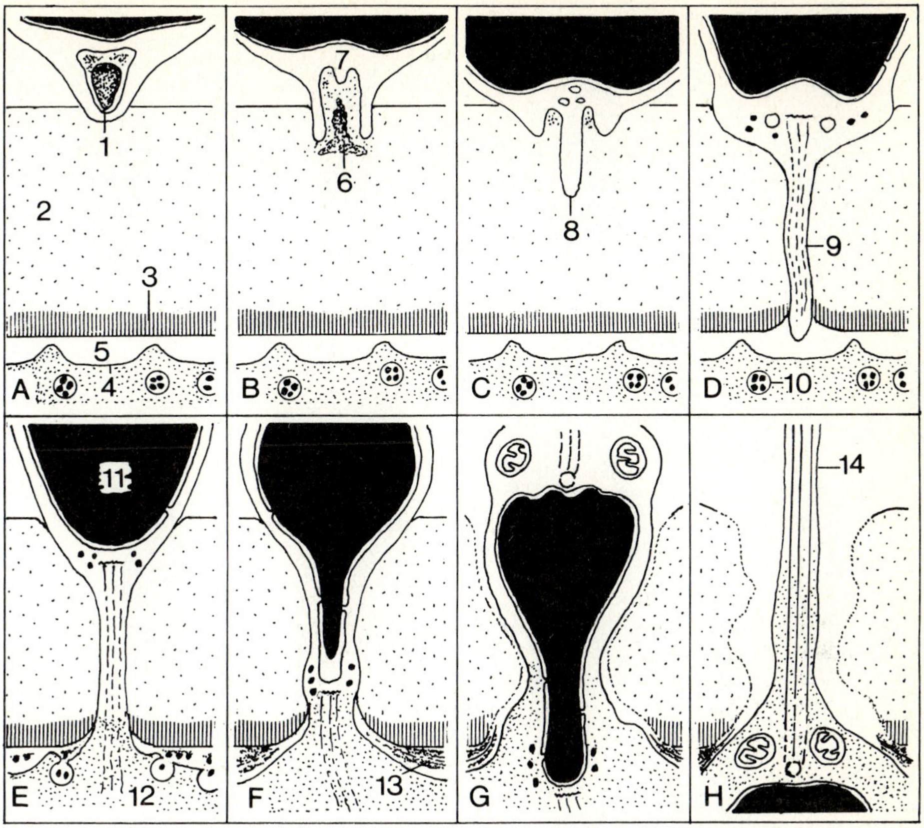

Fig. 30A–H. Schematic representation of the acrosome reaction in *Saccoglossus*. (After Colwin and Colwin).
1 = acrosome with acrosomal granule; 2 = outer egg membrane; 3 = inner egg membrane; 4 = cell membrane of the oocyte; 5 = perivitelline space; 6 = acrosome after having burst open with the granule in the process of dissolving; 7 = swelling from which the acrosome filaments arise; 8 = developing acrosome filament; 9 = acrosome filament shortly before it reaches the membrane of the oocyte; 10 = cortical granule; 11 = nucleus of the spermatozoon; 12 = cytoplasm of the oocyte; 13 = material released from the cortical granules, 14 = tail of the spermatozoon

binding sites on the membrane of the spermatozoon become freed or are unmasked by the removal of inhibitory or stabilising substances. Whichever is the case, the spermatozoon is apparently rendered reactive to substances which may be produced by the corona of granulosa cells surrounding the oocyte. Two different substances are thought to be involved: one ensures that the spermatozoon survives during this phase and another eventually leads to the acrosome reaction. What is certain is firstly, that the acrosome reaction can only occur after capacitation has taken place and secondly, that the spermatozoon can only penetrate the egg membranes when the acrosome reaction takes place. The latter reaction involves the fusion of the outer membrane of the acrosome with the plasma membrane at the front of the spermatozoon's head piece. Fusion of the head membrane of the spermatozoon with the outer acrosome

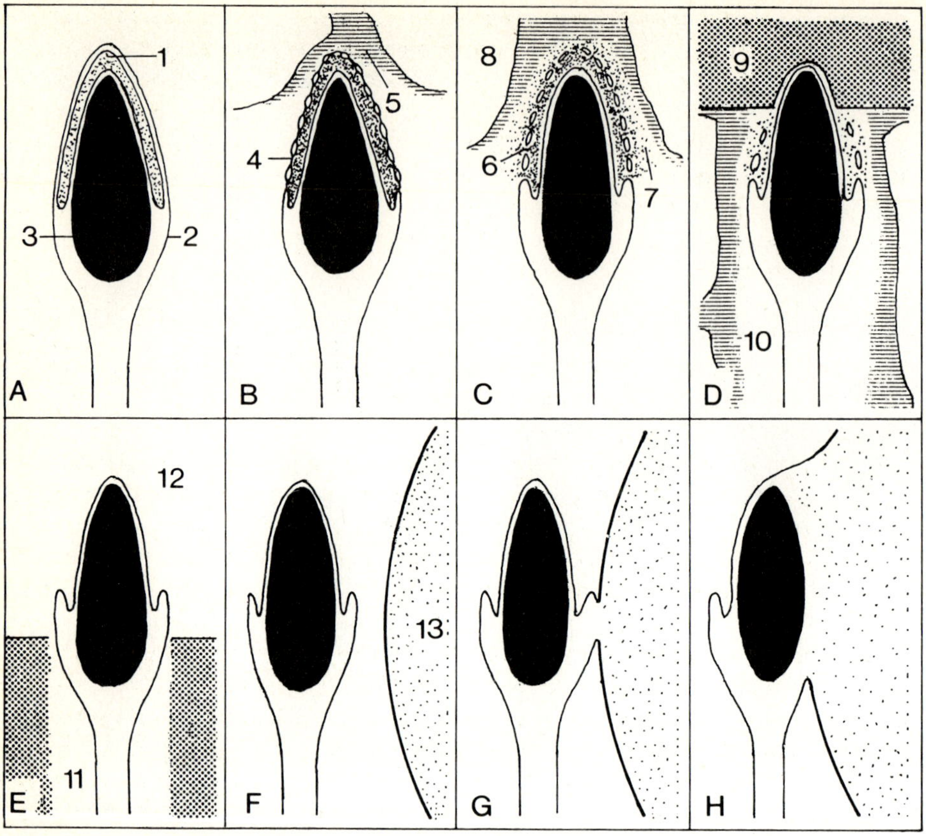

Fig. 31A–H. Schematic representation of the acrosome reaction and cytogamy in mammals. (After Austin, altered considerably).
1 = acrosome; 2 = membrane of the spermatozoon; 3 = nucleus of the spermatozoon; 4 = site of membrane fusion; 5 = intercellular substance between the cells of the corona radiata; 6 = acrosomal pore; 7 = liberated contents of the acrosome; 8 = cells of the corona radiata; 9 = zona pellucida; 10 = penetration canal through the corona radiata; 11 = continuation of 10 into the zona pellucida; 12 = perivitelline space; 13 = oocyte

membrane occurs at a number of sites, forming pores through which the contents of the acrosome are released (Fig. 31A–C). The ensuing penetration of the egg membranes occurs in two phases. Firstly, the hyaluronidase contained in the acrosome diffuses into and dissolves the substance cementing together the cells of the corona radiata, which is rich in hyaluronic acid. In this way the spermatozoon surmounts the first barrier. Secondly, the remains of the outer membrane disappear upon contact with the zona pellucida so that the front of the spermatozoon is then covered by the inner membrane of the acrosome whereas the rest of the spermatozoon is still covered by its original plasma membrane. Unlike the penetration of the corona radiata, penetration of the zona pellucida is probably not accomplished by the release of an enzyme but by a zona lysin bound to the inner membrane of the acrosome that renders the material of the zona pellucida fluid (Fig. 31D–F). In this way the spermatozoon tunnels through to the perivitelline space between the oocyte and the zona pellucida. The spermatozoon then contacts the plasma membrane of the oocyte and cytogamy begins, the spermatozoon being drawn sideways into the oocyte as their respective plasma membranes fuse together (Fig. 31G,H). In some species the tail of the spermatozoon remains outside the egg membrane or in the perivitelline space whereas in others it is taken into the oocyte, from which one may conclude that it probably does not play an important role in the further development of the zygote. The mitochondria of the spermatozoon, and also the axial filament in those cases where it is taken into the oocyte, are generally quickly broken down within the cytoplasm.

The acrosome reaction of mammals is thus characterised by the fact that no acrosomal filament is formed. To what extent the course of events described above for *Saccoglossus* is representative of submammalian species in general is a matter of speculation. Cyclostomata, Elasmobranchii, Amphibia (except Ranidae) and Aves, with the exception of the Passeriformes, all have spermatozoa which possess a distinct subacrosome composed of a fibrous material that in Amphibia and Aves takes the form of a rod extending from a depression in the nuclear membrane. A preformed acrosomal filament which extends past the rear of the nucleus is present in the case of the lamprey *Petromyzon fluviatilis* (see p. 92). The presence of such structures probably means that the acrosome reaction in these submammalians also involves the formation of an acrosomal filament.

The "wakening" of the oocyte by the penetration of the spermatozoon initiates a chain of reactions in the oocyte. Prior to this, when the oocyte enters the second metaphase of meiosis, its progress is arrested for a second time in that after ovulation its metabolism steadily decreases to a minimum level. The penetration of the spermatozoon causes the metabolism of the oocyte to increase again. This effect can also be produced by pricking the oocytes of some submammalians (e.g. the oocytes of Ranidae) with a fine glass needle or something similar. In favourable cases such eggs develop into tadpoles and may even metamorphose into frogs. Evidently, the trigger for the further development of the egg in these species is simply the mechanical stimulus of its being penetrated. This phenomenon is called egg activation, which refers to the initiation of embryonic development in general. The biochemical processes underlying this phenomenon have been studied primarily in invertebrates, mainly sea-urchins (Echinoidea). These studies show that the basic cellular machinery for the production of proteins and nucleic acids and for cell respiration are present within the oocyte prior to activation but that certain relevant enzymes are inhibited and that mRNA and the ribosomes are in some way "masked". The first effect of mechanical stimulation or of penetration by a spermatozoon is the release of proteases from lysosomes present in the oocyte which are thought to help bring about a disinhibition or demasking. It is noteworthy that the cell nucleus is apparently not absolutely necessary for the synthesis of protein which accompanies activation since activation also occurs after the nucleus has been removed with a micropipette. This implies that the requisite mRNA must have already been present in the cytoplasm. However, mechanical stimulation is not the only means of experimentally inducing activation of the egg. Treatment of frog's eggs with hypertonic or hypotonic solutions or even with a toxic solution of mercuric chloride can bring about an artificial parthenogenetic development, although this always results in so-called abortive cell divisions. The experiments mentioned above in which the oocyte is punctured by a glass needle only lead to the development of tadpoles or of adult frogs if the needle is "contaminated" with blood cells or cell debris from the same species. Numerous such experiments have been carried out on the eggs of sea-urchins and these demonstrate that activation is always associated with sublethal damage to the surface or cortex of the oocyte. This view is supported by the fact that activation generally begins soon after the spermatozoon contacts the oocyte and that the initial steps in this process take place in the cortical layer of the oocyte where the majority of the lysosomes mentioned above are located. This cortical reaction will be dealt with below.

Artificial parthenogenesis has also been induced in chickens (*Gallus domesticus*) and turkeys (*Meleagris gallopavo*) by simply incubating unfertilised eggs rather than by directly stimulating the eggs mechanically or chemically. Normal male animals develop in some cases. The oocytes of mammals can also be induced to undergo parthenogenetic development. Thus rabbit's eggs maintained in a tissue culture medium become activated after being incubated for 2 days. Low temperatures are similarly effective whereas treatment of the eggs with chemicals, while effective in the case of sea-urchins, is without affect in rabbit's eggs. Treated oocytes transferred into the tuba uterina of pseudo-pregnant females may continue to develop and, in a few reported cases, a young animal was born at full term, however improbable this may seem considering the size of the mammalian genome and the fact that mutations are often lethal and could not be compensated for by the male genome. This notwithstanding, reliable evidence indicates that over half of the pregnancies in rabbits and mice are the result of parthenogenetic development.

The penetration of the spermatozoon induces other changes in the oocyte besides activation, namely the cortical granule reaction (in those cases where these granules are present), the activation of mechanisms which protect the oocyte against being fertilised by more than one spermatozoon, the completion of meiosis involving the abscission of the second polar body, and finally the formation of the nucleus of the egg. In those species which possess cortical granules the first reaction of the oocyte following contact with the sperm is the release of the contents of the cortical granules into the perivitelline space where it liquifies and spreads over the surface of the oocyte. This process can be clearly seen in frog's eggs where the perivitelline space is narrow prior to fertilisation but considerably wider thereafter. Whereas in this case the egg membrane is in effect raised off of the egg, the corresponding process in the eggs of teleosts results in a "tanning" of the chorion, making it harder. In sea-urchins a so-called fertilisation membrane is formed but this does not occur in vertebrates. In those species of mammal which possess cortical granules the contents of the granules are similarly released into the perivitelline space, which can become somewhat wider as a result. This is evidently due not to the expansion of the zona pellucida but to the shrinkage of the oocyte as a result of the loss of fluid occasioned by the cortical reaction. Comparable morphological changes do not occur in species that do not possess cortical granules. The widening of the perivitelline space has been regarded as a simple physical mechanism preventing the penetration of other spermatozoa, but it is probable that other, more complex mechanisms are in fact responsible for this phenomenon. Although the details of these mechanisms are unclear, the basic process appears to involve substances which are released from the cortical granules and diffuse into the egg membrane and modify its nature. In mammals, for example, the cortical reaction leads to an almost simultaneous change in the permeability of the zona pellucida with the result that spermatozoa can no longer become attached to it: this is called the zona reaction. It is also possible that the structure of the oocyte's plasma membrane is altered by the contents of the cortical granules in such a way that it can no longer fuse with the membrane of a spermatozoon. Whatever is the case, the plasma membrane of the oocyte would nevertheless be altered by the insertion into it of the endomembrane bounding the cortical granules in the course of the cortical reaction.

The mechanisms preventing the penetration of more than one spermatozoon are reliable but not infallible. Thus monospermy, the penetration of the oocyte by a single spermatozoon, is normal in Teleostei, Ranidae and Mammalia but it can happen that two or, in rare cases, three spermatozoa penetrate the cell membrane of the oocyte despite the presence of preventive mechanisms. This is referred to as pathological polyspermy: the embryos do not survive since, in the simplest case, karyogamy produces a triploid set of chromosomes which is lethal. In mammals such embryos evidently develop normally but about half-way through their development they die from unknown causes. In Elasmobranchii, Urodela, Reptilia and Aves, which typically have eggs rich in yolk, it is normal that several spermatozoa penetrate the egg, this being physiological polyspermy. However, only one spermatozoon nucleus fuses with the nucleus of the egg, the remainder degenerate within the cytoplasm, although in some reptiles and birds they may first go through a number of abortive divisions. In the case of physiological polyspermy, the prevention of multiple karyogamy evidently lies within the oocyte. The reverse situation, of two female nuclei being present, is also possible when the separation of the second polar body is prevented by some pathological condition. In mammals the two egg nuclei fuse with the nucleus of a single spermatozoon, with the same outcome as pathological polyspermy. This phenomenon has sometimes been confusingly referred to as polygyny (see p. 29).

Two aspects of fertilisation remain to be dealt with, namely the fate of the spermatozoon up to and including the process of karyogamy and the changes which occur thereafter in the zygote up to the first cell division. Although the tail of the spermatozoon is taken into the oocyte in most cases, it does not appear to play any part in the subsequent course of development. The same applies to the middle piece, though not to the proximal centriole. The mitochondria become dispersed throughout the cytoplasm but it is not known whether they play an active role in development. In all vertebrates the oocyte only completes its maturation division, resulting in the abscission of the second polar body, after it has been penetrated by a spermatozoon and only when this division is complete can karyogamy begin. The essential parts of the spermatozoon, namely the nucleus and proximal centriole, separate themselves from the other elements of the spermatozoon and the highly condensed nucleus begins to increase in size by taking up fluid from the cytoplasm while the oocyte (or more precisely, the second order oocyte) goes through its second maturation division. The resulting haploid nucleus of the egg is called the female pronucleus and the enlarged head of the spermatozoon the male pronucleus. The two pronuclei then move, by some unknown mechanism, towards each other in the active cytoplasm of the animal pole of the egg while the proximal centriole from the spermatozoon divides and forms

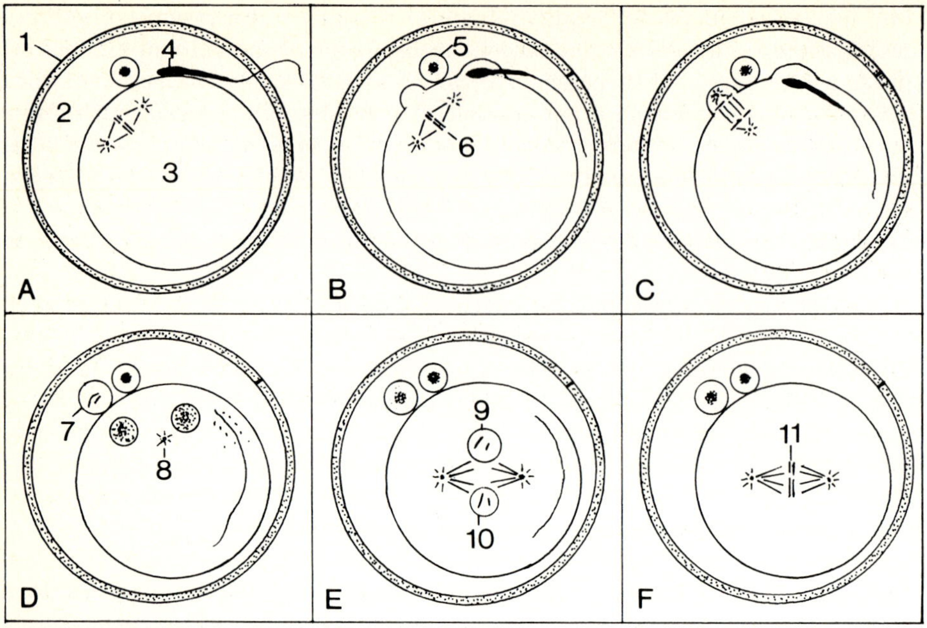

Fig. 32. A–F Highly simplified representation of the different stages in the process of fertilisation as it occurs in vertebrates.
1 = egg membrane (e.g. zona pellucida); 2 = perivitelline space (drawn much enlarged for convenience); 3 = second-order oocyte; 4 = spermatozoon; 5 = first polar body; 6 = spindle apparatus of the second maturation division; 7 = second polar body; 8 = proximal centriole of the spermatozoon; 9 = male pronucleus; 10 = female pronucleus; 11 = spindle apparatus of the first mitotic division

the spindle apparatus. Although true karyogamy – fusion of the nuclei and the formation of a new nuclear membrane – occurs in invertebrates such as the sea-urchin, karyogamy in vertebrates involves the dissolution of the membrane enveloping each of the two pronuclei followed directly by their respective chromosome sets becoming arranged along the equatorial plate of the first mitotic cleavage division (Fig. 32). This process, which represents the conclusion of fertilisation, is also called syngamy. After division a nuclear membrane reforms around each nucleus of the resulting two cell embryo.

The oocyte becomes reorganised in a number of different ways during the course of fertilisation (in the widest sense of the word) and the ensuing period up to the first cleavage division. The first of these, dealt with above, is the emptying of the cortical granules into the perivitelline space. As a consequence of this the membrane bounding the granules is integrated into the plasma membrane of the oocyte, thereby changing its nature. The other changes which occur have been primarily investigated in the eggs of Ranidae, which contain large amounts of pigment (see p. 106). Shortly after the penetration of the spermatozoon the superficial cytoplasm of the oocyte is displaced towards the animal pole, displacing also the pigment granules within it. At the same time the break caused by the pinching off of the primary polar body is sealed.

In *Rana temporaria* it has been found that about ten minutes after the spermatozoon has penetrated the oocyte the cortical layer of cytoplasm, representing the presumptive dorsal aspect of the egg, begins to rotate from the animal pole towards the vegatal pole on the opposite side of the egg. This displacement of the pigmented cortical layer reveals the underlying layer which is less heavily pigmented and gives rise to a grey crescent area which at a later stage of development is involved in the formation of the blastopore. The new distribution of the cortical cytoplasm also results in this area having new physiological characteristics. It has, for example, been found that prior to the formation of the grey crescent vital dyes are taken up uniformly over the entire surface of the oocyte whereas thereafter the permeability of the egg surface to these dyes decreases markedly over all but the area of the grey crescent. As a result of this redistribution of material the bilateral symmetry of the oocyte, formerly indistinct, becomes clearly evident, however the mechanisms underlying this phenomenon are largely unknown.

4.3 Cleavage Types and the Formation of Germ Layers

Karyogamy concludes the direct involvement of the parents or their gametes in reproduction and marks the beginning of the development of the new individual, or ontogenesis. In essence ontogenesis involves the controlled division of cells by mitosis which underlies the complex processes of differentiation that take place according to the genetically determined blueprint of the animal. It has four main stages: firstly the initial division of the zygote, which are called cleavages, secondly the formation of the germ layers, thirdly the formation of the different organs (organogenesis) and fourthly histological differentiation. The developing mammal is referred to as an embryo up to the end of the third stage but as a foetus thereafter. As mentioned at the beginning of this chapter, a detailed treatment of each of the four stages lies beyond the intended scope of this book which will deal with only the first, second and, more briefly, with the third stage of ontogenesis. The basic processes will be described first followed by a comparative account of early development in the individual classes of vertebrate.

4.3.1 Cleavage

After fertilisation the zygote divides mitotically into two daughter cells. Subsequent cleavages usually occur synchronously at more or less regular intervals and give rise to an even number of cells called blastomeres, 4, 8, 16 and so on. In the absence of any deleterious influence the cells always divide into equal halves and as a rule the plane of division is at right-angles to its previous orientation. The cytoplasm does not increase in volume, so that the blastomeres become increasingly smaller with each cleavage. The regularity of the intervals between each cleavage stems from the "behaviour" of the spindle apparatus, which always tends to be positioned in the middle of the cell. In fact, however, this is not always possible because the presence of yolk in the cell prevents it from assuming this position. The orientation of the plane of division also depends on the spindle. In a spherical, isolecithal egg the first spindle apparatus is normally orientated perpendicular to the polar axis through the animal and vegetal poles and hence the plane of division is perpendicular to the spindle axis so that the cell divides vertically (Fig. 33A). The spindle apparatus in each of the daughter cells remains in the horizontal plane so that they too divide vertically but at 90° to the plane of the first division. All the spindles then lie in the vertical plane so that the third division cleaves the four blastomeres horizontally into eight. This basic pattern is repeated in subsequnet divisions. As mentioned previously, the oocytes are giant cells (even those with little yolk) that possess a voluminous cytoplasm in relation to the size of their nucleus, but with each cleavage the nuclear/cytoplasmic ratio gradually approaches the "normal" value for body cells, which ideally is the same for every cell. This "normalisation" is furthered by the biosynthesis of DNA and proteins for the new cell nucleus from the reservoir of materials in the cytoplasm. At the same time the total surface area of the blastomeres grows in relation to that of the zygote. Local differences in the nature of the original plasma membrane of the zygote can be carried over to individual blastomeres or groups of blastomeres. The formation of daughter cells of equal size during the course of so-called equal cleavage is an ideal realised only in the case of isolecithal eggs (Fig. 33A).

In mesolecithal-telolecithal eggs the first cleavage spindle is shifted towards the animal pole by the yolk inclusions within the vegetal half of the zygote. Nevertheless, the first two cleavages usually give rise to four cells of equal size which are then cleaved into four smaller micromeres, representing the animal part, and four larger macromeres containing the yolk, this process being called unequal cleavage. In both equal and unequal types of cleavage the blastomeres divide completely to give rise to cells which are totally separate but still in contact with one another. This is referred to as total or holoblastic cleavage. There are therefore two versions of this, equal and unequal respectively, depending on the content and distribution of yolk in the egg. Among the Chordata equal cleavage occurs only in lancelets (genus *Branchiostoma*) and in mammals (with the exception of Monotremata). However, since the eggs of mammals have derived from those of reptiles they develop differently from those of other vertebrates with holoblastic cleavage. Unequal cleavage occurs in Petromyzontia, a few fishes (*Acipenser* and the Dipnoi for example) and amphibians (Fig. 33B).

In the macrolecithal-telolecithal eggs of Myxinoidea, Elasmobranchii, Teleostei and Sauropsida the cytoplasm of the animal pole divides but the yolk does not, hence this is referred to as partial cleavage. Since this type of cleavage gives rise to a germinal disc on top of the large vegetal portion of the egg, it is also called discoidal or meroblastic cleavage (Fig. 33C).

Other types of cleavage occur within members of the animal kingdom as a whole. The eggs of insects, for example, divide only on the surface whereas the yolk lying at their centre does not divide: accordingly, this is called superficial partial cleavage. Different types of total cleavage can also be recognised according to the arrangement of the blastomeres. Thus radial cleavage results in the blastomeres being arranged radially around the animal-vegetal axis whereas after spiral cleavage each blastomere is displaced by 45° in relation to the one beneath it. Bilateral cleavage results in a bilaterally symmetrical arrangement of the blastomeres within the embryo. Only this last type occurs in vertebrates. There is in addition an irregular type of cleavage that oc-

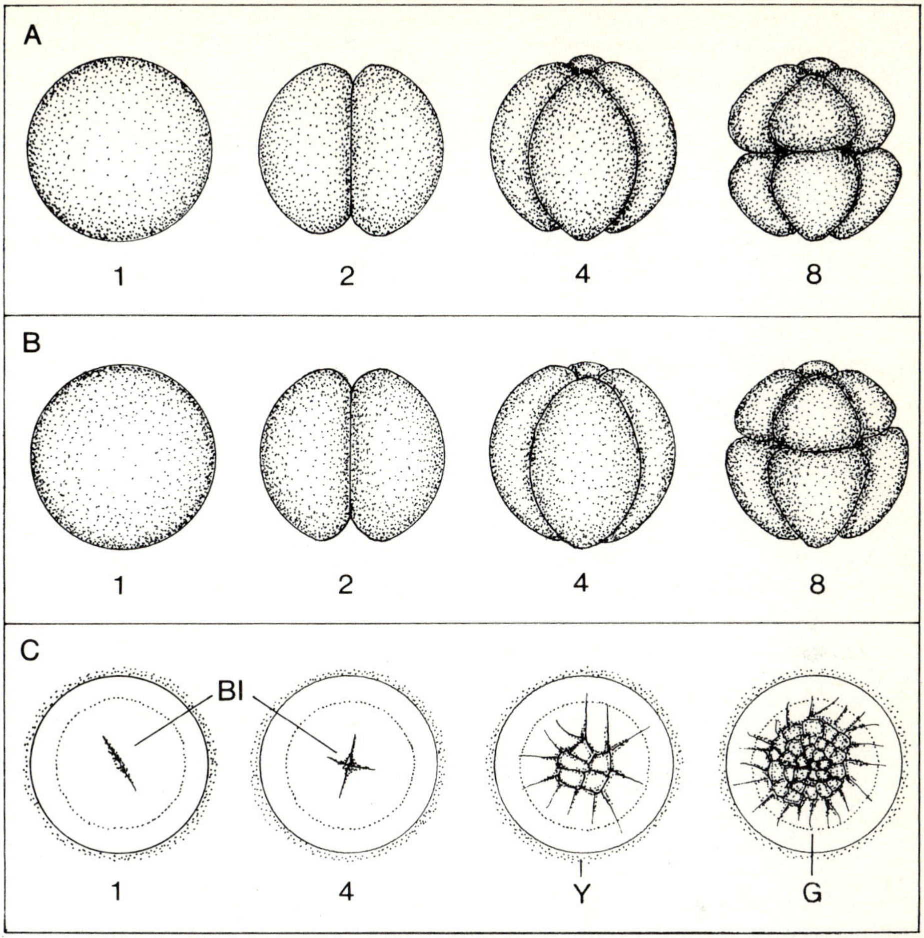

Fig. 33 A–C. Different types of cleavage. **A** Total equal cleavage. **B** Total unequal cleavage. **C** Partial discoidal cleavage (seen from above).
Bl – blastomeres; Y – yolk; G = germinal disc. Numbers indicate the number of cells at each stage

curs in invertebrates and so-called blastomeric anarchy which, among vertebrates, only occurs in Marsurpialia.

Total cleavage initially gives rise to a compact ball of cells called a morula because of its resemblance to a mulberry. As cell division continues a cavity is formed within the ball of cells, which is then referred to as a blastula. In the case of equal cleavage the blastomeres form a layer of cells of more or less uniform thickness, the blastoderm, around the central cavity, the blastocoel or primary body cavity (Fig. 34A). In the case of unequal cleavage the blastoderm of the vegetal region is thicker than elsewhere due to the large amount of yolk in the macromeres (Fig. 34B). In discoidal cleavage the early and late germinal disc correspond respectively to a morula and a blastula. During the latter stage dissolution of some of the yolk gives rise to a subgerminal cavity which in elasmobranchs and teleosts corresponds to the blastocoel

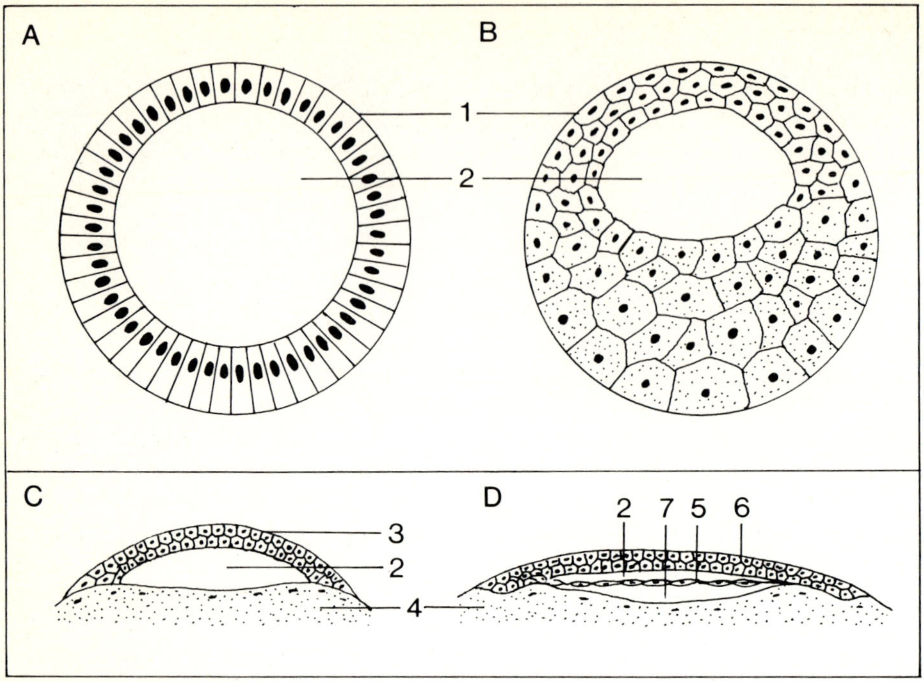

Fig. 34A–D. Different types of blastula. **A** Blastula formed by total equal cleavage. **B** Blastula formed by total unequal cleavage (both **A** and **B** after Balinsky). **C** Germinal disc of a teleost. **D** Discoblastula of a bird.
1 = blastoderm; 2 = blastocoel; 3 = germinal disc; 4 = yolk; 5 = hypoblast; 6 = epiblast; 7 = subgerminal cavity

(Fig. 34C). Contrastingly, in most reptiles and in birds a second, thin layer of cells is formed between the early germinal disc and the undivided yolk mass, the upper layer of cells being called the epiblast and the lower layer the hypoblast. The narrow space between these two layers corresponds to the remnant of a blastocoel whereas the space between the hypoplast and the yolk referred to as the subgerminal cavity (Fig. 34D). Such a germinal disc is also called a discoblastula.

The blastomeres of embryos which undergo total, equal cleavage are only of the same size in the case of the theoretical ideal. Normally, those of the vegetal pole are slightly larger than those of the animal pole. In general the blastomeres form a columnar epithelium. The situation is somewhat different in a blastula that arises by total but unequal cleavage. Apart from the fact that the blastocoel is eccentric because of the very different size of the blastomeres of the animal and vegetal poles, the micromeres and macromeres form a cuboidal epithelium. The cells of the outer layer of the blastula are cemented together by an extracellular substance whereas the inner cells are relatively loosely connected to one another. The process of cleavage ends with the formation of a blastula or, correspondingly, of a late germinal disc.

4.3.2 Formation of Germ Layers

The next steps in the development of the embryo lay the foundations for the subsequent formation of organs and can be subsumed under the general heading "formation of germ layers". This involves the formation of three different embryonic layers, each of which has a different developmental potential, namely the outer germ layer or ectoderm, the inner layer or endoderm, and the mesoderm which lies in between these two. As a crude generalisation one can say that each of these germ layers is fated to develop into specific tissues and organs. Thus the ectoderm forms the epithelial component of the skin and its derivatives (hair and feathers for example), the epithelia of the mouth and cloaca, and the nervous system, including the retina of the eye and the derivatives of the neural crests (the chromaffin tissue of the adrenal glands for example). The endoderm gives rise to the mucous membranes of the gut, the epithelial elements of the large digestive glands (liver and pancreas) and a considerable portion of the respiratory system (the gills or lungs). Finally, the mesoderm is responsible for the formation of the blood vessels, muscles, skeleton, connective tissue, the major part of the urogenital system, the lining of the body cavity, and the axial skeleton of the embryo: in short, the bulk of the new individual. This brief account of the fate or potency of the three germ layers is intended merely to help the reader find his or her way through the complex processes of development. A detailed account of which tissues stem from which germ layer was previously considered of great importance but since then abundant examples have been found which simply do not fit such schemes and which indicate that the whole phenomenon must be seen as being more flexible. For example, in the case of a holoblastic embryo which undergoes equal cleavage, the initial stage in the formation of the germ layers produces an inner and outer layer of cells corresponding to the endoderm and ectoderm respectively. This occurs in a number of different ways which will be dealt with later, but the result is always more or less the same. The mesoderm then develops between these two layers, and this also takes place in a variety of different ways. The mesoderm comes to form the bulk of the new organism. The idea that the formation of mesoderm follows a simple plan is confuted by the fact that different portions of mesoderm can develop at different times and at different places: a germ layer which develops in this way is said to be heterogeneous. On the other hand, ectoderm is potentially capable of forming structures which are normally of mesodermal origin: it can, for example, form elements of the musculature and the skeleton. Thus, in vertebrates, certain parts of the skull arise from both ectoderm and mesoderm and the pupillary muscles of the eye are formed solely from ectoderm. These examples indicate that the fate or competence of the germ layers given above cannot be regarded as being in any sense absolute but rather that some structures are formed by the co-operation and interaction of different germ layers. This will become clear in the sections dealing with the early development of the different classes of vertebrate.

In the animal kingdom as a whole, endoderm is formed in four different ways, one of which is a modification of one of the others. The first involves the invagination or involution of the blastoderm of the vegetal pole into the blastocoel to form a double layered, cup-shaped embryo. The blastoderm of the animal pole becomes the ectoderm and the invaginated blastoderm of the vegetal pole the endoderm. The endoderm forms the primitive gut whose lumen is the gastrocoel, or archenteron, which opens via the blastopore which is bordered by a pair of lips. At this stage the embryo is called a gastrula and the process by which it is formed gastrulation (Fig. 35A). The formation of endoderm by invagination occurs in *Branchiostoma*.

Among vertebrates invagination only occurs in a modified form in Petromyzontia and Amphibia. In this case the vegetal part of the blastula contains so much yolk that it cannot invaginate in a straightforward manner. Instead, the macromeres are overgrown by rapidly dividing micromeres of the animal pole. The dorsal lip of the blastopore is composed of rapidly growing micromeres which grow over the yolky macromeres, which retain their original position, to form the roof of the archenteron. In the course of this process the yolk macromeres are displaced so that they come to lie within the embryo where they later form the true primitive gut. This then is modified or "disturbed" invagination. In the case of gastrulation by epiboly the presence of a large mass of undivided yolk effectively prevents invagination so that gastrulation proceeds by the growth of cells around the mass of yolk from the animal pole. This type of gastrulation occurs in vertebrates with meroblastic cleavage, there being very little indication in such cases of any invagination. In Cyclostomata, holoblastic fish and amphibians gastrulation is a mixture of invagination and epiboly in which the relative contribution made by each of these processes varies from species to species according to the amount of yolk in the macromeres. Gastrulation in Myxinoidea, Elasmobranchii, Gymnophiona, Teleostei and Sauropsida is predominantly by epiboly.

Similarly gastrulation by immigration does not involve the invagination of the vegetal region of the blastula. Instead, cells migrate from this region into the blastocoel until it is more or less full of cells. A cavity then forms within this cell mass that subsequently opens to the outside when the blastopore is formed, i.e., secondarily. The cells lying in the blastocoel form the lining of the archenteron (Fig. 35C). Immigration, in the form described here, does not occur in vertebrates.

The final way in which endoderm can be formed is by delamination. In this case cells from all regions of the blastula migrate into and form a peripheral layer lining the blastocoel. Alternatively, the blastoderm divides into an ectodermal and an endodermal layer. In both cases the blastopore is formed secondarily by the blastocoel breaking through to the outside (Fig. 35D). As with immigration, this type of delamination does not occur in vertebrates.

The gastrocoel can therefore be formed in one of three different ways: both immigration and epiboly involve the internalisation of part of the external environment into the embryo to form the lumen of the archenteron which in the case of immigration arises secondarily by the formation of a cavity within a mass of cells whereas in the case of delamination the blastocoel becomes the gastrocoel by being bounded by a new wall of cells.

The process of endoderm formation is highly modified in those vertebrates which have meroblastic cleavage. Besides the process of epiboly one can recognise features of invagination in Elasmobranchii, of immigration in Teleostei and of delamination in Sauropsida. Details of these processes will be given below in the relevant sections dealing with development in the individual classes of vertebrate.

Mesoderm is formed in one of three ways, the most primitive of which is probably by the formation of a so-called enterocoel, as occurs in *Branchiostoma*. This involves

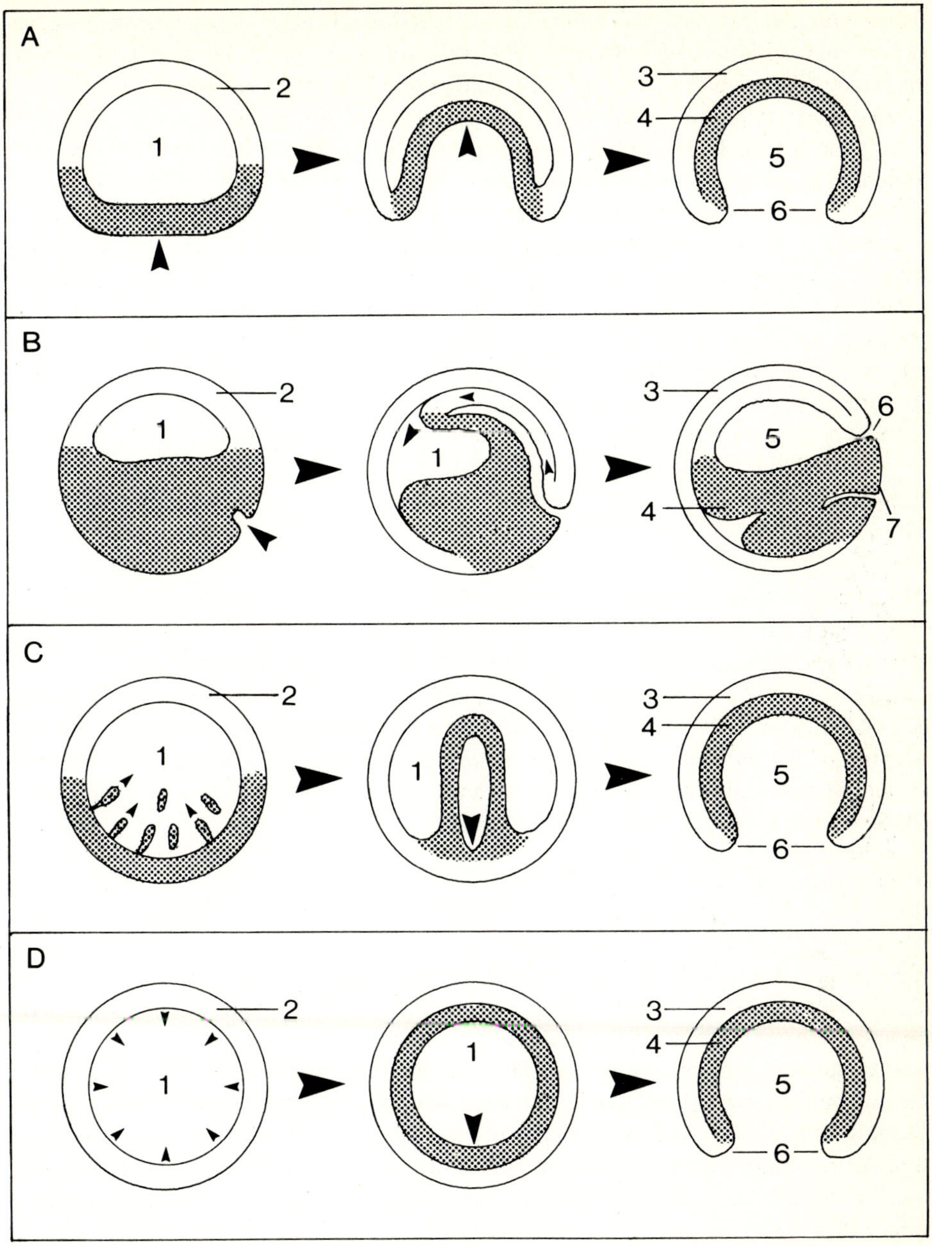

Fig. 35A–D. Different ways in which endoderm is formed. **A** Invagination. **B** Combination of invagination and epiboly. **C** Immigration. **D** Delamination.
1 = blastocoel; 2 = blastoderm; 3 = ectoderm; 4 = endoderm; 5 = gastrocoel (archenteron); 6 = blastopore (the lips of the primitive mouth)

the pinching off of bulges on either side of the archenteron to form cavities lying between the ectoderm and endoderm. The walls of these cavities constitute the mesoderm and the spaces they enclose the secondary body cavity, or coelom.

The term emigration refers to the formation of ridges or layers of cells on either side of the archenteron which then become hollowed out secondarily to form the coelom. This can also occur by individual cells migrating out of the archenteron and dispersing into the space between the ectoderm and endoderm where they begin to differentiate. This method of mesoderm formation is found in vertebrates and will be described in greater detail below.

The third way in which mesoderm forms is from primordial mesoderm cells, though this only occurs in invertebrates. Thus early in development cells arise in the vegetal region of the embryo which separate from the prospective endoderm and proliferate to form primordial streaks of mesoderm.

4.4 The Early Development of Acrania and Vertebrates

The early development of vertebrates will be presented as a comparative account using examples taken from the Petromyzontia, Elasmobranchii, Teleostei, Amphibia, Sauropsida and Mammalia. The acraniate *Branchiostoma lanceolatum* is included in this to extend the basis of comparison since it has a primitive type of development which can be regarded as the evolutionary antecedent of vertebrate ontogenesis. The development of *Branchiostoma* is therefore presented first, after which holoblastic then meroblastic development in vertebrates and finally the early ontogenesis of Theria are described.

4.4.1 The Early Development of *Branchiostoma*

The eggs of *Branchiostoma lanceolatum* are isolecithal and exhibit a bilateral symmetry after they have been fertilised. Cleavage is slightly unequal so that the blastomeres at the vegetal pole of the blastula are somewhat larger than those at the animal pole (Fig. 36A). The cells of the vegetal pole represent the prospective endoderm and are surrounded by a ring of cells that later form the embryonic mesoderm. The remaining cells of the blastula become the ectoderm. Gastrulation occurs by invagination of the dorsal area of mesoderm and results in the cells of the mesoderm and endoderm being drawn into the embryo. The ventral crescent of mesoderm remains where it is as the ventral lip of the blastopore. The roof of the archenteron formed by gastrulation is made up of the so-called chordamesoderm which later gives rise to the chorda dorsalis (also called the notochord), i.e. the axial skeleton (Fig. 36B,C). The gastrula then increases in length and the blastopore gradually shifts dorsally. Gastrulation is followed by neurulation, which involves the laying down of the anlagen of the nervous system and the formation of the neural tube dorsally within the embryo. The ingrowth of the chordamesoderm induces the dorsal ectoderm overlying it to differentiate into a plate of neuroectoderm. The neural plate then detaches from

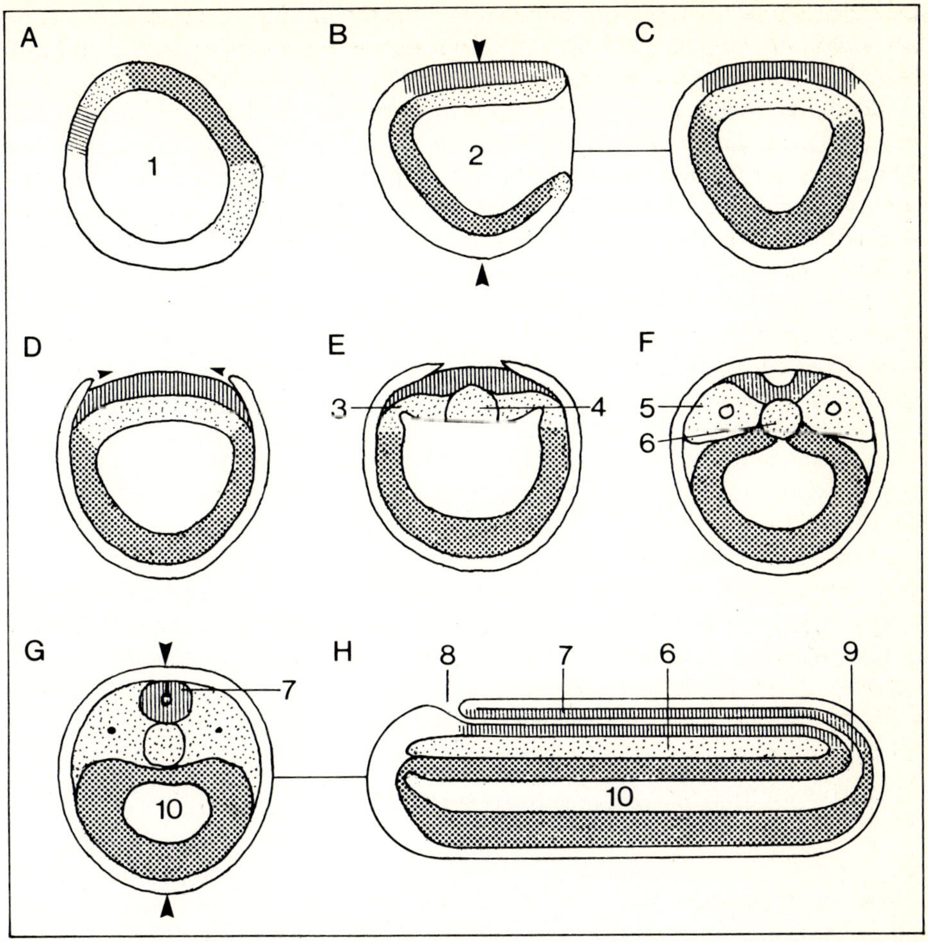

Fig. 36A–H. Early development of *Branchiostoma lanceolatum*. **A** Prospective germ layers of the blastula. **B** Gastrula. **C** Later stage gastrula sectioned in the plane indicated by the *arrowheads* in **B**. **D** Formation of the neural plate. **E** Growth of ectoderm over the neural plate and the initial differentiation of the chorda dorsalis and somites. **F** Closure of the ectoderm and formation of the chorda dorsalis and somites. **G** Completion of the neural tube and archenteron. **H** as **G** but sectioned in the sagittal plane indicated by the *arrowheads* in **G**.
Unshaded region: ectoderm; *hatched region*: neuroectoderm and later the neural plate (medullary plate) then neural tube; *light shading*: mesoderm; *heavy shading*: endoderm.
1 = blastocoel; 2 = gastrocoel; 3 = diverticulum of the archenteron; 4 = chordamesoderm; 5 = somite with coelomic cavity; 6 = chorda dorsalis; 7 = neural tube; 8 = neuropore; 9 = canalis neuroentericus; 10 = lumen of archenteron

the rest of the ectoderm, the margins of which then overgrow the neuroectoderm from both sides (Fig. 36D,E) so as to reform an unbroken layer of ectoderm (Fig. 36F). At the same time the neuroectoderm becomes folded in the mid-line forming a furrow which subsequently closes upon itself to form a neural tube. At the presumptive head end, that is the other end to that bearing the blastopore, the neural tube has a bulb-like expansion which later becomes the brain. At the blastopore end of the embryo the neural tube forms a small tail vesicle that opens to the outside on both sides of the embryo. However, the ectoderm does not only grow over the neural tube but also the blastopore so that a connection is established between the lumen of the archenteron and that of the neural tube, this being the canalis neuroentericus. The opening at the other end of the neural tube, the neuropore, persists during the course of development for some time (Fig. 36G,H). Parallel to neurulation there is a further development of the mesoderm involving the formation of the chorda dorsalis from the medial part of the roof of the archenteron. To either side of this the mesoderm bulges out towards the neuroectoderm to form furrow-like diverticula of the archenteron that later divide transversely to form, initially, 8 or 9 pairs of enterocoelous sacs of axial mesoderm. These sacs subsequently pinch off from the archenteron to give rise to the primitive segments, or somites (Fig. 36E,F). The formation of an enterocoel in this way does not occur in vertebrates. As development proceeds each somite divides into a dorsal myotome and a ventral splanchnotome. The myotome subsequently divides into four layers: the corium layer, the muscle layer, the fascial layer and a scleral layer. The myotomes remain in their original segmental arrangement throughout life. The layers of splanchnotome separate from the myotomes and divide into a parietal layer (somatopleure) associated with the ectoderm and a visceral layer (splanchnopleure) associated with the gut. The individual segments of splanchnotome eventually fuse to form a continuous body cavity, the dorsal part of the splanchnotome forming the mesenterium in which the gut hangs. Aside from the axial mesoderm, coelomic sacs of peristomal mesoderm are formed around the blastopore.

4.4.2 The Early Development of Petromyzontia

Petromyzontia have mesolecithal-telolecithal eggs that undergo total unequal cleavage to give rise, initially, to micromeres and macromeres of markedly different size. Subsequent cleavages do not occur synchronously and the difference in size between the vegetal and animal cells becomes less marked. The blastula has a relatively large blastocoel and, accordingly, gastrulation is mainly by invagination, although in principle it is a mixture of epiboly and invagination (Fig. 37A,B). Gastrulation results in the prospective mesoderm and the anlage of the chorda dorsalis invaginating over the dorsal ectoderm (Fig. 37B,C). This induces the subsequent transformation of the latter into neuroectoderm. The endoderm cells filling the blastocoel are initially disordered but those in the region underlying the neuroectoderm later become arranged into an epithelium bounding the relatively narrow lumen of the archenteron: the mesodermal portion forming the roof of the archenteron forms the anlage of the chorda. Bands of mesoderm are formed on either side of this anlage by the migration of endoderm cells, this being referred to as emigration. An enterocoel is not formed, although a connection does exist between the mesodermal bands and the lumen of the archenteron at the point where the latter joins the chorda dorsalis at the front end of the embryo. The coelomic cavity arises secondarily by the formation of a fissure within each flank of mesoderm (Fig. 37 D–F) which subsequently divide to form the somites. Concurrently the chorda dorsalis separates from the roof of the archenteron. By this stage

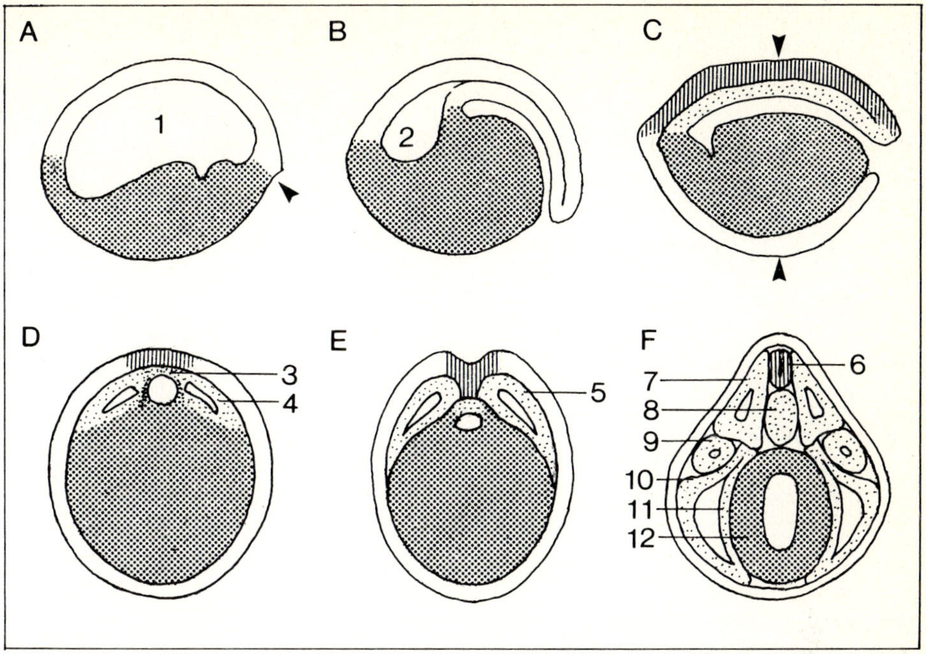

Fig. 37A–F. Early development of Petromyzontia. **A** Blastula. **B** Early gastrula. **C** Gastrula. **D** Late stage gastrula sectioned in the plane marked by the *arrowheads* in **C**. **E** Formation of somites and the neural plate. **F** Formation of the neural tube, closure of the archenteron and differentiation of mesoderm. *Shading* as in previous figure.
1 = blastocoel; 2 = gastrocoel; 3 = anlage of a somite; 4 = chordamesoderm; 5 = somite; 6 = neural tube; 7 = myotome; 8 = chorda dorsalis; 9 = pronephric duct; 10 = somatic layer of the lateral plate (somatopleure); 11 = visceral layer of the lateral plate (splanchnopleure); 12 = archenteron

there is already a solid cord of neuroectoderm in which a lumen is subsequently formed in the neural cord by cell dehiscence, and not as in *Branchiostoma*. The ectoderm finally closes over what is then the neural tube (Fig. 37E,F). Neurulation in Petromyzontia differs from that in *Branchiostoma* in a further respect: the neuroectoderm is overgrown by ectoderm not before the formation of the neural tube but thereafter.

4.4.3 The Early Development of Amphibia

Cleavage of the eggs of amphibians is total and unequal and gives rise to a blastula which has large, yolky cells in the vegetal region that have about the same volume as the blastocoel (Fig. 38A). The course of gastrulation is similar to that described for Petromyzontia except that epiboly plays a greater role. Gastrulation begins with the formation of a small invagination in the region where the blastopore forms. The region of the gray crescent (see p. 125) lying above the blastopore then grows over its dorsal lip into the developing gastrula. The prospective endoderm cells lying ventral to the blastopore grow over the ventral lip and into the embryo, thereby almost completely abliterating the blastocoel: a plug of yolk remains filling the blastopore

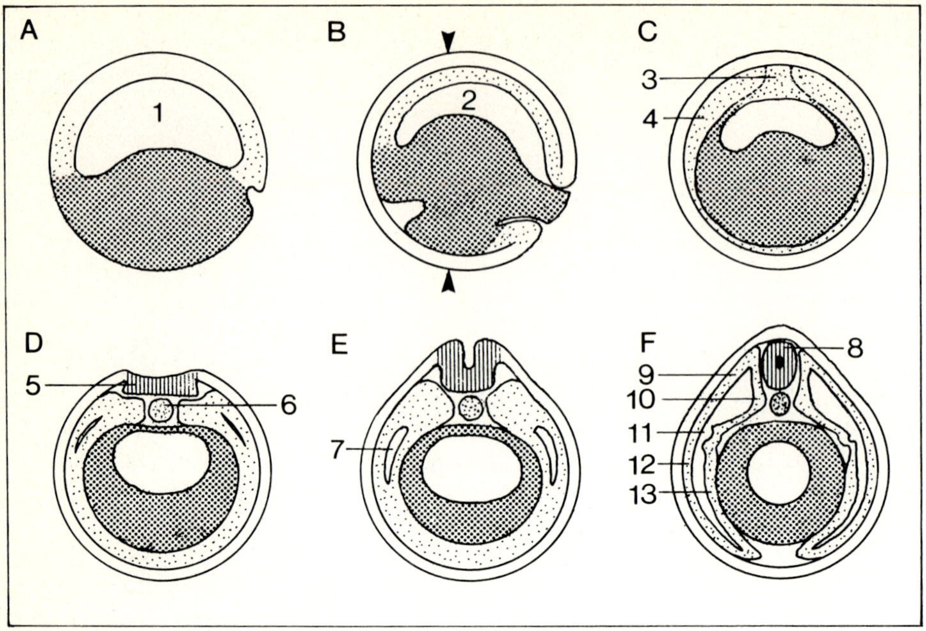

Fig. 38A–F. Early development of an anuran. **A** Blastula. **B** Gastrula. **C** Late stage gastrula sectioned in the plane indicated by the *arrowheads* in **B**. **D** Development of the chorda dorsalis, mesoderm and medullary plate. **E** Infolding of the neuroectoderm. **F** Formation of the neural tube and differentiation of the mesoderm. *Shading* as in the previous figure.
1 = blastocoel; 2 = gastrocoel; 3 = chordamesoderm; 4 = mesoderm; 5 = medullary plate; 6 = chorda dorsalis; 7 = coelom; 8 = neural tube; 9 = dermatome; 10 = myotome; 11 = nephrotome; 12 = somatopleure; 13 = splanchnopleure

(Fig. 38B). Thus a gastrocoel is formed whose dorsal lining includes the chordamesoderm, on each side of which is a mesodermal plate from which the somites arise (Fig. 38C). The dorsal ectoderm overlying the chordamesoderm differentiates into neuroectoderm and forms the neural plate, also called the medullary plate. This plate is bordered by neural folds which gradually approach the mid-line during neurulation to form a neural groove (Fig. 38D,E). This occurs shortly after the chorda dorsalis separates from the roof of the archenteron, which reforms beneath it. Subsequently the neural tube closes and is overgrown by ectoderm (Fig. 38D–F). A canalis neuroentericus is formed. Certain differences exist with respect to the formation of mesoderm in Anura and Urodela. In the former the dorsal region of the archenteron divides into two layers and the split between them subsequently extends to the ventral region until the innermost archenteron is encircled by a layer of mesoderm which is interrupted dorsally by the chordamesoderm (Fig. 38C). The latter forms a connection to the roof of the archenteron called the hypochordal plate. When the chorda dorsalis separates from the archenteron a gap is temporarily formed in the latter called the endodermal gap: this gap is most clearly to be seen in Urodela. Gastrulation in urodeles is much the same as it is in Anura except that the chorda dorsalis detaches from the endoderm together with the mesoderm so that the archenteron remains open dorsally for a much longer time than it does in Anura. At the same time mesoderm migrates from the ventral lip of the blastopore into the cleft between the archenteron and the ectoderm thereby forming a mantle of mesoderm around the archenteron. However, the latter is still contiguous with the anlage of the chorda dorsalis. This subsequently separates as the latter detaches from the archenteron and the resulting break is resealed.

The subsequent differentiation of the mesoderm in Amphibia and in all other vertebrates proceeds in a different way than it does in *Branchiostoma* and Petromyzontia. A distinct segmentation is evident quite early in the development of the latter animal whereas in Amphibia and other vertebrates the mesoderm detaches as an unsegmented mass, the first signs of segmentation appearing when the chorda dorsalis begins to differentiate. Somites are only formed dorsally however, the ventral mesoderm remaining undivided in the form of lateral plates that enclose the lateral plate coelom. The somites are connected to the lateral plates by so-called somite stalks (Fig. 39A). The somites then differentiate further, the medial portion next to the neural tube becoming myotome, each of which forms a segment of the trunk musculature. The lateral portion lying next to the ectoderm represents dermatome which is involved in the formation of the subepidermal connective tissue, i.e. the corium of the integument. In addition to this the mesoderm ventral to the myotome and adjacent to the chorda dorsalis differentiates into skerotome which contributes to the formation of the axial skeleton, that is the sheath of the notochord and, later, the vertebrae. The stalk of the somite becomes the nephrotome and gives rise to part of the excretory system (see p. 54). The lateral plates each divide into a visceral layer, the splanchnopleure, lying next to the gut and a somatic layer, the somatopleure, facing the ectoderm. Both of these later form the lining of the body cavity, the peritoneal epithelium, and the mesenterium in which the gut is suspended. At a more advanced stage of development the somites detach from the lateral plates and the dermatome and myotome spread out beneath the entire ectodermal covering of the embryo. In the case of the latter there are many regional differences, which are, however, too numerous to be detailed here.

It is pertinent to note that the development of holoblastic fish such as sturgeons (genus *Acipenser*) and lungfish (genus *Neoceratodus*) is similar to that of Amphibia. Cleavage of the egg is total and unequal and the blastula which is formed closely resembles that of amphibians. The early development of the blastula proceeds in much the same way as described for Anura.

To conclude this section on vertebrates with holoblastic cleavage mention must be made of the Gymnophiona since they provide a link to those animals with meroblastic cleavage. The Gymnophiona possess eggs very rich in yolk which can be classified as intermediate between the mesolecithal eggs of amphibians and the macrolecithal eggs of teleosts, for example. In Gymnophiona cleavage is partial, that is the nuclei within the yolk divide but the cytoplasm does not. As a result of this the blastula encloses only a very small blastocoel. Because of the large amount of yolk, invagination takes place close to the animal pole, beginning with the formation of a small slit-shaped invagination which breaks through to the blastocoel to form the archenteron. The roof of the archenteron differentiates into a plate of chordamesoderm which subsequently detaches as the chorda dorsalis thereby leaving a gap in the endoderm of the archen-

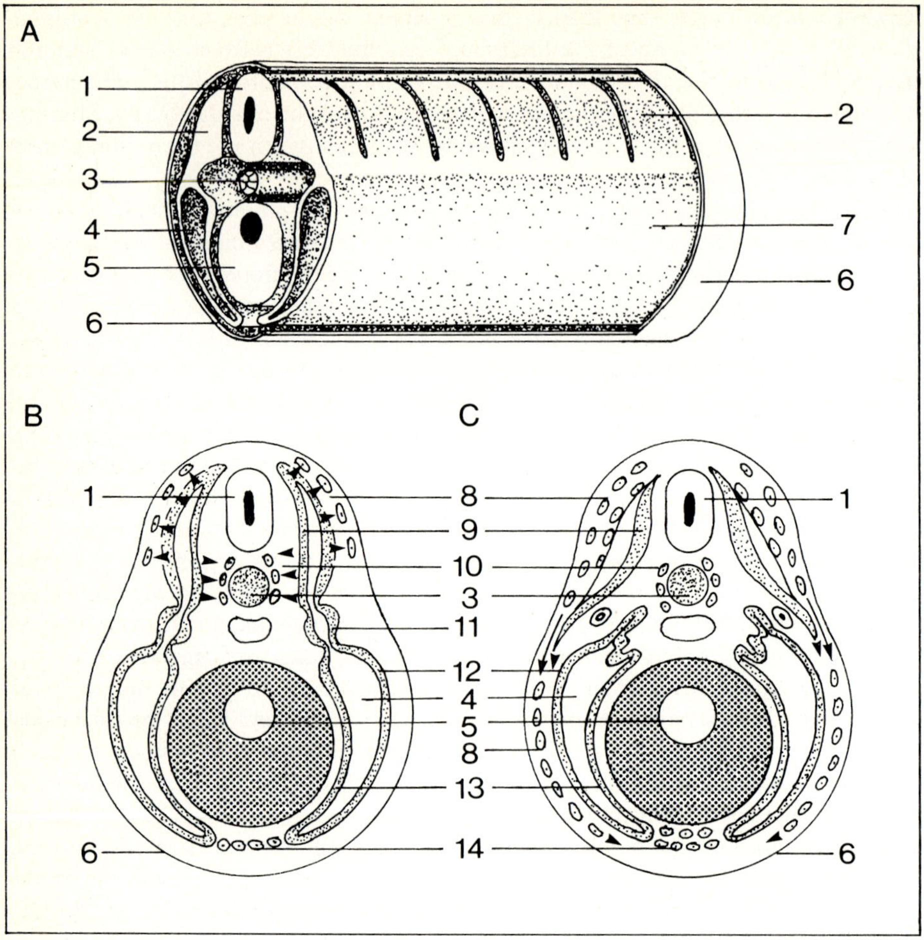

Fig. 39A–C. Differentiation of the mesoderm in a holoblastic vertebrate. **A** Stereogram of the middle section of an embryo (after removing part of the ectoderm). **B** Cross-section of the embryo at a stage prior to the separation of the lateral plates (both **A** and **B** after Portmann). **C** as **B** but after separation of the lateral plates.
1 = neural tube; 2 = somite; 3 = chorda dorsalis; 4 = lateral plate coelom; 5 = archenteron; 6 = ectoderm; 7 = lateral plate; 8 = dermatome; 9 = myotome; 10 = skerotome; 11 = stalk of the somite; 12 = somatopleure; 13 = splanchnopleure; 14 = cells from which blood vessels develop

teron that later closes as is the case in other amphibians dealt with previously. During the course of development the yolk becomes almost completely enclosed within the embryo, except for a plug of yolk that fills the blastopore. Only the dorsal lip of the blastopore participates in the formation of the mesoderm: the ventral lip only starts to develop after the yolk has been completely incorporated within the embryo. This type of development shows a fundamental similarity to that of meroblastic vertebrates, which will be dealt with next.

4.4.4 The Early Development of Elasmobranchii

The eggs of elasmobranchs are extremely rich in yolk and undergo discoidal cleavage to form a germinal disc, or blastodisc. However, the region of cytoplasm bordering the yolk mass does not cleave completely and, because of this, a syncytial boundary layer, the periblast, is formed which plays an important role in the provision of the embryo with nutrients from the yolk. The region of the periblast lying directly beneath the germinal disc is called the central periblast and the region outside this the peripheral periblast. The germinal disc is initially composed of a single layer of cells but later becomes multilayered as a result of tangential divisions. A subgerminal cavity corresponding to a blastocoel forms directly above the periblast (Fig. 40A). Gastrulation in elasmobranchs has a number of primitive features and resembles that described for Gymnophiona. In this context is should be remembered that sharks and rays are not the predecessors of teleosts, as might be supposed from their systematic position. Rather, both groups have undergone a parallel development from a common ancestor but the sharks and rays have evidently retained the more primitive form of ontogenesis.

The germinal disc of elasmobranchs is initially round or oval and has a thickened margin. Gastrulation begins at what will become the posterior end of the embryo by the disc rolling under itself in a process which can be interpreted as a greatly modified form of invagination. It results in a layer of prospective endoderm and mesoderm lying beneath the ectoderm (Fig. 40B,C). At the same time the anterior and lateral aspects of the disc gradually extend over more and more of the yolk, representing the extraembryonic part of the embryo, whereas growth of its posterior aspect is mainly directed towards growth of the embryo itself. The formation of endoderm and of mesoderm proceeds by the two prospective germ layers, which initially lie one behind the other, becoming arranged one on top of the other so that the mesoderm lies under the ectoderm and on top of the endoderm (Fig. 40D). Neuroectoderm differentiates above the chorda mesoderm and forms a neural groove. The gastrocoel is present, initially, only as a fissure between the periblast and the overlying layer of endoderm (Fig. 40 D–F); Mesoderm is formed at two sites: firstly, a sheet of axial mesoderm forms on each side of the medial chordamesoderm and secondly, peristomal mesoderm, also called marginal mesoderm, grows inwards from the peripheral margin of the blastodisc, which therefore corresponds to the lips of the blastopore. Both sheets of mesoderm finally fuse together to form a solid mass in which the coelom later develops (Fig. 40E,F).

After the development of the chorda dorsalis the neural groove fuses to form the neural tube, which is then covered by ectoderm. As development proceeds all three germ layers come to completely surround the yolk. Since the blastopore at the posterior of the embryo is overgrown by the folds of neuroectoderm, a canalis neuroentericus is formed and indeed persists for some time during the further course of development.

The embryo of meroblastic animals, unlike that of holoblastic animals, is divided into an embryonic and an extraembryonic region due to the large amount of yolk and these regions correspond respectively to the embryo, on the one hand, and the yolk sac and its stalk on the other. As already mentioned, in the early stages of develop-

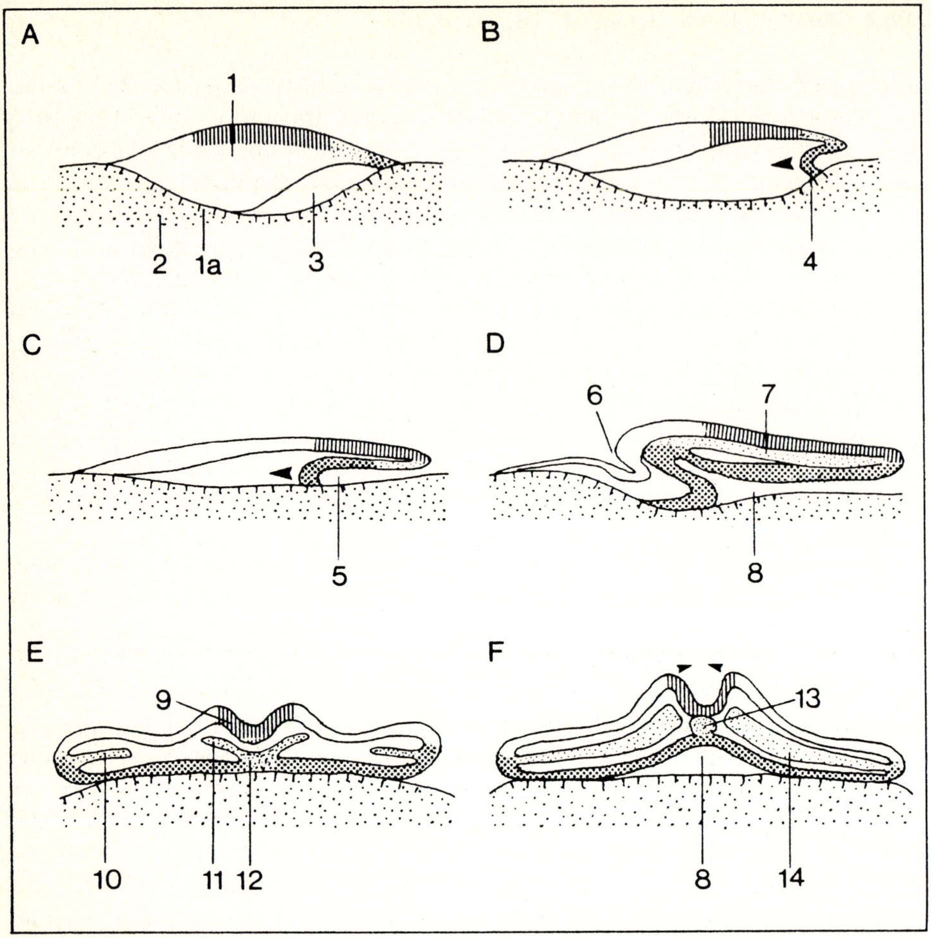

Fig. 40A–F. Early development of elasmobranchs. **A–D** Sagittal longitudinal sections of the embryo with the prospective head region on the left: **A** Germinal disc prior to gastrulation. **B** Beginning of gastrulation. **C** Invagination of the mesoderm. **D** Separation of the endoderm and mesoderm. **E** and **F** are cross-sections at the stage when mesoderm begins to form **(E)** and at a somewhat later stage **(F)**.
Arrowheads indicate the direction of growth. *Unshaded region*: ectoderm; *hatching:* neuroectoderm or the medullary plate; *light shading*: mesoderm; *medium shading*: yolk mass; *heavy shading*: endoderm.
1 = germinal disc; 1a = periblast; 2 = yolk; 3 = subgerminal cavity (blastocoel); 4 = locus at which gastrulation begins; 5 = gastrocoel; 6 = head fold; 7 = chordamesoderm; 8 = archenteron; 9 = medullary plate; 10 = lateral mesoderm; 11 = axial mesoderm; 12 = chordamesoderm; 13 = chorda dorsalis; 14 = continuous sheet of mesoderm

ment the endoderm is composed of a single layer of cells but does not form a continuous lining to the archenteron until later when the yolk has become enveloped by all three germ layers, with the endoderm lying adjacent to the yolk. Thus a large diverticulum of the archenteron is formed around the yolk which is continuous with the ar-

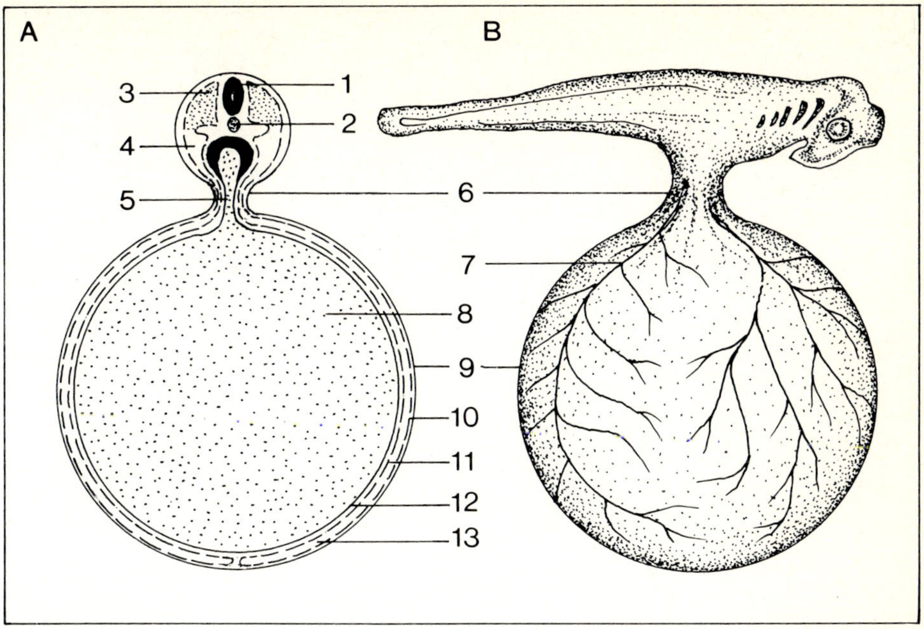

Fig. 41A,B. Structure of the yolk sac. **A** Diagram showing the three germ layers contributing to the formation of the yolk sac (cross-section). **B** Embryo of an elasmobranch showing the yolk sac and its stalk.
1 = neural tube; 2 = chorda dorsalis; 3 = somite; 4 = lateral plate enclosing the embryonic coelom; 5 = ductus vitello-intestinalis; 6 = stalk of the yolk sac; 7 = blood vessels of the yolk sac; 8 = yolk; 9 = yolk sac ectoderm; 10 = outer layer of yolk sac mesoderm (somatopleure); 11 = inner layer of yolk sac mesoderm (splanchnopleure); 12 = yolk sac endoderm; 13 = extraembryonic coelom (exocoel)

chenteron (sensu stricto) of the embryo via a yolk duct, or ductus vitellointestinalis (Fig. 41A). The yolk sac and its lumen are surrounded by four extraembryonic layers: outermost is the ectoderm, beneath which is the somatopleure then the splanchnopleure and finally the innermost layer is the endoderm which lies against the periblast. The coelomic space formed within these layers is termed the extraembryonic coelom, or exocoel for short. As the embryo develops it raises itself up from the yolk sac to a greater or lesser extent (Fig. 41B), this feature being particularly evident in elasmobranchs.

4.4.5 The Early Development of Teleostei

The eggs of teleosts are usually less rich in yolk than those of elasmobranchs but the blastodisc is nevertheless formed in essentially the same way. The blastodisc is round and has a thickened rim, especially posteriorly where it is referred to as the embryonic shield (Fig. 42A). The fate of the different areas of the blastodisc is known in the case of a number of teleosts. In some species there is no prospective endoderm on the

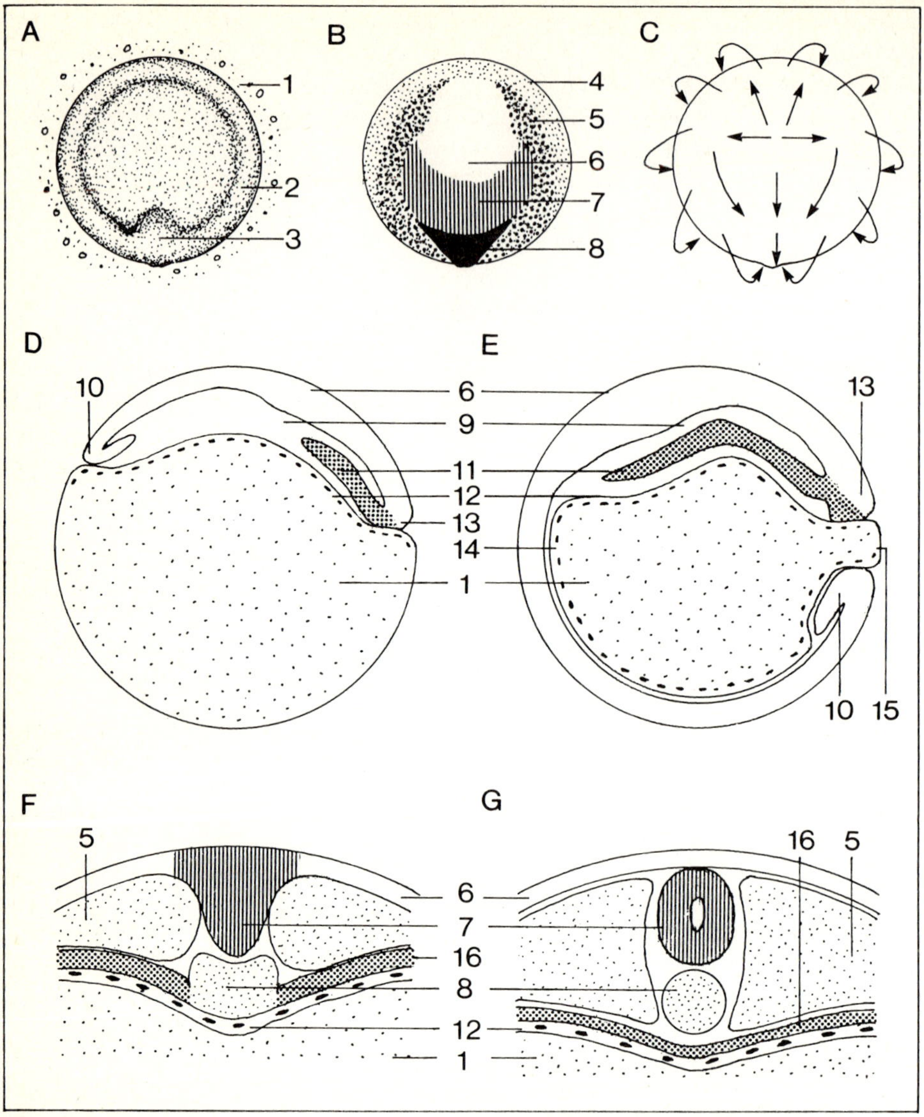

Fig. 42A–G. Early development of teleosts. **A** Surface view of the germinal disc of the trout. **B** Germinal disc showing the location of the prospective germ layers. **C** Schematic representation of the directions of cell growth (*arrows*) during gastrulation (**B** and **C** after Dalq and Pasteels). **D**, **E** Sagittal longitudinal sections of an early and late gastrula, respectively (after Wilson). **F**,**G** Cross-sections of the germinal disc after formation of the neural ridge and after neurulation, respectively.

1 = yolk; 2 = raised rim of the germinal disc; 3 = embryonic shield; 4 = lateral plate mesoderm; 5 = mesoderm of the somites; 6 = ectoderm; 7 = neuroectoderm; 8 = chordamesoderm; 9 = blastocoel; 10 = ventral lip of the blastopore; 11 = endomesoderm; 12 = central periblast; 13 = dorsal lip of the blastopore; 14 = periphery of the periblast; 15 = yolk plug; 16 = endoderm

surface of the blastodisc (Fig. 42B), as is typically the case in elasmobranchs, since this germ layer already lies beneath the surface of the posterior rim of the disc. A periblast is formed as in elasmobranchs. Gastrulation and neurulation proceed concurrently in teleosts and appear simplified in comparison to the same processes in elasmobranchs. The thickened rim of the blastodisc arises in the same way as it does in elasmobranchs, namely by the peripheral accumulation of cells produced by a centrifugally directed growth of cells within the blastodisc. Cell growth is most pronounced posteriorly, somewhat less so laterally and anteriorly, and least pronounced in the region which becomes the brain. The growth of cells from the margin of the blastodisc is directed both posteriorly, towards the embryonic shield, and beneath most pronounced posteriorly and becoming gradually less pronounced anteriorly. The cell layer which is thereby formed beneath the blastodisc does not separate immediately into endoderm and mesoderm and is therefore called endomesoderm. Two views may be found within the literature concerning this process of gastrulation. According to the first view the margin of the blastodisc which gives rise to the endomesoderm corresponds to the lips of the blastopore, as it does in elasmobranchs. The second and more probably view asserts that endomesoderm is formed by the rearrangement of the deeper lying layers of the blastodisc. Whichever is the case, the endomesoderm that arises first remains in contact with the margin of the blastodisc. According to the first view gastrulation is envisaged as being a consequence of displacement due to the growth of cells: this is illustrated schematically in Fig. 42C. Thus the actual process of gastrulation starts out with a blastodisc whose margin is tucked under itself, the tuck being largest beneath the posterior of the disc, and it is from the free edge of the latter that the sheet of endomesoderm then grows (Fig. 42D). Assuming that the margin of the blastodisc is homologous to the lips of the blastopore, the subsequent formation of the endoderm and mesoderm occurs from the equivalent of the dorsal lip only, unlike the situation in elasmobranchs and amphibians for example. At the same time the rest of the blastodisc continues to grow over the yolk, the ventral lip having to cover the most distance so to speak. Prior to its closure the blastopore is occluded by a plug of yolk which remains an uncleaved mass, in contrast to that of amphibians (Fig. 42E). The main bulk of endomesoderm arises not by the growth of cells beneath the blastodisc, in a process resembling invagination, but by the polar immigration of cells from the surface of the blastodisc. The formation of the true endoderm occurs at a later stage by the endomesoderm splitting into two distinct layers, one of endoderm that overlies the periblast and which initially also contains the chordamesoderm medially and another of mesoderm which takes the form of two lateral masses of axial mesoderm (Fig. 42F). At the same time the neuroectoderm above the anlage of the chorda develops into a solid mass of cells (Fig. 42G) and, as happens in Petromyzontia, a neural canal is formed secondarily by dehiscence: a canalis neuroentericus is therefore not formed. In teleosts the yolk is not completely enveloped by endoderm but is covered by the margin of the blastodisc (i.e. the lips of the blastopore which follows the gradual expansion of the peripheral periblast. The fully differentiated yolk sac of teleosts therefore partially consists only of an outer layer of ectoderm underlain by somatopleure, splanchnopleure, and finally a periblast layer adjacent to the yolk.

4.4.6 The Early Development of Sauropsida

The blastodisc of Sauropsida differs from that of elasmobranchs and teleosts in two respects. Firstly, the early blastodisc of teleosts arches over the blastocoel with only its margin being in contact with the yolk whereas in Sauropsida a greater area of the blastodisc contacts the yolk so that a subgerminal cavity exists only in the central area (Fig. 43D). This area appears somewhat lighter and is therefore called the area pellucida, the darker ring peripheral to it being the area opaca (Fig. 43A). The second difference has already been mentioned and is that the blastodisc is divided into an epiblast and hypoblast by a fissure representing the remains of a blastocoel and which becomes separated from the subgerminal cavity. The formation of the hypoblast can be interpreted as the first phase of gastrulation, which in this case involves immigration and delamination. The hypoblast forms the so-called yolk endoderm (also called the secondary endoderm) and the space between this and the yolk therefore represents the remnant of the gastrocoel. A periblast is formed beneath all but the central region of the blastodisc (Fig. 43D).

Development continues along a somewhat different course in reptiles and birds. In reptiles the aggregation of cells on the surface of the posterior third of the blastodisc forms a thickening called the embryonic shield (Fig. 43B). The middle of the shield then invaginates forming what is referred to as the primitive knot which, in effect, is a blastopore. Thus, strictly speaking, this constitutes the formation of a second archenteron whose roof is composed of mesoderm, chordamesoderm and endoderm. Since this invagination is considered to be a primitive feature, the endoderm is referred to as primary endoderm. The lumen of the second archenteron, which is the more primitive, is called the chorda canal. The formation of a secondary endoderm by delamination in Sauropsida has been interpreted as an adaptation that enables yolk to be quickly mobilised and supplied to the rapidly developing embryo (Fig. 43C,D). The formation of a second archenteron, though characteristic of reptiles, also occurs in some species of birds and mammals. The chorda canal subsequently breaks through to the yolk sac by fusing with the secondary endoderm (Fig. 43E–G). Thus the gut of the embryo is made up of secondary and primary endoderm which respectively constitute the major and minor portions of the gut. Gastrulation in most birds is different from but presumably related to that of reptiles.

An embryonic shield, such as is formed in reptilian species, is not formed in birds. Instead, a thickened zone, the primitive streak, develops in the centre of the blastodisc and in the middle of this a long primitive groove, which corresponds to a blastopore, is formed along the head-tail axis. Cells then migrate from the primitive groove into the deeper regions of the blastodisc where they later form the chorda dorsalis and the mesoderm between the epiblast and hypoblast. At the head end of the embryo the border of the primitive groove is thickened and is known as Hensen's node, which can be taken to correspond to the dorsal lip of a blastopore. Directly behind this node there is a shallow depression, the primitive pit, which represents a vestige of invagination (Fig. 44A,B). In a few species of bird, the duck for example, this can extend into the chordamesoderm as a vestigial chorda canal. The formation of mesoderm involves cells from the superficial layers of the blastodisc which migrate over the primitive groove, as if on a conveyor belt, to end up between the hypoblast and epiblast

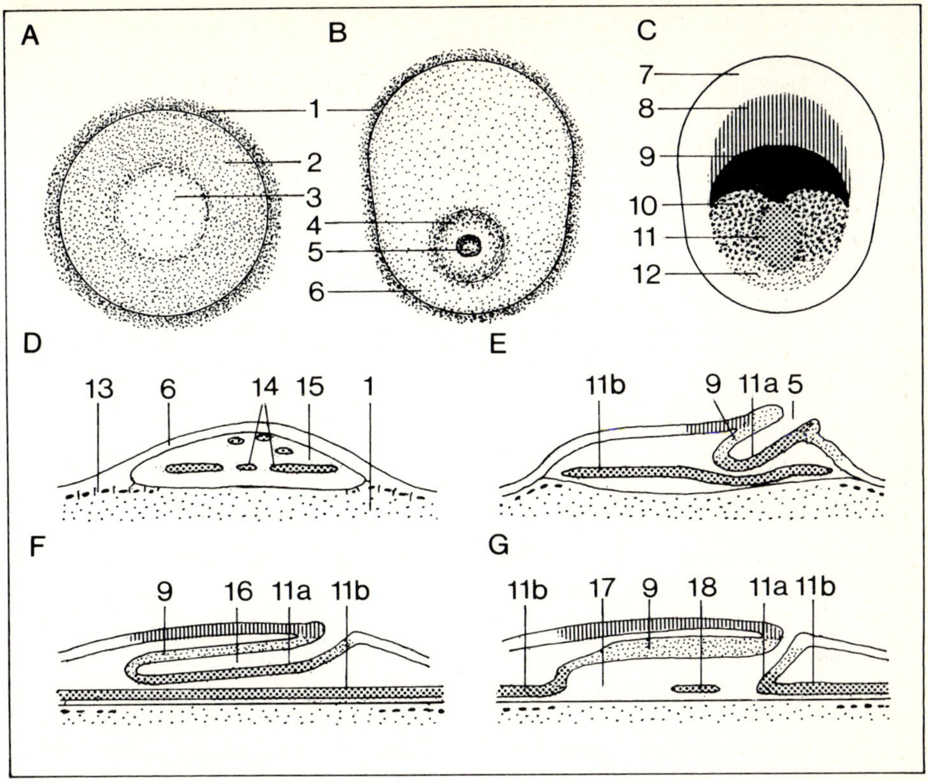

Fig. 43A–G. Early development of reptiles (except for **A** and **D**, after Will). **A** Surface view of the germinal disc of a sauropsid. **B** Surface view of an early gastrula. **C** Map of the germinal disc showing the prospective germ layers. **D–G** Sagittal longitudinal sections showing progressively later stages in the development of the germinal disc from the initial formation of the hypoblast (**D**), through the early (**E**) and late (**F**) gastrula stages, to that after the breakthrough of the chorda canal (**G**).
1 = yolk; 2 = area opaca; 3 = area pellucida; 4 = primitive node; 5 = blastopore; 6 = germinal disc; 7 = ectoderm; 8 = neuroectoderm; 9 = chordamesoderm; 10 = mesoderm of the somites; 11 = endoderm; 11a = primary endoderm; 11b = secondary endoderm; 12 = lateral plate mesoderm; 13 = periblast; 14 = hypoblast; 15 = subgerminal cavity; 16 = chorda canal; 17 = lumen of the archenteron; 18 = endoderm remnants

where they later develop into the somites and lateral plates (Fig. 44C). In contrast to the reptiles and the other vertebrates considered so far, the formation of the chorda in birds is effected in two stages. Firstly, the cells of the chordamesoderm congregate together beneath the node of Hensen where they form a small swelling called the head process. There they wait until the "conveyance" of the mesodermal cells beneath the surface has finished, whereupon the "blastopore" closes and the primitive groove gradually disappears starting from the head end. The cells of the chordamesoderm then undergo a second migration which results in the formation of the chorda dorsalis: the cells migrate between the two plates of axial mesoderm and subsequently develop into the axial skeleton of the embryo (Fig. 44 D–F).

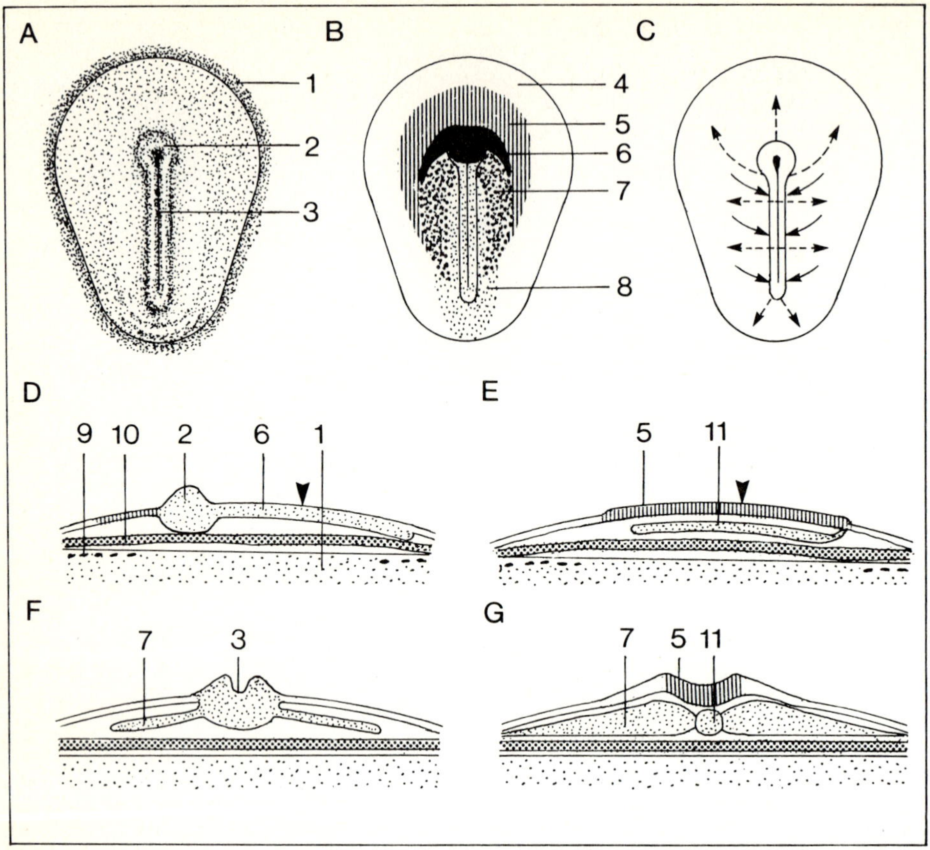

Fig. 44A–G. Early development of birds. (After Nelson, Hamburger and Hamilton; **B** and **G** have been modified). **A** Surface view of the gastrula. **B** Map of the prospective germ layers. **C** Diagram showing the directions of growth during gastrulation (surface movements indicated by full *arrows* and movement beneath the epiblast by *arrows* with *broken tails*). **D, E** Sagittal longitudinal sections of an early **(D)** and late **(E)** gastrula after the formation of the chorda. **F** and **G** are cross-sections at the level indicated by the arrowhead in **D** and **E** respectively.
1 = yolk; 2 = Hensen's node with a central primitive pit; 3 = primitive groove; 4 = ectoderm; 5 = neuroectoderm; 6 = chordamesoderm; 7 = mesoderm of the somites; 8 = lateral plate mesoderm; 9 = periblast; 10 = secondary endoderm; 11 = chorda dorsalis

The subsequent development of birds and of reptiles follows a similar path. As gastrulation comes to an end, the neural groove begins to form at the head end of the embryo. This subsequently closes to form the neural tube which lies above the chorda dorsalis. A yolk sac composed of ectoderm, the two layers of mesoderm and the endoderm is formed.

Although this description of the developing embryo has already reached the stage of organogenesis and thus the limit of its intended scope, one important aspect of organogenesis needs to be dealt with, one which also throws light on the development of mammals. Whereas the development of the animals considered prior to this section occurs in water or in the body of the mother, as in many elasmobranchs and a few

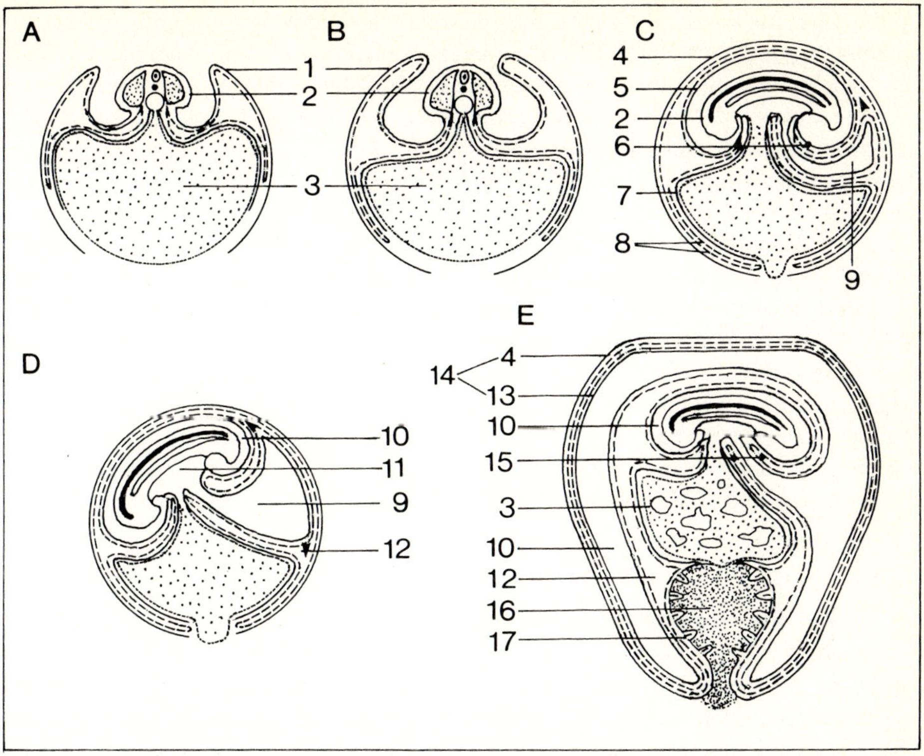

Fig. 45A–E. Formation of the amnion and allantois in birds. (After Weissenberg). **A, B** Cross-sections showing the formation of the amniotic folds. **C, D** Sagittal longitudinal sections at a stage when the amnion is complete and the allantois has begun to form (*arrowhead* indicates the direction in which the allantois expands). **E** Longitudinal section at a stage when the allantois is fully developed (the secondary egg membranes are not shown).
1 = amniotic fold; 2 = embryo; 3 = yolk; 4 = chorion (serosa); 5 = ectodermal lining of the amniotic cavity; 6 = umbilical cord; 7 = yolk sac endoderm; 8 = somatopleure and splanchnopleure; 9 = allantois; 10 = amniotic cavity; 11 = embryonic gut; 12 = exocoel; 13 = allantoic endoderm and mesoderm; 14 = chorioallantoic membrane; 15 = allantoic canal; 16 = albumen sac; 17 = chorionic villi

teleosts and amphibians, the embryonic development of most Sauropsida takes place in a dry environment: the few instances of ovoviviparity and viviparity which occur in reptiles can be regarded as a secondary adaptation. In reptiles and all higher vertebrates an extraembryonic organ, the amnion (see p. 21), is formed as an adaptation to the pressures of developing in a dry environment, hence these animal (viz., Sauropsida and mammals) are known as amniotes. The development of another important extraembryonic organ, the allantois, will also be dealt with in the following.

In front of the embryonic head is an area called the proamnion consisting of extraembryonic ectoderm underlain by mesoderm that develops into an amniotic fold. This fold grows both over the embryo and posteriorly until it eventually meets another fold that arises from the posterior of the embryo, and the folds fuse together to completely enclose the embryo which is bathed in the amniotic fluid (Fig. 45A–C). The wall of the amniotic cavity is therefore made up of an outer and an inner embryonic membrane, which lines the amniotic cavity, is similarly composed of ectoderm underlain by a layer of mesoderm (although as a result of the amniotic folds having folded over the embryo the mesodermal layer appears to overlie the ectodermal layer). The outer membrane is called the serosa, or chorion, and it not only encloses the embryo but also extends around the yolk sac and the allantois. The inner membrane lining the amniotic cavity is commonly referred to simply, though imprecisely, as the amnion. The allantois is a diverticulum of the embryonic hind gut and is therefore surrounded by splanchnic mesoderm. Its growth projects it further and further into the exocoel until it almost completely occludes this space, in the process of which it also surrounds the amniotic cavity. The outer layer of the allantois comes to lie against the chorion, thus forming the chorioallantois. The egg white remaining in the eggs of birds after the development of the allantois is enclosed within an albumen sac surrounded by parts of the chorion. The allantois functions in the exchange of gases to and from the embryo and may contribute to the formation of the placenta in viviparous reptiles (see p. 240), as does the yolk sac. Figure 45 shows the morphological organisation of the extraembryonic organs of birds.

4.4.7 The Early Development of Mammalia

The early development of egg-laying mammals, the Monotremata, follows the same course as that of Sauropsida. The blastodisc is somewhat larger than it is in Sauropsida and already envelops the yolk by the time gastrulation begins. During gastrulation endoderm is formed by the separating off of the hypoblast, which is composed of cells from the blastodisc that differ from those of the epiblast in their morphology and staining characteristics. Mesoderm develops in association with the formation of a primitive groove as it does in birds and leads, amongst other things, to the formation of a chorda canal.

Cleavage is total in placental mammals yet their development does not correspond to that of holoblastic vertebrates but exhibits a distinctly sauropsidian character. At first sight this would appear to be at odds with the fact that total cleavage in other vertebrates is not accompanied by the formation of a blastodisc and the development of extraembryonic regions of the embryo. An explanation for this can however be found by considering the development of Theria and particularly that of placental mammals, which exhibit many specialisations associated with caring for the young within the body. The term placental implies that close morphological and physiological relationships develop between the embryo and mother (as also exist in some other vertebrates; see Chap. 7) which maintain a supply of oxygen and food to the embryo and effect the elimination of carbon dioxide and other waste products (see Chap. 8). Because in mammals the mother takes over the care of the embryo after fertilisation, it is understandable that it is no longer necessary to provide the oocytes with yolk. However, since mammals evolved from reptilian ancestors it is highly likely that the eggs of mammals have developed from eggs similar to those of Sauropsida, and indeed a link between the two groups of animals is represented by the Monotremata. The fact that the eggs of Theria contain little yolk can therefore be seen as being secondary to this close evolutionary relationship. Furthermore, it has already been mentioned that the gene for vitellogenin has evidently been lost in the course of the phylogenetic development from the Prototheria to the Theria.

How it is that an egg which undergoes total cleavage nevertheless develops into a blastodisc is revealed by following the first steps in the development of mammalian embryo. The zygote undergoes total cleavage, equal or unequal, to give rise to a morula. This differentiates into a hollow ball of cells whose morphology correspond to that of a blastula, although strictly speaking it cannot be regarded as such according to the definition of the term used in this chapter. Opinions expressed in the literature differ as to whether or not it in fact corresponds to a blastula, with compelling arguments being presented on both sides, however the resolution of this problem is not of importance here. Whichever is the case, the embryo differentiates into two physiological distinct parts, the trophoblast (also called the trophoderm or trophectoderm) and the embryoblast, which corresponds to an early blastodisc. The embryoblast is either enclosed within a vesicle whose wall represents the trophoblast or it incorporated into the trophoblast as part of the vesicle wall, the whole structure being referred to as a blastocyst. The blastocyst represents the common starting point from which development subsequently proceeds in a number of different ways. In each case the blastocyst is formed by the time it reaches the tuba uterina of the female reproductive system. It then enters the uterus and, in a phenomenon referred to as placentation, establishes an intimate morphological and physiological relationship with the uterus.

Marsupials have relatively large eggs compared to those of other mammals, having a diameter of about 250 μm. The eggs are surrounded by a zona pellucida over which a protein layer and a shell membrane are laid down in the oviduct, evidently an inheritance from their reptilian-like ancestors. The subsequent formation of a blastocyst occurs within the confines of these layers. The eggs of marsupials contain a small amount of yolk within a yolk vacuole but this is eliminated from the egg during the first cleavage division, persisting for a while as a yolk body between the blastomeres until it eventually disintegrates and is phagocytosed by blastocyst cells. In a few species (e.g. of the genus *Didelphis*) the yolk is eliminated in the form of small particles, this being characteristic of Marsupialia but also of a number of Eutheria. Subsequent cleavages result in the cells separating completely from one another, i.e. blastomeric anarchy. Two sorts of cells arise, formative cells and trophoblast cells, which arrange themselves into a simple epithelium beneath the zona pellucida, the formative cells forming the embryoblast at the upper pole of a hollow ball formed by the cells of the trophoblast (Fig. 46A–E). The blastocyst therefore consists of two regions, the formative embryoblast and the trophoblast, which corresponds to the extraembryonic ectoderm formed in the case of meroblastic development. Endoderm is formed in more or less the same way as it is in Sauropsida by cells migrating out of the formative region (corresponding to the blastodisc) to form a second layer of cells: these two layers thus correspond to an epiblast and a hypoblast respectively. The innermost hypoblast layer then extends beneath the wall of the entire blastocyst to form a yolk sac which, however, contains no yolk (Fig. 46F–H). The subsequent development of Marsupialia corresponds to that of birds and Monotremata and therefore need not be considered further other than to note that the amnion develops from amniotic folds. The

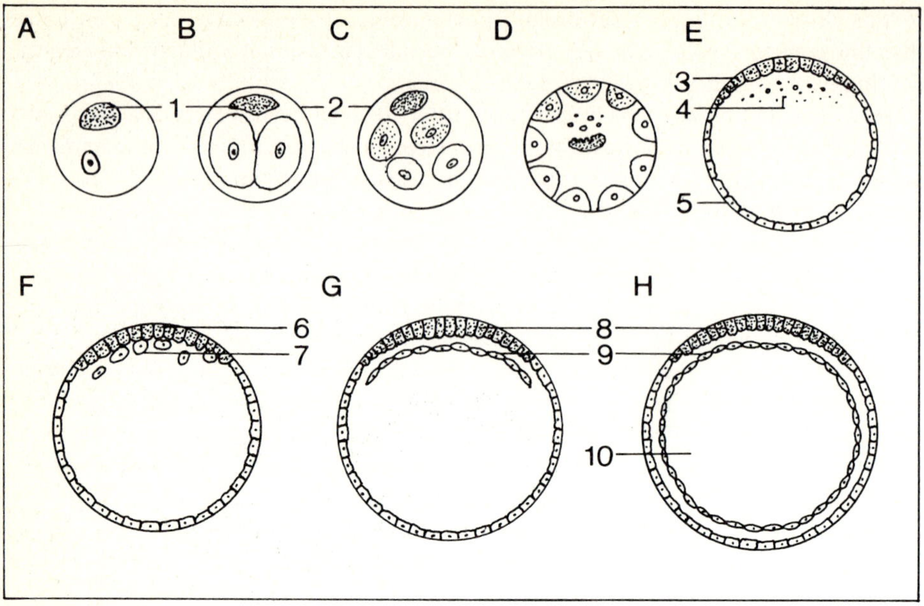

Fig. 46A–H. Early development of mammals: Marsupialia. (After Hartmann). **A–E** Elimination of the yolk and formation of the blastocyst. **F–H** Formation of the endoderm.
1 = yolk mass; 2 = zona pellucida; 3 = embryoblast; 4 = remains of the yolk mass; 5 = trophoblast; 6 = embryonic knob; 7 = hypoblast cells; 8 = epiblast cells; 9 = hypoblast (endoderm); 10 = yolk sac. The secondary egg membranes are not shown

changes in the trophoblast which take place after it has "hatched" out of the zona pellucida are dealt with in Chapter 8.

It is appropriate to mention here that placentation in marsupials typically involves the formation of a yolk sac placenta (omphaloplacenta or omphalogenic placenta) which represents a primitive condition that occurs in elasmobranchs and reptiles. The more "advanced" allantoic placenta occurs in only a few marsupial species.

Cleavage in Eutheria differs from that in Marsupialia in that the blastomeres do not separate from each other completely. Thus, a morula is formed whose peripheral cell layer becomes the trophoblast whereas the cells at the centre congregate at one pole of the blastocyst to form the embryoblast, which is therefore completely surrounded by the trophoblast: this phenomenon is referred to as entypy (Fig. 47A–C). In some species the embryoblast takes the form of a compact embryonic knob, in others it forms a hollow embryocyst. Cells then migrate out of the embryoblast to form an epiblast and hypoblast similar to that of marsupials. The cells of the hypoblast give rise to a yolk sac that does not contain any yolk. Development continues in different ways in different mammals, the differences relating principally to the formation of the early blastodisc and amnion and the fate of the trophoblast and yolk sac. The development of the embryo itself proceeds as it does in Sauropsida with the formation of a primitive streak, an associated primitive groove and Hensen's node, as described earlier.

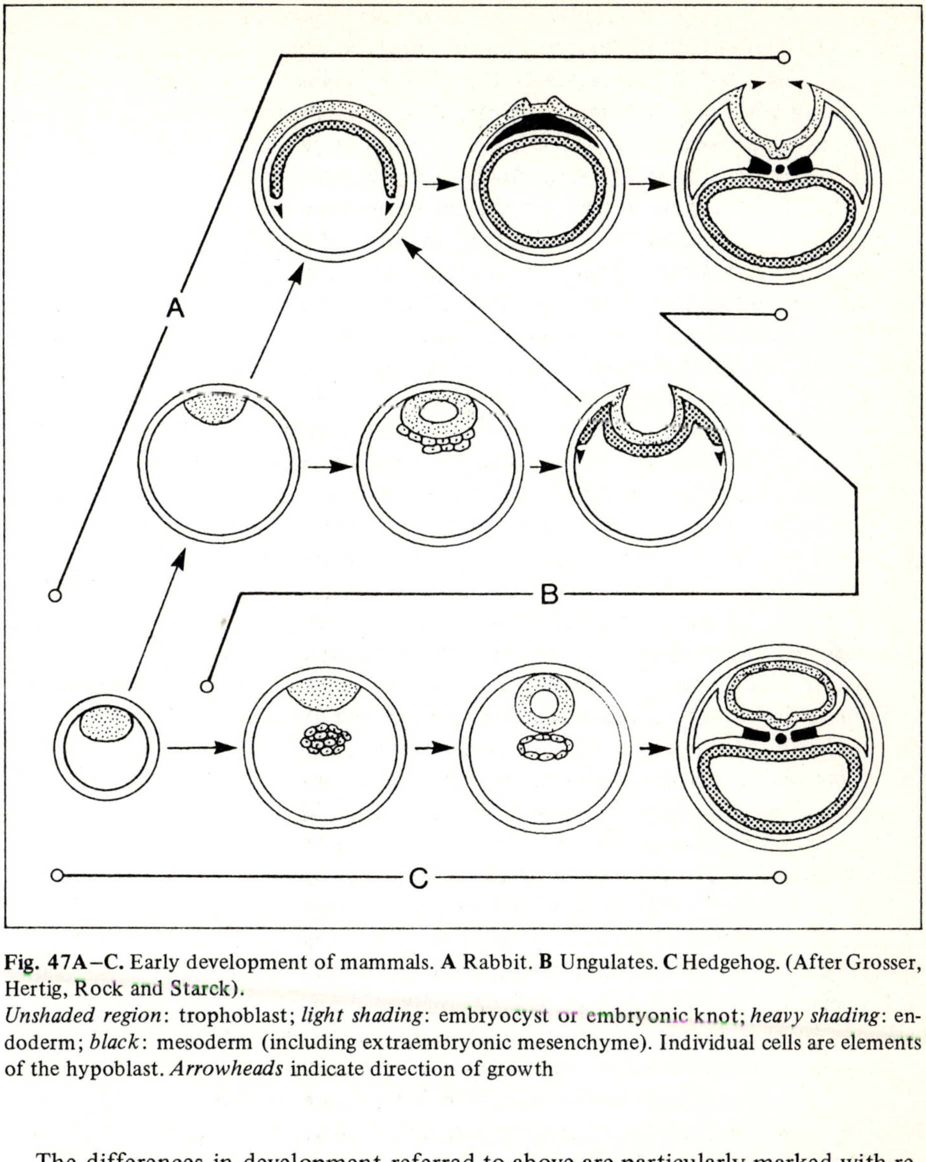

Fig. 47A–C. Early development of mammals. **A** Rabbit. **B** Ungulates. **C** Hedgehog. (After Grosser, Hertig, Rock and Starck).
Unshaded region: trophoblast; *light shading*: embryocyst or embryonic knot; *heavy shading*: endoderm; *black*: mesoderm (including extraembryonic mesenchyme). Individual cells are elements of the hypoblast. *Arrowheads* indicate direction of growth

The differences in development referred to above are particularly marked with respect to the formation of the amnion. There are basically two types of amnion. The first is formed from amniotic folds, as is that of Sauropsida and Marsupialia described above, and will be referred to here simply as an amnion (*Faltamnion* in German). In the case of the second type the lumen of the embryocyst, which arises by dehiscence, becomes the amniotic cavity: such a cavity is referred to as a cavitation cavity and the amnion so formed as a cavitation amnion.

One can identify five variations of the two basic types of amnion formation which can be seen as a series representing the progressive abbreviation or simplification of development. The first variation occurs in, for example, rabbits and, in a sligthly different form, in carnivores. Thus a blastocyst with an embryonic knob is formed, out of which hypoblast cells migrate to form the yolk sac whereas the epithelium of the embryonic knob becomes incorporated into the trophoblast. A primitive streak with a primitive groove and a node of Hensen then develops followed by the formation of an amnion from amniotic folds which results in the embryo again being completely surrounded by the trophoblast (Fig. 47A). In the course of this process the blastocyst implants in the uterus (see Chap. 8). The allantois forms during the development of the amnion and makes contact, via its outer covering of mesoderm, with the placental region of the chorion. The peripheral part of the yolk sac and the wall of the trophoblast overlying the yolk sac later degenerate.

The second variation occurs in Ungulata and, although appearing more complicated than the first variation, can nevertheless be regarded as a step towards the third and simpler variation. As before, a compact embryonic knob is formed within the trophoblast but in this case the embryocyst arises by dehiscence, producing a lumen within the embryonic knob. This lumen then becomes open to the outside and the embryocyst merges into the surface of the trophoblast. The yolk sac is formed by hypoblast cells that have migrated out of the embryocyst. As in the former variation, the amnion develops from amniotic folds after the completion of the second phase of gastrulation (Fig. 47B). This is no longer the case in the third variation where the amnion is formed by entypy. This type of development occurs in the hedgehog (*Erinaceus europaeus*) and begins with the transformation of the embryonic knob into an embryocyst. The hypoblast cells which have previously become separated from the embryonic knob build a compact clump of cells in which the lumen of the yolk sac is then formed by deshiscence, which represents a much simpler course of development than those described above. This also applies to the way the amnion is formed in that the lumen of the embryocyst does not open to the outside, it simply becomes the amniotic cavity. The primitive groove then develops from the floor of this cavity and gives rise to intraembryonic and extraembryonic mesoderm as well as the chorda dorsalis: it subsequently forms the neural tube (Fig. 47C). Other types of development which exist can be considered as transitional between the second and third variations, although this may not have any phylogenetic significance. For example, in Microchiroptera (insectivorous bats) the amnion is formed in the same way as in ungulates and Megachiroptera, although the latter animals have a cavitation amnion.

The fourth variation occurs in the Muridae. It appears more complex than the other variations but in fact represents a simplification of development. Thus one pole of the blastocyst becomes enlarged into a growth called the carrier (or ectoplacental cone) whose base incorporates the embryonic knob. This latter gives rise to hypoblast cells which form the yolk sac (Fig. 48A,B). Cavities develop within the carrier which fuse to form a single carrier cavity. Concurrently, the trophoblast epithelium overlying the yolk sac degenerates, as does the peripheral layer of the yolk sac at a later stage of development. That part of the carrier cavity which is bounded by the embryonic knob then becomes sealed off to form the amniotic cavity. Before this occurs, however, there is a growth of extraembryonic mesenchyme between the remaining wall of the yolk sac and the embryonic knob, this being at a stage of development preceding the onset of mesoderm formation within the embryo (Fig. 48C,D). The significance of this is interpreted as being that it provides the early embryo with a source of

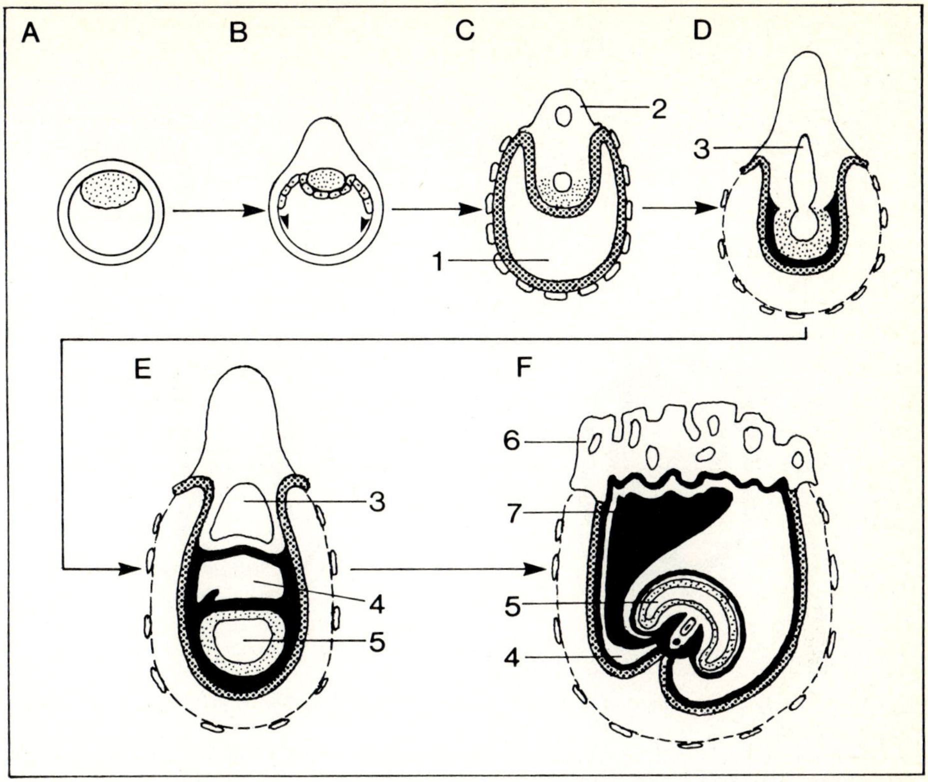

Fig. 48A–F. Early development of mammals: Muridae. (After Grosser, Hertig, Rock and Starck). *Shading* as in Fig. 47.
1 = yolk sac; 2 = carrier; 3 = cavity of the carrier; 4 = exocoel; 5 = amniotic cavity; 6 = placental region; 7 = allantoic mesenchyme

nourishment since blood vessels quickly develop within the mesenchyme. The carrier becomes the region where placentation takes place. A sheet of extraembryonic mesenchyme develops, as it does in other mammals, and extends over the inner suface of the exocoel to cover the region of placentation. Although this layer has no associated diverticulum of the archenteron, it is nevertheless interpreted as an allantois because it is vascularised and forms a connection between the embryo and the placenta. The carrier becomes flattened where it contacts the uterine epithelium and the carrier cavity divides up into a number of small cavities (Fig. 48E,F).

The fifth and final variation occurs in primates, including man. It involves not only the formation of a yolk sac from a compact clump of cells by dehiscence, as in the hedgehog, but also the early formation of large amounts of extraembryonic mesenchyme. Initially, the mesenchyme encloses a primary exocoel but in the course of development cavities arise within the mesenchyme and merge together to form a secondary exocoel that surrounds the yolk sac and amnion (whose cavity represents the lumen of the embryocyst). The primary exocoel gradually becomes smaller and small-

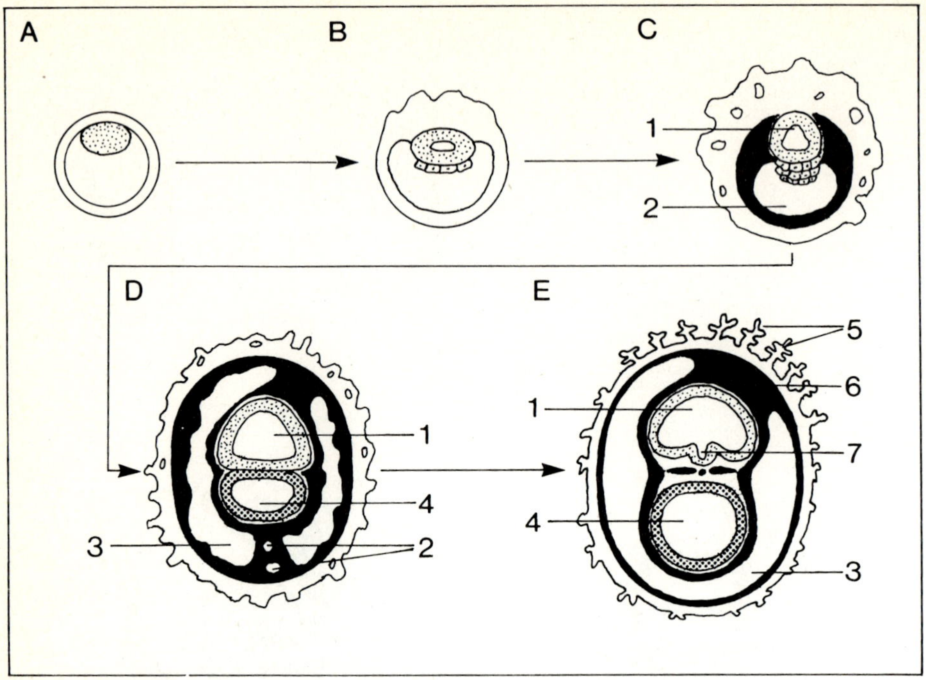

Fig. 49A–E. Early development of mammals: primates. (After Grosser, Hertig, Rock and Starck). *Shading* as in Fig. 47.
1 = amniotic cavity; 2 = primary exocoel; 3 = secondary exocoel; 4 = yolk sac; 5 = chorionic villi in the placental region; 6 = body stalk; 7 = neural groove

er as the secondary exocoel enlarges until the amnion and yolk sac are connected to the wall of the trophoblast only by a stalk, the so-called body stalk. At this stage the secondary exocoel is referred to as the chorionic cavity. The body stalk joins the embryo to the placental region of the trophoblast in which the chorionic villi are most well-developed. The epiblast forming the floor of the amniotic cavity gives rise to the primitive streak, primitive groove, node of Hensen and, later, the chorda canal. Unlike the situation in Muridae, a rudiment of the allantois extends into the body stalk (Fig. 49D,E). Later, when the embryo and its enveloping amnion come to fill the chorion cavity, what remains of the yolk sak is also incorporated into the body stalk which eventually becomes the umbilical cord. Stages in the development of a mammalian embryo up to the formation of a chorda canal are illustrated in Fig. 50, which depicts the development of the human embryo.

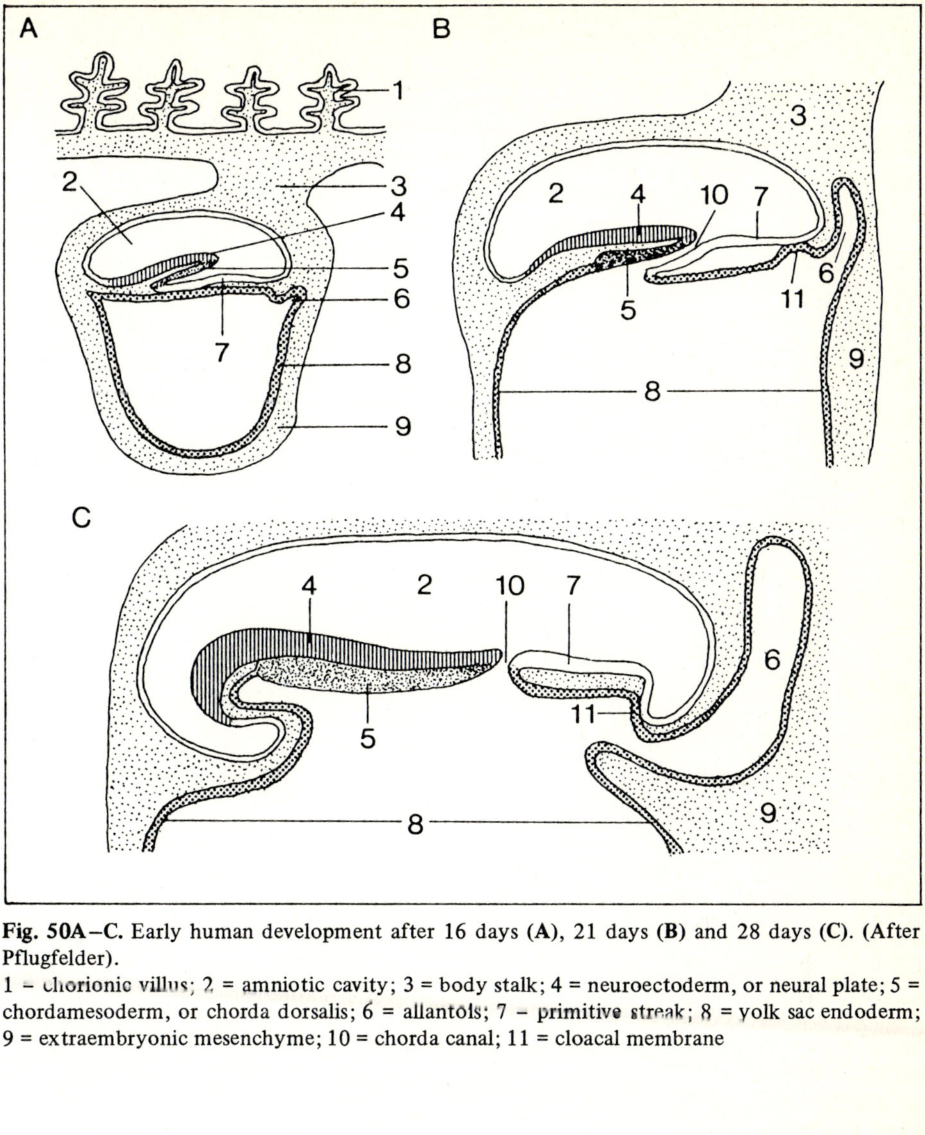

Fig. 50A–C. Early human development after 16 days (**A**), 21 days (**B**) and 28 days (**C**). (After Pflugfelder).

1 = chorionic villus; 2 = amniotic cavity; 3 = body stalk; 4 = neuroectoderm, or neural plate; 5 = chordamesoderm, or chorda dorsalis; 6 = allantois; 7 = primitive streak; 8 = yolk sac endoderm; 9 = extraembryonic mesenchyme; 10 = chorda canal; 11 = cloacal membrane

CHAPTER 5

The Regulation of Reproduction

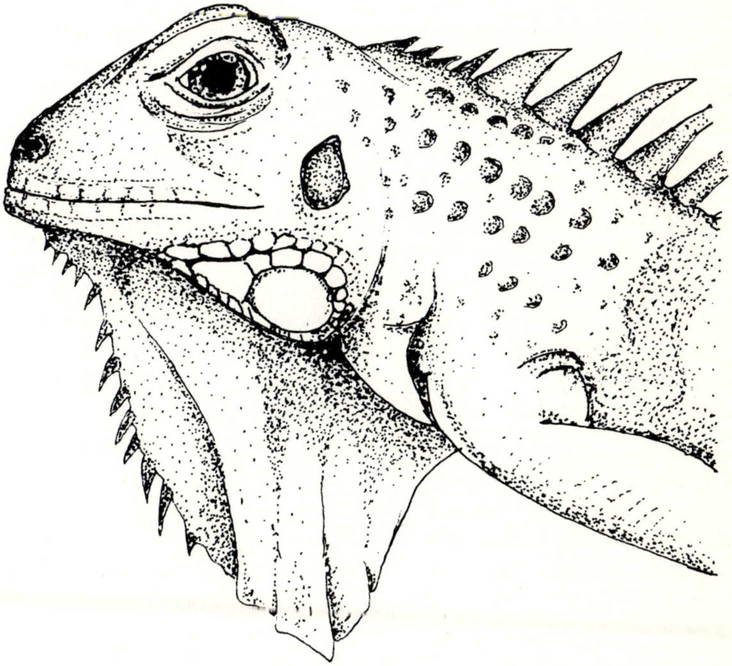

A male green iguana *(Iguana iguana)* with throat appendage

5.1 Characteristics of the Integrative Systems

From the foregoing chapters it will be apparent that reproduction is an extremely complex physiological phenomenon which includes the development and maturation of the gonads, the production of functional gametes, the development of the reproductive ducts and the maintenance of their physiological condition, the development of primary and secondary sex characteristics, reproductive behaviour, and associated phenomenon such as the production of attractants or scents and the nourishment of the body in order that the parent animal be able to care for its young. All these components not only have their own strictly defined form but must also be precisely coordinated which each other in space and time. This is achieved by two integrative systems, the nervous system and the endocrine system. The two systems function, each in accordance with its specific characteristics, in part upon identical, in part upon quite different physiological principles. Both systems convey signals from one part of the body to another where the signals elicit a reaction in a target organ.

The nervous system transmits information by means of chains of nerve cells, the path over which information is transmitted being structurally defined by the physical pattern of interconnection. This chain stretches from a receptor organ to a target organ between which several levels of integration may be interposed. Within the nerve cells information is transmitted as electrical energy whereas a biologically active chemical (neurotransmitter) usually carries the signal from one nerve cell to the next and finally to the target organ. Because the routes over which information is conveyed are fixed and the electrical impulses are conducted along the nerves at great speed, the nervous system operates exceedingly quickly and with great specificity. Accordingly, it is found to control rapid processes along predetermined paths. The endocrine system functions according to a different principle. Certain structures produce biologically active substances, hormones, and release them directly into the circulation whereby they are carried to all parts of the body and trigger a physiological response in certain target organs. The hormone thus appears to act as a chemical messenger carrying information from the structure that produced it, via the blood to the target organ. This implies that the hormone molecule has an information content, which it does not in fact possess in the usual sense of the term. It merely conveys the information that certain physiological processes have been initiated and continue to operate. The qualitative information concerning the nature of these processes lies in the cells within the target organ which have, for example, a specific complement of enzymes. However, a precondition for this model of action is that the cells of the specific target organ "recognise" the hormone and can "understand" its message. This precondition is met by the existence of hormone receptors, these being proteins located either on the outside of the cell membrane or in the cell cytoplasm which bind a specific hormone and in one way or another initiate a physiological process. A certain number of receptors must be "occupied" for the process to be initiated, just as they must for the process to continue. This means that an adequate level of hormone must be built up in the blood, against its continual degradation in the liver and kidneys, and sustained over what is usually a long period of time. The establishment of this level takes a relatively long time and correspondingly the physiological processes which are initiated in this way normally continue for a long time. Thus the characteristics of the endocrine system are well adapted to the regulation of long-term processes within the animal.

Despite their differences the two systems are fundamentally the same in one respect: they act upon the target organ by the release of a biologically active chemical, whereby the endocrine system makes a "detour" via the circulatory system whereas the nervous system goes all the way to the "door" of the target organ and releases the active substance "personally", so to speak. That electrical energy is involved in the transmission of information in the latter case does not alter the fact that the final link to the target organ is chemical in nature.

Hormones play an essential role in the regulation and integration of the complex physiological processes of reproduction. Vertebrates have acquired specialised systems in the course of their evolution whose sole function is in the service of reproduction: other systems merely collaborate in this or are appropriated for this purpose via the co-operative agency of the nervous and endocrine systems.

5.2 Types of Hormone and Their Sites of Production

Three types of hormone can be distinguished in vertebrates. The first type, neurohormones, represent a link between the nervous system and the endocrine system. They are formed in modified nerve cells called neurosecretory neurones which receive signals from "normal" nerve cells. The neurohormone is produced within the cell body and packaged into membrane-bound granules approximately 150 nm in diameter (type A granules) which are transported into the axonal process of the cell. The neurohormone is stored in the axon terminals. The latter form specialised synapses on a blood vessel into which the neurohormone is released upon nervous command. The unit formed by the nerve terminal, or terminals, and a blood vessel is called a neurohaemal organ (Fig. 51A). Neurosecretory neurones are, phylogenetically speaking, the oldest sites of hormone production and they occur in even very "simple" invertebrates. The neurosecretory neurone therefore represents an ideal link between the nervous and endocrine systems, a position which it in fact occupies.

The second type of hormone are the tissue hormones. These are produced by cells which lie in a tissue that has a different primary function (e.g. hormone-producing cells in the mucous membrane of the gut) and are released directly into the blood. These hormones have no direct involvement in the regulation of reproduction.

The third type of hormone is referred to as a glandular hormone and is produced in a specialised endocrine gland. Such glands are typically composed of sheets or strings of cells arranged like an epithelium around a blood vessel. The hormone is produced within the cells, stored in the form of granules, and released directly into the blood upon appropriate stimulation of the cells (Fig. 51B). The thyroid gland is the only exception to the rule that the hormone is produced and stored within the cells of the gland. This gland is composed of follicles which are surrounded by a simple

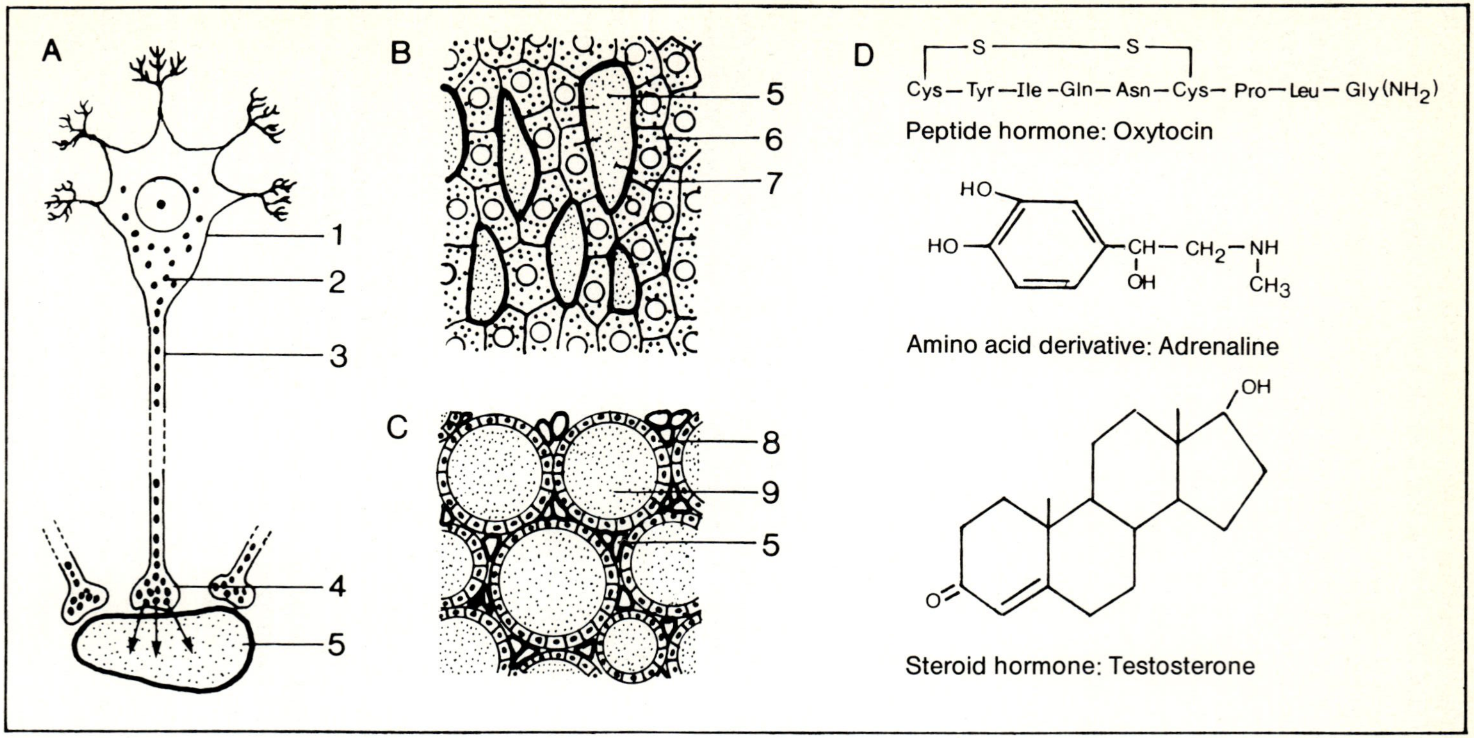

Fig. 51A–D. Types of hormone-producing structures and types of hormone. Basic structure of a neurohaemal organ (**A**), an endocrine gland (**B**), the thyroid gland (**C**), and three classes of hormone (**D**).
1 = cell body of a neurosecretory neurone; 2 = granules of neurohormone; 3 = axon of the neurosecretory neurone; 4 = terminal bulb containing granules of neurohormone; 5 = blood vessel; 6 = endocrine gland cell; 7 = granules of hormone; 8 = follicle of the thyroid gland; 9 = colloid

epithelium and filled with colloid. It is in the colloid that the actual production and storage of the thyroid hormones occurs (Fig. 51C). The thyroid is only indirectly involved in the regulation of reproduction but is of great relevance in this respect in its role a synergist of metabolism in general.

Hormones are chemically heterogeneous, with those of vertebrates falling into three main groups: the polypeptide or proteohormones, amino acid derivatives and the steroid hormones. The first group consists of polypeptides or relatively small proteins with a maximum molecular weight of about 40,000 Dalton and in some cases possessing a carbohydrate moiety. Some of the most important reproductive hormones belong to this group. The second group derive from amino acids and accordingly have a low molecular weight: adrenalin and the thyroid hormones are examples of such hormones, although these are only of indirect relevance to reproduction. The steroid hormones all have a characteristic chemical structure which derives from the saturated tetracyclic form of cyclopentanoperhydrophenanthrene (or gonan). The male and female sex hormones are, without exception, steroids. An example from each of the three groups of hormone is illustrated in Fig. 51D.

5.3 Molecular Mechanisms of Hormone Action

As mentioned at the beginning of this chapter, the cells of a target organ capture or bind molecules of hormone by means of specific receptor molecules and this leads to the initiation of a particular physiological process. This never happens directly. In those cases where the receptors are bound to the cell membrane, cyclic 3′5′ adenosine monophosphate (cAMP) plays an important role as an internal transmitter between the receptors and the metabolism of the cell. Thus the hormone is first bound to the outside of the cell membrane by a specific receptor according to the so-called "lock and key principle". The enzyme adenylate cyclase on the inside of the membrane is then activated via a transducer and a protein, which binds cyclic guanosine monophosphate. Adenylate cyclase catalyses the conversion of adenosine triphosphate (ATP) into cAMP and pyrophosphate. The cAMP then activates certain enzymes, primarily protein kinases, and thereby specific processes within the cell. Thus cAMP functions as a second messenger mechanism. An overview of the processes associated with the formation of cAMP are shown schematically in Fig. 52A. Calcium ions frequently play an important role in the functioning of cAMP, and in this context a recently identified substance called calmodulin has been found to be of increasing significance. The second messenger mechanism is engaged primarily by proteohormones and hormones derived from amino acids. It is perhaps surprising that different hormones can, via such a simple messenger mechanism, induce very different effects. However, two factors reveal how this comes about. Firstly, the hormone receptors in the membrane are highly selective and each reacts with only one particular hormone and secondly, the physiological response of the cell is determined by the specific complement of enzymes it contains. A phosphodiesterase rapidly inactivates cAMP by converting it into 5′ AMP. Receptors also exist which, when occupied by a hormone molecule, lead to the inhibition of cAMP production.

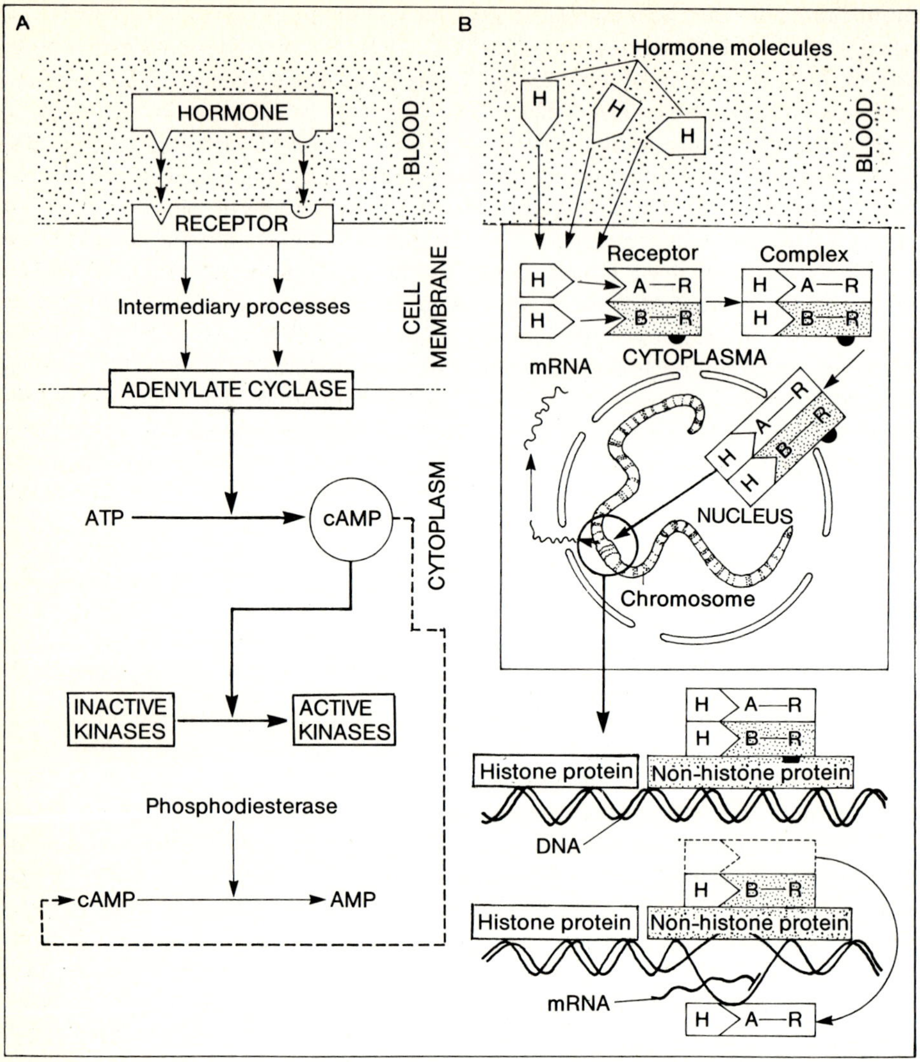

Fig. 52A,B. Molecular mechanisms of hormone action. **A** The "second messenger" mechanism. **B** The gene activation mechanism.
H = hormone molecule; A–R = the A subunit of the cytosolic receptor; B–R = the B subunit of the cytosolic receptor

Steroid hormones engage a different molecular mechansim. The cells of the target organ have receptor molecules within their cytoplasm which are composed of two subunits (A and B). Although the subunits do not differ physically, they bind differently to the chromatin of the chromosomes in the cell nucleus. The hormone is believed to diffuse through the cell membrane, which is probable owing to the lipophilic nature of steroid hormones, and into the cytoplasm where two molecules of hormone bind to each receptor molecule, one on each subunit. It is supposed that the bonding of the hormone molecules to the subunits of the receptor activates the receptor in some way. This hormone-receptor complex then diffuses through the pores in the nuclear membrane into the nucleus and reacts with the proteins associated with the chromosomes. These proteins are of two main types: alkaline histones and non-histone proteins. The hormone-receptor complex reacts with a particular fraction of the latter type of protein. The hormone-receptor complex binds to the non-histone protein via the B subunit and in doing so subunit A, together with its associated hormone molecule, dissociates itself from the complex and goes through some unknown sequence of events whereby it becomes associated with the chromosomal DNA. This results in the "activation" of a section of the DNA and the initiation of messenger RNA synthesis, probably by the "binding" of RNA polymerase to certain initiation sites along DNA where gene transcription begins. A change in the number of initiation sites per cell results in a like change in the number of hormone-receptor complexes present. The mRNA synthesised leads finally to the biosynthesis of a protein (e.g. an enzyme) on the ribosomes in the cytoplasm of the cell. A schematic representation of hormonal gene activation is shown in Fig. 52B. Certain details of this mechanism are still unclear and it should therefore be regarded as hypothetical. New results indicate that the hormone-receptor complex also may interact directly with the DNA.

5.4 The Neuroendocrine Reflex Arc: Interconnections Between the Two Systems

As mentioned, the nervous and endocrine systems regulate and co-ordinate the complex physiological processes of reproduction, each according to its own specific characteristics. The neuroendocrine reflex arc represents one aspect of the co-operation between these two systems. In general, this involves the reception of a stimulus (long-lasting or otherwise) by the nervous system which then directly or indirectly stimulates the endocrine system. The latter responds by producing a hormone which circulates in the blood and elicits a physiological response in a target organ (Fig. 53). Such a mechanism underlies the maturation of the gonads in birds in response to increasing day length and the discharge of milk from the alveolae into the ducts of the mammary gland in response to the suckling of the young. The different characteristics of the nervous and endocrine systems and the part each system plays in a neuroendocrine reflex arc are particularly well illustrated in the case of ovulation in the rabbit. Mature follicles in the female rabbit do not rupture as a consequence of an endogenous mechanism but as a result of the stimulation caused by the male's penis being introduced into the female's vagina during copulation. This act stimulates receptors in the wall of the vagina which then stimulate an afferent nerve fibre that conveys the stimulus via the spinal cord to the brain where the release of neurohormone from certain neurosecretory neurones in the hypothalamus is triggered. In this case the hormone is a so-called releasing hormone that specifically promotes the release of luteininsing hormone (LH) from the adenohypophysis of the pituitary. The blood stream then carries the LH to the ovaries where it causes the mature follicles to rupture, setting free oocytes which can then be fertilised in the tuba uterina (Fig. 53). The neural

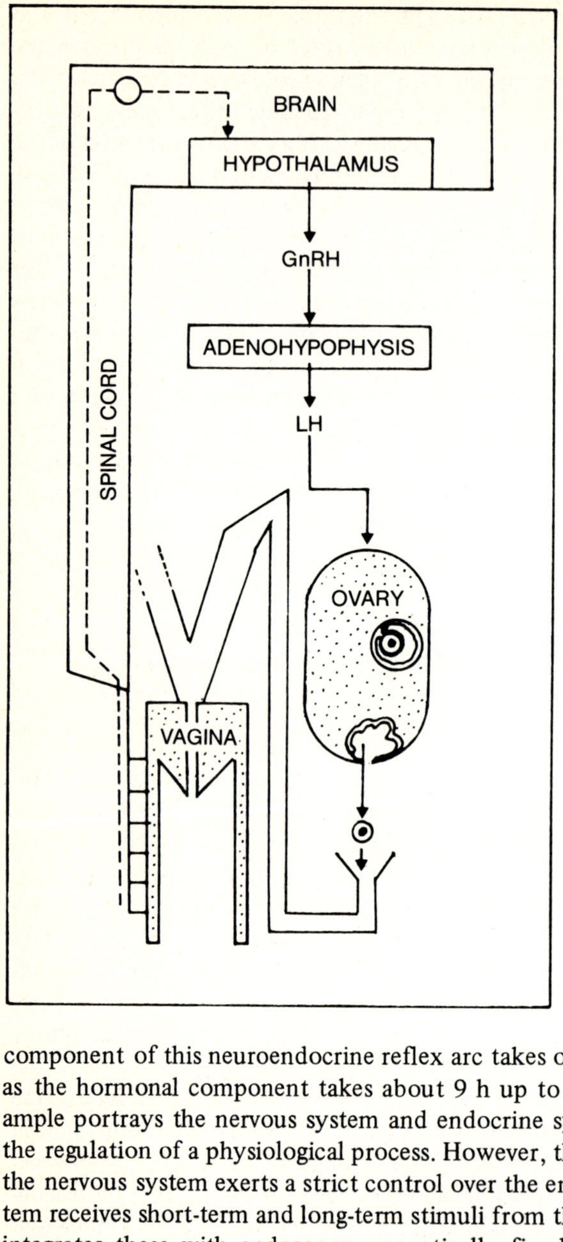

Fig. 53. Schematic representation of the neuroendocrine reflex arc responsible for ovulation in the rabbit

component of this neuroendocrine reflex arc takes only a fraction of a second whereas the hormonal component takes about 9 h up to the point of ovulation. This example portrays the nervous system and endocrine system acting as equal partners in the regulation of a physiological process. However, this equality disguises the fact that the nervous system exerts a strict control over the endocrine system. The nervous system receives short-term and long-term stimuli from the environment and in some cases integrates these with endogenous, genetically fixed rhythms. The hypothalamus in the mid-brain is the actual "command centre" where nervous commands are passed on to neurosecretory neurones which act as mediators between the nervous and endocrine systems, as has already been mentioned. The neurosecretory neurones produce either releasing hormones (RH) or inhibiting hormones (IH) that are released into the blood and influence the release of hormone from a first order endocrine gland, either promoting or inhibiting it, respectively. The adenohypophysis mentioned above is one

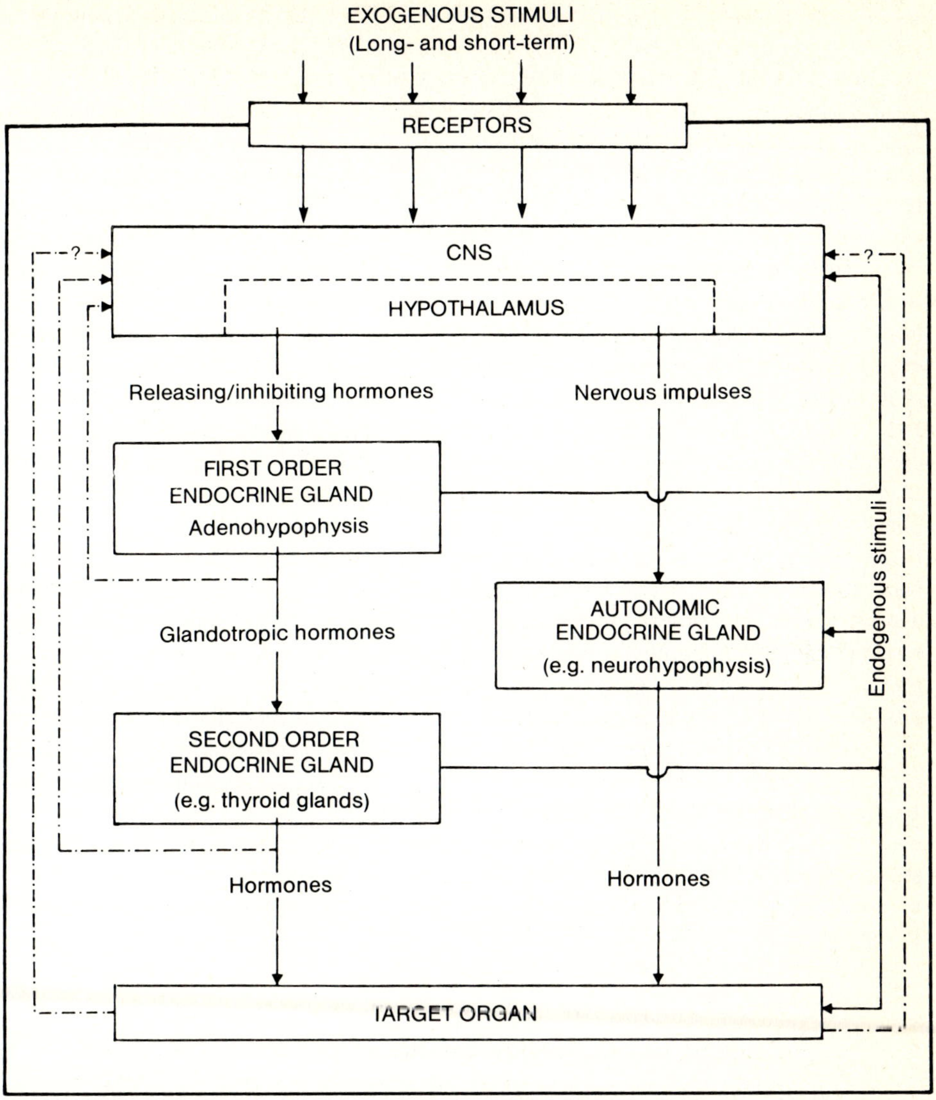

Fig. 54. Scheme showing the organisation of the vertebrate hormone system

example of a first order endocrine gland and for every hormone produced in it there is a specific RH and/or IH. The first order endocrine glands then release hormones into the blood that promote the production of hormones in a second order endocrine gland: the former are referred to as glandotropic or endocrinokinetic hormones. In their turn the hormones of the second order endocrine glands, the gonadal hormones for example, circulate in the blood and influence specific target organs. First order endocrine glands may also produce hormones that act directly upon a target organ. In addition, there are so-called autonomous endocrine glands which are not under the control of releasing or inhibiting hormones or glandotropic hormones. Such glands are, however, probably stimulated or inhibited by the products of other endocrine glands or of the target organ. Endogenous stimuli of various kinds can act as modulators of activity at all levels within the hierarchy of hormone-producing tissues.

The chain of command just described is an open one, meaning that commands flow in only one direction, from top to bottom. Such a system is not capable of controlling one or more physiological parameters so that homoeostasis is achieved, this requiring an antagonistic system. There are only a few instances of such systems in vertebrates, namely in the regulation of the blood sugar level and calcium homoeostasis. In such cases one speaks of a bipolar control. However, a self-regulating system requires information concerning the deviation of the parameter under control (e.g. the level of a target organ product in the blood) from a certain "normal" or set value. For example, if the level of an adenohypophyseal hormone is below the "required" level, more must be released and this is achieved by releasing more of the corresponding RH (or less of the IH) from the hypothalamus. This requires that the level of the adenohypophyseal hormone is "measured" and that this information is fed back to the central nervous system which then initiates a counteracting response. This phenomenon is known as negative feedback. Negative feedback loops can operate between any of the levels of the hierarchy outlined above. For example, they have been shown to exist between the hypothalamus and endocrine glands of the first and of the second order, these being referred to as short and long feedback loops respectively. The hierarchy is made up of individual components, each of which represents a closed control loop, ordered one above the other and interacting one with another. Regulation within such a cybernetic system can logically only be achieved by negative feedback loops. Strictly speaking, positive feedback loops, such as have often been described in the endocrinological literature, are not viable since they lead to the self-destruction of the system: the chain reaction leading to the explosion of an atom bomb is an example of positive feedback. Such phenomena usually involve stimulation or the so-called positive phase of a system. The organisation of the paths of interaction between the nervous system and the endocrine system is summarised in Fig. 54.

5.5 The Release of Hormones into the Blood

Hormones are released from their site of production in two different ways. Firstly, they may be released continuously over a long period of time. A certain level of hormone in the blood is thus established, against the biological degradation of the hormone, which is then maintained at a set level by the regulatory system described above. The second mode of release is either in a single pulse or in a series of pulses that occur at more or less regular intervals, this being called pulsatory or episodic release. The interval between pulses can be anything from a few minutes to about 8 h. The "profile" of a hormone, which graphs the change in its concentration in the blood with time, resembles the teeth of a saw in some cases. Each peak corresponds to the hormone concentration directly after a pulse is released. The concentration then sinks, due to dilution of the hormone in the blood and to it being "used up" or degraded, until the concentration reaches a minimum value, at which point the next pulse occurs leading to a sharp increase in the concentration of hormone. Iteration of this process gives rise to an oscillating level of hormone that fluctuates around a mean value. The minimum value reached depends of the speed with which the hormone is eliminated from the blood and the amount of hormone released in each pulse. A change in the mean value is brought about by a change in the pulse frequency. The time required for the hormone concentration in the blood to sink to 50% of its maximum value is called the half-time. Thus, when the interval between pulses is shorter than the half-time the hormone concentration inevitably increases so that after a few pulses a new mean value is established. Contrarily, a pulse frequency longer than the half-time results in a lowering of the mean hormone concentration. The level of hormone in the blood can be adjusted within a certain period of time to meet the prevailing physiological conditions by "resetting" the set value, but how the frequency of hormone release is changed remains largely unknown. An important aspect of the regulation of reproduction in mammals is the pulsatory release of LH from the hypophysis. For example, in sheep whose ovaries have been excised it has been found that during the month of May LH is released in pulses at intervals of from 50–100 min. For each individual animal the interval was more or less constant, each having a characteristic minimum and mean level of LH in the blood. The results of other experiments suggest that the hypothalamic-releasing hormone for LH (LHRH) is also released in pulses. Thus rams taken outside of the breeding season and injected with pulses of LHRH (a 1 min pulse every 2 h) show an increase in the size of their testes which otherwise only occurs during the breeding season.

One effect of the aforementioned resetting of the hormone level is the cyclical release of hormone whereby so-called biological clocks in the central nervous system give rise to neural "programmes" which regulate either the setting of the hormone level in the blood or the release of hormone. Exogenous factors such as day length frequently act as a trigger or a modulating factor in this process. In the case of diurnal or circadian rhythms the pattern of hormonal release repeats itself every day at the same time. For example, in the Japanese quail (*Coturnix coturnix japonica*) gonadotropic hormones that influence the gonads are released from the hypophysis for a period of 4 h every day, beginning 16 h after the day begins. However, this only happens when the animals live under "long day" conditions (20 h of light and 4 h of darkness). A circannual rhythm, in contrast, refers to a cyclical pattern in the changing level of hormone in the blood which repeats itself every year, and these rhythms are of great importance in the regulation of reproduction. A large number of the vertebrates which live in temperate regions are monocyclic and reproduce only once a year at a time when the conditions are optimal for either the birth, development or hatching of the young. In such cases the relevant physiological processes are regulated according to a genetically fixed "programme" of hormone release which may be considerably modified by exogenous factors such as day length and temperature. In the case of polycyclic species, several periods of breeding occur successively in the course of a year followed by a period of rest or, alternatively, the cycle repeats itself uninterruptedly throughout the year. In every cycle the pattern of changing hormone levels is repeated, a classic example of which is the menstrual cycle of women that proceeds autonomously according to a programme which is largely endogenous. The regulation of cyclical reproduction will be dealt with further in a later section (see p. 189).

5.6 Hormone Systems Involved in Reproduction

The hormone systems directly involved with reproduction are, for the most part, components of the so-called hypothalamic-hypophyseal-gonadal axis. Thus gonadotropin-releasing hormones (GRH) are released from neurosecretory neurones in the hypothalamus as a result of endogenous rhythms or an exogenous trigger. In the adenohypophysis the GRH causes the release of hormones which constitute the components of the gonadotropin system. In tetrapods, with the exception of some squamate reptiles, this system has two components, follicle-stimulating hormone (FSH) and luteininsing hormone (LH), which affect the gonads. The gonads respond in two different ways, namely by producing functional gametes and by producing gonadal hormones: the ovaries produce the female sex hormones (oestrogen and progesterone, or gestagen), the testes the male sex hormones (androgens). These hormones in turn influence a range of phenomena: the development of primary and secondary sex characteristics, reproductive behaviour, the maintenance of functional wolffian and müllerian ducts, pregnancy, the production of scents and of attractants and the nutrition of the parent so that it can provide for the offspring. Another hormone of the adenohypophysis, prolactin, is of particular relevance in the context of parental care: it has sometimes been regarded as a gonadotropin although strictly speaking it is not. The hormones stored in the neural part of the hypophysis include oxytocin, or an analogue thereof, which is thought to be involved in triggering birth contractions in mammals and the release of gametes in vertebrates in general. In mammals, additional hormones are produced by the maternal and foetal elements of the placenta whose functions are related to gestation. The different hormones involved in reproduction and their most important regulatory functions are summarised in Fig. 55.

Most of the other hormone systems in vertebrates are also involved, indirectly, with the events of reproduction. Thus the steroidogenic component of the interrenal tissue, referred to as the adrenal glands in mammals, produces steroid hormones that play an important role in the metabolism of carbohydrates and minerals. Steroid sex hormones are also synthesised by the interrenal tissue. This tissue is, in part, under the control of adrenocorticotropic hormone (ACTH) form the adenohypophysis. The thyroid glands are under the control of thyroid-stimulating hormone (TSH = thyrotropin) and produce the hormones triiodothyronine (T_3) and thyroxine (T_4, T_x) which are mainly involved in the processes of growth and differentiation as well as metabolism in general. The participation of the thyroid glands in the regulation of gonadal function has long been a matter of debate for although there is evidence indicating that they are in fact involved, no-one has yet been able to clearly define or demonstrate such an action. The parathyroid glands and the ultimobranchial bodies act antagonistically to maintain calcium and phosphate homoeostasis. The hormones insulin and glucagon from the endocrine component of the pancreas, the Islets of Langerhans, serve to keep the level of glucose in the blood constant. The hormone adrenalin from the catecholaminergic tissue of the adrenal gland (the adrenal medulla in mammals) also plays a role in this. Adrenalin is further involved in "emergency situations", causing the rapid "activations" of the circulation and mobilisation of carbohydrate reserves. The gastrointestinal hormones are tissue hormones of the gut

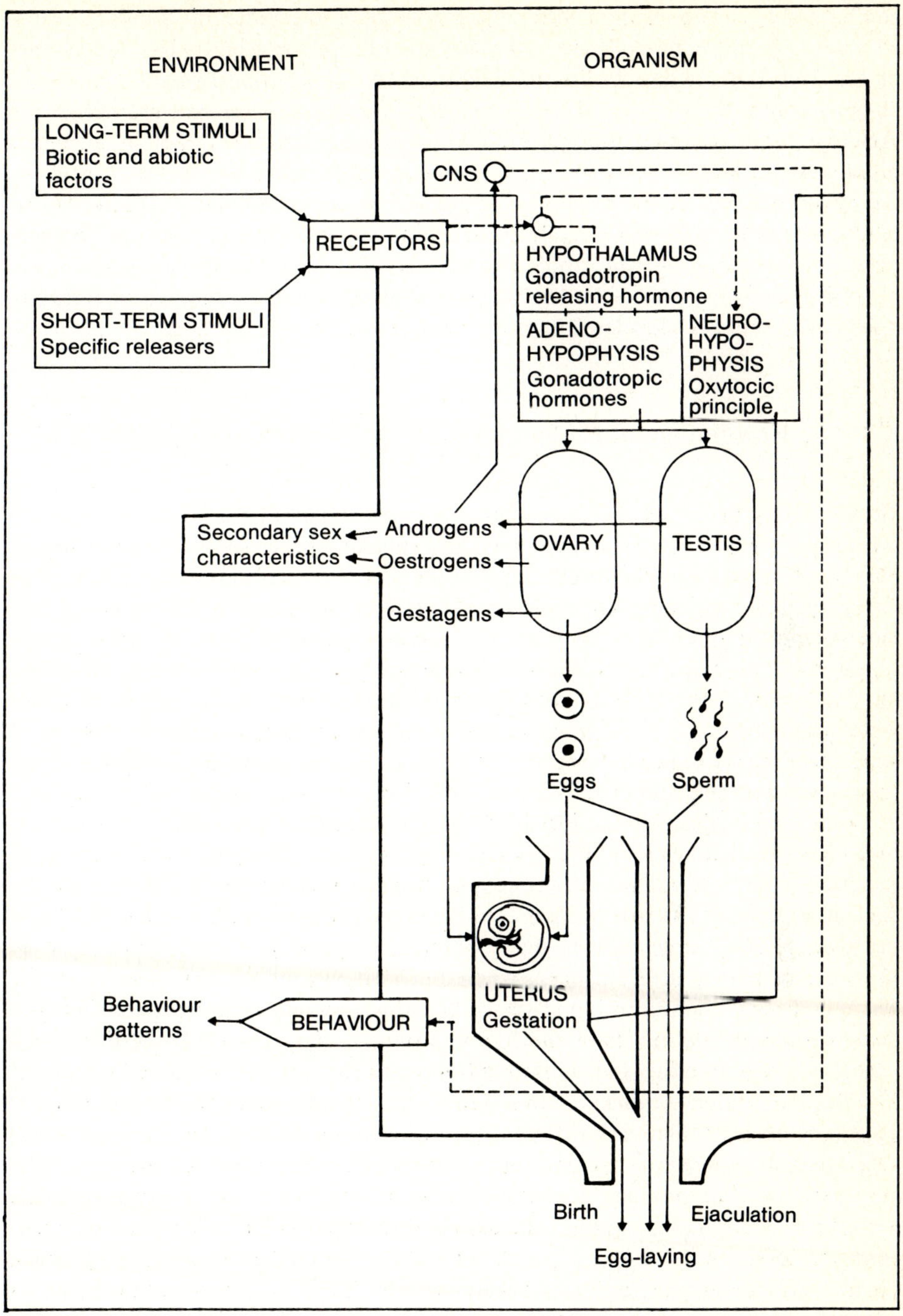

Fig. 55. Schematic representation of the hormone systems associated with reproduction in vertebrates

and have the least connection with reproduction of all the endocrine glands mentioned since they regulate exclusively the activity of digestive glands. The adenohypophysis also produces growth hormone (also called somatotropin) which, alone or in co-operation with the thyroid hormones, regulates metabolism and growth of the skeleton. It also plays an important role in the complex processes whereby mammals produce milk. Melanocyte-stimulating hormone (MSH) from the pars intermedia of the hypophysis causes the dispersion of pigment in the melanophores of lower vertebrates. The last source of hormone to be mentioned is the epiphysis which, in mammals, can exert an antigonadotropic influence on the gonads but also a progonadotropic influence in the case of birds. The physiological role of the epiphysis is not fully known: this organ is considered in more detail in a subsequent section.

5.7 The Morphology of Endocrine Systems Involved in Reproduction

The hypothalamus is a part of the unpaired diencephalon that lies behind the paired fore-brain (telencephalon). Behind the diencephalon is the mid-brain (mesencephalon) which is only paired dorsally. The hind-brain (metencephalon and myelencephalon) is unpaired and merges into the spinal column (Fig. 56A). The diencephalon contains a cavity referred to as the third ventricle that is bounded by three areas: dorsally, the roof is formed by the epithalamus, lateral to the third ventricle are the thalami, and ventral is the bowl-shaped hypothalamus. The most ventral part of the latter is the infundibulum, in front of which (i.e. rostral) is the median eminence: both these structures play an important part in the formation of the neurohypophysis. The hypothalamus is not only the neural "conductor of the endocrine orchestra" but is also a two-fold source of hormones. Firstly, it contains collections or nuclei of neurosecretory neurones that belong to the so-called magnocellular system. In lower vertebrates the most significant nucleus is the preoptic whereas in Amniota it is the supraoptic nucleus and paraventricular nucleus. The hormone-producing cells of these nuclei give rise to axons that unite to form the hypothalamo-hypophyseal tract which extends to the neural part of the hypophysis where the hormones are stored in Herring bodies. These bodies junction onto blood vessels thereby forming a neurohaemal organ. The second source of hormones in the hypothalamus is the parvicellular system which synthesises the releasing and inhibiting hormones that affect the adenohypophysis. The neurosecretory neurones of this system are not aggregated together in discrete ganglionic nuclei, rather each different type of neurone is localised within a particular area which may, however, overlap that of another neurone type. All the hypothalamic neurones that produce a neurohormone are innervated by catecholaminergic nerve fibres from higher centres. The axons of the parvicellular system end on a network of blood vessels (the primary plexus) within the median eminence (Fig. 56B). These vessels are supplied by the hypophyseal artery. The venous side of the plexus is connected to a system of portal veins in the adenohypophysis so that the releasing and inhibitory hormones released into the primary plexus are rapidly conveyed directly to the adenohypophysis. Such a venous network of capillaries exists in all vertebrates except the Cyclostomata and the Teleostei.

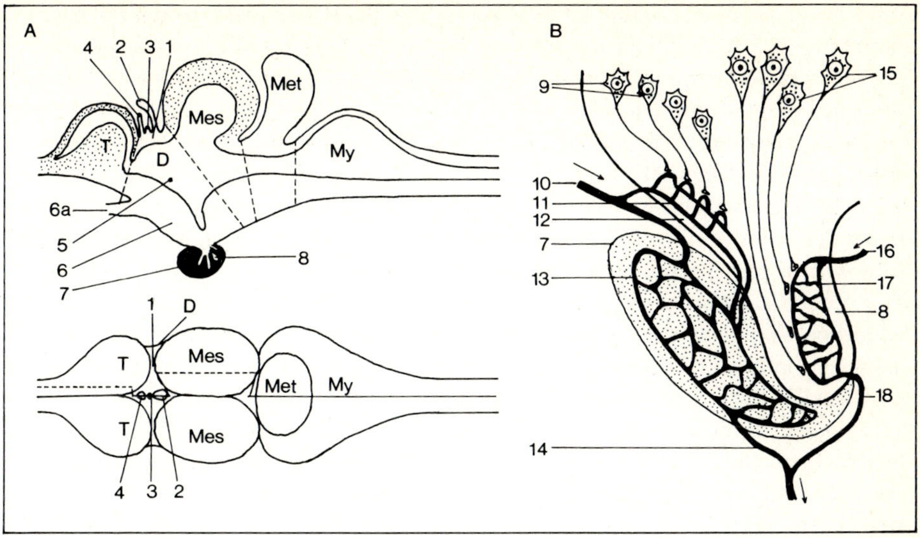

Fig. 56A,B. Diagram of the vertebrate brain and of the hypophysis of tetrapods showing its blood supply. **A** Brain of a teleost: *upper figure* shows a longitudinal section cutting the brain medially (*unshaded area*) and paramedially (*shaded area*); *lower figure* shows the brain as seen from above; *continuous line* indicates the medial plane, the *broken line* the paramedial plane (but only dorsally in the region of the mesencephalon). **B** Vascularisation of the median eminence and hypophysis of mammals.
T = telencephalon; D = diencephalon; Mes = mesencephalon; Met = metencephalon; My = myelencephalon.
1 = epithalamus; 2 = epiphysis (pineal organ); 3 = parietal organ; 4 = paraphysis; 5 = thalamus; 6 = hypothalamus; 6a = optic chiasma; 7 = adenohypophysis; 8 = neurohypophysis; 9 = neurosecretory neurones of the parvicellular system; 10 = hypophyseal artery; 11 = primary plexus; 12 = eminentia mediana; 13 = portal system of the adenohypophysis; 14 = adenohypophyseal vein; 15 = neurosecretory neurones of the magnocellular system; 16 = neurohypophyseal artery; 17 = capillary network of the neurohypophysis; 18 = venous return from the neurohypophysis

The roof of the mid-brain has a complex structure and is also a source of hormones. Three outgrowths develop from the epithalamus: the paraphysis is situated cranially, in front of the parietal organ and the caudally situated epiphysis, or pineal organ (Fig. 56A). The latter organ is laid down early in development in all vertebrates and in most cases is also present in the adult animal. It is an endocrine gland. Most vertebrates do not have a parietal organ, it being present only in a few reptiles. The paraphysis is supposedly a part of the telencephalon. It is connected to the paraventricular nucleus by neurosecretory fibres, but its function is unknown. The pineal and parietal organs are rudimentary sense organs for the reception of light. Photoreceptors can be found in the epiphysis of all vertebrates up to the phylogenetic level of the reptiles.

The hypophysis is a neuroepithelial organ of which the neural part is formed from the infundibular region of the hypothalamus during ontogenesis. The epithelial part stems from the ectodermal roof of the mouth as an outgrowth which is either hollow (called Rathke's pouch: as in Elasmobranchii, some primitive fish, and Amniota) or solid (as in Amphibia and most fish) and grows dorsally to make contact with the infundibulum. The epithelial part usually separates from the oral ectoderm at a later stage of development to form the adenohypophysis, the associated infundibular part forming the neurohypophysis.

The hypophysis of fishes other than the Myxinoidea and Dipnoi has a largely uniform structure. The neurohypophysis is made up of the infundibulum and the median eminence which lies rostral to the latter. The infundibulum gives rise to more or less well-developed finger-like processes that extend into and establish a close association with a part of the adenohypophysis called the pars intermedia: this region is therefore also referred to as the neurointermediate lobe. In many cases the median eminence also gives rise to processes that penetrate into the adenohypophysis. The adenohypophysis is typically composed of two parts, the pars intermedia and the pars distalis. Each type of adenohypophyseal hormone is produced by a different cell type, and in fishes' the cells of each type are often situated together in groups resembling an epithelium. Thus the rostral part of the adenohypophysis contains primarily the cells which produce prolactin (PRL), adrenocorticotropic hormone (ACTH) and thyroid-stimulating hormone (TSH = thyrotropic hormone) whereas the cells producing gonadotropic hormones lie between this part and the pars intermedia. The pars distalis of fishes is therefore usually divided into a rostral and a proximal part (Fig. 57A). The pars intermedia of all vertebrates produces solely the melanocyte-stimulating hormone (MSH). In elasmobranchs the adenohypophysis has, in addition, a pars ventralis in which the cells producing the gonadotropins are located, although the adenohypophyseal portal system in this group of fishes does not extend to this part of the hypophysis.

The hypophysis of tetrapods differs from that of fishes in two respects. Firstly, there is an additional lobe to the adenohypophysis, the pars infundibularis (= pars tuberalis) and secondly, a large part of the neurohypophysis is formed by a so-called pars nervosa that arises from the caudal region of the infundibulum and which has its own blood supply (Fig. 56B). The typical tetrapod hypophysis therefore consists of the following elements. A neurohypophysis composed of the median eminence, infundibulum (hypophyseal stalk) and pars nervosa into which the third ventricle projects (as the hypophyseal recess) to a greater or lesser extent. Partly encasing the cranioventral aspect of the pars nervosa is the pars intermedia, outside of which is the pars distalis. The hypothalamic border of the adenohypophysis is formed by the pars infundibularis which can form a collar around the infundibulum (Fig. 57B). In many species there is a cleft between the pars intermedia and the pars distalis (the hypophyseal cleft) which represents a relict of the lumen of Rathke's pouch. A further difference between the hypophysis of fishes and that of tetrapods is the fact that in the latter the different cell types are not each clustered in particular areas but are all mixed together.

The basic structure of the hypophysis is not invariant however. That of birds and some mammals, whales and elephants for example, has no pars intermedia, and in many reptiles the pars infundibularis is lacking. In most vertebrates the hypophysis lies at the base of the skull in a depression of the sphenoid bone called the sella turcica. Despite its great physiological significance the hypophysis is a relatively small endocrine gland which in man is about as big as a pea and weighs little more than

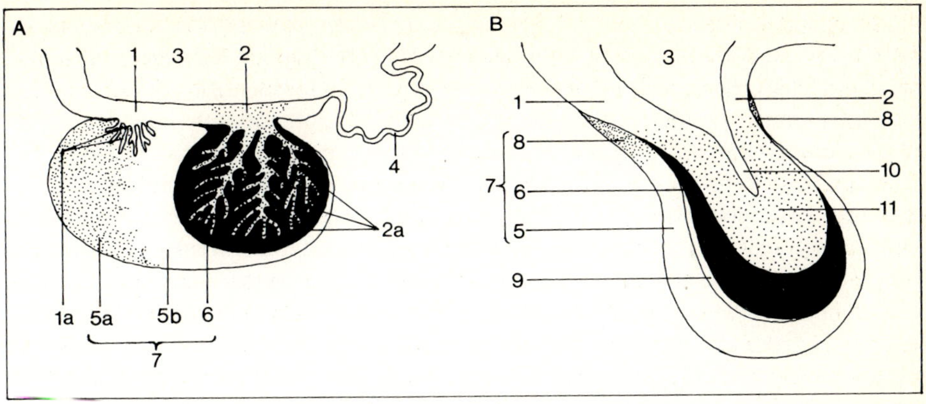

Fig. 57A,B. Diagram of the hypophysis of teleosts (**A**) and tetrapods (**B**).
1 = eminentia mediana; 1a = processes of the eminentia mediana; 2 = infundibulum; 2a = extension of the infundibulum (neurohypophysis); 3 = third ventricle; 4 = saccus vasculosus; 5 = pars distalis; 5a = rostral pars distalis; 5b = proximal pars distalis; 6 = pars intermedia; 7 = adenohypophysis; 8 = pars infundibularis; 9 = hypophyseal fissure; 10 = recessus hypophyseus; 11 = pars nervosa

half a gramm. The adenohypophysis of mammals produces eight hormones and a further two hormones are stored in the neurohypophysis. The term frontal lobe hormones, as they are referred to in some textbooks, derives from a terminology according to which the frontal lobe is made up of the pars infundibularis together with the pars distalis whereas the intermediate lobe represents the pars intermedia. Since the pars infundibularis evidently does not produce hormones, the frontal lobe hormones are those produced in the pars distalis, but do not include MSH produced in the pars intermedia.

It has already been mentioned that the gonads are bifunctional organs responsible for the formation of gametes as well as the production of the male and female gonadal hormones. According to one school of thought, it is the Leydig cells, also called interstitial cells, which are the endocrine structures of the testes of all tetrapods except the Urodela. These cells lie in the interstices between the seminiferous tubules in association with blood vessels and typically contain lipid droplets during periods of heightened physiological activity. Exceptionally, Leydig cells also occur in some fishes and in Petromyzontia, although the hormone-producing structures of most fishes and Urodela are the so-called boundary cells. These cells form a sort of epithelium around the tubules or ampullae of the testes and are postulated to be the homologues of the Leydig cells. The only somatic cells within the lumen of the seminiferous tubules or ampullae are the Sertoli cells and, so far as is known, they probably occur in all vertebrates, either as Sertoli cells sensu stricto or as cyst cells. For a long time they were interpreted as being nurse cells which "administer" to the developing sperm. However, no evidence has yet been presented to support this view: on the contrary, the evidence indicates that they have an endocrine function. The hypothesis that animals can be divided into those with boundary cells and those with Leydig cells has recently been seriously questioned with respect to teleosts. According to light and electron microscopical studies the Leydig cells appear to be the real endocrine components outside the seminiferous tubules of teleosts. The presumptive boundary cells that have been observed in a range of species were evidently misidentified Sertoli cells lying in the lumen of the tubules. According to the hypothesis, true boundary cells, such as those supposedly present in the testes of teleosts, are not endocrine cells. However, these cells do not completely surround the tubules, are completely absent in some regions, and contain myofilaments. They therefore stand better comparison with the myoepithelial cells which lie around the seminiferous tubules of mammals. To what extent these objections also apply to Urodela remains to be seen.

The endocrine structures of the ovary, which have already been mentioned briefly (see p. 47), are more complex than those of the testes. In some cases they form an integral part of the follicle wall, this being particularly well illustrated in the case of the mammalian ovary, which therefore will be now considered in some detail. It is surrounded by a single layer of coelomic epithelium that is covered on its inner surface by a thin layer of connective tissue, the tunica albuginea. The ovarian stroma is divided into a cortical layer and the medulla. The former contains primary follicles and corpora lutea at different stages of development which are surrounded by connective tissue whereas the latter contains blood vessels, lymphatic vessels and nerves (Fig. 58A). First order oocytes are initially invested in a single layer of granulosa cells, forming a primary follicle. As the oocyte grows in size the granulosa becomes multilayered and the developing zona pellucida (the primary egg membrane; see p. 109) comes to separate it from the oocyte. A thin layer of connective tissue, the theca folliculi, also forms around the granulosa, at which stage in development the whole structure is referred to as a secondary follicle. The formation of a tertiary follicle involves firstly, the development of a cavity, the antrum folliculi, filled with liquor folliculi in which the oocyte is suspended by a conical mass of granulosa cells (the cumulus oophorus). Secondly, a portion of the follicle wall differentiates into an endocrine gland. In addition, a basal membrane forms between the granulosa and the theca. The latter becomes bilaminar, the outermost theca externa consisting of connective tissue whereas the innermost theca interna is cellular and becomes richly vascularised (Fig. 58B). The theca interna is the primary source of female sex hormone. At this stage of development the follicle is called a Graafian follicle and when fully mature the wall of the follicle and the overlying tissues rupture and the oocyte is ovulated. The oocyte, surrounded by a corona of granulosa cells (the corona radiata), is then taken up by the oviduct in which it goes through the final stages of maturation division. The empty follicle then undergoes a series of major changes in its morphology and physiology. The ruptured granulosa becomes progressively thicker until its cells almost fill the lumen of the follicle. Blood vessels subsequently grow into the granulosa from the theca interna and the granulosa cells start to produce hormones. These cells possess lipid droplets within their cytoplasm which contain a yellow lipochrome that imparts a yellow colour to the whole structure, hence its name, the corpus luteum (yellow body). The modified granulosa cells, now referred to as granulosa lutein cells, produce a progestin called progesterone. Cells of the theca interna may also form part of the periphery of the corpus luteum, although they continue to synthesise oestrogen. The final phase in the formation of a corpus luteum varies greatly from species to species. In those cases where there is an active luteal phase the corpus

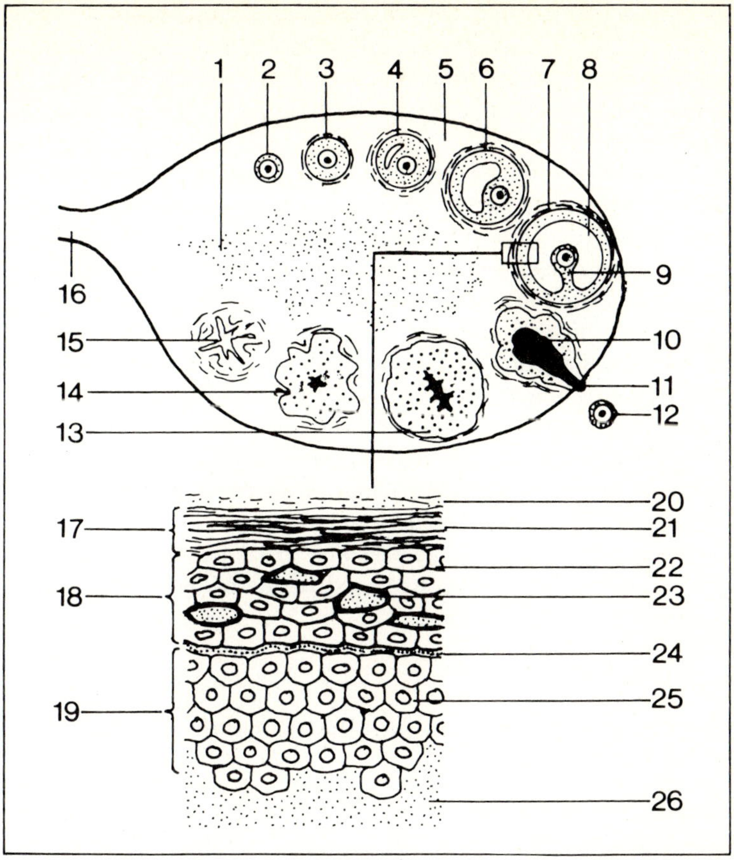

Fig. 58A,B. Structure of the mammalian ovary.
1 = medulla; 2 = primary follicle; 3 = secondary follicle; 4 = intermediate stage between a secondary and tertiary follicle; 5 = cortex; 6 = young tertiary follicle; 7 = Graafian follicle; 8 = antrum folliculi; 9 = cumulus oophorus; 10 = blood filling the lumen of the follicle; 11 = blood spot; 12 = oocyte surrounded by the corona radiata; 13 = corpus luteum; 14 = degenerating corpus luteum; 15 = corpus albicans; 16 = mesovarium; 17 = theca externa; 18 = theca interna; 19 = granulosa; 20 = ovarian stroma; 21 = connective tissue fibres; 22 = endocrine cells of the theca; 23 = capillaries; 24 = basal membrane; 25 = granulosa cells; 26 = liquor folliculi

luteum is formed prior to the beginning of a pregnancy (e.g. the corpus luteum menstruationis of women: see p. 196) whereas in other cases it is formed thereafter. This will be dealt with later in greater detail. Whichever is the case, after a certain length of time the corpus luteum is broken down, the granulosa lutein cells degenerate and there is an ingrowth of connective tissue from the theca. This results in a scar within the ovary called the corpus albicans which is eventually also degraded. The overall structure of the mammalian ovary, the processes of follicle maturation and the formation of the corpus luteum are illustrated schematically in Fig. 58A. By far the largest number of follicles present in a mammalian ovary do not mature but are degraded in a process known as follicular atresia involving the breakdown of the granulosa and the proliferation of theca interna cells, which temporarily form a thecal gland that produces oestrogen. A comparable phenomena also occurs in submammalians, as described earlier (see p. 48), in which case the degenerate follicle is referred to as a corpora atretica or corpora lutea praeovulatoria, although whether if functions as an endocrine organ is doubtful.

Comparatively little is known about the sources of oestrogens and progesterone in submammalians. A theca interna similar to that found in mammals is absent in every case, although cells are present between the connective tissue fibres which are not fibroblasts and which have been called specialised thecal cells. In teleosts such cells have been ascribed the function of producing oestrogen and gestagen but the granulosa cells are also contenders for this role. Corpora lutea occur in all vertebrates except birds and are presumed to be the source of progesterone, although this has never been irrefutably demonstrated: progesterone is found in members of all the vertebrate classes. Evidence shows, for example, that steroid biosynthesis occurs in the corpora lutea of viviparous teleosts and that these structures are responsible for the production of progesterone in reptiles. On the other hand, progesterone is also found in birds, although no corpus luteum is formed in these animals.

In mammals the uterus and both sides of the placenta, foetal and maternal, are important sources of hormones during gestation. The structure of these organs is described in detail in Chapter 8.

5.8 Chemistry and Physiology of Hormones that Regulate Reproduction

Among the hypothalamic neurohormones which function as a releasing hormone (RH) or an inhibitory hormone (IH), the first to be considered here is the releasing hormone for hypophyseal gonadotropins (GRH), so-called gonadoliberin. This is probably a single substance which causes the release of FSH as well as of LH and is, therefore, referred to as FSH-RH, LH-RH, or FSH/LH-RH, or GRH. It is a decapeptide with the following sequence of amino acids:

pyro Glu-His-Try-Ser-Tyr-Gly-Leu-Arg-Pro-Gly NH_2.

There are, it seems, no species specific differences in this molecule, at least with respect to mammals. Gonadoliberins have now been found in most classes of vertebrate. Thus immunological and biochemical studies have shown that the gonadoliberin molecules of mammals and amphibians are very similar, if not identical, whereas those of birds, reptiles, teleosts and Chondrichthyes differ slightly from these, by only two amino acids (at positions seven and eight) in the case of one species of salmon. A synthetic form of GRH is also available which is a potent as the natural substance. The release of gonadotropins in Elasmobranchii, and perhaps in Cyclostomata as well, is evidently not under hypothalamic control. Aside from stimulating the release of gonadotropins from the cells that produce them, GRH also stimulates their synthesis. The half-time of GRH is about 4 min in mammals.

A topic that will be taken up again later in greater detail relates to the fact that the role of FSH and LH as regulators of the reproductive cycle requires that they be released at different times and in different amounts, so that the existence of only a single releasing hormone presents certain conceptual difficulties. However, these difficulties are mitigated by evidence which indicates that the response of the relevant adenohypophyseal cells is subject to certain modulating influences. It must, nevertheless, also be said that there is strong circumstantial evidence for the existence of two separate hypothalamic hormones which regulate the release of FSH and LH respectively. Gonadoliberin also occurs outside the hypothalamus, having been found, for example, in the fore-brain of the rainbow trout (*Salmo gairdneri*) and in the epiphysis and extrahypothalamic regions of the brain of mammals. Nothing is known of the function it may fulfil at such sites: a neurotransmitter role has been mooted.

In mammals the secretion of prolactin from the adenohypophysis is controlled by an inhibitory hormone referred to as prolactostatin or prolactin-inhibiting hormone (PIH). Its precise chemical nature is not yet known but it is probably dopamine. There are also a number of findings indicating the existence of a prolactoliberin, i.e. a prolactin-releasing hormone (PRH), however its status is still uncertain, particularly since the releasing hormone for thyrotrophic hormone (TRH) apparently also causes the release of prolactin. Whether in fact PRH exists in mammals and whether or not TRH plays a role in the secretion of prolactin is far from certain. In contrast to these equivocal results, it is certain that the release of prolactin in birds is stimulated by a hypothalamic prolactoliberin. Contrarily, in amphibians and bony fish the hypothalamus appears to exert an inhibitory influence, although serious doubts have been expressed as to whether this is indeed so in bony fish.

Mention has already been made of one other hypothalamic factor, namely TRH. This is a tripeptide (pyro Glu-His-Pro NH_2) which has been found to be present in species from every class of vertebrate. A somatostatin which inhibits the release of growth hormone has also been isolated from mammals. It is a teradecapeptide but its full structure is not known. The release of growth hormone in all other vertebrates is also subject to some inhibitory influence. A releasing hormone for adrenocorticotropic hormone (CRH) has been postulated to exist in all vertebrates. It appears to be a very unstable polypeptide which probably differs greatly from species to species.

The neurohypophysis of mammals stores two neurohormones from the magnocellular system of the hypothalamus, namely oxytocin (OT), which plays a role in reproduction, and antidiuretic hormone (ADH: also referred to as vasopressin), which promotes the retention of water in the kidneys. Homolgous substances occur in members of other vertebrate classes. All these neurohypophyseal hormones are cyclic nona- or octapeptides that differ from each other at only three positions within the molecule (Fig. 59). The neurohypophyseal hormones which have so far been identified in vertebrates are listed in Table 2.

Cyclostomata possess only arginine-vasotocin (AVT) whereas Elasmobranchii also possess glumitocin (GLT), valitocin (VAT) and aspartocin (AST). In addition to these Holocephala possess OT. AVT and isotocin (ICT) are found in Actinopterygii whereas AVT and mesotocin (MST) are found in Amphibia. The neurohypophyseal hormones of submammalian amniotes are similarly AVT and MST. Monotremata and Marsupialia have arginine-vasopressin (AVP) and OT. The same applies to the Eutheria, with the exception of Suiformes which have lysine-vasopressin (LVP) instead of AVP. It can be seen from the above that all submammalians possess AVT, which evidently

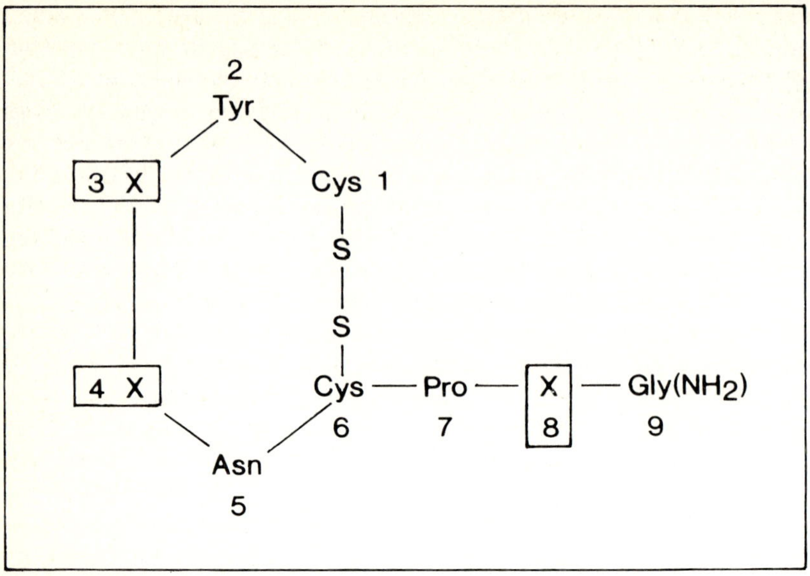

Fig. 59. General structure of the neurohypophyseal peptides. Positions at which the amino acid residues vary are marked with an X

Table 2. The neurohypophyseal hormones of vertebrates. The hormones differ in which amino acid is substituted at positions 3, 4 and 8 (X) in their common structure

	1 2 3 4 5 6 7 8 9	
General structure	Cys-Tyr-(X)-(X)-Asn-Cys-Pro-(X)-Gly(NH_2)	
Basic polypeptides		
Arginine-Vasopressin (AVP)	-Phe-Gln-	-Arg-
Lysine-Vasopressin (LVP)	-Phe-Gln-	-Lys-
Arginine-Vasotocin (AVT)	-Ile-Gln-	-Arg-
Neutral polypeptides		
Oxytocin (OT, OXY)	-Ile-Gln-	-Leu-
Mesotocin (MST)	-Ile-Gln-	-Ile-
Isotocin, Ichthyotocin (ICT)	-Ile-Ser-	-Ile-
Glumitocin (GLT)	-Ile-Ser-	-Gln-
Valitocin (VAT)	-Ile-Gln-	-Val-
Aspartocin (AST)	-Ile-Asn-	-Leu-

represents a primitive phylogenetic feature that has been retained for a long time in the course of evolution.

The only neurohypophyseal hormone in mammals which is of significance to reproduction is OT in that it firstly, enhances the contractility of the uterus and secondly, plays a role in the expression of milk (see p. 202). The first effect is evidently due to the resting potential of the smooth muscle cells of the uterus being lowered. OT also increases the frequency of contraction and the muscle tone. It is nevertheless still not clear whether or not this hormone is essential for the process of birth in mammals. There is also evidence that OT facilitates the passage of spermatozoa within the female genital tract and may, in males, be involved in the ejaculation of semen. Little is known concerning the affects of neurohypophyseal hormones on reproduction in submammalians. It is known, for example, that AVT induces contractions of the oviduct in birds, reptiles and amphibians which force the egg to be laid, and this effect is essentially the same as that of OT on the uterus of mammals. It has also been shown that injections of AVT into male birds temporarily raises the sexual activity of the birds. Another, dramatic effect of this hormone has been observed in the teleost *Fundulus heteroclitus* in which it induces the so-called spawning reflex response. This response involves both partners making characteristic movements as they release their gametes. The same effect is induced by OT and ICT, though they are not as potent as AVT. Similarly, injection of OT into female and male labyrinth fish (*Macropodus opercularis*) has been found to cause a characteristic flexure of the body, although the gametes are not released in this case. Another case of interest is provided by the seahorse (*Hippocampus* sp.) in which injections of OT or of ICT lead to contractions of the brood pouch on the male's abdomen (see p. 235) which are similar to the normal birth contractions. In ovoviviparous teleosts (see p. 251) AVT causes contractions of the ovarian muscles, which become more sensitive to AVT as gestation takes its course. It is possible that all the effects of OT and AVT reduce to the common ability of these hormones to stimulate the smooth muscles of certain structures of the urogenital system, which respond selectively by virtue of their possessing a specific complement of hormone receptors.

The second effect produced by OT in mammals has been largely elucidated. OT represents part of a neuroendocrine reflex arc whereby suckling on the teat (or milking) excites an afferent nerve fibre which, via the hypothalamus, induces the release of OT from the neurohypophysis, the entire process taking only a fraction of a second. The OT released is conveyed by the blood to the mammary glands where it causes the contraction of the myoepithelial cells surrounding the alveoli and thus the expression of the milk formed therein.

The gonadotropins from the hypophysis of mammals, follicle-stimulating hormone (FSH) and luteinising hormone (LH), together with thyroid-stimulating hormone (TSH = thyrotropic hormone) all belong to the family of glycoprotein hormones, i.e. the basic protein molecule is substituted with carbohydrates. All these hormones consist of an α and a β subunit which are not covalently bound together. Their α subunits are virtually identical whereas their β subunits differ. The former can be exchanged for one another experimentally. For example, when TSH-α is coupled with LH-β the effect of the complex is that of LH. In other words, it is the β subunit that determines the biological activity of the hormone. However, the β subunit alone is without effect: it only exhibits activity when coupled with the α subunit. The subunits of related hormones from vertebrates belonging to different classes can similarly be combined to form active hormone chimeras: reptilian LH-α and mammalian LH-β, for example. In this instance, however, the hormone chimera is less effective in reptiles than is the native hormone and there are also other quantitative differences in its activity. Thus the specific activity of a mammalian α subunit coupled with a mammalian β subunit is greater than when coupled to a non-mammalian β subunit. Furthermore, such experiments show that the nature of the α subunit is evidently also important, perhaps with regard to the binding of the hormone to the receptor. Both FSH and LH are synthesided by a single cell type in the adenohypophysis.

Human FSH has a molecular weight of 34,000 Dalton and is evidently a relatively unstable molecule. Its amino acid sequence has been determined and the α subunit found to consist of 89 amino acid residues which are identical to those of the α subunit of human LH. The β subunit has 115 amino acid residues of which 49 are in the same sequence as those in TSH-β, but only 39 of which are in identical positions to those of LH-β. The β subunit of FSH consists of 18% carbohydrate, a proportion of which is sialic acid. Splitting this group off with a neuraminidase leads to the loss of the subunit's biological activity. FSH affects the growth of follicles in the ovary, promotes growth of the testes, and regulates spermiohistogenesis and possibly other processes: details will be given later (see p. 185).

The α and β subunits of mammalian LH from sheep and cattle are made up of 96 and 119 amino acid residues respectively, and together have a molecular weight of about 30,000 Dalton. Carbohydrates constitute between 13% and 16% of the molecule and include hexose, hexosamine and fucose. Sialic acid, which is characteristic of FSH, is present only in very small amounts. In the few mammalian species which have been studied, only minor differences have been found between the amino acid sequence of the two subunits. The physiological effects of LH are primarily the triggering of ovulation in the ovary and the control of steroid production in the gonads of both sexes.

It should be noted however that in many cases it is very difficult to distinguish between the effects of FSH and LH: in many instances these hormones appear to work together and indeed are perhaps best regarded as acting synergistically.

Birds also possess a dual system of gonadotropins, as described for mammals. Among reptiles, two gonadotropins have been found in Chelonia and Crocodilia whereas members of some squamate families have only one whose activity resembles neither that of FSH nor LH. A pair of gonadotropins have been identified in two species of anuran whereas in teleosts, the sturgeon and an elasmobranch only one has been isolated. All these hormones are composed of two subunits. The gonadotropin of carp has a molecular weight of about 28,000 Dalton and its subunits are similar in structure to those of mammalian gonadotropins with the β subunit of carp more closely resembling that of mammalian LH than that of FSH. An additional hormone has been isolated from the hypophysis of a number of teleost species which promotes vitellogenesis within the ovary. Although it has been referred to as a gonadotropin, it should not be included within the group of hormones considered here since it is atypical in not being a glycoprotein.

Lastly, it remains only to mention that the TSH of mammals has a molecular weight of about 28,000 Dalton and is very similar to LH. It functions as a glandotropic hormone regulating the production of hormones by the thyroid gland.

In man, both the foetal and maternal sides of the placenta produce a chorionic gonadotropin (so-called human chorionic gonadotropin, or HCG) during pregnancy. This hormone has a molecular weight of about 37,000 Dalton and consists of approximately 30% carbohydrate. It is therefore a glycoprotein and, like the hypophyseal gonadotropins, is composed of an α and β subunit. The former is very similar to the α subunit of TSH and LH, with which it can be substituted. The α and β subunits are made up of 92 and 136 amino acid residues, respectively. The proline content of HCG is relatively high. The effects of this hormone are similar to those of LH. Pregnant women excrete HCG in the urine and its detection in this fluid can be used as a proof of pregnancy. Other primates also produce this hormone.

Prolactin (PRL), also called luteotropic hormone (LTH), occurs in vertebrates of every class and is only absent in the Cyclostomata within the class Pisces. The molecular weight of mammalian PRL is approximately 22,500 Dalton. The primary structure of the molecule has been elucidated in the case of cattle and sheep. It is a pure protein which has no substituted carbohydrate residues and has three disulphide bridges within it. There is a close similarity between the PRL molecule and that of the growth hormone (STH), although large differences exist between the prolactins of the different classes of vertebrate, that of teleosts differing to such an extent that it is not capable of inducing any of the typical effects of prolactin in higher vertebrates. Prolactin has been isolated from two species of bird and several species of teleost. Turkey PRL has a molecular weight of about 26,000 Dalton whereas the corresponding value for the PRL of the teleost *Sarotherodon mossambica* is 19,400 Dalton. The physiological effects of prolactin are diverse, its primary role with respect to reproduction being the hormonal regulation of certain aspects of parental care. In general it brings about the manufacture of secretions for the nutrition of the offspring, this being the lactogenic effect, a specific instance of which is lactation in mammals (see p. 199) from which this hormone gains its name. In pigeons PRL induces and regulates the production of the so-called crop milk which is fed to the young and which is formed by the proliferation of the fatty epithelial cells of the crop: this effect is the basis of the standard international assay for PRL. A comparable phenomenon occurs in members of the teleost genus *Symphysodon* where PRL causes an increase in the production of surface mucus and a thickening of the epidermis in both the male and female parent animal. The secretion, together with the epidermal scurf, constitutes the so-called discus milk which the young eat off the surface of the parent's body. Prolactin in birds also induces the formation of a highly vascularised, oedematous and featherless area of the skin of the body called the brood patch, by virtue of which heat is efficiently transferred to the eggs during brooding. This effect of prolactin is promoted synergistically by oestrogens or androgens respectively, according to whether the female or the male broods the eggs. Prolactin also affects certain behaviour associated with parental care (see p. 219). The amount of PRL in the hypophysis of a variety of species of bird has been found to increase during the brooding period. The injection of PRL into hens causes them to sit on their eggs and induces clucking behaviour directed towards their chicks. It is possible that progesterone also plays a role in this, perhaps by stimulating the secretion of PRL. The injection of PRL into a range of bird species has been observed to stimulate brooding behaviour. Related phenomena occur in teleosts. Thus PRL induces fanning of the brood (see p. 220) to a degree that is dependent on the amount of PRL administered; in mouthbreeding cichlids it is responsible for the parent holding the eggs in its mouth but not eating them (see p. 219), and it is evidently essential for the production of oral mucus in Anabantidae during nest building.

The alternative name for prolactin, luteotropic hormone, derives from the ability of this hormone to maintain the secretion of progesterone from the corpus luteum in the rat, mouse and hamster. Whether this is also the case in other mammals is not clear: PRL is evidently not involved in the production of progesterone in primates, cows, pigs, rabbits and guinea-pigs. In other species it is postulated to act synergistically together with one or both gonadotropins as a "luteotropic complex". In some vertebrates PRL has an antigonadotropic effect, that is it causes the regression of mature gonads or inhibits the development of immature gonads. In several species of birds the gonads of both sexes have been found to regress under the influence of prolactin: this also involves the atrophy of structures which are stimulated by the action of the gonadal hormones. A similar phenomenon occurs in the great crested newt and in male teleosts of the family Cichlidae. In one species of reptile prolactin evidently acts as an antigonadotropin only in the female sex.

As already mentioned, the effects of PRL are diverse and include important functions not related to reproduction. Thus it promotes the growth of certain structures in a manner that is comparable to that of STH. In teleosts prolactin is essential to survival in freshwater since it promotes the retention of sodium ions, and it evidently retains this ability even in mammals.

STH is involved in the regulation of lactation. This hormone is a protein whose sequence of amino acids has been elucidated in the case of man, cattle and sheep. In the two former species the protein is composed of 191 amino acid residues, has a molecular weight of approximately 22,000 Dalton, and contains two disulphide bridges. In general STH promotes growth, particularly of the skeleton, in that it induces the formation of cartilage in the epiphyses. In addition STH has a range of effects on metabolism which are mainly related to the promotion of protein biosynthesis.

A hormone which is very similar to STH and PRL is produced in humans by both the foetal and maternal elements of the placenta during pregnany: it is referred to as human choriosomatomammotropin (HCS), or human placental lactogen (HPL). It is a protein with a molecular weight of about 21,000 Dalton and contains 190 amino acid residues, no less than 160 of which have the same position as in human STH.

There is also a close similarity between HCS and prolactin. Thus the effects of HCS include the promotion of growth, affects on the mammary glands and on the crop of pigeons, as well as a luteotropic effect. The somatotropic activity of HCS is, however, much less than that of STH from the hypophysis. The physiological role of HCS is nevertheless still uncertain. It has been suggested that it serves to prepare the mammary glands for the production of milk after birth. A comparable hormone also exists in other primates. Of the adenohypophyseal hormones not directly involved in reproduction, adrenocorticotropic hormone (ACTH) remains to be considered. It is a polypeptide composed of only 39 amino acid residues whose sequence is known in the case of several mammals. ACTH regulates the secretion from the steroidogenic interrenal tissue of so-called glucocorticoids whose main effect in mammals is the promotion of gluconeogenesis.

The gonadal hormones are steroids, the principle male sex hormone or androgen being testosterone (T). The testosterone molecule contains 19 carbon atoms and has a structural formula as illustrated in Fig. 60A. The most active androgen in dioecious teleosts is probably 11-oxo testosterone. With respect to reproduction, testosterone has a wide spectrum of action: it promotes the growth and development of the external genitalia and regulates the functioning of the accessory glands of the wolffian

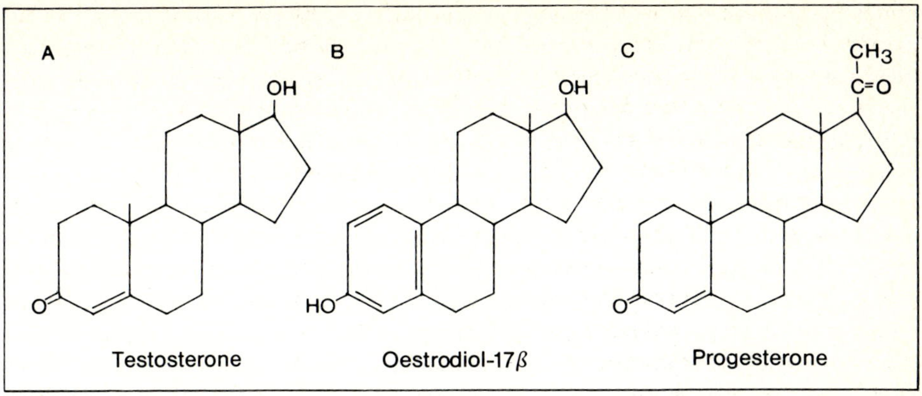

Fig. 60A–C. The structural formula of testosterone, oestrodiol-17 β and progesterone

duct. In addition, the development of the male secondary sex characteristics (see p. 35) is due to the influence of testosterone. Such characteristics may be either irreversible or reversible. Examples of the former include the breaking of the voice and changes in the structure of the skeleton in man, and the development of the "sword" on the tail fin of the teleost *Xiphophorus helleri* whereas the second group includes the development of the comb of cockerels and of the helmet in males of certain species from the family Blenniidae (Teleostei). The development of secondary sex characteristics in vertebrates is frequently dependent on androgens, as for example is the growth of antlers in Cervidae, the coloration of the bill of birds, the development of head and throat crests in male lizards and of the "wedding dress" of urodeles and teleosts. Nevertheless, other secondary sex characteristics are determined purely by genetic factors, for example the pterygopodia of Elasmobranchii and Holocephala. A further example of this is the display plumage of many male birds, the development of which in females is suppressed by oestrogens. Testosterone also promotes sexual behaviour in males, as it does in primate females to some extent. In addition, testosterone plays an important role in the processes of differentiation during the course of ontogenesis (see p. 202). With respect to metabolism, testosterone has been found to have a strong anabolic effect on proteins in that it promotes the growth of muscles, in part in association with the formation of structures characteristic of males, such as the arms of frogs and toads and the musculature associated with chewing in rodents.

Finally, in mammals it is possible that the active principle which affects the target organ is dihydrotestosterone (DHT), a metabolite of testosterone which is formed within the target organ by the action of an enzyme. Once formed, DHT can end up in the blood as a circulating androgen: it is in fact released into the blood from the testes, the proportion of testosterone to DHT being approximately 10:1 in man. In light of this, it is possible that testosterone is merely a precursor of the actual active principle which is DHT. Other vertebrates, however, have been little investigated in this respect. The involvement of testosterone in spermatogenesis will be dealt with later (see p. 185).

Ovarian oestrogens, the most important of which are oestrodiol-17 β and oestrone, have 18 carbon atoms in their molecular structure (Fig. 60B). In most of the vertebrates which have been investigated oestrone occurs together with oestrodiol 17 β but merely represents a precursor of the latter. The physiological role of the oestrogens is highly complex and relates primarily to the female urogenital system, although other structures are also affected. Broadly speaking the main function of oestrogens is to maintain the physiological condition of the müllerian ducts: this will be dealt with in detail later. Oestrogens are also involved in the development of the female reproductive tract during sexual maturation and influence, in an analogous way to androgens, the development of the external genitalia and secondary sex characteristics of females, such as the mammary glands and body form etc. In birds, oestrogens are involved in the formation of the calcerous egg shell, in so far as they serve to maintain a high level of calcium in the blood by promoting the breakdown of a special kind of bone (medullary bone) in the tibia and femur: its initial development is under the influence of both oestrogens and androgen. One important aspect of oestrogen function is that they stimulate the production in the liver of yolk proteins which are transported in the blood to the ovary where they are taken up by the oocytes (see p. 103). The effects of oestrogens on metabolism are complex. For example, oestrogens antagonise the effects of growth hormone and thereby bring growth to a stop, with the result that females tend to be smaller than males. A further effect of oestrogens is that they promote sexual behaviour in females.

Progesterone is the only biologically active gestagen (progestin) produced by the ovaries. Its molecular structure is made up of 21 carbon atoms (Fig. 60C). In mammals progesterone is responsible for the maintenance of pregnancy, as will be explained later (see p. 196). The effects of progesterone are usually manifest at the same time as those of oestrogens, or thereafter. The predominant effect of oestrogens is growth whereas progesterone usually promotes differentiation, as can be seen from their respective effects on the mucosa of the human uterus (see p. 194). In birds progesterone induces the production of the protein avidin, a component of the egg white. The gelatinous secondary membrane of toad eggs is similarly produced under the influence of progesterone. It is probable that progesterone also plays a role in ovulation in birds whereas in ovoviviparous reptiles the development of the ovary and ovulation is impeded by progesterone, as is also the case in Chelonia. Progesterone evidently plays a part in the triggering of certain types of sexual behaviour in rodents, as does oestrogen.

Oestrogens and progesterone are produced during pregnancy by the human trophoblast, in addition to the proteohormones already mentioned. Oestrogens are in fact produced by all mammals that have a gestation period of longer than 70 days. However, there is considerable variation among mammals with regard to the production of progesterone in the placenta. Large amounts of progesterone are produced in both the maternal and foetal parts of the human placenta whereas in the rat and rabbit the placenta produces only a small proportion of the total amount of this hormone present in the blood, the greater part stemming from the ovary. The biosynthesis of both oestrogens and progesterone within the placenta evidently proceeds autonomously, free of any controlling influence of the hypophyseal gonadotropins.

It is possible that the ovaries of mammals produce another hormone, namely relaxin. This is a polypeptide resembling insulin with 48 amino acid residues which form two chains linked by disulphide bridges. It is produced primarily during pregnancy and causes a softening of the ligamentum pubis and the joints of the pelvis and reduces the contractility of the uterus.

The prostaglandins are a further group of substances of relevance to reproduction. These have been grouped together with the so-called parahormones since they do not fully meet the criteria of being hormones. Their name derives from the fact that they were first found in the seminal fluid of mammals and were thought to be produced by the glandula prostatica. In fact they occur in practicallyall tissues and body fluids. They are polyunsaturated fatty acids built up from 20 carbon atoms which are coupled with a cyclopentane ring and which contain to several functional hydroxyl or keto groups. The preovulatory follicle of mammals produces large amounts of prostaglandin E_2 and $F_{2\alpha}$, the latter also being produced in the uterus. Prostaglandin E_2 is thought to play a role in ovulation whereas in many species (e.g. horse, cow, pig) $F_{2\alpha}$ reaches the ovary via a counter-flow mechanism where it causes the regression of the corpus luteum. It is also possible that the same effect is brought about by the local synthesis of prostaglandins within the ovary of primates.

5.9 Hormonal Regulation of the Testes

In all vertebrates the functioning of the gonads is under hypophyseal control. This control is usually mediated by the complex of gonadotropic hormones released from the hypophysis in response to a hypothalamic releasing hormone, although the latter are apparently lacking in the case of the Cyclostomata and Chondrichthyes. The gonadotropins in turn promote not only gametogenesis but also the production and release of gonadal hormones.

The functioning of the testes in mammals is regulated by FSH, LH and androgens. LH stimulates the Leydig cells of the interstitial tissue to produce or release testosterone and is therefore also referred to as interstitial cell stimulating hormone (ICTH). In general, FSH causes growth of the testes and is, amongst other things, necessary for the occurrence of spermiohistogenesis. Testosterone plays a similarly important role in spermatogenesis in that it induces mitotic division. In the rat it has been found that the early prenatal and postnatal development of the germ cells and the meiotic division of the spermatocytes are conditional on the presence of testosterone. However, this is evidently not the case in all mammals, although there are insufficient data available to discern any pattern in the stages of spermatogenesis that are dependent on the involvement of hormones. The role of FSH in spermatogenesis is difficult to define. As has been mentioned, it is evidently necessary for the occurrence of spermiohistogenesis but it is not clear whether it acts directly. FSH receptors have been identified on the Sertoli cells (and also on the spermatogonia) within the seminiferous tubules of the rat which, when activated, bring about the production of a specific androgen binding protein (ABP) that is secreted into the lumen of the tubule. On the other hand, the Leydig cells outside the tubules produce testosterone, the majority of which is bound to a carrier protein and transported throughout the body in the blood. Some testosterone nevertheless finds its way into the lumen of the tubule where it is probably bound by ABP, resulting in its accumulation, and it is possible

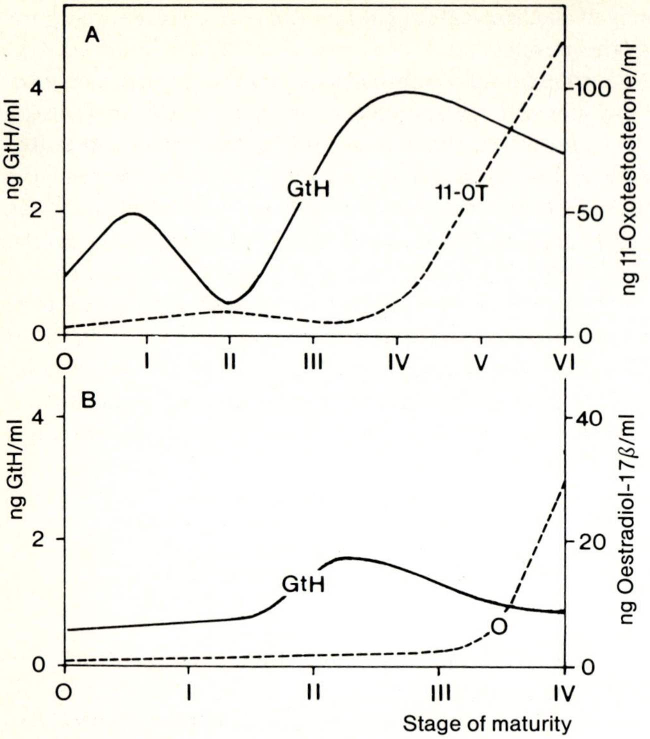

Fig. 61A,B. Concentration of gonadotropin and sex steroids in the serum during the course of gonadal maturation in male (**A**) and female (**B**) rainbow trout (*Salmo gairdneri*). The immature gonad at stage 0 is quiescent. The testis reaches full maturity at stage VI whereas the ovary reaches full maturity at stage IV, prior to ovulation. *Unbroken line:* gonadotropic hormone (GtH); *broken line*: sex steroids (as indicated on the *right*)

that these ABP-testosterone complexes act upon the germ cells. With respect to Sertoli cells it has been found that those of mammals can produce oestrodiol-17 β in vitro under the influence of FSH, probably by the conversion of androgen. It has also been suggested that Sertoli cells produce testosterone. On the other hand, LH also has an indirect influence on spermiogenesis via its effect on the production of testosterone by the Leydig cells. High levels of testosterone act via a long feedback loop so as to inhibit the secretion of gonadotropins from the hypophysis, primarily LH in the case of man. However, such an action appears to mitigate against the two gonadotropins exercising a differential effect on the testes since it does not provide for the possibility of specifically inhibiting the release of FSH alone. Nevertheless, a water-soluble substance called inhibin with a molecular weight of between 10,000 and 20,000 Dalton has been found in the testes which has exactly this effect. It is probable that FSH itself stimulates the production of inhibin. On the other hand, inhibin has not yet been detected in the blood of male animals. Aside from this, a medium in which granulosa cells have been cultured has been found to possess an inhibin-like activity. Finally, it has been shown that Sertoli cells also possess gonadoliberin receptors.

The testes of submammalians, excepting amphibians, are under the same type of regulatory control – via gonadotropin(s) and androgen – as those of mammals. In amphibians a gonadotropin similar to LH stimulates not only the production of androgen but also spermatogenesis, although whether or not androgens play a role in spermatogenesis is uncertain, especially since testosterone suppresses the development of spermatozoa. In all other vertebrates androgen appears to be necessary for the completion of spermatogenesis. Birds and most reptiles have two gonadotropins, but it is even harder to distinguish the specific action of FSH and LH in these animal than it is in mammals. The testes of reptiles appear to be influenced by a gonadotropin which in most respects resembles FSH. In Chondrichthyes gonadotropin is necessary, in either a direct or an indirect way, for the completion of the late meiotic divisions of the germ cells in the testes whereas in Cyclostomata the final stages of spermiogenesis are not under the control of the hypophysis. The changes in the serum concentration of gonadotropin and androgen in relation to the maturation of the testes of a teleost are shown in Fig. 61A.

As a broad generalisation one can say that the production of androgens in the testes of all vertebrates is dependent on gonadotropin but which stages of spermatogenesis are hormone-dependent remains unresolved – it appears to be the later stages in fish, reptiles, birds and mammals.

5.10 Hormonal Regulation of the Ovaries

The maturation of the ovaries, though more complex than the maturation of the testes, is primarily controlled by gonadotropins. In animals which do not suckle their young, ovarian maturation involves not only the mitotic and meiotic divisions of the germ cells and the production of hormones but also the phenomenon of vitellogenesis. In mammals, and very probably in some ovoviviparous and viviparous submammalians, the function of the ovaries as a source of gametes and of oestrogens does not come to an end at ovulation but continues with the formation of the corpus luteum.

Relatively little is known about the regulation of ovarian function in submammalians, with the exception of birds. It is known that in Cyclostomata and Chondrichthyes the hypophysis exerts a control over the ovary but is itself independent of the hypothalamus. Both corpora lutea and progesterone occur in Chondrichthyes but to what extent they are essential for the maintenance of gestation in ovoviviparous and viviparous species in unknown. The development of oocytes, vitellogenesis, and the production of oestrogen in teleosts are all dependent on gonadotropin. However, the early development of the oocyte is apparently not influenced by gonadotropin as the level of this hormone does not rise until the stage of so-called endogenous vitellogenesis is reached when the precursors of the cortical granules are formed. Nevertheless, the subsequent stage of exogenous vitellogenesis is influenced by oestrogen. This hormone stimulates the liver to produce yolk protein which is transported as vitellogenin in the blood to the ovary where it is taken up by the oocytes. A vitellogenic hormone from the hypophysis is probably also involved. The level of gonadotropin in the blood serum increases sharply at the time of ovulation and it is probable that steroids are involved in this process. The molecular structure of these steroids is made up from 21 carbon atoms. An example of the types of changes that occur in the levels of gonadotropin and oestrogen during the development of the ovary is illustrated in Fig. 61B.

The development of amphibian oocytes is under hypophyseal control, although it is not known whether the gonadotropins involved in this act directly. This control system also influences the production of oestrogens and progesterone but no evidence of negative feedback of these steroids on the hypothalamus has yet been found. As with all vertebrates that produce yolk a vitellogenin is formed in the liver of amphibians under the influence of oestrogen. Vitellogenin is subsequently taken up by the oocyte in response to its being stimulated by gonadotropin. The subsequent maturation of the oocyte is promoted by a maturation agent, probably progesterone. The production of progesterone is, in turn, mediated by a gonadotropin that also induces ovulation. This latter effect forms the basis of a pregnancy test which has long been used in human medicine whereby the HCG in the urine of a pregnant woman induces ovulation in mature female clawed toads (*Xenopus laevis*). Of the two gonadotropins identified in anurans, the one resembling LH plays the leading role in the various processes described above.

Most reptilians have two gonadotropins, others only one. However, in every case it is apparently a gonadotropin similar to FSH that controls the development of the ovaries, the production of ovarian steroids and ovulation. The müllerian ducts of reptiles are not only structurally more highly differentiated than those of lower vertebrates but are also influenced by oestrogens. These hormones appear to exert a negative feedback on the hypothalamus. Following ovulation, corpora lutea and progesterone are formed in the ovaries of reptiles and this is of significance physiologically in ovoviviparous and viviparous species (see p. 256) in that they inhibit the development of the ovary during gestation. As in other animals, the production of yolk proteins in the liver is stimulated by oestrogens, the effects of which may possibly be promoted synergistically by STH.

The ovaries of birds mature due to the influence of FSH and LH, though exactly what each hormone does has not yet been resolved. The developing follicles release oestrogens which affect the müllerian ducts. Ovulation either occurs spontaneously, as for example in fowl, ducks and geese, or is triggered by a specific releaser, usually the appearance of a male. Ovulation in the domestic chicken is dependent on LH but the mechanism whereby this hormone is released is unknown. It has been found that the presence of an egg in the oviduct inhibits the following ovulation via a neural mechanism that probably blocks the release of LH from the adenohypophysis. The rise in the level of LH which occurs a few hours before ovulation is accompanied by a rise in the concentration of progesterone in the blood and, since injection of progesterone promotes ovulation, one may assume that this hormone also plays some as yet unknown role in triggering ovulation. Despite the fact that ovulation is inhibited by the injection of oestrogen, a rise in the concentration of oestrogen in the blood serum can be observed shortly before the peak in the level of LH associated with ovulation. Yet it is known that in birds oestrogens are incapable of inducing the release LH from the hypophysis: in short, it is not at all clear how ovulation is controlled. It is possible that oestrogens sensitise the hypothalamus to progesterone which in turn induces the release of LH. The source of progesterone in birds remains similarly obscure. Since functional corpora lutea are not produced in birds, the possible sources that come into question are the follicle cells themselves, the interstitial tissue of the ovary, or perhaps the corpora atretica. Much of the physiological and endocrinological evidence concerning reproduction has been gained from domesticated birds, primarily chickens, and therefore it is not known to what extent they may apply to birds in general.

The growth and differentiation of the follicle in mammals, including the formation of the antrum, is dependent on FSH. The granulosa cells bear receptors for FSH and may be able to convert testosterone into oestrogens. The principle effect of LH is on the theca interna, where it regulates the production of oestrogens. In many species LH is also necessary for the formation and maintenance of the corpus luteum. This topic is taken up again in the following section.

Ovoviviparity and viviparity occur in a range of Elasmobranchii, Teleostei, Amphibia and Reptilia and involve the parent animal carrying the developing embryos or larvae for a short or long period of time. In association with this there are complex structural and functional adaptations by means of which the offspring are provided for in either the uterus, ovary, or secondary oviducts. There is a growing body of evidence demonstrating that hormones control various aspects of the development and physiological condition of specialisations associated with ovoviviparity and viviparity. Specific instances thereof will be mentioned in Chap. 7 where such specialisations are described in detail.

5.11 Hormonal Regulation of the Female Genital System of Mammals

Mammals, with the exception of the Monotremata, are viviparous. In most Marsupialia the yolk sac placenta represents a relatively simple and loose connection between the foetus and the mother animal whereas the corresponding association in Placentalia is considerably more complex and may involve profound structural and physiological changes, as will be described in full in Chap. 8. Such changes are controlled by an array of hormones eminating from the hypophysis, the ovary and the foetal and maternal elements of the placenta and it is appropriate to deal with these control processes here in the wider context of the hormonal regulation of the female reproductive tract as a whole rather than in Chap. 8.

Very little is known concerning the hormonal control of reproduction in the egg-laying Monotremata. They presumably possess the hypophyseal gonadotropins FSH and LH but the only established fact is that progesterone is present in females, indeed its concentration in the blood is much higher than in gravid Placentalia.

The hypophysis of Marsupialia is a source of both FSH and LH whereas the follicles produce oestrogens and the corpora lutea progesterone. The activity of the corpora lutea, which have been shown to possess PRL receptors, is inhibited by PRL. The oestrus cycle of marsupials has a period of from 28–32 days and can be divided into five phases. During the luteal phase the activity of the corpus luteum, which arises from the follicle as a result of ovulation, serves to prepare the uterus for implantation. In the absence of any inhibitory influences (occasioned by pregnancy) the corpus luteum slowly regresses during the ensuing postluteal phase. In the proöestrus phase a new follicle develops in the ovary which did not give rise to the previous corpus luteum and the uterine epithelium becomes thicker due to the influence of oestrogen. After about the fifteenth day of the cycle the level of progesterone in the blood steadily increases then falls again over the next few days with the approach of the oestrus phase during which ovulation takes place. What causes ovulation is still unknown, except that it is independent of any exogenous stimulus. During the following postoestrus phase a new corpus lutem is formed and the cycle begins again.

A characteristic of marsupials is that the pregnancy following successful copulation does not inhibit the oestrus cycle, which proceeds as normal, at least at first. The developing blastocyst implants into the limb of the uterus on the same side as the ovary from which the oocyte originally came and development proceeds. At the same time the oestrus cycle continues in the other ovary. An oocyte that matures during pregnancy is ovulated at different times (in relation to the time of birth) in different animals. For example, the group of kangaroos (Macropodidae) to which the wallaby *Wallabia bicolor* belongs ovulate and can copulate, resulting in the development of a blastocyst in the other limb of the uterus, before the birth of the first foetus: in the species named the period of the cycle is 32 days and gestation lasts 35 days. As soon as the new-born foetus reaches the pouch and suckles on the teat, thereby initiating lactation, the blastocyst is "put to sleep" and develops no further; i.e. implantation is delayed (see p. 34). The onset of lactation also blocks the continuation of the oestrus cycle. The development of the blastocyst begins again when the baby kangaroo has left the pouch. In another group of kangaroos ovulation and copulation, as well as the formation of a blastocyst, occur shortly after birth. Again in this case, the development of the blastocyst is suppressed so long as the baby kangaroo in the pouch continues to suckle. It is a curious fact that even after it has left the pouch the first offspring continues to take milk from one teat while its younger sibling suckles at the other. In this case each mammary gland produces milk of a different composition. The phenomenon is all the more remarkable when one considers that both mammary glands are subject to the same hormonal influences from the blood of the mother. In another group of marsupials, in which species other than kangaroos predominate, the gestation time is considerably shorter than the period of the oestrus cycle. In this case the stimulus of the offspring suckling on the teat inhibits the next ovulation and effectively freezes the oestrus cycle. In some species of this group of marsupials the level of progesterone in the blood during gestation is somewhat higher than normal whereas in other species little or no change can be detected. In kangaroos the development of the blastocyst is suppressed after it has reached the 80-cell stage, at which stage it is surrounded by a secondary egg membrane and a layer of albumin. The blastocyst can survive for several months in this "frozen" state, and this in fact occurs if the weather conditions are unfavourable: an 11-month dormancy has been reported in the case of *Protemnodon euginii*. The block on the oestrus cycle is usually lifted when the young stop suckling. The suppression of blastocyst development and of ovulation is evidently mediated by an as yet unidentified neuroendocrine reflex arc. One can waken the sleeping blastocyst experimentally by injecting either oestrogen or progesterone. During the period of lactation the corpus luteum is relatively inactive but after suckling has ceased it develops rapidly and produces a greater amount of progesterone. This hormone promotes the development of the dormant blastocyst and at the same time initiates the luteal phase of the cycle which prepares the uterus for the implantation of another blastocyst. There is some evidence to show that oxytocin is released from the neurohypophysis in response to the stimulation caused by suckling: it is possible that oxytocin inhibits the development of the corpus luteum. In one species of kangaroo suckling also results in the release of PRL, which could similarly suppress the development of the corpus luteum.

The suppression of the oestrus cycle by lactation is a characteristic of all marsupials. It is effected in two different ways, both of which involve the stimulatory affect of the young suckling in the pouch. In marsupials other than Macropodidae the nervous stimulus appears to affect the hypophysis directly, causing the release of an inhibitory hormone whose nature is unknown. On the other hand, in most Macropodidae the suppression of the ovary during lactation depends on the corpus luteum. Again, the stimulus of the suckling young leads to the release from the hypophysis of an unknown hormone which is neither LH nor prolactin. The hormone acts on the corpus luteum, which in turn produces a hormone that inhibits the ovary. The latter effect may be due to a low level of progesterone.

The process of birth in marsupials may also be controlled by hormones, possibly oestrogens and progesterone. However, the interplay of the different factors involved seems to vary considerably from species to species. There is some evidence to indicate that oxytocin is involved in the process of birth and it is noteworthy that after birth it stimulates the expression of milk.

The oestrus cycle of placental mammals differs from that of marsupials with respect to the formation and persistence of the corpus luteum. Three basic types can be recognised. In the first type, represented by the sheep, cow and pig, the corpus luteum is present and active for almost the entire duration of the cycle. The second type is represented by the menstrual cycle of humans and other primates in which the corpus luteum is present only during the second half of the cycle. The rat provides an example of the third type which is distinguished by the small size and low level of activity of the corpora lutea.

The oestrus cycle of the sheep will be taken here as an example of the first type of cycle. As a rule this begins in autumn after an inactive summer phase. The first cycle is triggered by the shortening day length at this time of year and lasts 16 or 17 days. If a pregnancy does not follow the cycle repeats itself, continuing on into the next spring. For convenience, the beginning of oestrus is taken as a reference point representing the start of the cycle (i.e. day 0). Oestrus itself lasts almost exactly 24 h and ends with ovulation. A corpus luteum develops from the empty follicle and reaches its maximum size on the eighth day. Ovulation is induced by a sudden rise in the concentration of LH in the serum. Thus LH in fact acts as a luteinising hormone by inducing the formation of the corpus luteum. The low level of this hormone present is nevertheless sufficient, in terms of its luteotropic activity, to stimulate the production of progesterone by the corpus luteum. The concentration of progesterone in the serum therefore rises steeply until day 11 or 12 then on day 13 begins to fall again, reaching its original level after 3 or 4 days. The decrease in the level of progesterone reflects the regression of the corpus luteum, which becomes markedly less sensitive to LH. Parallel to the events described so far, a new follicle matures due to the influence of FSH and oestrogens: production of the latter in the follicle wall is stimulated

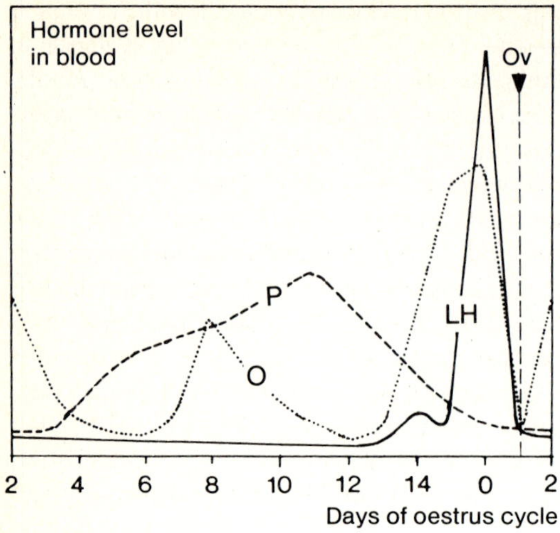

Fig. 62. Concentration of luteinising hormone (LH), oestrogens (O) and progesterone (P) in the serum during the oestrus cycle of the sheep.
Ov = ovulation. (After Hausel and Echternkamp)

by LH. The concentration of oestrogens in the serum is maximal on day 0 of the cycle but falls rapidly during the first day only to rise again to reach a peak on the second day. The level sinks again thereafter, reaching a minimum on day 6. On day 8 a second peak follows which similarly sinks to a second minimum on day 12. From then on the level rises to reach its maximum on day 0. This periodically changing level stands in sharp contrast to the single peak in the concentration of progesterone. Both oestrogen and progesterone affect the vagina and uterus in such a way that the conditions for fertilisation and implantation are optimal. The low serum concentration of LH between day 1 and day 15 is produced by a negative feedback mechanism involving progesterone whereas the rapid rise in LH prior to ovulation is a consequence of the stimulatory effect of the high oestrogen concentration on the hypothalamus. The oestrogen concentration in the serum falls rapidly during the oestrus phase. It is possible that the behaviour of sheep during oestrus is regulated by a combination of oestrogens, an androgen (androstendione) and, to a small extent, by progesterone. The changing levels of the different hormones underlying the oestrus cycle of sheep are illustrated in Fig. 62. The timing of the cycle is regulated by the corpus luteum. The rapid degeneration of the corpus luteum is probably caused by a so-called luteolysin produced by the non-gravid uterus. There is some evidence to indicate that this substance is a prostaglandin which is conveyed from the uterus to the ovary via a special system of blood vessels.

The second type of oestrus cycle is that represented by the menstrual cycle of humans, the most prominent characteristic of which is the cyclical occurrence of menses. These periods of bleeding, menstruation, usually last 4 to 5 days. Humans have a permanent oestrus cycle, that is each cycle follows on directly from the previous one and this continues throughout the course of the year without any intervening inactive phase. Menses only occur in primates. The bleeding that female dogs show when on heat comes from blood vessels in the vagina and is known as a diapedesis. The theoretical period of the human cycle is 28 days but the average duration of the cycle in a large number of women between the ages of 20 and 39 has been found to be 29.5 days. The first day of menstruation has been arbitrarily chosen as the start of the cycle (Fig. 63A). Each cycle is broken down into two phases, the follicular and the luteal phase, there being no marked oestrus phase during which the likelyhood of mating is greater. During the follicular phase the follicle matures under the influence of FSH, LH and oestrogens. On the first day the concentration of FSH in the blood increases slightly, as does that of LH, causing the growing follicle a start producing oestrogens, whose concentration in the blood thus rises to reach a maximum on about the thirteenth day. The oestrogens affect the follicle itself, primarily causing an increase in its reactivity to LH, although they also promote its growth. The changes in the concentration of PRL in the blood more or less follow those of the oestrogens. The concentration of FSH falls slightly towards the middle of the cycle. From the eleventh day on the concentrations of FSH and LH in the blood rise sharply to reach their maximum levels in the middle of the cycle on day 14 and initiating the second phase of the cycle. During the first half of the cycle, the so-called proliferative phase, the oestrogens stimulate the growth of the inner layer of the uterus wall (endometrium). Ovulation is induced by LH, although it is uncertain whether FSH may not also be involved in this. It is conceivable that FSH is released from the hypophysis together with LH simply as a consequence of their having a common releasing hormone, even though ovulation may not depend on FSH. The release of LH is due to the high oestrogen concentration in the blood exerting a stimulatory effect on the hypothalamus, an effect which has also been interpreted as part of a positive feedback loop. Ovulation occurs 16 to 24 h after the concentration of LH has reached a maximum. LH also stimulates the subsequent development of the corpus luteum, which takes 8 days to complete. This development is accompanied by a steep rise in the concentration of progesterone in the blood which reaches a maximum on the 18th day and remains more or less at this level until day 21. It falls thereafter and attains its original low level towards the end of the cycle. This decline is due to the process of luteolysis, that is the regression in the structure and function of the corpus luteum, which takes place during the last days of the cycle. In the second half of the cycle the concentration of oestrogens in the blood varies between its maximum level prior to ovulation and its original minimum level: this equally describes the changes in the level of PRL. The progesterone produced by the corpus luteum during the luteal phase stimulates secretory activity (initiating a secretory phase) within the endometrium that has been built up under the influence of oestrogen. However, this activity can no longer be maintained as the level of progesterone falls towards the end of the cycle. In primates the endometrium is penetrated by so-called spiral arteries and these contract during the phase of low progesterone, resulting in an ischemia that dissolves the intercellular substance and the breakdown of the structural integrity of the tissue. This process leads to the detachment, or desquamation, of the inner layers of the uterus, the mucosa (see p. 260), and the tearing of the spiral arteries. The resulting menstrual blood, which does not coagulate, flushes out the desquamated tissue debris. At the same time a new follicle in the ovary begins to develop and produce oestrogens that initiate the next proliferative phase. The role prolactin plays in the menstrual cycle is largely obscure. Contrary to earlier opinions, the function of the corpus luteum in man is not dependent on prolactin.

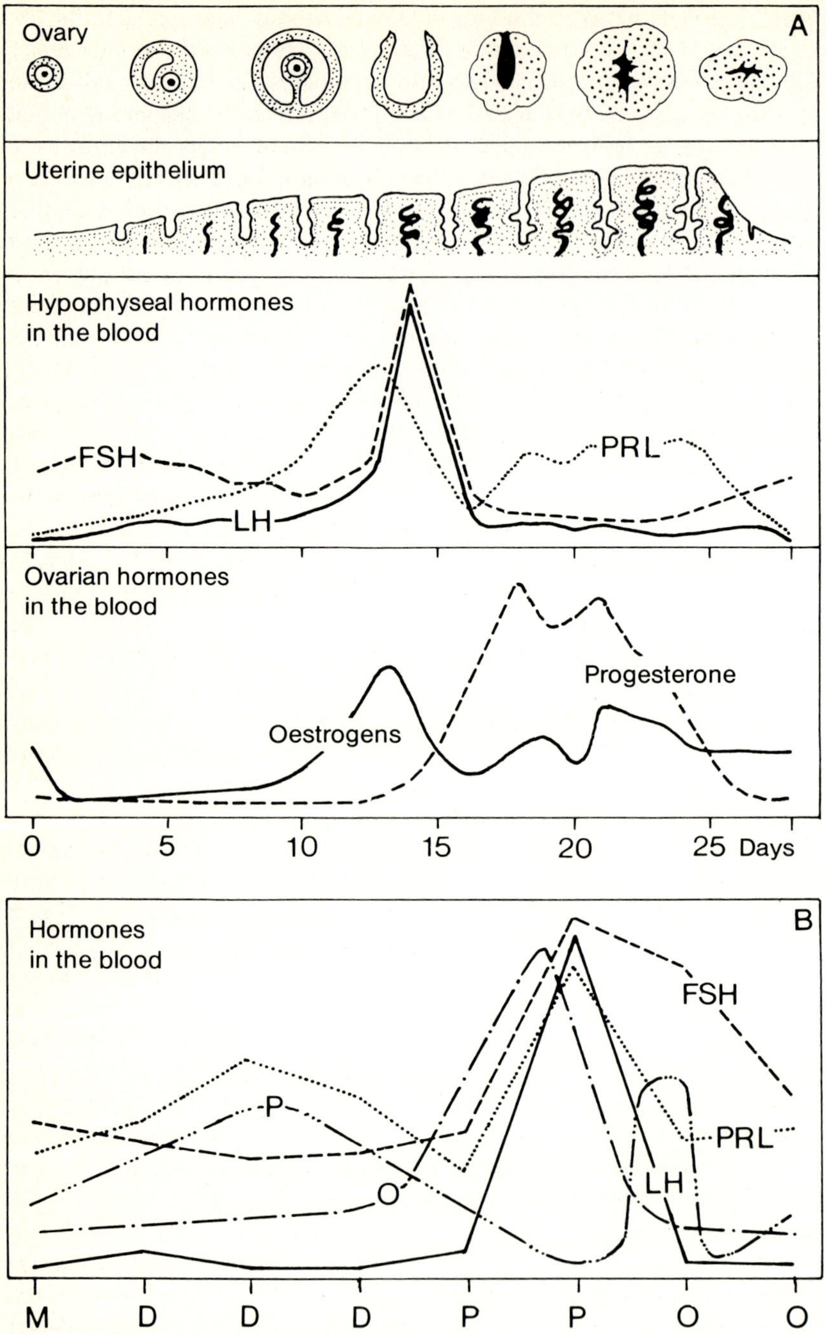

Fig. 63A,B. Endocrinology of the human menstrual cycle (**A**) and the oestrus cycle of the rat (**B**). (After Kalra et al. and Bentley, respectively).
M = metoestrus; D = dioestrus; P = proöestrus; O = oestrus; FSH = follicle stimulating hormone; LH = luteinising hormone; PRL = prolactin; P = progesterone; O = oestrogens

The oestrus cycle of the rat will be taken here as an example of the third type of cycle. It has a period of about 4 days and can be divided into 4 phases. Thus oestrus takes from 9 to 15 h and is followed by a metoestrus phase lasting from 10 to 14 h. The ensuing dioestrus phase, with a duration of from 60 to 70 h, is the longest part of the cycle and is followed by a proöestrus phase. The cycle is evidently subject to a diurnal rhythm, ovulation always occurring shortly after midnight. During metoestrus and dioestrus there is a low level of oestrogens in the blood that stimulate the release of FSH which, together with LH, induces the production of oestrogen in the developing follicle. During metoestrus the uteri are small and in the ovary one can find corpora lutea and small follicles. The corpora lutea regress during dioestrus and the small uteri are only poorly supplied with blood and are only weakly contractile. In the proöestrus phase there is a sharp increase in the concentration of LH in the blood brought about by the stimulatory effect of the relatively high concentration of oestrogens in the blood on the third day of the cycle. At the same time there is a similarly sharp increase in the levels of FSH and PRL. Ovulation is induced by LH, which luteinises 10–12 follicles. At the beginning of the oestrus phase the concentrations of PRL and LH in the blood fall back to their original levels. During the proöestrus phase the "old" corpora lutea degenerate completely and the uteri become filled with fluid and also contractile. The latter are fully filled by the time of oestrus, in which condition the uptake of blastocysts is most efficacious. The level of FSH falls slowly during oestrus. One particular characteristic of the rat's oestrus cycle that clearly distinguishes it from the two other types of cycle described is the way in which the level of progesterone changes. Thus the progesterone concentration rises to a maximum during metoestrus and early dioestrus then falls again during the first two-thirds of the proöestrus phase. After the LH, FSH and PRL peaks the progesterone level again rises rapidly to a second maximum to fall equally rapidly shortly after the beginning of oestrus. However, a corpus luteum is not present at this point in time, possibly due to a luteotropic effect of LH on the interstitial tissue of the ovary. PRL evidently has a luteotropic effect in the rat. The corpus luteum of the rat clearly does not function as a timer as it does in the oestrus cycle of the sheep since it is only small in size and does not persist for any length of time. The changes in the concentration of the different hormones involved in the oestrus cycle of the rat are shown in Fig. 63B. As in males, the effects of FSH and LH on the ovary cannot always be distinguished from each other. Indeed, it is highly probable that these gonadotropins do not themselves act directly on the oocyte. As has already been mentioned, FSH promotes the growth of the granulosa cells whereas LH induces these cells to secrete the liquor folliculi and stimulates the production of oestrogens in the theca interna. Thus FSH and LH can be regarded as acting together synergistically as a gonadotropin complex, as they do in the male sex.

In each of the three examples of the oestrus cycle described ovulation occurs spontaneously under the influence of LH. The release of LH in the rat and the cow, for example, is induced by a high level of oestrogens, which act in two different ways: they cause the release of LH-releasing hormone from the hypothalamus, and at the same time sensitise the LH cells of the adenohypophysis to this hormone. In the rhesus monkey it has been found that oestrogens can also elicit the release of gonadotropin directly from the hypophysis when this organ is sufficiently stimulated by the LH-releasing hormone. Spontaneous ovulation occurs in a wide variety of mammals.

Ovulation in some other animals is, in contrast, not spontaneous but induced by some form of sexual stimulation or the act of copulation itself. Induced ovulation is underlain by a neuroendocrine reflex arc, an example of which (that of the rabbit) has already been described in detail (see p. 163). The domestic cat, the ferret, raccoon and mink are all induced ovulators. But as with spontaneous ovulation, what actually triggers ovulation is LH. Thus the maturing follicle produces oestrogens which elicit sexual behaviour that leads to copulation. Due to neural stimulation of the hypothalamus copulation results in the release of a pulse of LH which triggers ovulation. It is supposed that the mature follicle becomes turgid and that proteolytic enzymes digest the follicle wall so that the follicle eventually ruptures. Prostaglandins may participate in this by causing the microfilaments in the theca to contract. These substances are formed in the follicle under the influence of LH and may act as mediators between the LH stimulus and the rupturing of the follicle.

After oestrus the uterus of mammals undergoes changes that prepare it for the reception of blastocysts. These changes are brought about by the co-operative action of the ovarian steroids oestrogen and progesterone, which respectively promote the growth and differentiation of the tissue, progesterone being the dominant factor. After fertilisation has taken place, the developing blastocyst in some cases embeds itself into the wall of the uterus and develops further. The profound structural and physiological changes associated with this highly complex process of placentation are all directed towards providing the developing embryo with foodstuffs and oxygen etc. and the removal of waste products and gases: placentation will be dealt with thoroughly in Chap. 8. Placentation can however only take place when the uterus remains in an appropriate condition and this is achieved by a complex system of regulatory hormones. This system has been intensively investigated in humans. The most important precondition for a successful pregnancy is a sufficiently high level of progesterone in the blood so that the blastocyst can implant securely. In man, when fertilisation has taken place, the "normal" corpus luteum transforms itself into a corpus luteum graviditatis and continues to produce progesterone. At the same time the trophoblast begins to produce HCG, which can be detected in the urine of pregnant women in the second week of pregnancy. The concentration of HCG increases rapidly and reaches a maximum at the end of the second month of pregnancy, after which it falls steadily to remain at a relatively constant low level from the end of the fourth month onwards. It is possible that HCG stimulates the production of progesterone and oestrogen in the placenta. Whatever is the case, HCG is necessary for the development and functioning of the corpus luteum graviditatis during the first third of pregnancy since the secretion of gonadotropin from the hypophysis is blocked via a negative feedback effect of the increasing concentration of oestrogen and progesterone. Consequently, no new follicles mature in the ovary and therefore the menstrual cycle remains blocked. From about the third month on the concentration of HCS in the blood rises steadily to reach a maximum in the ninth month. Nevertheless, the function of this hormone is obscure. It is possible that it has no direct role in pregnancy but serves primarily to prepare the mammary glands for lactation. The concentrations of oestrogen and progesterone similarly increase steadily during the course of pregnancy, that of the former showing an additional increase shortly before birth. At about this time the muscles of the uterus are activated by prostaglandins and the concentration of pro-

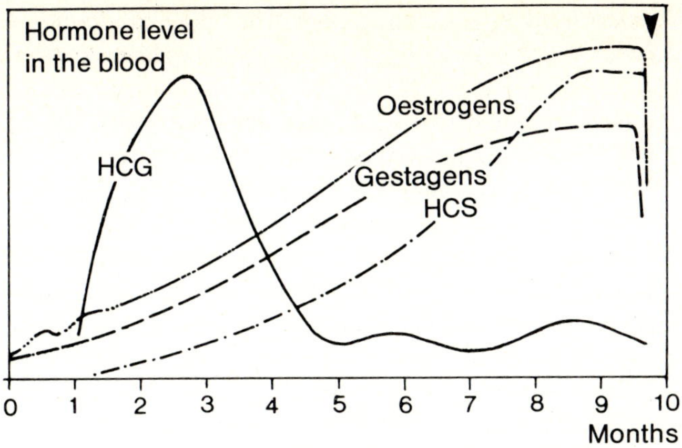

Fig. 64. Level of hormones in the blood during pregnancy in women (simplified). *Arrowhead* marks the time of birth. HCG = human chorionic gonadotropin; HCS = human choriosomatomammotropin

gesterone in the blood falls rapidly followed shortly thereafter by that of the oestrogens. As a result the wall of the uterus becomes sensitised to the action of oxytocin whose affect is then unimpeded by other influences. The embryo exerts a pressure on the cervical region of the uterus which, via a neuroendocrine reflex arc, stimulates the release of oxytocin from the neurohypophysis. Oxytocin then acts on the muscles of the uterus causing contractions that lead on to labour. As already mentioned, the role of oxytocin with respect to parturition in man is uncertain. The birth of the child is followed by the expulsion of the afterbirth, namely the placenta with its attached umbilical cord, and with this the hormones HCG and HCS rapidly disappear from the maternal blood. The changing levels of hormones throughout the course of human pregnancy are shown in Fig. 64.

The progesterone required during pregnancy in humans is thus provided by the corpus luteum graviditatis and the placenta. In other species, such as the rabbit and goat, the placenta produces neither steroids nor gonadotropin, the sole source of progesterone being the corpora lutea graviditates. The uterine luteolysins are evidently inhibited in such species. The number of corpora lutea formed is normally the same as the number of oocytes ovulated, or the number of embryos in the uterus, but in some species, the horse for example, additional "accessory" corpora lutea graviditates are formed which supplement the supply of progesterone. The same result is produced by a reduction in the rate at which progesterone is degraded biologically within the body of the gravid female by progesterone being coupled to binding proteins which in effect shield the hormone. In some species, large quantities of these binding proteins are produced in the liver during pregnancy, probably in response to the influence of oestrogens. The foetus also supplies steroids to the mother since the steroidogenic tissue of the adrenal glands produces two steroids that are metabolites of oestrogen and progesterone, namely dehydroepiandrosterone sulphate and pregnenolone sulphate. These metabolites pass from the foetal to the maternal blood and are converted in the placenta into oestrogen and progesterone, respectively. There are, however, a number of differences between different species. Thus the placenta of the pig can synthesise only oestrogens whereas that of the sheep produces large amounts of progesterone.

The level of progesterone in the blood before birth is not always as high as it is in man. Indeed, in the sheep and ferret it is relatively low. The role of oestrogens prior to birth can be seen to be primarily that of inducing the formation of oxytocin receptors in the muscles of the uterus. Oxytocin appears to be an essential factor in the process of birth in a range of species although its role in man is still unresolved. The hormone relaxin, mentioned briefly above (see p. 184), is very probably also involved in the process of birth by virtue of the fact that it softens the ligaments of the pubic symphysis and so allows the body of the foetus to pass through. In addition to their effect on the uterine muscles in humans mentioned above, prostaglandins also cause a softening of the cervix uteri and are responsible for the breakdown of the corpus luteum graviditatis, in those species in which this structure persists, in the manner described (see p. 185): that is, they have a luteolytic effect.

The foetus also plays a part in determining the time at which it is born in that it produces certain corticosteroids which induce the production of an enzyme that changes the metabolic path taken by progesterone so that it is converted into oestrogen, thereby resulting in a more rapid decrease in the level of progesterone in the blood.

5.12 The Role of the Epiphysis

One organ that remains to be mentioned in connection with the hormonal regulation of the gonads, one which may exert a modulating effect, is the epiphysis. This organ produces several biologically active substances including noradrenalin, serotonin and melatonin, the last of which is regarded as its true hormone. The content of melatonin in the epiphysis of mammals and birds exhibits a daily fluctuation related to the rhythm of light and darkness, it being higher at night than during the day. This is evidently not due to a build up of melatonin during darkness but rather is a reflection of an increased biosynthesis of melatonin, as are the corresponding changes which can be detected in the level of melatonin in the peripheral blood. Thus pulses of melatonin are released whose duration and amplitude are positively correlated with the occurrence of darkness. In other words, the rhythm of day and night is represented by the melatonin pulses. It has been suggested that this representation underlies the ability of animals to adjust to different daily rhythms and their ability to "perceive" and respond appropriately to the changes in the photoperiod that occur during the course of a year by, for example, becoming reproductively active. Indeed, the epiphysis does appear to have an antigonadotropic effect. Thus, if rats are held in continuous light, which must result in the almost complete "shut-down" of the epiphysis according to the above mentioned observations, the weight of their ovaries increases and they enter a permanent oestrus. Again, the testes of Syrian golden hamsters regress when the animals are taken from "long-day" conditions and set in continuous darkness. This effect can also be elicited by injections of melatonin. Similarly, the administration of melatonin results in ovarian atrophy in female mice. Comparable findings have been obtained from reptiles and goldfish, in which case pinealectomy prevents the regression of the gonads that normally occurs as the days get shorter, or results in the gonads increasing in weight. All these observations indicate that melatonin has an antigonadotropic effect. However, the situation is clouded by the fact that the epiphysis of mammals also contains a substance that acts as an antigonadotropin but which is not melatonin. In cattle this so-called pineal antigonadotropin (PAG) is a water-soluble substance with a molecular weight of between 700 and 1000 Dalton which causes a reduction in the concentration of LH in the serum, probably by inhibiting the LH-releasing hormone.

On the other hand, the epiphysis is also reported to have a progonadotropic effect. For example, pinealectomy prevents sexual maturation in Petromyzontia and in sheep the epiphysis appears to inhibit the secretion of PRL. Moreover, the rate of PRL secretion in the sheep is normally low in winter but pinealectomy results in a significant increase in the serum concentration of PRL.

One is still far from being able to define the precise role of the epiphysis in reproduction. Nevertheless, it appears to represent an intermediary organ whereby diurnal or seasonal changes in the photoperiod are transformed into endogenous humoral signals which in turn mediate long-term physiological changes in the organsim.

5.13 The Hormonal Regulation of Lactation

Lactation – the structural and physiological processes associated with the production and expression of milk – is a characteristic of mammals. Among other vertebrates two comparable phenomena are known. Firstly, a nutritious secretion is produced by the skin of teleosts of the genus *Symphysodon* and secondly pigeons produce a so-called crop milk. However, neither of these processes has a level of complexity approaching that of lactation in mammals. The mammary glands are not capable of producing milk early in life but must first go through a process of maturation which is controlled by hormones, whereby one must distinguish between their prepubescent development and the processes occurring after puberty which precede the production of milk. The latter are highly complex metabolic processes, as indicated by the complex composition of milk which includes fats, carbohydrates, proteins, minerals and other essential substances. These metabolic processes are regulated by the co-operative action of different hormones. Consideration of how these processes are controlled necessitates that the development and structure of the mammary glands are also mentioned.

The mammary glands, or mammae, after which Mammalia are named, are derivatives of the integument, i.e., they are epidermal glands that represent modified hair follicles and sweat glands. The mammary glands of Marsupialia play a vital role in reproduction since placentation in these animals lasts only a short time and the subsequent development of the young depends almost solely on the production of milk. On the other hand, the mammary glands of Monotremata are of lesser importance since the young live for a long time off the large reserves of yolk in the egg. Compared with those of Marsupialia, the mammary glands of Eutheria also play a minor role as their young are already relatively well developed when they come into the world: young guinea-pigs, for example, can survive without their mother's milk. The macroscopic anatomy of the mammary glands varies considerably among different mammals. Thus, instead of teats Monotremata have milk fields in the abdominal region of their body which are composed of about two hundred tubular epidermal glands, each of which opens to the base of a hair follicle. The latter also have an associated sebaceous gland. In this case the milk is a fatty secretion which the young lick from the milk fields. The mammary glands of Marsupialia and Eutheria, which bear teats, vary in number. Many Rodentia and the rabbit have 5 or 6 pairs of glands situated between the shoulders and the groin whereas those of Carnivora and the pig lie in the thoraco-abdominal region. Several pairs of mammary glands are laid down early in development in primates, elephants and bats but only one pair persists in later life. The mammary glands of Ruminantia fuse to form a specialised gland in the abdominal region, the udder. This topic will be taken up again in Chap. 9 (see p. 330).

Mammary glands are laid down during embryonic development in a process which is probably independent of hormonal influence. The teat, or nipple, and the primary duct are then formed by the outgrowth and ingrowth of the epidermis. Early in development a system of ducts develops within a richly vascularised region of fatty tissue. These ducts have rounded ends that exhibit a high mitotic activity. Branches to the ducts then arise, in a similar way as in the root-tip meristem of higher plants, forming a highly branched system of ducts, the so-called milk tree. Alveoli subsequently develop at the ends of the ducts and these cluster together during the course of development to form lobules. This then is the typical structure of the mammary glands of a gravid mammal. The terminal lobules are the main sites of secretory activity although secretion is also produced in the terminal sections of the ducts and the alveoli stemming from the larger ducts. Early in pregnancy the alveoli have a small lumen enclosed within a relatively thick epithelium and they secrete a pre-milk consisting of a watery secretion that contains lipid. At a later stage the milk of mammals contains two other characteristic components, namely the protein casein and the carbohydrate lactose, which change the pre-milk into a yellowy white, in many cases, opaque milk. The composition of milk varies greatly among different mammals. Its fat content lies between 0.3% (rhinoceros) and 49% (dolphin), its lactose content between 0.3% (whale) and 7% (man), and its protein content between 1.2% (man) and 13% (whale). At a late stage in pregnancy the alveoli are swollen but do not as yet produce any milk. Fat droplets are present both intacellularly and extracellularly. Later, when lactation is well under way, the lumen of the alveoli are even more voluminous and the epithelium becomes flattened. The developing and physiologically active alveoli are entwined in myoepithelial basket cells. In the virginal mammary gland there is a large amount of fatty tissue present which is progressively replaced by glandular parenchyme during the course of development.

The development of the mammary glands can be divided into five steps: the development of functional teats or nipples, of the duct system, of the alveoli, the production of milk and lastly the expression of milk. All these processes are controlled by hormones.

The development of the teats is independent of other processes and is controlled solely by oestrogens, which promote their full development.

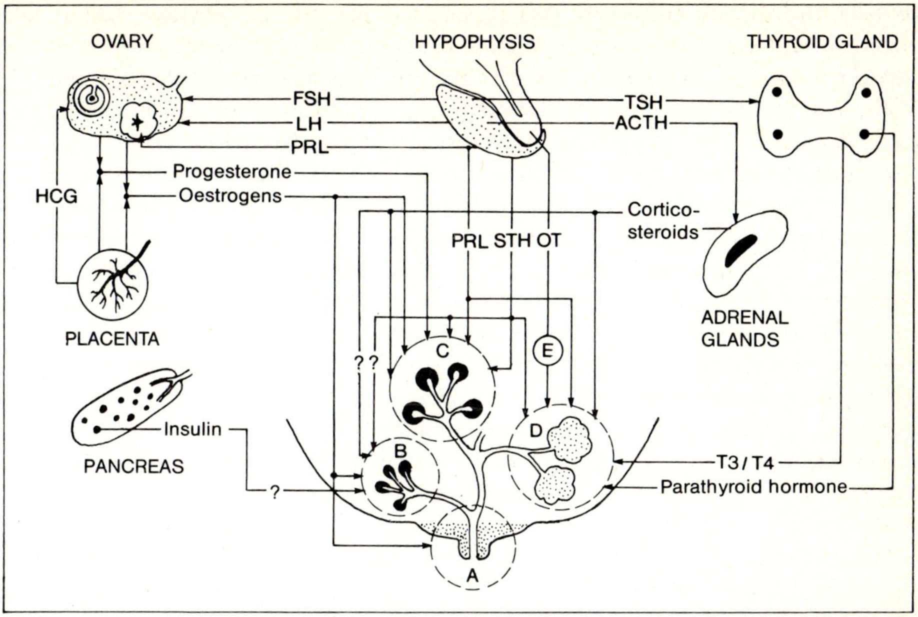

Fig. 65. Schematic representation of the hormonal regulation of lactation in mammals. A = formation of teats, or nipples; B = formation of a ductal system; C = formation of the alveolae; D = milk production (lactogenesis and galactopoesis); E = release of milk; TSH = thyroid stimulating hormone; ACTH = adrenocorticotropic hormone; HCG = human chorionic gonadotropin; PRL = prolactin; STH = somatotropic hormone; OT = oxytocin; T_3/T_4 = triiodo- and tetraiodothyronine

The development of the duct system similarly occurs primarily under the influence of oestrogens. What other influences operate is difficult to say: PRL is one possible influence since it appears to act synergistically with the oestrogens on the proliferation of the duct epithelium: there is also evidence that corticosteroids and insulin are involved.

The control of alveolar development differs in different species. In the mouse, which has a very short oestrus cycle without a functional luteal phase, there is a slight development of the alveoli even in virgin animals, the hormones involved being a combination of oestrogens, STH and corticosteroids. In species with an active luteal phase, however, alveolar development in virgin animals is brought about by progesterone and PRL. Similar changes in the mammary glands also occur during the course of the oestrus cycle in such species. During early pregnancy the development of the alveoli is stimulated synergistically by oestrogens, PRL and progesterone, and possibly also by STH and corticosteroids. Progesterone is referred to as having an alveoli-mammogenetic effect on the mammary glands whereas oestrogen exerts a ductal mammogenetic effect.

The production of milk can be divided into the process of lactogenesis, the intial production of milk, and galactopoesis, the continuous production of milk. Lactogenesis is initiated by PRL but only when it acts together synergistically with corticosteroids, thyroid hormones and STH. Notably, oestrogens and progesterone are not involved, necessarily so if milk production is to proceed since these hormones inhibit lactogenesis and are responsible for the drying up of the mammary glands at the end of the lactation period, although oestrogens alone are necessary to bring this about in some species. The hormonal regulation of galactopoesis is difficult to distinguish from that of lactogenesis. The primary agent is again PRL but there is evidence indicating the involvement of thyroid hormones, corticoids, parathyroid hormone (a hormone from the parathyroid glands involved in calcium metabolism) and hormones of the pancreas. There are also differences between species with respect to the control of lactogenesis. Thus in the case of the rabbit it is induced by PRL alone but PRL, STH, corticosteroids and the thyroid hormone triiodothyronine are required in the case of the goat. In the rat and mouse the combination of ACTH and PRL is effective in initiating lactogenesis. Such groups of hormones acting synergistically to bring about lactogenesis are referred to as lactogenic complexes.

The last process to be considered is the expression of milk, which is regulated via a neuroendocrine reflex arc. The stimulus of the young animal suckling or pressing on the teat excites an afferent nerve fibre and leads to the release of oxytocin from the neurohypophysis. The oxytocin is then conveyed in the blood to the mammary glands causing the myoepithelial cells to contract, thereby squeezing milk out of the alveoli into the ducts. The nervous impulse in this reflex arc travels from the spinal column to the mid-brain and thence to the hypothalamus. Suckling is not the only stimulus that can trigger the expression of milk: it can also occur in response to visual, tactile or acoustic stimuli, in which case it is a conditioned reflex. The continuation of lactation is dependent upon the continued expression of milk from the mammary glands since, as a rule, when milk is no longer expelled from the alveoli the glands atrophy. In Eutheria the stimulation of the teat also inhibits ovulation, probably by β-endorphin causing a decrease in the secretion of gonadotropin-releasing hormone. Furthermore, this stimulus induces the release of PRL from the hypophysis.

The mammary glands are capable of synthesising triglyceride, lactose and proteins, although small amounts of the latter may also come from the blood. The mammary glands can only do this if the animal as a whole is capable of supplying the necessary metabolites. Therefore, with respect to the hormonal regulation of the development and activity of the mammary glands, one must distinguish between those hormones which affect the glands directly and those which act upon the general metabolism of the body. Indeed, the phenomenon of lactation is the most comprehensive of all the multi-hormonal processes which occur within the vertebrate body. The control of lactation is shown schematically in Fig. 65.

5.14 The Role of Hormones in Sexual Differentiation

Although sex is evidently genetically determined in all vertebrates, hormones can influence the differentiation of the gonads in lower vertebrates and the development of the reproductive ducts in mammals and, in certain circumstances, even cause the appearance of a feature characteristic of a sex that does not correspond to the genetically fixed sex. Moreover, various species of teleost undergo a natural sex reversal (see p. 32) that can be mimicked experimentally by treating them with oestrogens or androgens. Such treatment can also cause sex reversal in a few species of amphibian, although only in juvenile animals. However, it is not known to what extent the sex steroids are in fact involved in the differentiation of the gonads or the natural reversal of sex.

The hormonal regulation of the development of the characteristic reproductive ducts of each sex has been investigated extensively in mammals. For experimental work one starts with gonads at a stage of development at which it is already clear whether they are ovaries or testes. The genital tract of the foetus initially develops in the direction of female irrespective of the genetic sex. In genetic males, however, the foetal testes produce a hormone that suppresses the development of the existing female ducts and androgens that promote the development of male ducts. The former hormone, so-called factor X, is responsible for suppressing the development of the müllerian ducts: it is also referred to as anti-müllerian duct hormone (or AMH). This hormone is produced early on in development by the Sertoli cells of the foetus and is probably a protein or glycoprotein with a molecular weight of between 200,000 and 300,000 Dalton. The differentiation of the reproductive ducts in the direction of male or female is determined only by the presence or absence of androgens and factor X: oestrogens and gestagens are not involved. When both androgens and factor X are present the ducts are male but when both are absent the ducts are female. If androgens are present but not factor X then müllerian ducts develop in addition to a complete male urogenital system (Fig. 66). Most of these observations relate to rabbits castrated while still in the uterus but, so far as is known, the basic principles of regulation they illustrate also appear to hold for other mammals. Among mammals, factor X is evidently not specific to each species. It also occurs in chickens but whereas chicken AMH is effective in both chickens and mammals, mammalian AMH is not active in chickens. The fact that chickens possess a factor X does not mean however that sexual differentiation in birds is regulated by the same hormonal mechanism as in mammals.

The initial development of mammals towards being female irrespective of their genetic sex involves not only the reproductive tract but also the central nervous system. Thus the cyclical release of gonadotropins from the hypophysis is much more pronounced in females than it is in males. The release of hormones is regulated by a neural element within the hypothalamus which can be either "male" or "female". Moreover, the neural centres responsible for sex-specific behaviour are also developed to a different degree and have different patterns of synaptic connection in the two sexes. During a certain phase of foetal development, or during the first day of life in those species with a short gestation, these areas of the brain become sexually "imprinted" on one sex or the other. This always occurs after the differentiation of the genital system. As in the case of the reproductive ducts, the only determining factors are the androgens produced by the foetal testes, which change a pre-existing female principle into a male one. If the brain is not subjected to the influence of androgen it develops in a feminine direction irrespective of the genetic sex of the animal. Alternatively, testosterone from the foetal testes reaches the brain and diffuses into the nerve cells, though curiously it is not itself the active principle but is converted in the cytoplasm

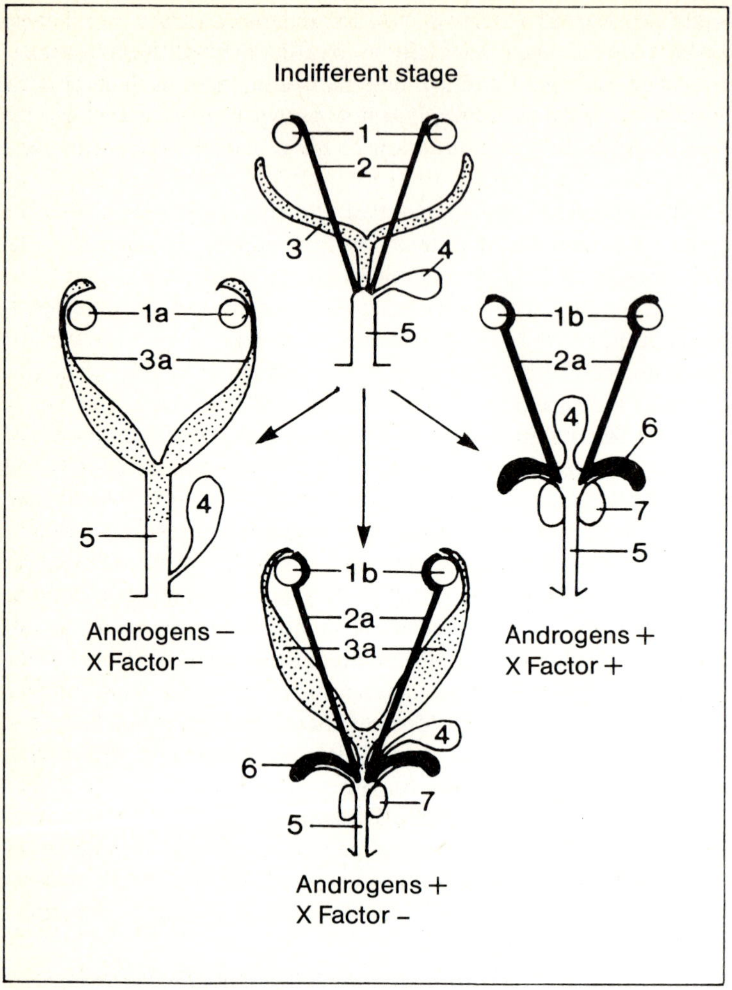

Fig. 66. Differentiation of the mammalian genital system in the presence (+) and absence (–) of androgens and factor X. (After Neumann et al.).
1 = gonad; 1a = ovary; 1b = testis; 2 = undifferentiated wolffian duct; 2a = differentiated wolffiand duct; 3 = undifferentiated müllerian duct; 3a = differentiated müllerian duct; 4 = vesica urinaria; 5 = sinus urogenitalis; 6 = glandula vesicularis; 7 = glandula prostatica

into oestrodiol-17 β which is then bound by a receptor in the cytosol. This complex passes into the nucleus and activates chromosomal DNA, leading to the formation of mRNA and thus of certain proteins which are evidently necessary for the nerve cells to develop in the masculine direction. This appears all the more extraordinary since the female foetus also produces oestrogens (and receives them from the maternal blood) which logically would automatically lead to the reversal of the imprinted sex. However, there is a highly selective mechansim that prevents this happening. The liver of the foetus produces so-called α-foetoprotein which, in rats, can be detected in the liquor cerebrospinalis up to the third week post partum and which specifically binds oestrogens but not testosterone. Thus the oestrogens are "captured" and rendered ineffective before they can reach the brain cells. Most of the investigations into the regulation of the sexual differentiation of neurones have been performed on the rat but the findings are probably applicable to mammals in general. Results relating to members of the other vertebrate classes are indecisive: the observations made so far on lower vertebrates indicate that the processes of reproduction are controlled by neural elements that are not sex-specific.

5.15 Pheromones and Reproduction

Aside from hormones, which play an integrative role within the animal, there is another group of chemical signal substances, the pheromones, whose function in co-ordinating events extends beyond the individual. These substances are produced by one animal and released into the environment in some way. Another animal of the same species then perceives the pheromone by either olfaction or gustation and makes some response to it. In this way pheromones can co-ordinate the actions of the individual members of a population. One can recognise two different types of pheromone, signal pheromones and primer pheromones, although they cannot be clearly distinguished from one another in many cases. The former, as their name suggests, convey a signal to another individual which responds to it within a short span of time by performing some behavioural reaction. In the context of reproduction signal pheromones serve as sexual attractants. For example, the scent of a sexually mature male is used to attract a female, or vice versa. Other signal pheromones are only indirectly involved in reproduction in that they serve to mark out a territory or as a signal whereby sexual rivals can be recognised. Signal pheromones are therefore characterised by the fact that they cause a relatively rapid reaction in the central nervous system of the receiving individual. Primer pheromones, in contrast, act on the receiver's hormonal system and bring about long-term physiological changes by activating different systems of hormones via a neuroendocrine reflex arc. In mammals, several examples of this kind of phenomenon have been investigated in depth.

Pheromones have a highly heterogeneous chemical nature, although only relatively few have been studied in detail. For example, it is known that a mixture of short-chain fatty acids in the vaginal secretion of the rhesus monkey and in the secretion of the anal gland of mongooses (*Herpestes*) acts as a pheromone. In many other cases this mixture includes valeric acid as an additional component of the pheromone, as for example in the secretion from the anal gland of the fox and the lion. Musk-like substances such as muscon and zebeton are common: these are macrocyclic compounds resembling steroids. Indeed steroids can also act as pheromones, as does a group of Δ16-steroids in the boar, for example.

Signal pheromones play a role in the reproduction of species from all classes of vertebrate. Female allice shad (*Alosa* sp.) produce a pheromone that attracts the male of the species and this may also be the case in Salmonidae. In Blenniidae there are glands associated with the first and second spines of the anal fin that produce a pheromone which attracts females. Such a substance is evidently present in the ovarian fluid of goldfish and it signals to the male that the eggs have been laid. A comparable pheromone is produced by females of the genus *Bathygobius* which promotes courtship behaviour in males. This phenomenon has been made use of commercially by fishermen in the case of *Ictalurus punctatus* in that a sexually mature female is used as bait to lure large numbers of males. The action of the pheromone of *Brachydanio rerio* is more complex since its effect also depends on the amount of pheromone released. The low concentration of pheromone released by a single male attracts other males that produce the same substance, thereby increasing its concentration in the water. Only then are the females drawn by the pheromone to the scene of events. The females in turn release their own pheromone which serves to attract both males and females. However, when the concentration of this substance reaches a certain level additional females arriving on the scene are repelled and swim away.

Pheromones are common among amphibians, at least within Urodela, the only group for which evidence is available. The cave salamander (*Proteus anguineus*) and salamanders of the genus *Triturus* will serve here as examples. Male cave salamanders can distinguish sexually mature females from males according to their pheromones and can track rivals back to their holes. Pheromones are of particular importance in this species since they have lost the sense of sight. Males of the genus *Triturus* release pheromones during mating which stimulate the female.

Pheromones evidently play an important role in the reproduction of reptilians. Thus *Testudo radiata* can recognise the sex and degree of sexual maturity of a conspecific from its cloacal scent: the smell of the faeces is also important. The anal sac of lizards produces a strongly smelling secretion which similarly facilitates recognition of the sexes. Crocodiles have glandular sacs in the throat region which release a secretion that smells strongly of musk. This scent, together with that produced by the cloacal glands, plays a role in the mating behaviour of crocodiles. The so-called thigh pores of lacertilians probably produce a pheromone. Its production appears to depend on a hormone in the case of the sand lizard since the organ atrophies after castration.

Very little is known about whether signal pheromones are involved in the reproduction of birds except that ducks produce a secretion in their rump glands whose composition during the reproductive period is different in males and females. This secretion supposedly acts as a sexual pheromone in the domestic duck (*Anas platyrhynchos*).

Sexual attractants are common among mammals. Female rats produce scents during oestrus which serve to attract males and which probably originate from a musky smelling secretion of the female's glandulae praeputiales. In males these glands also appear to secrete a pheromone. The vaginal secretion of female hamsters contains an attractant that induces the male to go through the motions of copulation, this being an innate behaviour pattern released in response to the pheromone even when all other relevant stimuli are lacking. However, if their sense of smell is blocked the males fail to react in this way. The anal glands of beavers secrete an attractant that serves to stimulate the male, as do the secretions of the vaginal and anal glands of female dogs. The urine of female cattle, horses and camels apparently contains a pheromone that acts as an aphrodisiac on the males, as does the vaginal secretion of the rhesus monkey. A group of Δ16-steroids produced by the Leydig cells in the testes of the boar have been identified as pheromones. It is noteworthy that the biosynthesis of these substances takes a different course to that of androgens although in each case pregnenolone and progesterone are their precursors. The release of these so-called odour steroids into the blood is evidently dependent upon LH. From the blood they pass into the salivary glands and into fat tissue where they are stored. They are mainly secreted with the saliva and via the sweat glands. Their effect on the sow, who becomes highly excited when ready to mate, is to elicit the so-called patience reflex whereby she stays relatively still, for if she did not copulation would not be possible. The phenomenon has important economic consequences since the pheromones are responsible for the musky smell of the boar and not only is this unpleasant to the human nose but the stores of these substances in the fatty tissues renders the meat of the boar practically unmarketable. One of these steroids (5Δ-androst-16-en-3-one) also occurs in the serum of men and in considerably higher concentrations than in the serum of women. It is excreted in the sweat from the axillae of men.

The examples mentioned above are intended merely to convey a general impression of the roles played by signal pheromones in reproduction – a thorough treatment of this topic would go well beyond the scope of this book. It is nevertheless pertinent to note that these substances also play an important role in territoriality and social behaviour which in some cases is closely related to their rôle in reproduction, and this is especially so in the case of the recognition of individuals. Thus the so-called parent or maternity pheromones are an important aspect of parental care since they enable the parent animals to identify their young. An example of this is provided by a species of teleost from the genus *Cichlasoma* in which case the nurturing parents can recognise their young by their smell. Similarly, young rats are evidently drawn to the mother rat to suckle by a scent contained in her faecal pellets and which she produces up until the young are about 33 days old. There is some evidence to indicate that the production of this scent depends on PRL.

The only primer pheromones known to be involved in reproduction occur in mammals and, understandably, the bulk of the relevant observations have been made on laboratory animals. For example, juvenile female mice become sexually mature earlier when in the presence of a sexually mature male due to a pheromone in the urine of the male. Experimentally this hormone has been found to increase the release of LH from the adenohypophysis. This presumably results from an induced elevation in the level of gonadotropin in the blood. The pheromone of the sexually mature male mouse can also "normalise" and synchronise the oestrus cycle of female mice, even if it is irregular. This stimulatory effect only lasts for one cycle but can be prolonged by constantly changing the male. A similar phenomenon is known to occur in rats. A further effect of the primer pheromone found in the urine of sexually mature male mice and rats (as well as guinea-pigs and possibly sheep and goats) is that it causes a shortening of the female's oestrus cycle by up to 25%.

The synchronisation of the annual breeding period of *Lemur catta* is very probably regulated by a pheromone. The smell of the pheromone causes the progesterone level in the female's blood to fall thereby precipitating the development of a new follicle and initiating a new oestrus cycle. Ovulation in the vole *Microtus agrestris* is triggered by a pheromone from the male which is only effective over a very short dis-

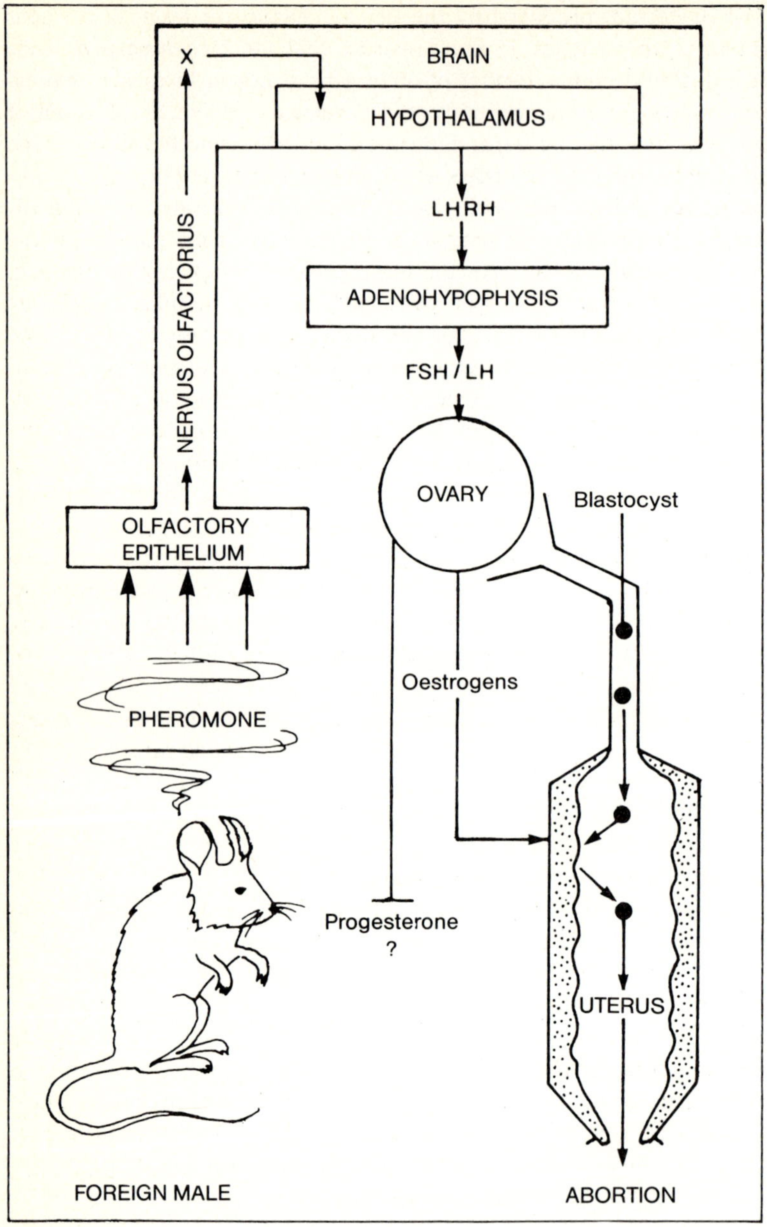

Fig. 67. Schematic representation of the neuroendocrine reflex underlying the BRUCE effect in the mouse

tance: it is smelled by the female during mating. There is a reciprocal effect of one female mouse on the oestrus cycles of other mice, and vice versa, which is evidently mediated by a pheromone. Thus spontaneous pseudo-pregnancies frequently occur in conditions of low population density whereas above a certain density the female mice have irregular cycles with long anoestrus periods. Blocking the sense of smell abolishes these effects. Contrarily, there is evidence that close contact between women can lead to a synchronisation of their oestrus cycles. A study carried out in a women's college has shown that close friends and room-mates show a significant synchronisation of their menstrual cycles during the course of a semester.

Perhaps the most spectacular and therefore the most studied effect of a primer pheromone is the so-called BRUCE effect: that is, when female mice are brought together with a male of another strain during early pregnancy (before implantation of the blastocyst) the pregnancy is frequently aborted. This effect is induced by a pheromone in the male's urine which the female smells, evoking a nervous impulse that travels along the olfactory nerve to the telencephalon and thence to the hypothalamus where it causes the release of LH-releasing hormone. This in turn results in the release of gonadotropin from the adenohypophysis. The ensuing production of oestrogens in the ovary disrupts the development of the uterine epithelium so that it is no longer in an optimal state for the implantation of the blastocyst and consequently there is a good chance of the pregnancy being aborted (Fig. 67). If the female is treated with PRL or progesterone during the critical phase the pheromon has no effect. It has therefore been suggested that the hormonal component of this neuroendocrine reflex arc normally involves an inhibition of the release of prolactin or progesterone. The BRUCE effect has also been demonstrated in the New World deer mouse *Peromyscus maniculatus bairdi.*

CHAPTER 6

Reproductive Behaviour

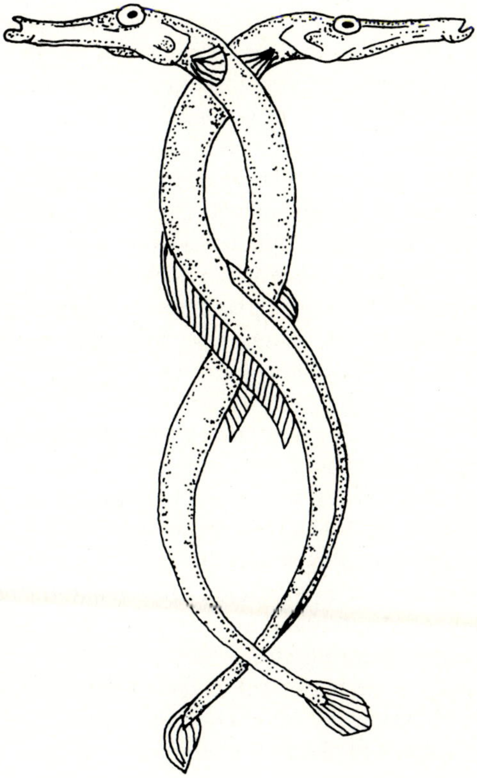

Sea needles of the species *Siphonostoma floridae* in the act of spawning

6.1 General Definitions

The term behaviour can be broadly defined as the adaptive response of an organism to changes in the environment. The response may be internal or external whereas the environmental change may be biotic or abiotic. Behaviour is manifest as a pattern of movements which are made up of a series of muscle contractions co-ordinated in space and time. These activities range from a part of the body being moved once to a number of individual movements, which may be the same or different from one another, being performed in a protracted sequence. Movement may indeed cease and the animal remain completely still, playing dead. Changes in body colour, the activity of glands, or the production of sound can also be interpreted as behaviour, in its widest sense, but not "movements" which result from the process of growth.

As intimated above, behaviour can be triggered not only by exogenous factors but also by endogenous factors. Thus behaviour can be elicited in the absence of any external stimulation from the environment by endogenous stimuli, such behaviour being referred to as spontaneous behaviour (e.g. waking from sleep) or as an "action". In contrast, behaviour triggered by exogenous stimuli can be referred to as a response or a "reaction" which is triggered at a particular point in time by a definable cause.

There are basically two types of behaviour: that which is genetically fixed and thus innate and that which is learned and founded in experience. In the former case all the components (e.g. the pattern of synaptic connectivity in the central nervous system) are genetically fixed and thus follow a predetermined programme (e.g. movements associated with breathing). In the second case, such predetermined programmes are "mixed" with the engrams of earlier experiences, whereby the latter components may predominate.

In higher animals the simplest stimulus-response unit is a monosynaptic reflex consisting of a sensory and a motor neurone and an associated effector organ (e.g. the knee-jerk reflex). Although a single reflex action can be regarded as behaviour, it is usually only one component of a more complex sequence of behaviour. The greater the number of neurones mediating a reflex action, the more complex it becomes. Behaviour effected by one or several reflexes is referred to as stereotype behaviour. As a rule the behavioural repertoire of invertebrates includes more reflex actions – stereotype behaviour – than does that of vertebrates, which is based more on experience: experience becomes increasingly more important as one climbs the evolutionary ladder. Nevertheless, the sexual behaviour of vertebrates also includes reflexes. In man, for example, erection of the penis and ejaculation are spinal reflexes. The former involves the stimulation of mechanoreceptors in the glans penis. Nervous impulses are then conveyed to the spinal column via afferent fibres resulting in the excitation of efferent fibres which cause the dilation of arterioles in the erectile tissues and the production of secretion in the accessory glands. Ejaculation is induced in a similar or identical way, and may involve oxytocin from the neurohypophysis (see p. 179). Thus tactile stimulation of the glans penis leads, via a neuroendocrine reflex arc, to rhythmical contrations of the muscles of the ductus deferens and the base of the penis. The former muscles force semen through the canalis urogenitalis, expelling it in spurts. The contractions of the muscles at the base of the penis occur automatically in a pattern which is controlled by neurones. These two reflexes can however be influenced via their connections to higher centres in the nervous system. For example, visual stimuli can trigger erection whereas ejaculation can be brought on or delayed at will.

Fixed action patterns are sequences of reflexes or of impulses in motor nerves which are induced by a specific external stimulus. In some cases the control of a behavioural sequence requires that the stimulus be continuously repeated whereas in other cases the whole sequence is triggered by a single stimulus event. The basic sequence in which the muscles contract is however pre-programmed within the central nervous system. In some species of birds the sight of an egg which has rolled out of the nest triggers an egg-rolling response in the parent bird in which it pushes the egg back into the nest with forward and sideways motions of the beak. The forwards and backwards motion with the beak is triggered solely by the sight of the egg and may continue for a short time even after the egg is removed. On the other hand, the sideways movements are performed independently of the forwards-backwards motion and are triggered only as long as the egg continues to wobble.

Taxes are automatic movements which orientate the body towards some particular factor in the environment. They are common among invertebrates but also play a role in the reproductive behaviour of certain vertebrates, such as behaviour elicited by scents (see p. 205).

Much of the behaviour associated with reproduction is motivated. Motivated behaviour is initiated by endogenous stimuli and is largely under the control of the hypothalamus and the telencephalon (in mammals the cerebral hemispheres). The different types of motivated behaviour are only triggered by "adequate" external stimuli and are highly specific to particular states of the animal: they include behaviour related to feeding, drinking, courtship, copulation, parental care, and territoriality etc. Although the hypothalamus plays the major role in the regulation of such behaviour via either facilitatory or inhibitory centres, the limbic system and the cerebral hemispheres also play a part. Motivated behaviour is also influenced by other factors: for example, previous experience can lead to the inhibition or discontinuance of such behaviour and certain hormones can have an influence. Reproductive behaviour can include very complex elements which are performed one after the other, each of which may be motivated variously by aggression, flight, and sexuality, and the same applies to mating behaviour although in this case the sexual component predominates. Copulation is typically motivated by the sexual drive alone, whereas behaviour associated with parental care is usually motivated by parental drive and, to a varying degree, aggression. Such drives are due to the effect of hormones on the structures in the central nervous system responsible for generating reproductive behaviour. The hormones may serve to induce or to modulate the performance of a particular behaviour, which is commonly triggered by a highly specific exogenous factor. Other feedback signals indicating the consumation of a drive can also induce the start of a new phase of behaviour.

6.2 Phases of Reproductive Behaviour

The different types of behaviour associated with reproduction normally occur in cycles having a circadian, seasonal or circannual rhythm which are often closely correlated with certain environmental factors, although they can also be manifest in the absence of such factors. General physiological states such as sleep and wakefulness are usually subject to a circadian rhythm. The reproductive cycles of vertebrates often demonstrate a distinct periodicity, escpecially in the case of female mammals. As mentioned in Chap. 1, some species only have one phase of reproduction per year (monocyclic species) whereas others have several (polycyclic species). The intervening period may be only a matter of days in the latter species. The reproductive cycle is often entrained by environmental factors, but there are also those which run purely according to an endogenous, genetically fixed rhythm. In the former case the environmental rhythms are frequently superimposed on the endogenous rhythms. The continuous oestrous cycle of many female mammals is regulated primarily by a fixed, endogenous rhythm effected by a mechanism (see p. 189) involving the central nervous system and hormonal components, which also influence reproductive behaviour.

Reproductive behaviour can thus be seen as a sequence of behaviour patterns controlled by a complex interplay of several factors. The animal receives various environmental stimuli through its receptor organs which are conveyed to the central nervous system. In the simplest case this directly triggers a response, which takes the form of a reflex or a chain or reflexes. Hormones may also be involved in this as intermediaries subordinate to the controlling influence of the hypothalamus which regulates their production or release according to both exogenous and endogenous stimuli. The hormones then activate or change the threshold of elements within the central nervous system responsible for a particular behaviour, thereby initiating, facilitating or inhibiting its performance. The performance of a particular behaviour may in turn have a feedback effect, via the higher nervous centres, on the endocrine system, causing an increase or decrease in the production or release of hormones. It is also possible that such neuroendocrine systems are co-ordinated within a particular time-frame by endogenous or exogenous rhythms. Endogenous stimuli that do not arise from either the central nervous system or the endocrine system can similarly act as modulating factors at each of the different levels of this scheme. However, a distinction between innate and learned behaviour has yet to be made. This is often a difficult distinction to make, particularly in those cases where the pattern of behaviour is subject to ontogenetic change, i.e. maturation. The problem is most accute in highly evolved species in which the ability to learn is most pronounced. Innate patterns of behaviour can even be suppressed in such animals as a consequence of previous experience so that the behaviour of the animal is, in a sense, a sham. This is particularly true in the case of human behaviour.

A further distinction must also be made in connection with the above. It has been shown that sensory-motor co-ordination involved in the performance of a behavioural pattern can be perfected by practice, hence the old adage "practice makes perfect". Thus experience and practice can be seen as influences whereby a pattern of behaviour can be perfected. Such practice can be regarded as an accumulation of experience by individual animals. An exception to this is imitative behaviour, or learning from others. The development of the song pattern in birds is in most cases a combination of practice and imitation. Other than to note that learning implies the capacity to store experiences and to recall them again, the complexities of the phenomenon of learning will not be enlarged upon here.

According to both ethological and physiological criteria the reproductive behaviour of the majority of vertebrates has two phases, a sexual phase and a phase of parental care. The latter phase may be of only short duration or indeed, in many lower vertebrates, may not be in evidence at all. In general the objective of the sexual phase is the release of gametes or fertilisation, i.e. spawning or copulation. The sexual phase can consist of several subphases. In many cases it begins, in the male sex, with the demarcation and defense of a mating ground or territory which may be held permanently or temporarily and vary considerably in size. This territorial phase can be taken to be the first subdivision of the sexual phase. A characteristic behaviour during this phase is the marking out of the boundary of the occupied area. In addition, increased aggressiveness is usually shown towards intruders and neighbouring animals, as evidenced by boundary conflicts and conflicts with sexual rivales. These are frequently ritualised conflicts which are triggered by a particular signal. Territoriality can change from being an aspect of reproduction to being associated with, for example, the maintenance of an adequate food supply There are many species that are not territorial.

The phase of establishing and defending a mating ground leads on to a second phase of nest building and courtship. One or more females are wooed into the territory by the display of certain signals. Prior to this or at the same time, one or both of the sexual partners often build either a nest for the eggs or brood or performs some equivalent activity. For example, some species of bony fish clean the site where the eggs are to be laid. In some cases a part of the body takes on the function of a nest, for example the honeycomb structure on the back of Surinam toads (genus *Pipa*) and the brood pouch of the male sea-horse (see p. 235). The courtship behaviour of the male is usually the adequate releaser for eliciting the sexual behaviour of the female, just as some aspect of the female usually acts as a releaser of certain behaviour patterns in the male. Although mutual role-play is the rule there are exceptions where it is the female which plays the active role in courtship (e.g. Syngnathidae). These activities culminate in spawning or copulation, which represent the third and final subphase of the sexual phase, its climax so to speak. The sexual partners normally release their gametes, or copulate, after performing a series of characteristic actions, i.e. courtship. The release of gametes and copulation similarly involve characteristic patterns of behaviour. The final subphase may be repeated once or several times together with the second phase. This concludes the sexual phase of reproduction. All the activities associated with the following phase of parental care are directed towards caring and providing for the offspring. One sexual partner, often the male, may leave and take no further part in the reproductive cycle but it is more usual that both parents participate in caring for their young. The phase of parental care is the second longest phase in the reproductive cycle after the sexual phase with its three subdivisions. As mentioned above, some species do not have such a phase. The simplest types of parental care occur in the various ovuliparous and oviparous species of vertebrate. It begins with the care of the fertilised eggs. Many species of bony fish which spawn on the substrate clean their eggs by sucking them into the mouth thereby covering them with mucus, which may possibly have an antibacterial effect. The parent fish also fan oxygenated water over the developing embryos with characteristic movements of their pectoral fins. Other examples include midwife toads which carry their

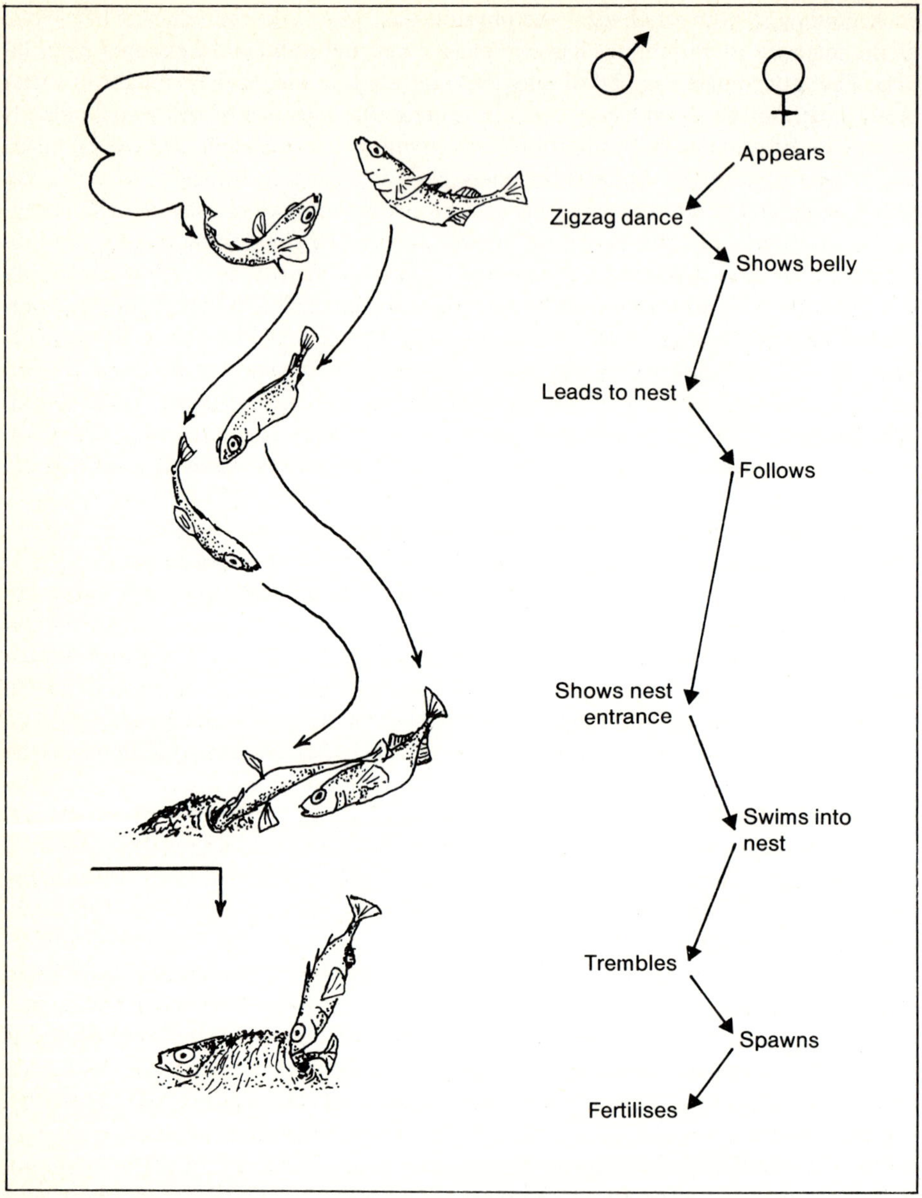

Fig. 68. Representation of the reciprocating mating behaviour of male and female *Gasterosteus aculeatus* in which each responds to a specific releaser provided by the other. (After Tinbergen)

eggs on their body, crocodiles which watch over their clutch of eggs and birds which incubate their eggs. The young are often cared for after they have hatched by being fed or led around. Bodily secretions are sometimes used by non-mammalians to feed the offspring: for example, South American cichlids of the genus *Symphysodon* feed their young a mucus secretion of the skin and pigeons produce a kind of milk from their crop (see p. 181). In ovoviviparous and viviparous species the phase of parental care follows on from gestation, during which time there are usually no characteristic patterns of behaviour. Birth also marks the beginning of a period of parental care for mammals which includes activities such as preening, feeding, and leading the young etc. The course of the parental care phase is in many cases determined by whether the young remain in the nest (altricial) and require prolonged care or leave the guardianship of their parents soon after birth and fly the nest (precocial). This topic is dealt with in detail in Chap. 9.

The different phases of the reproductive cycle are most distinct, and therefore more easily studied and analysed, in species whose behavioural repertoire is relatively primitive and based on genetically fixed, instinctive behaviour which is little, if at all, affected or changed by experience. A particularly clear example of this is the reproductive behaviour of the three-spined stickleback (*Gasterosteus aculeatus*: see Fig. 68). The male establishes a territory and builds a nest out of plant material glued together with a secretion from the kidneys. He is then ready to mate. If a sexually mature female appears the characteristic shape of her body acts on the male as a specific releaser of courtship behaviour. The female then displays her swollen belly to the male who responds with his so-called zig-zag dance in which he swims a zig-zag course leading the female to his nest. He then shows her the entrance to the nest and she swims in, whereupon he thrusts his snout rhythmically against her flank in the region of the genital opening, stimulating her to lay her eggs. The female then leaves the nest and the male swims in and fertilises the eggs. Later he tends the eggs by fanning them with his pectoral fins. This reproductive cycle thus contains all the typical phases and subphases.

6.3 The Physiological Basis of Reproductive Behaviour

The phenomenological treatment of behaviour presented above provides a number of points which serve as a basis from which one can consider the physiological regulation of reproductive behaviour. The division of the reproductive cycle into a sexual phase and a phase of parental care indicated by the results of ethological studies suggests that the controlling mechanisms are correspondingly distinct. It will be recalled that behaviour can be regarded as consisting essentially of muscle contractions which occur in accordance with a pattern of neural commands. The functional interconnections between structures in the central nervous system responsible for mediating such commands are for the most part genetically fixed and this is particularly so with regard to reproductive behaviour. This necessarily results in a behavioural repertoire which is genetically fixed in the sense that it is restricted to instinctive behaviour patterns. The region of the central nervous system responsible for instinctive behaviour is associated with a centrally located learning centre which can add to, perfect, promote, inhibit or cancel the predetermined pattern of behaviour. Both agonistic and antagonistic interactions are possible between these two centres, which theoretically are sufficient to generate the appropriate behaviour in response to an adequate exter-

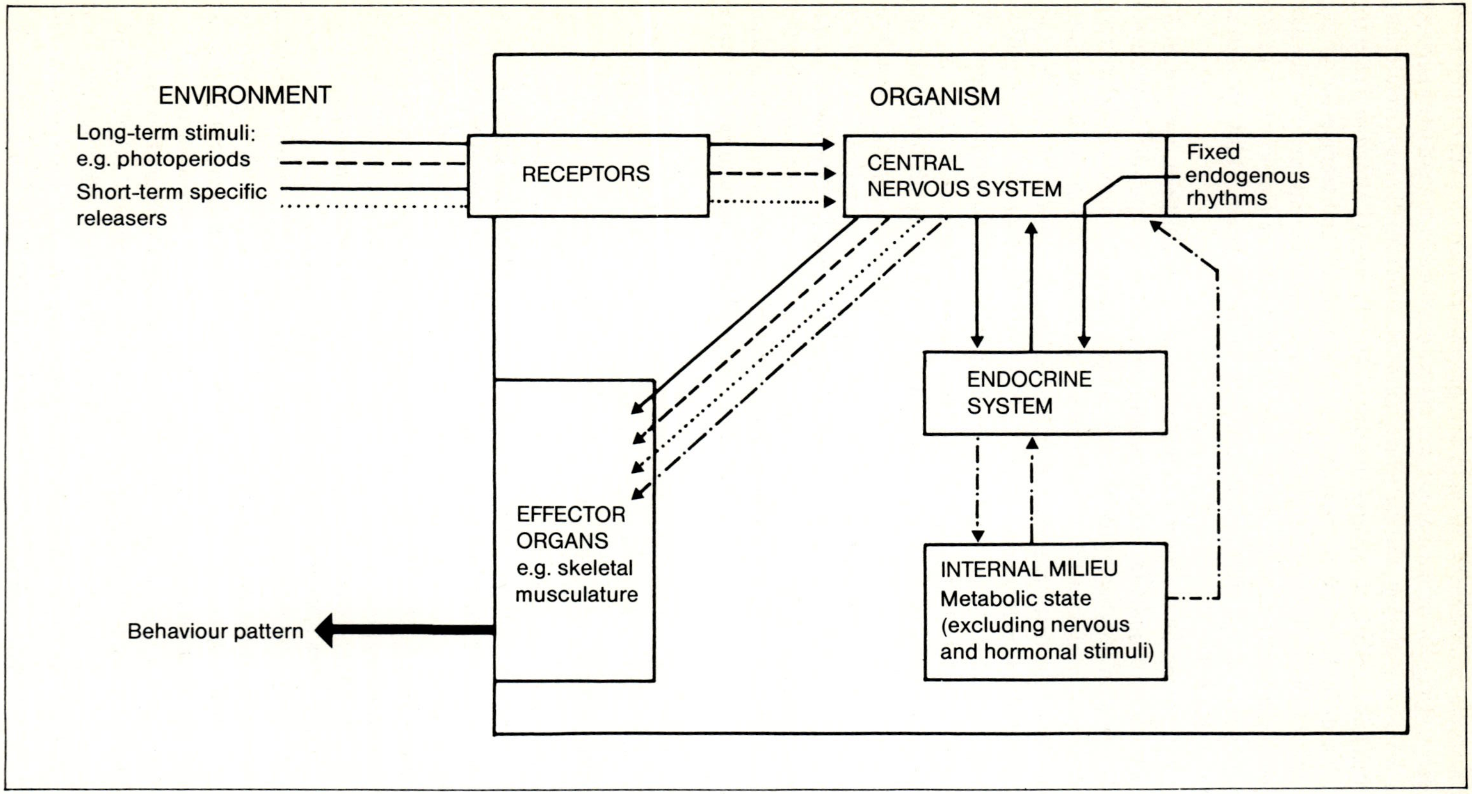

Fig. 69. Highly simplified scheme showing the involvement of hormones in the regulation of reproductive behaviour

nal stimulus. However, another factor is necessary, particularly with respect to reproductive behaviour, namely an underlying drive which results in specific types of behaviour, such as sexual behaviour, being brought to the fore. In other words, a degree of motivation within the central nervous system is required which influences the threshold to adequate stimuli so that it becomes lower with increasing degrees of motivation. In this way it is possible to explain why many animals respond little or not at all to a sexual partner outside of the sexual phase. The level of motivation often increases slowly over a long period of time. Hormones are the pre-eminent agents of such changes in the context of reproduction. Thus both long-term and short-term stimuli from the environment, often in combination with endogenous rhythms, act via the hypothalamus to bring about the stimulation of the endocrine system (see p. 168). The increasing level of certain hormones directly activates or changes the threshold of certain centres in the nervous system leading to an increase in the level of motivation (Fig. 69). The hormones may act at various levels within the nervous system (see p. 225). It can be safely assumed that the division based on ethological studies of the reproductive cycle into a sexual phase and a phase of parental care can also be explained in physiological terms. For example, there is mounting evidence that hypophyseal gonadotropic hormones act via the androgens, oestrogens or progestin produced in the gonads to influence behaviour during the sexual phase (see p. 182). Furthermore, sexual motivation in males and females is commonly induced by their respective sex hormones and can, if the threshold is lowered sufficiently, reach such a high level that sexual behaviour occurs in response to relatively weak stimuli. This has been particularly well studied in bony fish; for example, in the three-spined stickleback (*Gasterosteus aculeatus*). Castrated males treated with androgens not only exhibit the full sequence of sexual behaviour but they also develop the characteristic courtship coloration. Females which have had their ovaries removed perform the courtship display of the male when injected with androgens, with the exception of the zig-zag dance which is evidently not laid down in the female's genetic code. The basic principle demonstrated by these experiments probably applies to all vertebrates. Thus when oestrogens are applied to certain centres in the hypothalamus of cats which have had their ovaries removed the animals exhibit the typical behaviour of an oestrous female. According to current understanding only androgens, oestrogens and in some cases progestins are involved in the control of behaviour during the sexual phase: gonadotropins appear to have no direct effect in this respect. Behaviour associated with the phase of parental care is, in contrast, evidently controlled primarily by prolactin and progesterone from the hypophysis, and this seems to be the case for most vertebrates. Thus prolactin, aside from its tropic affect on, for example, the skin of cichlids of the genus *Symphysodon*, the crop of pigeons and mammary glands, can also affect behaviour. Fanning behaviour, which is an aspect of parental care, can be specifically induced by prolactin in certain species of cichlid (e.g. *Symphysodon* and *Pterophyllum*) according to a characteristic dose-response curve (Fig. 70A). In the mouth-breeding cichlids prolactin, together with oestrogens, help to ensure that the eggs or young are held in the mouth but not eaten. Cocks can be made to cluck like a hen by treating them with prolactin whereas, in pigeons, brooding of the eggs can be induced experimentally by injecting the birds with progesterone. The physiological basis of the change over from the sexual phase to the phase of parental care is

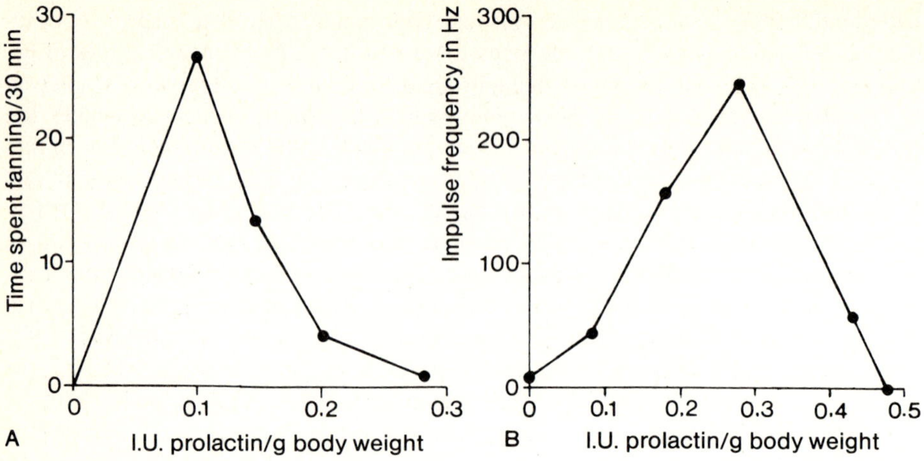

Fig. 70A,B. Relationship between prolactin dosage and fanning behaviour (**A**) induced by prolactin in prepubescent pompadour fish (*Symphysodon aequifasciata axelrodi*), and the response to injected prolactin of certain neurones (**B**) in the area dorsalis pars dorsalis of the telencephalon of the sunfish *Lepomis gibbosus*

largely unknown. It is possible that it involves oxytocin, or an analogue thereof, since if oxytocin is released when the gametes are shed, as probably occurs in some species, it could stimulate the secretion of prolactin from the hypophysis. Other, exogenous stimuli such as the sight of the eggs undoubtedly also play a role. The higher up the evolutionary ladder of vertebrates, the more difficult it becomes to discern the specific role that hormones play in the regulation of behaviour, beyond the recognition that they motivate some aspect of sexual behaviour or behaviour associated with parental care. In such cases the dominant role is played by the central nervous system, especially the learning centres. Nevertheless, the above scheme can probably still be applied: accordingly, gonadal steroids and prolactin activate centres in the central nervous system by modifying their threshold to environmental stimuli in a manner that depends directly on their level in the blood. The threshold can, by the appropriate application of hormone, be experimentally reduced to such an extent that the relevant behaviour is performed without apparently having been triggered by any external stimulus.

6.4 Variation in Reproductive Behaviour

The scheme presented above in which the reproductive cycle is devided into phases, each characterised by particular types of behaviour, provides a rough framework within which the widely different types of behaviour exhibited by vertebrates can be ordered. It is of particular significance that phases which are characterised by a weak motivation, such as the territorial phase, can occur outside of the context of reproduction. Ethological studies enable one to recognise several different types of behaviour, which will be considered more closely in the following.

6.4.1 Territorial Behaviour

Territories can be established exclusively for the purpose of reproduction or for the purpose of obtaining sufficient food, or both. In the case of a feeding territory, not only are rivals of the same species driven out or fought with but also members of other species. A territory may be lived in temporarily or permanently. In the latter case the territory often includes specialised compartments for sleeping, eating, drinking, defaecating, storing food supplies etc. The boundary of the territory is often marked out with a scent, or with urine or droppings. Calling signals can also be used to demarcate the territory, which is then as extensive as the voice can reach.

Territories solely for the purpose of reproduction are usually vigorously defended by the resident or ruling individual, who is the only one to reproduce since rivals beaten in territorial conflicts normally leave the territory and subsequently often quickly lose their state of reproductive readiness. Such animals are sometimes incorporated into a hierarchical social structure. The aggressiveness of the holder of a territory is usually greatest at its centre and decreases gradually the further the animal is away from the centre. Evidently the animal is more ready to engage in conflict in familiar surroundings and for this reason even intruders which are much stronger than the holder of the territory are driven out. Heightened aggressiveness within the territory is undoubtedly of considerable selective advantage. This also applies with respect to caring for the young, although during this phase the territory is often restricted to an area immediately surrounding the brood or nest.

6.4.2 Finding a Partner, Courtship and Mating

The main problem of reproduction is that of two partners of different sexes being at the same place at the same time in the same state of reproductive readiness. For this to occur it is necessary that the two partners first find each other and recognise each other for what they are. The search for a partner is made easier by the use of sexual attractants. For example, attractants make it possible for a sexually mature female to communicate to a male that she is prepared to mate and is not a male rival. In those species which are not sexually dimorphic, individuals must identify a potential partner firstly as a conspecific and secondly as a sexual partner, and this is often based on the reactions of the other. For example, during the spawning period male toads leap onto all individuals of the same species and attempt to cling onto them. If another male is approached in this way it gives out an acoustic signal which, in effect, says that he is not a female. In general the initial contact between sexual partners is frequently accompanied by aggression on the part of one of the partners, whose subsequent behaviour depends on the response of the other. A rival usually responds in kind whereas a female ready to mate responds with behaviour that preludes courtship. Three major sources of motivation influence such behaviour: firstly aggression, secondly sexuality and thirdly flight, if the rival is the stronger or the female unprepared to mate. This often results in a tripartite conflict between aggression, sexuality and flight which is manifest in different kinds of behaviour according to the changing balance between these types of motivation. The behaviour associated with the highest level of sexual motivation and which precedes the actual shedding of gametes or copulation is referred to overall as courtship, although this term more precisely refers only to the specific act of courtship. Among vertebrates the male is the active partner as a rule, though there are exceptions to this. The behaviour of the male serves to prepare the female for mating and to neutralise any behaviour on the part of the female which could hinder this. The females of some vertebrate species are very aggressive and hesitant, especially if they inhabit their own feeding ground. In such cases the male often emits sounds that mimic those of the young and which have the effect of pacifying the female. Aggressive behaviour by the male can also serve as a form of courtship in certain species of bird (e.g. mallards) where the males defend a tiny display ground in a ritualised tournament as the females look on. After this so-called group courtship, the defended display ground loses its significance as such. In many cases group courtship breaks up and the birds court each other individually. Behaviour patterns similar to those of courtship can also occur after mating as a form of play.

Although gametes can be shed spontaneously, courtship behaviour appears to be a necessary preliminary to this event in many cases. For example, in many species of frog the females only spawn if they are grasped by a male of the same species. As the female begins to lay her eggs she throws her head back thereby signalling the male to release his semen. This is essentially a chain of action and reaction in which each partner triggers an instinctive response in the other. It frequently occurs that males can only copulate in response to the female performing a particular type of behaviour. Copulation leads on to fertilisation, the consumation of the sexual phase, although in many cases fertilisation takes place some time after copulation. In bony fish and Gymnophiona, the males of which have specialised copulatory organs (see p. 68), fertilisation may take place some considerable time after copulation. Male urodeles deposit a spermatophore during mating which the female then takes up in her cloaca. Male reptiles possess a penis with a complex structure which is usually inserted into the cloaca of the female during copulation. Copulation in birds normally involves the partners pressing their cloaca's together: some birds have a penis; for example, ducks and ostriches. Mammals copulate exclusively by the insertion of the male's penis into the cloaca, or vagina, of the female. Only in a relatively few species of mammal is copulation accompanied by an orgasm, that is, a pleasurable sensation: humans and female cats for example. In the latter orgasm occurs some while after copulation (post coital orgasm). With respect to the physiological basis of sexual behaviour it is noteworthy that the behaviour of the partner can also induce the production and release of hormone.

In many species of vertebrate the sexual partnership is restricted to a single reproductive period or to the period of courtship or copulation. Other species, in particular bony fish and birds, form a partnership ("marriage") which is often life-long. Geese even have a period of engagement and those which have been widowed early usually remain partnerless for several reproductive periods. Such marriages would be impossible without a highly developed mechanism for the recognition of individuals. In some species the male only marries one female (e.g. the gibbon, cichlids and birds) whereas in others he takes several wives (apes, seals, red deer).

6.4.3 Parental Care

The young may be cared for by one or both parents; the mother in the case of ducks, the father in the case of sticklebacks, syngnathids and ostriches, and both in many species of song-bird. The different types of behaviour associated with parental care are essentially directed towards three goals: protecting and defending the young, providing them with food and educating them. The first of these can also occur indirectly as a consequence of territorial defence, however it encompasses other patterns of behaviour such as carrying the young back to the nest, the young remaining motionless while they are being carried (e.g. lions). The young may also hold onto the mother's fur with their jaws and so be carried around (e.g. the shrew). All types of feeding behaviour are relevant to the ways in which the young are provided with food, the gathering of food being of prime importance. Food is fed to the young in response to them performing a particular type of begging behaviour. In most cases this involves highly specific, innate releasing mechanisms. The behaviour of the parent animal in feeding its young similarly involves specialised mechanisms, for example the vomiting or choking up of predigested food from the crop or gastrointestinal tract. Such behaviour includes all those whereby the young are fed with a bodily secretion, namely that from the skin of *Symphysodon* (see p. 219), the crop of pigeons, and the mammary glands of mammals, in which case the innate behaviour mechanisms are supplemented by learning. Thus searching for the teat is innate in young mammals, though they do learn its position, often with the help of the mother holding the young to the teat.

Many birds and mammals can recognise their young by acoustic signals (sea-lions), optical and olfactory signals (mammals), among others. The young often imprint on and thereby establish a bond with the mother during a sensitive period. The timing of this period is genetically fixed but imprinting itself is a process of learning, features of the head and face often serving as points of identification. In many species the mother emits different calls that either warn the young, call them back to her, or prompt them to follow her. Alternatively she may make some corresponding gesture, or do both these things.

In higher mammals parental care also involves the education of the young. In the simplest case the parent animals play with the young thereby encouraging them to imitate and perfect certain paterns of behaviour. This leads on to the parents actually instructing their young: for example, apes teach their young how to climb. In such cases one or both of the parents actively encourages the young to learn, a relationship that takes on immense importance in the case of humans. In man a significant role is also played by non-biological factors such as ideology, religion etc., to the extent that patterns of innate behaviour and, more generally, the motivation underlying behaviour can be completely suppressed and this applies in particular to reproductive behaviour.

6.5 Reproductive Behaviour as Social Behaviour

Social behaviour by definition involves at least two individuals as does reproductive behaviour, including that associated with the parental care of the young. The common end of all these types of behaviour is the survival of the species. There are however aspects of social behaviour which are related directly or indirectly to reproduction. In some cases animals form groups, schools of fish and flocks of birds for example, the members of which do not know each other individually. The groups are held together by the members maintaining visual or acoustic contact, although olfactory stimuli (e.g. the smell of the nest) are sometimes involved. There is usually no established hierarchy between the members of such groups. This is well illustrated by the swarms formed by the brown rat (*Rattus norvegicus*) which may consist of several hundred animals living together without any sign of there being a hierarchical structure. Other relevant features are that females on heat are served by several males; no special area is established for the young; there are no sexual conflicts between the males; and the young of several females are reared together. Well developed systems of communication within the group exist whereby its members are, for example, informed about a poisoned source of food. Over and above this facet of social interaction, small groups may form between individuals which recognise each other to some extent, as is evident in colonies of birds and large families of mammals that include adults and grown youngsters. However, this behaviour must evidently result from an innate disposition: rearing animals in captivity or holding young animals together which do not socialise never leads to the formation of groups. Membership of an individual to a particular group of animals commonly determines its behaviour. Thus members are tolerated whereas non-members are fought with or even killed. The latter behaviour may change during the reproductive period: for example, lionesses will tolerate a strange lion, and vice versa, during the mating season. The formation of groups undoubtedly represents a considerable selective advantage to its members. In many cases the members of a group become organised hierarchically and perform specific roles. For example, the leader is usually the most alert member of the group who warns others in the group of danger and who is followed by the group. Similarly, older members of the group force back young animals which have strayed from the group and may settle conflicts between members by simply separating the animals. Cases are also known of one individual helping another: dolphins reportedly do this. The collective attack on a predator and the development of traditions within the group can also be mentioned in this context.

A "pecking order" is usually established within the social structure of the group. Conflicts over food or a resting place usually lead to the animals becoming more aggressive and a fight results, the outcome of which is learned. In effect the subordinate animal is taught a lesson. As a rule the loser then recognises the winner as its superior and fighting between them ceases for some time. Such fighting then takes place between other, weaker members of the group, in the course of which they come to recognise each other, and this leads eventually to the establishment of a pecking order. The dominant animal, or leader, may be male (as for example in wild horses, lions and baboons) or female (as in the case of red deer not on heat and the onager, or wild ass). An important consequence of such hierarchical social structures with respect to reproduction is that it is frequently only the dominant animal which reproduces. Packs of wolves in captivity have a dominant male and a dominant female which mate with each other. The dominant animal is generally referred to as the alpha animal, its subordinate as the beta animal, and so on. Dominance is typically manifest in the amount of freedom given to those animals occupying the lower positions in the hierarchy. The degree of differentiation of a hierarchy is variable and depends on factors such as the amount of space and food available. Pecking orders are usually not rigid but subject to change, as determined by the outcome of renewed fighting. The position of an individual in a peck order depends on its size, strength, readiness to fight, skill and perhaps also on its intelligence. Hormones may also play a role in so far as they influence the level of motivation in the manner described above. In some cases the age of an animal determines its position in the pecking order. Old deer and anthropoid apes for example can occupy the alpha position despite being physically weaker than subordinate members of the group – their experience evidently counts for more in terms of selective advantage when they occupy the alpha position than any lower position in the hierarchy. Apes occupying such a position develop markings distinctive of old age. Members of a group may also have a displaced social status; thus a female may be promoted to the status of her mate (as occurs with jackdaws and apes) or one of the young may stand in for its mother (ungulates and apes) or father. Comparable phenomena can be recognised in human hierarchies, for instance the succession of the throne and the social status of experienced old men in the council of elders etc.

6.6 Physiological Aspects Underlying Reproductive Behaviour

The regulation of the reproductive cycles of vertebrates by hypophyseal gonadotropins, gonadal steroids and prolactin has already been explained in Chap. 5, from which it was evident that hormones are involved in the long-term regulation of reproduction. And it will be evident from what has already been said in this chapter that hormones also play a fundamental role in the long-term regulation of reproductive behaviour in that they induce long-term changes in motivation. It is of particular significance that the role of hormones in this respect is restricted to the generation or the modification of motivational states whereas the central nervous system alone regulates the co-ordination of behaviour in space and time. The question then is, in which ways do hormones affect the nervous system. According to our present understanding hormones can have an effect in five possible ways, both direct and indirect. Firstly, they can activate centres in the central nervous system, as has already been mentioned. Thus experiments have shown that after injecting radioactive sex hormones into gonadectomised animals radioactivity is concentrated temporarily in the cell nuclei of certain populations of cerebral neurones. In male collared turtle doves, after injection of labelled testosterone the highest level of radioactivity is found in the hypothalamus with lower levels being present in the forebrain. Similarly, after injecting radioactively labelled oestrodiol into female rats, the levels of radioactivity are highest in the hypothalamus, intermediate in the hippocampal region of the cerebral hemispheres and lowest in the cerebral cortex. On the other hand, implantation of microcrystals of testosterone into the brain of pigeons promotes different elements of the courtship behaviour according to the site of implantation. Testosterone implanted into the lateral diencephalon of castrated male cats results in increased aggression but copulatory behaviour if implanted into the preoptic region. Simultaneous implantation at both sites faciliatates the activation of courtship behaviour, providing physiological support for the view that the motivation underlying this behaviour is not of a single kind. It has also been shown that the sites at which radioactively labelled sex hormones accumulate in the brain of collared turtle doves, rats and cats coincide closely with the sites at which hormone implants stimulate elements of sexual behaviour. Correspondingly detailed results relating to lower vertebrates are not available excepting the observation that in cichlid species whose behavioural repertoire includes fanning of the brood a population of neurones in the forebrain can be stimulated specifically by systemic injection of prolactin. The dose-response relationship characterising this effect is similar to that relating to fanning induced experimentally by prolactin (Fig. 70B). Species which do not fan the brood do not respond to prolactin in this way, this being particularly well illustrated by two species of the genus *Tilapia: T. mariae*, a fanning substrate breeder, reacts to prolactin in the manner described above whereas the mouth-breeding species *T. leucosticta* does not.

The determining effect of hormones on the central nervous system has been described in detail in Chap. 5 (see p. 203) in connection with the differentiation of male and female structures within the brain: the description given there includes details of the mechanism underlying this process in mammals. Hormonal determination not only affects the setting or suppression of endogenous cycles but is also the means by which specifically male or female behaviour patterns are developed. Such determination can usually only occur during a specific phase in the ontogenetic development of the nervous system, which may begin prior to birth: it may be referred to as hormonal imprinting by analogy to the process of imprinting. At present only testosterone is known to have such a determining effect in mammals. Thus, after they are born, male dogs castrated in utero by treatment with hormones urinate in the same way as females do. When rats castrated shortly after birth become adults they behave like females towards other males. Treating pregnant guinea-pigs and rats with androgen results in the partial or complete suppression of female sexual behaviour in the offspring, and treating the female offspring can even lead to them exhibiting male sexual behaviour. Data relating to lower vertebrates are scarce and only indirectly of relevance. The basic tendency for the brain of mammals to differentiate in the direction of female, independent of the genetic sex (see p. 17), appears to extend to the realm of behaviour. Thus the behavioural repertoire laid down genetically in the central nervous system corresponds to that of the female: male behaviour results only after the central nervous system has become hormonally imprinted on testosterone. It is possible that these early processes in ontogeny are connected with the later, transitory effects of androgens and oestrogens on elements of the central nervous system in that testosterone brings about a permanent change which manifests itself later as an increase in the affinity and sensitivity of neural elements to testosterone. Logically, if testosterone were not present early on, one would subsequently expect to find that the sensitivity to oestrogens is increased. Whether or not this is in fact the case is unknown but it is consistent with the known ability of certain neurones to produce specific receptors for testosterone or oestrogens and is supported by the observation mentioned above that exogenous radioactive gonadal steroids accumulate in the cell nuclei of certain populations of neurones.

Hormones are known to affect the threshold characteristics of receptor cells thereby modifying the information these cells supply to the central nervous system. Testosterone exerts just such an effect on mechanoreceptors in the skin of the rat's penis, raising their sensitivity so that they give a stronger response which increases the likelyhood of sexual behaviour being activated. Similarly, the tactile sensitivity of the brood patch in many species of bird and the skin of the pigeon's crop is increased by prolactin thereby facilitating the activation of brooding and feeding behaviour respectively. In general, such mechanisms have been little studied.

Whereas the modes of hormone action dealt with so far have been direct, hormones can also stimulate behaviour indirectly via the central nervous system and the endocrine system. Thus it has been found that radioactively labelled oestrogens accumulate in the cats brain relatively quickly then disperse and that an associated change in sexual behaviour only occurs several days later. This delayed effect is probably due to the stimulation of neurones in the hypothalamus producing gonadotropin-releasing hormone (see p. 176) leading to an increase in the amount of gonadotropin released from the anterior lobe of the hypophysis which in turn causes an increase in the level of oestrogens in the blood. In this case the specific behaviour pattern is activated by one of the mechanisms which have already been described.

Finally, hormones can facilitate the triggering of sexual behaviour by the partner, that is the hormones which act within the sexual partner and not those of the responding individual. This is because sexual hormones play a decisive role in the development of secondary sex characteristics (see p. 182) and that these features frequently function as signals or as specific releasers of behaviour so that the more well-developed these features are, as a consequence of hormonal stimulation, the greater or more intense is their effect in releasing a specific pattern of behaviour in the sexual partner. Such a mechanism is common among vertebrates. Examples of features which act as releasers include the many different kinds of courtship coloration, such as the red colour of the male stickleback's belly during the reproductive period. In general these features are either permanent characteristics of the animal or they are only developed during the period of reproductive activity.

Each of the five mechanisms mentioned above can act singly or in combination to change the motivation underlying reproductive behaviour (Fig. 71). Importantly, the behaviour of an animal can influence its own endocrine system or that of another animal of the same species, such as its offspring or sexual partner. This effect of behaviour is mediated by the central nervous system and in particular by the centres in the hypothalamus that produce releasing hormones. For example, a suckling rat presents an array of stimuli to the mother which induce a release of prolactin from the adenohypophysis within 30 min. The details of this process are given in Chap. 5 (see p. 202). In many species of bird, the sight and feel of the eggs while brooding similarly leads to the release of prolactin. As a rule the behaviour exhibited by the sexual partner during courtship also stimulates the release of hormones. Thus the courtship behaviour of male collared turtly doves acts sequentially via the eyes and ears, brain,

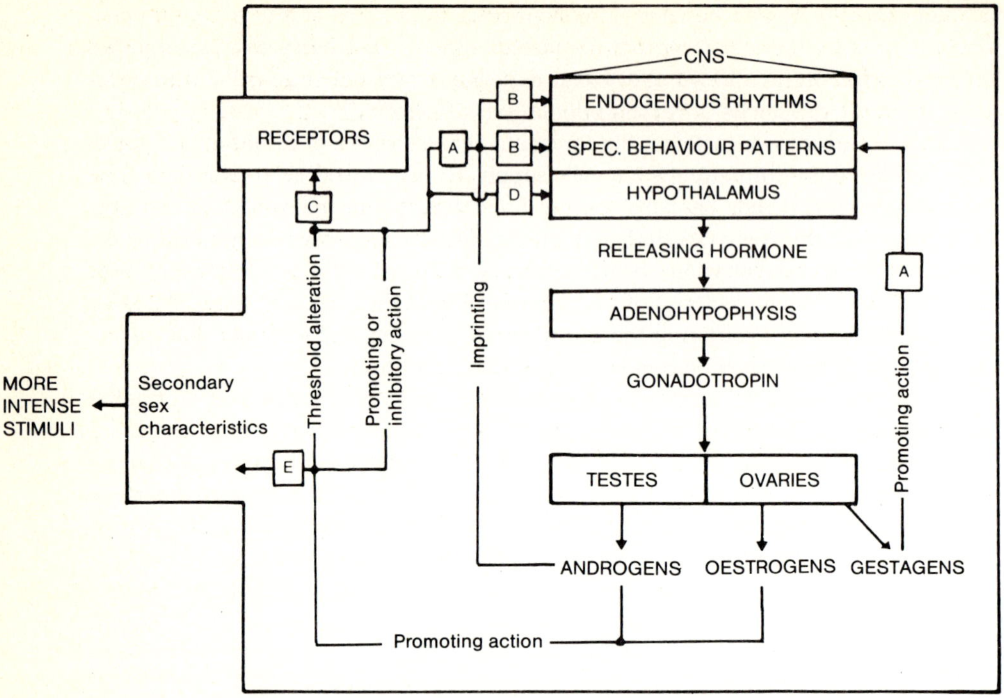

Fig. 71. Scheme illustrating the various ways in which hormones can influence the motivation of reproductive behaviour: *A* Activation of centres in the central nervous system (CNS). *B* By determination of brain structures. *C* Alteration of receptor thresholds. *D* Feedback on the endocrine system. *E* By making the secondary sex characteristics more effective releasers of behaviour

and hypophysis of a conspecific female to cause the release of oestrogens from the ovaries which promote nest building behaviour. This behaviour in turn acts via the hypothalamus to induce the release of luteinising hormone which causes ovulation and the subsequent synthesis and release of progesterone. Finally in this process, progesterone causes the bird to sit on the eggs and brood them. In a few cases such mechanisms operate between conspecifics of the same sex. The song of the male canary, for example, can induce the testes of other males to develop. The various kinds of interaction between hormones and behaviour are often difficult to demonstrate experimentally yet they play an integral role in the overall phenomenon of reproduction.

The role played by the central nervous system with respect to reproductive behaviour accords with its nature as an integrative system, in that it serves primarily to regulate short-term processes. Aside from having the function of receiving and integrating stimuli, of controlling motor responses, and of establishing memory engrams, the central nervous system is also responsible for the vegetative regulation and "orchestration" of the endocrine system. To go into these different roles in any detail lies far beyond the scope of this chapter but one may note that despite its high rank in the hierarchy of integrative systems (see p. 163) and the controlling influence it exerts on the endocrine system, the central nervous system also acts as an equal partner together with the endocrine system in regulating reproductive behaviour for both systems are essential for the generation and modification of motivational states within the animal. Although this equilibrium can be upset by experimental manipulation, under physiological conditions both systems are in a finely balanced state of homeostasis in which hormones regulate the behavioural readiness of the animal to mate and, on the other hand, make it possible for the central nervous system to "make sense" of the exogenous and endogenous stimuli associated with mating and to coordinate specific behaviour patterns in space and time. Finally, it must be stressed that science is still a long way from being able to offer a comprehensive explanation of the full role played by the central nervous system in regulating reproductive behaviour.

CHAPTER 7

Care of the Young Within the Body: Submammalians

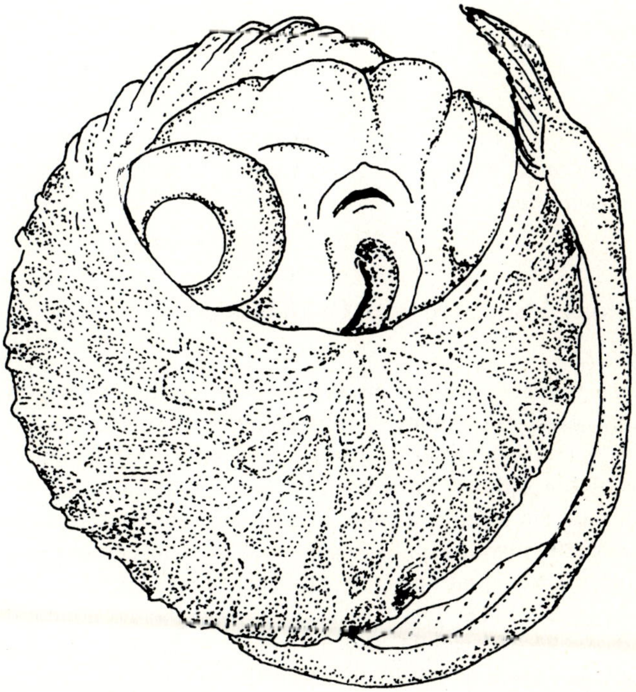

Embryo of a live-bearing cyprinodont of the genus *Poeciliopsis* with the pericardial sac drawn over its head

Most non-mammalian species lay their eggs directly in the environment, that is they are either oviparous or ovuliparous according to whether the eggs are fertilised or unfertilised when laid. In many cases the parent animals take care of their eggs to some extent and of their young after they have hatched out, these being aspects of parental care (see p. 223). However, in many species of Elasmobranchii, Teleostei, Amphibia and Reptilia various methods have arisen during the course of evolution whereby the embryos, larvae or brood are kept within the body of the parent so as to protect them whilst they develop. This is usually accompanied by a drastic reduction in the number of eggs produced. In some cases the adaptations associated with this are highly specialised, as they are in mammals, and serve to supply the embryos or larvae with oxygen and nutrients derived from the body of the parent. Three grades or forms of caring for the young within the body can be recognised:

1. adaptations associated with ovuliparity,
2. ovoviviparity,
3. viviparity.

The first of these encompasses a range of complex phenomena which are undoubtedly not of common origin whereas the second and third forms are often difficult to distinguish from each other.

7.1 Adaptations Associated with Ovuliparity

The adaptations whereby ovuliparous species care for their young within the body exhibit different levels of complexity, though all these species have in common the fact that the eggs are only fertilised after, even if only a short time after, they have left the body of the mother. Subsequently the eggs are taken up into some cavity within the body of the parental animal, though never one that is part of the urogenital system, where the offspring undergo part or all of their development.

The most primitive adaptation is very probably mouth-breeding, which occurs in a range of teleost species. This does not usually involve the development of specialised structures, instead the eggs simply develop within the mouth or gill cavities of either the male or female parent. Typically however, the eggs laid by the female are fertilised by the male who then takes them into his mouth. Sometimes both the eggs and the sperm are taken into the mouth, where fertilisation then takes place. Embryonic development also takes place in the mouth and, after hatching out, the free-swimming young of many species are taken back into the mouth at moments when danger threatens or during the night. Such behaviour is also exhibited by substrate breeders, from which mouth-breeders are thought to have evolved. Mouth-breeders do not feed while brooding and in some cases the oesophagus may close and the empty gut shrink in upon itself. For part of their lives the embryos live off the store of yolk in the egg. The young are not only given protection by the brooding parent but also provided with oxygen by the constant flow of water through the gills. A few species possess specialised structures which prevent the eggs falling out of the mouth. For example, female cichlids of the species *Pelmatochromis lateralis* have a flap of skin behind the teeth of the upper jaw. Among teleosts, true mouth-breeding occurs in the Siluridae, Cyprinodontidae and Serranidae and is particularly common in Cichlidae. The Amblyosidae, in contrast, take their eggs into the gill chamber. The genital opening of adult females of this family lies so far forward that the eggs can pass directly from the oviduct into the gill chamber. In the past these fish were erroneously described as giving birth to live young. There is evidence that in the mouth-breeding cichlid *Sarotherodon mossambica* the behaviour of the female in caring for her young is controlled by the co-operative action of oestrogen and prolactin: the motivation causing the animal to take the eggs into her mouth is induced by the former, whereas the hormone prolactin is responsible for the inhibition of eating and inducing a state of brooding readiness which results in the animal simply holding the eggs in her mouth and not eating them.

Phenomena comparable to those just described are rare in vertebrates other than the Teleostei. True mouth-breeding has only been observed in one species of frog, *Hylambates brevirostris*, in which case the female takes the fertilised eggs into her mouth where she holds them until they hatch out after metamorphosing into small frogs. She herself does not feed during this time.

A more specialised form of parental care is shown by the South American bell frog (*Rhinoderma darwinii*). A single female lays between 20 and 40 eggs which are fertilised then guarded by a number of males. After 10–20 days, when the development of the embryos is already well advanced, each male takes between 5 and 15 eggs into his mouth and then shifts them into his voluminous throat sac. This sac otherwise serves as an air reserve and resonating chamber for the production of sounds (Fig. 72A). The embryos develop into tadpoles and hatch out within the throat sac, remaining there during the larval stage until they metamorphose into small frogs and finally leave the male's body. The embryos live off the yolk reserves within the egg but how the tadpoles feed themselves is not fully known. It is possible that they feed off a secretion of the throat sac epithelium in a manner similar to embryotroph (see p. 239). The oxygen requirements of the larvae, which have no external gills, are probably met by oxygen diffusing through the skin and the relatively large, flat and richly vascularised tail. This tail, which is lost after metamorphosis, may also play a role in the uptake of nutrients. Strictly speaking, this way of caring for the young is not mouth-breeding, although it could well represent a development thereof.

The Australian frog *Rheobatrachus silus* has a unique and highly specialised way of caring for its young. This species lives in running water and only occurs in one particular area of Queensland. The female swallows the eggs soon after they have been fertilised and retains them in her stomach where they grow and finally metamorphose. The small frogs are then "born" orally (see title picture to Preface). The females evidently do not feed during this period of "gestation", in which the stomach is much distended and has only an extremely thin wall.

The fourth way in which the young are cared for outside of the urogenital system probably arose from the eggs simply being carried about on the body of the parent, such as occurs in teleosts and amphibians (see p. 309, 314). The development of specialised structures in association with this is particularly well illustrated by species of the teleost family Syngnathidae. In these fish the male is always the one who cares for the young. Protective structures develop on the ventral aspect of the body, either in

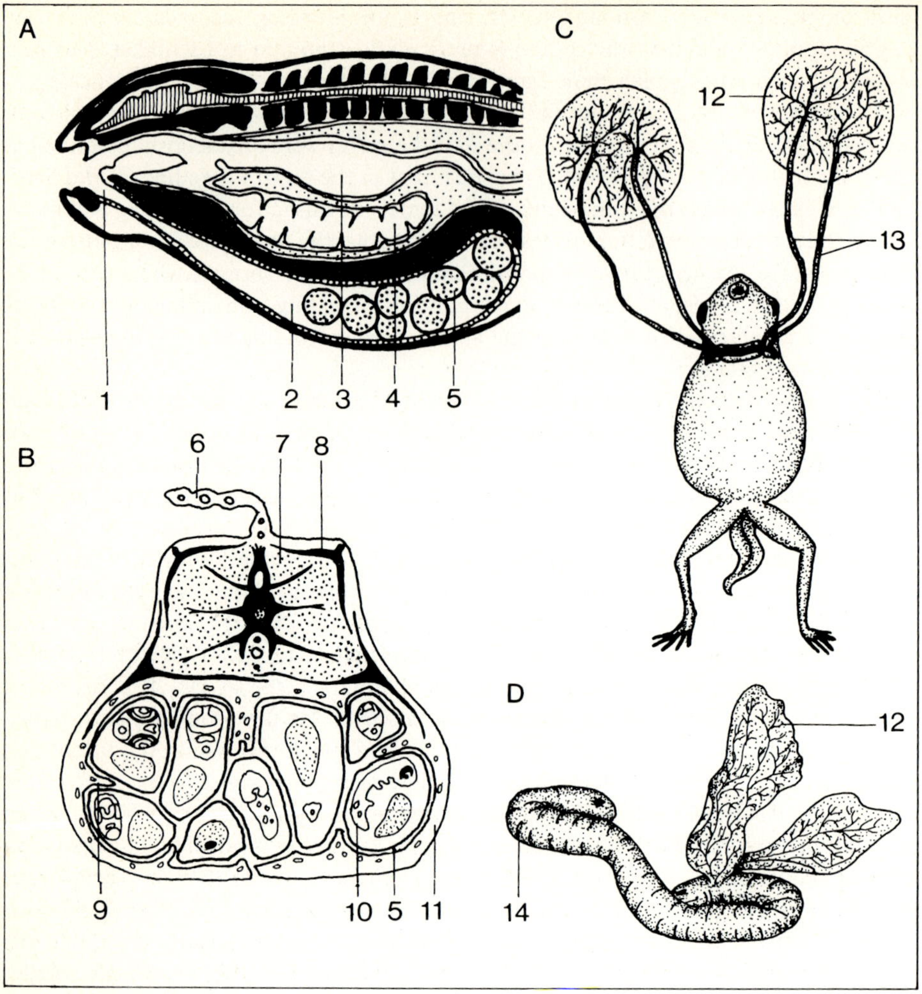

Fig. 72. A Sagittal longitudinal section through part of the body of the frog *Rhinoderma darwinii*. (After Wiedersheim). **B** Cross-section through the body of a male *Syngnathus dumerilii* carrying eggs. (After Huot). **C** Larva from the brood pouch of the marsupial frog *Gastrotheca oviferum* bearing plate-like gills. (After Brandes and Schoenichen). **D** Larva of *Typhlonectes compressicauda* bearing large leaf-like gills. (After Sarasin).
1 = entrance of the throat sac; 2 = lumen of the throat sac; 3 = stomach; 4 = lung; 5 = egg enclosed in the egg membranes; 6 = dorsal fin; 7 = muscle; 8 = plates of dermal skeleton; 9 = vascularised wall of the comb; 10 = embryo; 11 = fold of skin enclosing the eggs; 12 = gill; 13 = blood vessels; 14 = body

the abdominal region (Gastrophori) or under the tail in the case of tail breeders (Urophori). In sea adders of the genus *Entellurus* and *Nerophis* the eggs are simply stuck to the skin of the male's abdomen and carried about by him. In Solenognathina the skin of the ventral surface is sculptured into a honeycomb-like structure whose walls surround the eggs. Dorichthyna and Syngnathina have, in addition, developed protective organs on both sides of the body which are formed from either plates of the dermal skeleton (protective plates), epidermal duplicatures (protective folds), or both. After the eggs have been laid the protective folds may unite along the mid-line to form an enclosing chamber (Fig. 72B), but protective plates never unite in this way. The folds separate again when the brood hatch out. The most elaborate form of this kind of parental care is exhibited by sea-horses (genus *Hippocampus*). In these animals the lateral folds join to form a brood pouch, or marsupium, which has a small opening rostrally that can be closed by a sphincter. Over a period of several days a number of females deposit their eggs in this pouch, where they are then fertilised. The embryonic and larval development of the young takes place within the pouch, after which they are born by being pumped out of the pouch by what appear to be birth contractions. There is evidence indicating that this process is triggered by the release from the neurohypophysis of isotocin, and analogue of oxytocin (see p. 179). The young are very probably supplied with oxygen and nutrients from the father's body during the late embryonic and larval stages, as probably also occurs in other syngnathids. In *Syngnathus typhle*, for example, the epidermal cells of the father's skin are reported to give rise to villus-like projections which penetrate the egg membranes. Moreover, the eggs lie in close proximity to simple, pocket-like glands situated in a richly vascularised area of the abdominal skin. The young die if they are exposed to sea water by opening the brood chamber experimentally. In many syngnathids the provision of the young with oxygen and nutrients has been attributed to a honeycomb layer in the brood pouch which is reportedly expelled as an after-birth in the case of sea-horses. In general, young syngnathids are born at a stage of development which correlates with the degree of specialisation exhibited by the protective structures. In those cases where the protective structures are simple, the young are born with fins that still have the undivided embryonic form, whereas the young of species possessing highly developed protective structures are born with a small habitus which is characteristic of the adult animal. The structures which are associated with parental care in syngnathids are first formed at the onset of sexual maturity. Protective plates, on the other hand, are usually formed early in development and persist throughout life. Protective folds are fully developed during the spawning period but regress to a greater or lesser degree thereafter, whereas the brood pouch of sea-horses, once formed, does not regress.

The ways in which amphibians care for their young are similar to those already described. The simplest method is, as in teleosts, where the eggs are carried on the body of the parent (see p. 309) and in a few species simple structures are developed for the protection of the eggs. Thus female frogs of the species *Hyla evansi* carry about 20 large eggs on their backs, the skin of which forms a low wall around the eggs so that it appears like a flat honeycomb. In *Flectonotus goeldei* and a few other South-American frogs the eggs are lodged in a bowl-shaped depression formed in a hump on the back of the animal (see p. 315): the floor of the depression is richly vascularised. Most of these species hatch out of the eggs as small frogs but some hatch out as larvae that possess specialised gills. The young are cared for not on the body but fully within it in the case of pouched frogs of the genus *Gastrotheca*. Contrary to the implication of their name the females have a marsupium on their back which opens caudally in a slit. During the brood period the pouch contains from 4–50 eggs, depending on

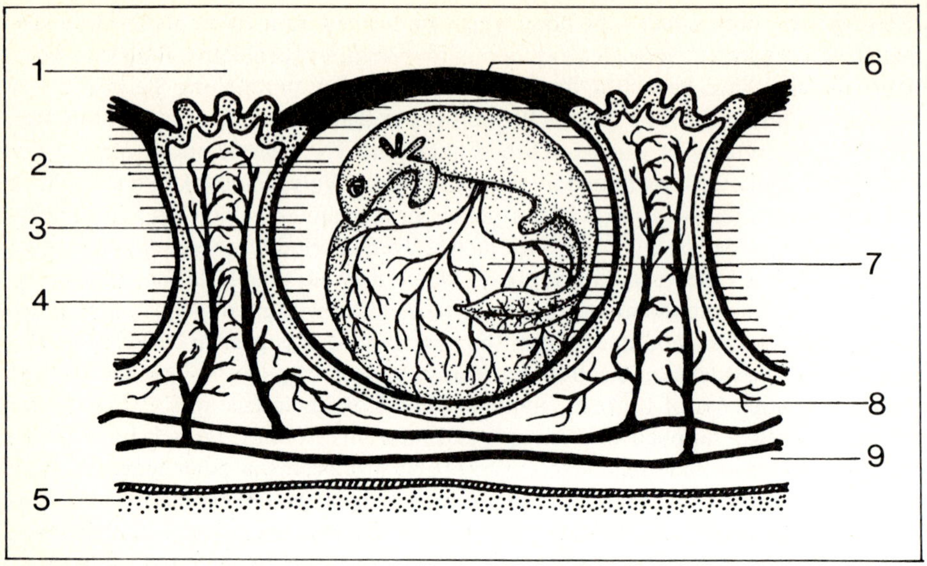

Fig. 73. Section through the comb of a toad of the genus *Pipa*. (After Wiedersheim). 1 = epidermis; 2 = larva; 3 = gelatinous material; 4 = capillary; 5 = subepidermal lymph space; 6 = cap to brood chamber; 7 = yolk sac; 8 = blood vessel; 9 = corium

the species, which hatch out as free-swimming tadpoles or as fully metamorphosed young frogs. In the latter case the larvae have large leaf-like or bowl-shaped gills which are connected to the body via an afferent and efferent blood vessel (Fig. 72C). The gills make contact with richly vascularised folds in the epithelium of the brood pouch from which the tadpoles are probably provided with oxygen by diffusion. Nothing is known concerning the nutritional inter-relationships between the mother and larvae, nor is it known how the pouch is formed; it probably arises from longitudinal duplicatures of skin on either side of the back which partly fuse together medially. In some species the female deposits the eggs in the marsupium herself by repositioning her body whereas in others the male assists by using his legs. In general, the larger the brood the earlier they hatch out, i.e. large broods hatch out as tadpoles, small broods as young frogs.

The eggs of South American surinam toads of the genus *Protopipa* and *Pipa* develop in a honeycomb-like structure on the back of the animal which forms after the eggs have been laid. Spawning and fertilisation of the eggs involves a complicated pattern of behaviour in *Pipa*, in which the sexual partners somersault and thereby transfer the eggs onto the female's back. After the eggs have been laid they become embedded in the skin of the back as the honeycomb structure develops, each chamber of which is then sealed by a cap. The walls of the honeycomb structure are richly vascularised (Fig. 73). Both the embryonic and larval development takes place within this comb. During the final stages of the latter process the larva is surrounded by a gelatinous substance. The developing young are probably supplied through the walls of the comb with oxygen from the mother's blood. The tadpoles have a long, flattened and richly vascularised tail which probably serves as a respiratory organ. The embryos and larvae live off yolk reserves, though whether or not this is supplemented by nutrients from the mother is unknown. In the genus *Protopipa* the chambers of the comb are less numerous and structurally more simple than they are in *Pipa*.

7.2 Ovoviviparity

The term ovoviviparity applies when the embryos within their egg membranes develop within some part of the female urogenital system and by definition excludes those cases in which the embryo is supplied with nutrients from the mother, although not the possibility that gases and water diffuse through the egg membranes. Thus the mother's body primarily affords the developing young protection. Development in some species takes place in the oviduct, more exactly a specialised section thereof, whereas in others it occurs in the uterus or in the ovary, in which case it may occur inside or outside the follicle. Ovoviviparity is often extremely difficult to distinguish from viviparity since various kinds of intermediate forms exist. It will therefore be dealt with only briefly: certain relevant details are discussed in the more extensive coverage of viviparity.

Ovoviviparity occurs in Elasmobranchii, Teleostei, Amphibia and Reptilia. Many sharks and rays give birth to live young, there being relatively few oviparous families (viz., Scyllidae, Heterodontidae, Rajidae). Fertilisation is internal in all elasmobranchs (see p. 114). Only two families of shark are reported in the literature to be truly viviparous (the Carcharhinidae and Sphyrnidae), all other live-bearing forms being ovoviviparous. In this case the distinction between ovoviviparity and viviparity according to the definition of ovoviviparity employed here breaks down since in many species various structures and mechanisms exist which in some way support the development of the young. Furthermore, many intermediate forms exist. Ovoviviparous species occur, for example, in the Squalidae, Chlamydoselachiidae, Orectolobidae and Trikaidae. The contrast between ovoviviparous and viviparous forms can be seen with particular clarity within the genus *Mustelus*. The smooth dogfish (*Mustelus mustelus*) is ovoviviparous whereas the shark *Mustelus laevis* first goes through a phase during which it does not receive nutrients from the mother, then has a phase in which it does. In all ovoviviparous elasmobranchs the eggs are carried in the uterus and as a rule the young hatch out shortly before they are born. In some cases only a small amount of the yolk in the yolk sac is enclosed within the egg membranes, the rest becomes a milky fluid which at first is resorbed through the filamentous gills of the developing young and later is taken in through the mouth.

Ovoviviparity and viviparity in Osteichthyes are similarly often difficult to distinguish from each other. In many cases it is not known whether the embryos developing within the egg membranes live only from the reserves of yolk or receive nourishment from the mother and, as stated above, the distinction is not always a clear cut one. Sometimes it is purely a question of definition, for example, as to whether or not the function of the follicular epithelium is to supply nutrients to the embryo. In general, the eggs may develop either within the follicle or within the ovarian cavity, this being referred to respectively as follicular and ovarian gestation. Ovoviviparity occurs within the follicle in members of the genus *Poecilia* but in the ovarian lumen in members of the genus *Sebastodes* (family Scorpaeniidae). This will be dealt with further when considering viviparity in teleosts. Fertilisation is, necessarily, internal in ovoviviparous and viviparous species of teleost.

The only living crossopterygian, *Latimeria chalumnae*, is evidently ovoviviparous since in one specimen several larvae, each possessing a yolk sac, were found in the oviduct though there was no trace of any structure which might serve to supply nutrients to the larvae.

Ovoviviparity occurs in all three orders of Amphibia. Among anurans, only two species of toad from the genus *Nectophrynoides* are known to be ovoviviparous: *N. tornieri* and *N. vivipara*. In the latter species over one hundred embryos sometimes develop in the uteri. The larvae have a long, thin and richly vascularised tail which probably serves as a respiratory organ. They live from the reserves of yolk in the egg. Similar features occur in *N. tornieri*, the young of which still have external gills when they hatch out. Although in this case, as in all other ovoviviparous and viviparous species, fertilisation is presumably internal, this has yet to be demonstrated. Notably, members of the species *Nectophrynoides* do not possess copulatory organs.

Gymnophiona of the genera *Gymnopis, Chthonherpeton* and *Typhlonectes* give birth to fully formed young, development having taken place in the oviducts. In the latter genera the respiratory organs of the larvae are large, lobular gills (see Fig. 72D). Fertilisation in Gymnophiona is internal, being effected by the insertion of the male's copulatory organ into the cloaca of the female. Our knowledge of the physiological processes associated with gestation is fragmentary and our understanding correspondingly based largely on conjecture. It is therefore not always possible to decide whether a particular case represents ovoviviparity or viviparity.

Among urodeles, the spotted salamander (*Salamandra salamandra salamandra*) is ovoviviparous. The eggs are fertilised in the uterus of the female by a spermatophore which is deposited by the male and then picked up by the female. The eggs take up to 10 months to develop into larvae, which then hatch out within the mother's body and are finally born, at which stage they possess external gills. Animals whose young are born as larvae are referred to as larviparous. Ovoviviparity in the cave salamander *Proteus anguinus* is facultative, for although it normally lays eggs, under certain conditions the eggs are retained in the oviduct to be born later as fully developed young salamanders.

Fertilisation in reptiles is internal and in some species leads on to ovoviviparity which ends by the rupturing of the egg membranes while the embryo is in the uterus, as for example occurs in *Anguis fragilis*. As with other groups of animals, it is often difficult in the case of reptiles to draw the line between ovoviviparity and viviparity. Moreover, some species regress to being oviparous. *Lacerta vivipara* for example, normally develops a true placenta but can become oviparous at temperature above 25°C. Neither ovoviviparity nor viviparity occur in Chelonia or crocodiles.

7.3 Viviparity

Internal fertilisation, whether by copulation or the transfer of a spermatophore, is a prerequisite of viviparity, as of ovoviviparity. In essence, viviparity involves the embryos being held within some cavity in the urogenital system and being provided with substances, which are essential for their survival, up until the time of birth. The many different adaptations which have evolved to meet this end extend beyond providing the embryo with nutrients to include aspects of gas exchange and excretion. Three basic mechanisms can be distinguished, which can occur individually or in combination:

1. oophagy, or adelphophagy,
2. the production of an embryotroph,
3. placentation.

It must be stressed that the definitions put forward in this chapter are to some extent arbitrary and do not always agree with those to be found elsewhere in the literature to which however this caveat can also be applied. With respect to lower vertebrates for example, reference is sometimes made only to two forms of ovoviviparity, placental and aplacental. True viviparity is, in contradistinction, where the embryo is supplied exclusively from the maternal blood via some form of placenta. The term placenta in general denotes a physiological interdependence between the embryo and the maternal organism and does not specify the nature of the structures involved. Another terminology in use completely avoids the term ovoviviparity and recognises only placental and aplacental viviparity, which is certainly justified on etymological grounds. On the other hand, the term ovoviviparity is well established in the literature and its use should therefore not be discontinued here. The definitions used in this chapter accord with a system of classification within which most cases can be accomodated without problem. Nevertheless, no matter which definitions are employed, there are always some cases which either represent intermediate forms that cannot be classified exactly or exhibit the characteristics of both types of reproduction during the course of gestation, starting out as ovoviviparous and becoming viviparous. Such is the case with *Mustelus laevis*. In the first months of gestation the embryo of this shark is enclosed within a relatively thick egg membrane and lives solely from its reserves of yolk. As these reserves become depleted a functional placenta forms between the yolk sac and the uterine epithelium. Thus the terms ovoviviparity and viviparity are used in this chapter according to the definition each has been given at the start of the relevant sections of the text and in light of the points just mentioned.

In the case of oophagy the embryo lives off the substance of other eggs. Only one or a few of the eggs produced actually develop, the remainder degenerate into a nutritious "soup" which is taken up by the surviving embryos. The consumption of sibling embryos which are at an advanced stage of development is referred to as adelphophagy or embryonic cannibalism.

The term embryotroph generally describes nutritory fluids which are secreted by the wall of the uterus or ovary and with which the embryo is provided. It usually also contains leucocytes and tissue debris. Various specialisations are commonly associated

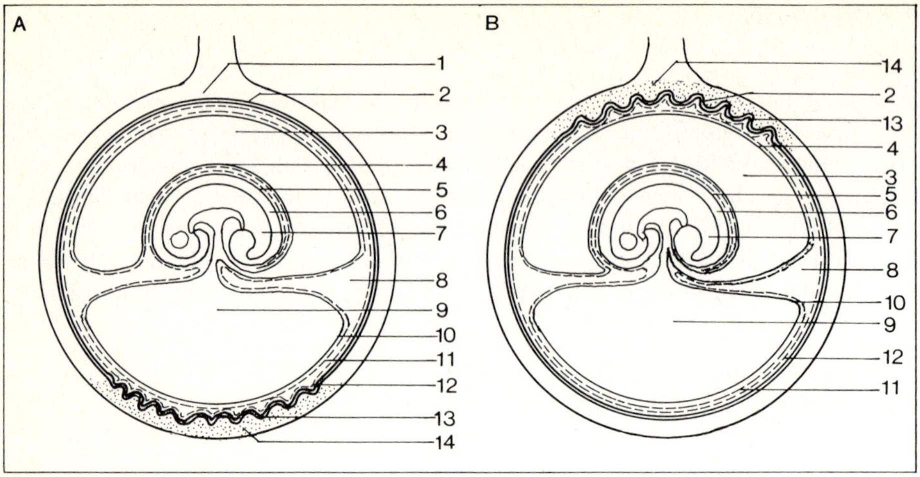

Fig. 74A,B. Diagrams of the two main forms of placenta which occur in submammalians, namely the omphaloplacenta (**A**) and allantoic placenta (**B**).
1 = wall of the uterus; 2 = uterine epithelium; 3 = allantois; 4 = endoderm of the allantois; 5 = amnion; 6 = amniotic cavity; 7 = embryo; 8 = extraembryonic coelom; 9 = yolk sac; 10 = yolk sac endoderm; 11 = splanchnopleure; 12 = somatopleure; 13 = chorionic ectoderm; 14 = placental region (*shaded*)

with this means of providing for the embryo. Thus in some cases the wall of the uterus develops secretory villi, so-called trophonemata, which are thread-like processes that extend through a spiracle or gill slit of the embryo to the oesophagus where they release a secretion. In other cases it is the embryo which develops specialised structures which facilitate the uptake or resorption of nutrients, examples of these being the so-called trophotaenia and the development of superficial capillary networks.

Placentation involves the formation of specialised structures whereby a close association or fusion between the maternal and embryonic tissues is established which facilitates the physiological exchange of substances between the maternal blood and the embryo. The two main types of placenta in vertebrates are named after the embryonic element which makes contact with the maternal tissues of the uterus wall. In the case of an omphaloplacenta (yolk sac placenta) this is the region of the chorion overlying the yolk sac, whereas it is the region overlying the allantois in the case of the allantoic placenta (Fig. 74). The former is undoubtedly the more primitive type and occurs in elasmobranchs and reptiles: it is also referred to as a preplacenta. The structural and physiological associations which occur in teleosts between the wall of the ovary or the follicle epithelium and structures of embryonic origin also have the character of a placenta but fall outside the rather narrow definition of that organ given above. A range of other specialisations also exist, including a follicular pseudoplacenta. Whether the omphaloplacenta has the function of supplying nutrients to the embryo is questionable in many cases, even though the appropriate structures are present. In such cases the placenta has been ascribed the function of providing the embryo with oxygen and water. This seems justified in the case of elasmobranchs and reptiles whose eggs contain a large amount of yolk, particularly in the former animals where the embryo is frequently also supplied by embryotroph. The uptake of water appears to be of particular importance for mobilising the reserves of foodstuffs bound up in the yolk. However, there is also good evidence to show that the omphaloplacenta of some species of elasmobranch is involved in supplying nutrients to the embryo. Among submammalians, only the reptiles have an allantoic placenta. In some of these species it is probably involved to a limited extent in providing the embryo with nutrients, aside from providing it with water and oxygen.

7.3.1 Viviparity in Elasmobranchs

The embryos of viviparous elasmobranchs are provided not only with water but also with oxygen and nutrients. Elasmobranchs reproduce either once every year or once every 2 years and have a period of gestation of between 3 and 22 months. The uterine epithelium of viviparous species usually forms ridges, papillae or villus-like structures. The uterus of placental species, and of a few aplacental species, exhibits a particular specialisation, namely that during the course of gestation several chambers are formed, each of which contains an embryo. As a rule, both uteri are functional and only exceptionally is one used, which can be either the right or the left uterus.

Oophagy occurs in the shark *Lamna cornubica*. The reserves of yolk are fully depleted during the early development of the embryo. Thereafter, the embryo, or foetus, consumes immature eggs and degenerating ovarian tissue which passes down the oviduct. This material is "packaged" in a secretion of the shell gland that is evidently nutritious (see p. 109). The embryo of *Lamna nasus* takes up large amounts of "yolk" which differs from that formed in the egg. This yolk collects in the stomach, thereby distending it into a so-called yolk stomach. This also occurs in the case of the mako *Isurus oxyrhyncus*. It has been reported that the foetuses of the tiger shark *Carcharias taurus* make spontaneous eating movements within the uterus and can even bite the experimenter's fingers if the uterus is opened. Oophagy is not common among elasmobranchs.

Embryotrophy plays an important nutritional role in elasmobranchs and, one suspects, may even be of significance at one time or another during gestation in all viviparous species. The term embryotroph is used here as a collective term encompassing all nutrient fluids produced by the uterus and includes the phenomenon referred to in some publications as histiotroph, in which case the fluid also contains cells. As a rule, the fluid is a secretory product of the uterine epithelium, which may develop specialised secretory folds or villi. The composition of the embryotroph differs from species to species, as does its content of organic substances which, for example, is 1.3% in the electric ray *Torpedo marmorata* and 13% in the stingray *Dasyatis violacea* of which 8% is lipid (Fig. 75). The mucus associated with embryotrophs is mostly glycoprotein, together with reduced sugars and urea. The fluid within the uterus may also contain proteases which are however inactive owing to the pH-value of this fluid. It has been suggested that they become active digestive enzymes once they reach the embryo's gastrointestinal tract. This could be of particular relevance in those species in which an embryotroph is digested in the stomach, whereas the contents of the yolk sac are digested in the intestines. The egg membrane, which is as much in evidence in

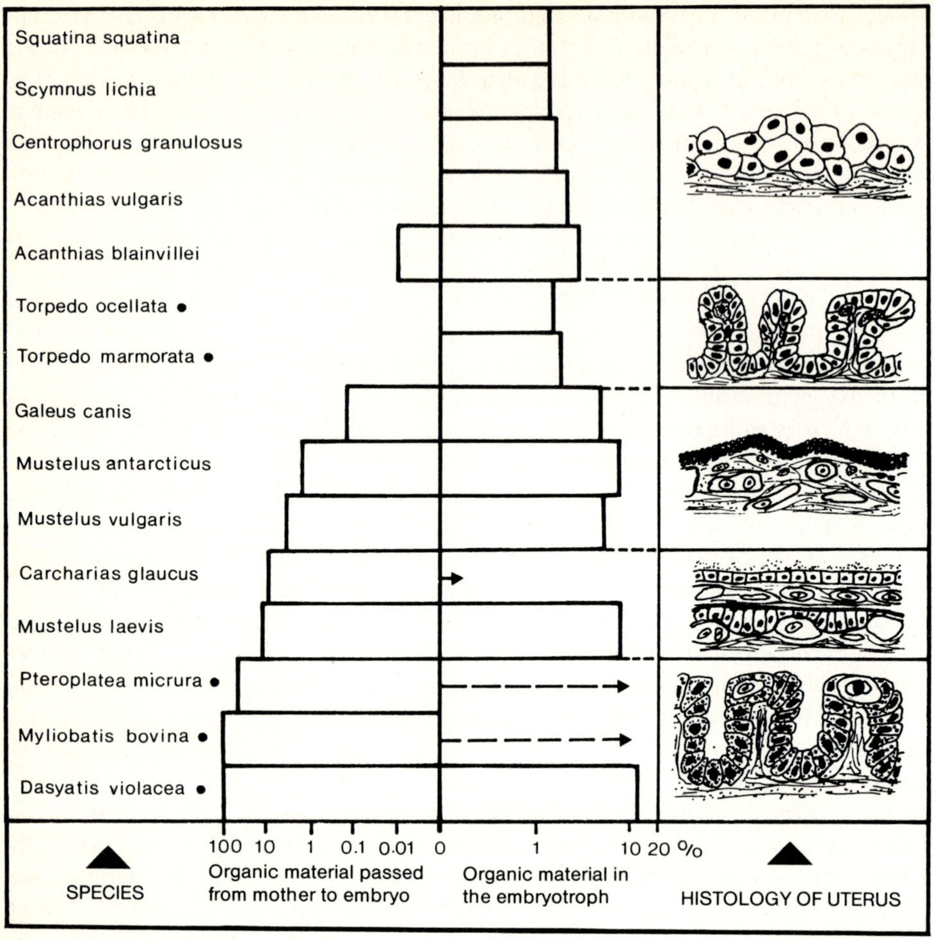

Fig. 75. Summary diagram of the foeto-maternal relationship in Elasmobranchii. (After Ranzi and Needham from Hoar and Randall; considerably modified). • indicates species of ray, all others are sharks

viviparous species as it is in non-viviparous species, must also be permeable to the constituents of the embryotroph or it must possess specialised pores.

One can distinguish three basic types of uterine secretion, namely serous, mucous and fatty secretions, the nature of the secretion often reflecting the physiological state of the uterine epithelium. Serous secretions are produced by species of *Torpedo* and by *Scymnorhinus licha*, which has been investigated more thoroughly. In this animal the wall of the gravid uterus bears a number of villi, each of which is provided with an ascending and descending arteriole between which extends a subepidermal network of capillaries. Whereas it is possible that substances simply diffuse from the capillaries into the lumen of the uterus, it has been found that the cells of the uterine epithelium produce a secretion which may involve the discharge of whole cells. Leucocytes containing lipid droplets are found in the subepithelial capillaries and these can penetrate the uterine epithelium and become a component of the embryotroph. The serous embryotrophs of different species vary in their lipid content but, understandably, do not usually have a mucus component. In general, serous embryotrophs and their production are characterised by the following features: the diffusion of substances out of the villar capillaries; certain cell types of the uterine epithelium are secreted; a low lipid content (i.e. relatively few immigrating leucocytes); the lack of a mucus component; a low total content of organic substances (Fig. 75). The mucous type of embryotroph is characteristic of placental species and aplacental species with a chambered uterus. Studies on the smooth dogfish *Mustelus mustelus* have shown that the uterine epithelium is underlain by a dense network of capillaries from which large amounts of material is released into the lumen of the uterus, particularly towards the end of gestation. In addition to this, a large amount of mucous material and a very small amount of fats is secreted from the uterine epithelium. The total amount of organic substances in this type of embryotroph is very high (Fig. 75). Species of the genera *Dasyatis, Myliobatis* and *Pteroplatea* produce a lipid embryotroph. In *Dasyatis violacea* the epithelium of the functional left uterus bears long villi that are provided with tubular glands. During gestation the villi grow to a length of about 20 mm, not by cell proliferation but by the cells increasing in size. The villi are richly vascularised. The nearer the glands of the villi are to the end of the villus, the more active they are. The glands secrete large amounts of a very fatty secretion, as do other cells of the uterine epithelium. In association with this, cells dislodged from the epithelium become suspended in the fluid within the uterus. Lipid may be produced either as lipid droplets by the epithelial cells themselves or by a leucocyte penetrating into and completely degenerating within an inactive epithelial cell whereupon the latter begins to produce large amounts of protein and lipid. The characteristics of lipid embryotrophs are as follows: the uterine epithelium secretes small amounts of a serous secretion; substances diffuse out of the subepithelial capillary network; the uterine epithelium secretes lipid; the uterine fluid contains numerous cells rich in lipid; there is no secretion of mucus; lipid embryotrophs are rich in organic substances (Fig. 75). The secretory villi may be of considerable length. For example, in *Pteroplatea micrura* they extend through the spiracle and into the oesophagus of the foetus where the lipid embryotroph is secreted from the tubular glands (Fig. 76A,B), i.e. they function as trophonemata. In *Dasyatis* and *Myliobatis* however, the villi are shorter but more numerous and produce a secretion which is taken up by the embryo mainly through the spiracles.

Placental viviparity occurs in only a few individual species of elasmobranch, namely *Mustelus laevis, Carcharias glaucus, Carcharias melanopterus, Carcharrhinus falciformis, Carcharrhinus plumbeus, Sphyrna tiburo, Scoliodon sarrakowah* and *Scoliodon palassorah*. Of these placentation has been studied most thoroughly in *Mustelus laevis*, the animal in which Aristotle was the first to recognise the nutritional interrelationships between mother and foetus. Placentation in *Mustelus laevis* begins in the chambered uterus when the embryo has a length of between 10–16 cm. The yolk sac is connected to the body of the embryo by an umbilicus which can reach a length of about 20 cm by the end of the gestation. The yolk sac is differentiated into a proximal and distal portion (with respect to the embryo). The former, which lies outside the uterine epithelium and is surrounded by the uterine fluid, is much folded and still

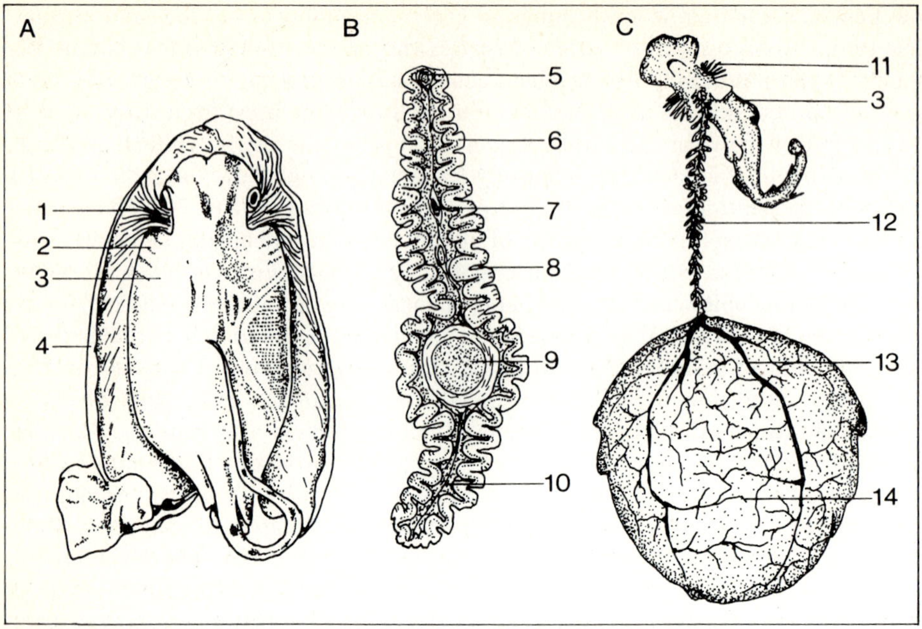

Fig. 76A–C. Foetus of *Pteroplatea micrura* showing the trophonemata entering the spiracles (**A**). Cross-section of a trophonematum (**B**). (After Wood-Mason and Alcock). Foetus of *Sphyrna tiburo* attached to the yolk sac by the umbilicus (**C**). (After Leuckhart).
1 = trophonemata; 2 = spiraculum; 3 = foetus; 4 = cut wall of the uterus; 5 = artery; 6 = uterine epithelium; 7 = capillary; 8 = tubular gland; 9 = vein; 10 = connective tissue; 11 = external gills; 12 = umbilicus with appendiculata; 13 = blood vessel; 14 = yolk sac

contains yolk during the early stages of placentation. Its wall is composed of a layer of yolk sac endoderm underlying the splanchnopleure, which is separated from the first by an extraembryonic coelom. The distal portion of the yolk sac contacts the wall of the uterus and both these elements together form the yolk sac placenta. The uterine epithelium goes through three stages during gestation. In the first it secretes various mucoid substances which are resorbed by the ectoderm of the distal portion of the yolk sac. Thereafter the uterine epithelium comes into contact with the wall of the yolk sac and this is accompanied by a sharp reduction in the secretion of mucus. At this stage the maternal and foetal tissues are separated by the egg membrane which lines the entire surface of the uterine chamber. During the last stage blood vessels and capillaries grow into the uterine epithelium and come into close contact with the yolk sac whose ectodermal epithelium gradually becomes less extensive during the course of development until if finally disappears so that the capillaries within the mesoderm covering the yolk sac are only separated from the uterine epithelium by the egg membrane, which has by this stage become considerably thinner. The egg membrane is impregnated with mucous material and is presumably permeable to nutrients and respiratory gases. The omphaloplacenta of *Mustelus laevis* is an example of an epithelio-chorial type of placenta (see p. 263). Two structural features of the placental region which are very probably involved in supplying nutrients to the embryo remain to be mentioned. The first is the mucus secreting chamber that forms within the wall of the uterus beneath the region of placentation. It is penetrated by long, foetal diverticula which evidently resorb mucus produced by the uterine epithelium: it is possible that the mucus not taken up by the embryo is stored within the chamber. The second structure is a glandular sac lined with mucus producing cells that extends parallel to and beneath the surface of the uterine wall. Its function is unknown. The basic structure of the omphaloplacenta of *Mustelus laevis* (Fig. 77) is essentially the same as that in other species of placental, viviparous species of shark. The most highly developed placenta is that of *Scoliodon sarrakowah*. In this case the yolk sac no longer contains any yolk but a network of tissue strands pervaded with blood vessels. In the fully developed placenta the yolk sac is no longer confluent with the foetal gut so that they are connected only by blood vessels, as is the case in mammals.

As has already been mentioned, the foetus is connected to the yolk sac or placenta by an umbilicus. This cord is ensheathed by a thin layer of ectoderm overlying a layer of somatopleure which encloses the arteria umbilicalis (or arteria vitellina), the vena umbilicalis (vena vitellina), and between these two the ductus vitello-intestinalis. The extraembryonic coelom extends between the somatopleure and the splanchnopleure (Fig. 77). In some species (e.g. *Sphyrna tiburo* and *Scoliodon terra-novae*) the umbilicus bears appendages along its entire length, the so-called appendiculata (Figs. 76C and 78). These are composed of thin cords of connective tissue covered by a single epithelial layer which is sometimes ciliated. Numerous capillaries usually lie beneath the epithelium and therefore the appendiculata have been implicated in the resorption of embryotroph. Such an umbilicus is typical of those species with a primitive placenta, but is also found in the aplacental species *Carcharrhinus dussumieri*.

To summarise briefly, the maternal organism alone is responsible for providing the embryos of viviparous elasmobranchs with water, mineral salts, oxygen and organic substances from the yolk within the yolk sac, from other eggs, from uterine milk (embryotroph) and from the maternal blood via an omphaloplacenta. The nourishment of the embryo from reserves in the yolk sac is of particular importance during the early stages of embryonic development. These nutrients reach the embryonic intestine via the ductus vitello-intestinalis. In many cases an additional, inner yolk sac connected to the ductus vitello-intestinalis is formed in which the yolk is digested. Exactly how the component substances of the embryotroph are resorbed is, in many cases, not known but it is thought to involve the uptake of material over the embryonic ectoderm and yolk sac epithelium and perhaps the filamentous gills of the embryo, supplementary to their role in gas exchange. A further important means of nutrition is the uptake of uterine milk through the mouth and spiracles or oesophagus via the trophonemata. The appendiculata of the umbilicus are probably also organs of resorption. In placental species the foetus is additionally provided with material from the maternal blood and this is reflected structurally in the association between the blood vessels of the placenta and the foetal circulation, as is illustrated with particular clarity in the case of *Carcharias glaucus*: the arteria umbilicalis arises from the arteria intestinalis and the vena umbilicalis leads back into the vena portae hepatica. It is noteworthy that in no case is the foetus of a viviparous elasmobranch supplied by only one of the routes mentioned above, rather the foetus is always supplied with nutrients

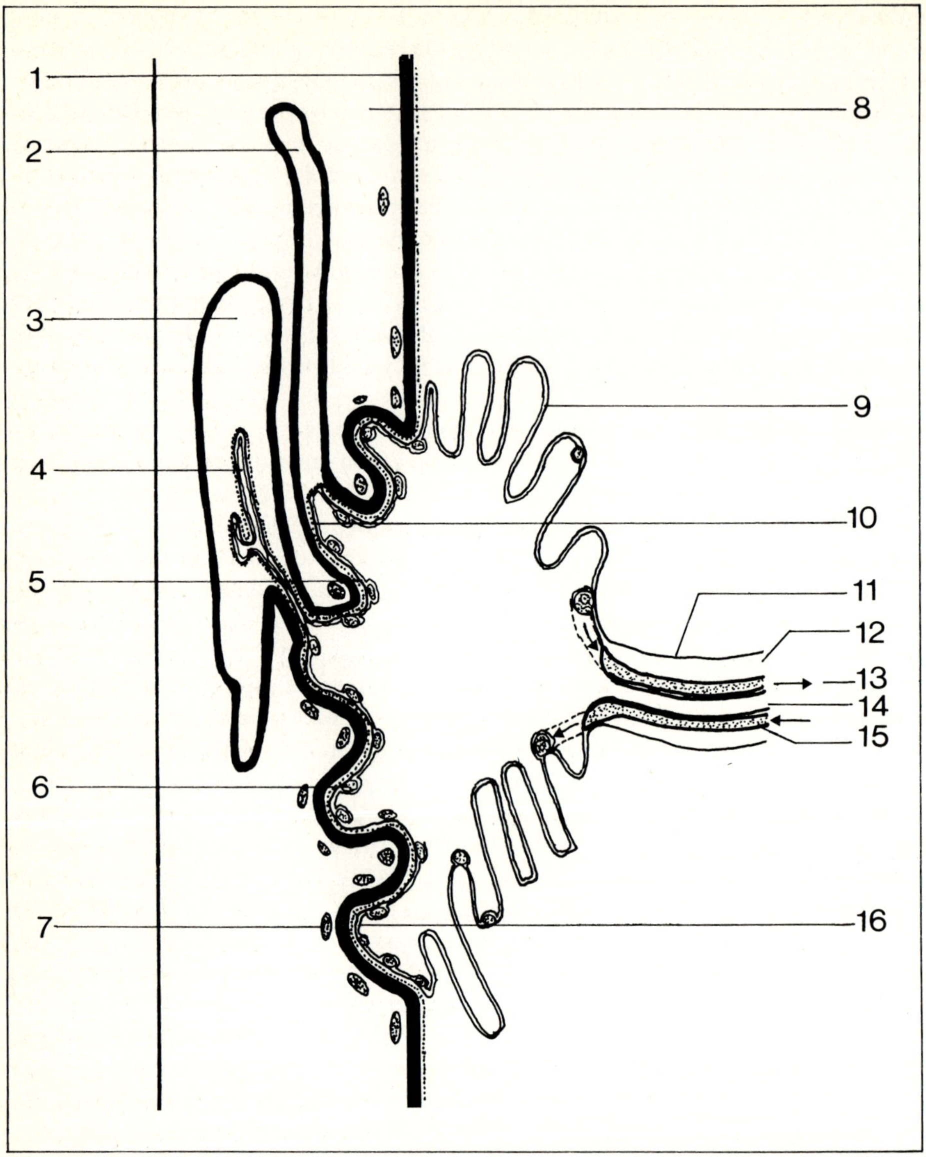

Fig. 77. Diagram of the omphaloplacenta of *Mustelus laevis*. (After Ranzi, greatly modified). 1 = uterine epithelium; 2 = glandular sac; 3 = mucous secretion chamber; 4 = foetal diverticulum; 5 = distal (placental) portion of the yolk sac; 6 = blood vessel of the yolk sac (placenta); 7 = uterine blood vessel (placental); 8 = wall of the uterus; 9 = proximal portion of the yolk sac; 10 = egg membrane; 11 = umbilicus; 12 = extraembryonic coelom; 13 = vena umbilicalis (vena vitellina); 14 = ductus vitello-intestinalis; 15 = ateria umbilicalis (arteria vitellina); 16 = placental region of the yolk sac

from the egg yolk and from at least one other of the sources mentioned. A comparison of the total expenditure on the part of the maternal organism on behalf of the foetus over and above that represented by the reserves of yolk in the egg is shown for a number of different species in Fig. 75. From this figure it can be seen that viviparous species which are placental do not have the highest level of expenditure: the species of rays represented in the figure which produce a lipid embryotroph are significantly superior in this respect.

Little is known concerning the hormonal regulation of gestation in elasmobranchs. Hypophysectomy does not appear to terminate pregnancy if performed during the first 3 months and also does not prevent ovulation in oophagous species. The ovaries of elasmobranchs produce both of the typical types of corpora lutea, namely preovulatory corpora lutea atretica (CLA), which arise from atretic eggs, and postovulatory corpora lutea ovulatoria (CLO). In some species both these structures exert an influence on the activity of the shell glands and in ovoviviparous and viviparous species they suppress ovulation and reduce the contractility of the uterus during gestation. Other than this they evidently inhibit the laying of eggs in pairs, which is a characteristic of many oviparous species. There is no evidence to show that they affect the attachment of the placenta to the uterus or the secretion of embryotroph. It is noteworthy however, that in some species CLA become more numerous during gestation and when folds develop in the uterus in the course of the normal oestrus cycle. Most then disappear again shortly after the birth of the young. These changes resemble those undergone by the accessory corpora lutea in some mammals. Although it has been demonstrated that oestrogens and progesterone are present in the ovarian tissues of a few species of elasmobranch and that CLA as well as CLO are capable of producing steroids, there is no conclusive evidence implicating steroids in the endocrine control of gestation. More general considerations suggest that steroids inhibit ovulation and reduce the contractility of the uterus.

7.3.2 Viviparity in Teleosts

True viviparity occurs in a range of teleost species but in a form which differs in one important respect from that in all other vertebrates, namely that gestation never takes place in the oviduct or uterus but in the ovary. Teleosts do not possess müllerian ducts and the oviducts (see p. 65) are unsuitable for the containment and provision of embryos. As mentioned in connection with ovoviviparity, gestation in such animals can take place either in the follicle (follicular gestation) or in the ovarian cavity (ovarian gestation). Fertilisation is internal in all viviparous teleosts, to which end the males of the most species possess copulatory organs which have a complex structure (see p. 68). The eggs are normally fertilised when they are in the follicle and although one might suppose that the more primitive form of fertilisation is where the ovulated eggs are fertilised in the lumen of the ovary, it is very probable that this only occurs in the case of the viviparous *Zoarces vivipara*. The relative timing of ovulation, hatching of the young, and of birth differs between species. However, three basic types of gestation can be distinguished which can be represented as follows:

Zoarces type O-F. .H. .B
Jenynsia type F . . . . .O. . . . . H . . . . .B
Gambusia type F . . . . O-H-B

where F stands for fertilisation, O for ovulation, H for hatching and B for birth. Viviparity only occurs in two orders of bony fish, the Microcyprini and the Perciformes, within the families Goodeidae, Poeciliidae, Jenynsiidae and Anablepidae in the former order and the Embiotocidae, Clinidae, Zoarcidae, Brotulidae, Scorpaenidae and Cottidae in the latter. As mentioned above, it is difficult to distinguish between ovoviviparity and viviparity in teleosts and even by comparing the weight of the eggs and the birth weight of different species the distinction cannot be made with any certainty. Nevertheless, it is currently accepted that the embryos or foetuses of most live-bearing species are supplied in one way or another with nutrients, water, mineral salts and oxygen within the ovary. They are therefore viviparous according to the definition given in this chapter and it is merely playing with words to regard some species as, for example, pseudo-ovoviviparous or as partially ovoviviparous. A variety of mechanisms for the uptake of material by the embryo have evolved within the members of the families mentioned above. Viviparous teleosts include both marine and freshwater species.

The surf perch *Cymatogaster aggregata* (Embiotocidae), a teleost in which gestation occurs within the ovarian cavity, has been studied in some depth. Following copulation the spermatozoa of this species are stored in the ovary for about 6 months, after which time fertilisation takes place within the follicle. Ovulation occurs during the early stages of cleavage and the subsequent development, which takes 10–12 months, takes place in the ovarian cavity. The young are born at a very advanced stage of ontogenetic development, some males already being sexually mature. The young are provided for during gestation by mechanisms which are particularly complex. Thus, at a stage when the larvae are about 15 mm long, richly vascularised, lobular processes which serve as organs for the resorption of nutrients and gases develop between the rays of the dorsal fin and persist until shortly before birth. During the late stages of gestation long villi develop in the intestines which are evidently involved in the digestion and resorption of embryotroph ingested through the mouth. Furthermore, the embryonic gills become associated with the tissues of the ovary. This type of gestation is, nevertheless, simple compared with others. The Goodeidae, Jenynsiidae, Embiotocidae and Zoarcidae possess a range of more complicated structures for the resorption of embryotroph. In all cases the embryos or larvae, both while they are contained within the egg membrane and after they hatch out, are bathed in an ovarian fluid which has a complex composition that includes dissolved proteins and salts, glycogen granules, tissue remnants, leucocytes and oil droplets. This embryotroph is secreted by the wall of the ovary, the epithelium of which becomes specialised during the course of gestation: for example, it may become folded or lacunae may form within it or it may be invaded by leucocytes. It may also become more richly vascularised and begin to secrete actively.

The embryo or foetus resorbs the embryotroph in a variety of ways. Firstly, material is taken up through the skin, in particular that of the yolk sac and gills. This includes the uptake not only of dissolved substances but also of emulsified lipid droplets, which may be mediated by leucocytes. Active transport is probably also involved.

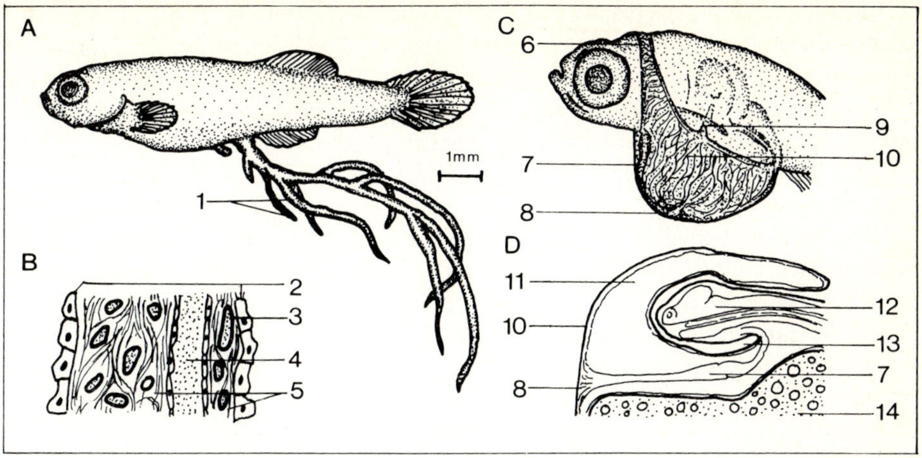

Fig. 78A–D. Well developed embryo of *Zoogoneticus quitzeoensis* with trophotaenia (**A**). Longitudinal section of a trophotaenium (**B**). (After Turner). **C** Embryo of *Lebistes reticulatus* to show the portal system in the wall of the yolk sac. (After Turner). **D** Sagittal longitudinal section through the front end of an embryo of a species of *Poeciliopsis*. (After Turner).
1 = trophotaenia; 2 = epithelia; 3 = capillary; 4 = blood vessel; 5 = mesenchyme; 6 = neck band; 7 = heart; 8 = sinus venosus; 9 = ductus cuvieri; 10 = embryonic portal system; 11 = pericardial sac; 12 = embryo; 13 = pseudo-amniotic cavity; 14 = yolk sac

The uptake of material by the vascularised processes of the dorsal fin of Embiotocidae which was mentioned above can be included here, although these structures are thought to be primarily involved in the exchange of gases. Towards the end of gestation the intestine may be responsible for the resorption of a significant amount of embryotroph taken in through the mouth or gill slits. In Embiotocidae long villi are formed in the hind gut. Similarly, in Goodeidae long processes called trophotaenia develop around the anus after hatching, their form being different in different species. They are made up of skeins of connective tissue, lymph lacunae and numerous blood vessels which are all in part covered with foetal ectoderm and in part with intestinal endoderm in the form of an isoprismatic epithelium (Fig. 78A,B). One species of Brotulid also has trophotaenia, but in this case they are covered only with foetal ectoderm. In a few Goodeidae and in Jenynsiidae a pericardial sac develops during the phase of development prior to the outgrowth of the trophotaenia: this structure also has the function of resorbing embryotroph.

Among those teleosts with ovarian gestation are species which are also oophagous or adelphophagous. In brotulid species for example, the embryo takes up material from clumps of small eggs within the lumen of the ovary. Adelphophagy occurs in Goodeidae and Jenynsiidae and in some cases up to three quarters of the brood of eggs serve as food for the remaining quarter. It has been suggested that adelphophagy could only have evolved in association with ovarian gestation because the embryos are not separated from each other, as they are in the case of follicular gestation. True adelphophagy is often difficult to distinguish from a case where the embryos are nourished by an embryotroph composed of the products of other, degenerating embryos.

Highly specialised mechanisms have evolved in the Jenynsiidae for the provision of the embryos. Ovulation normally follows shortly after fertilisation and during the early stages of its development the embryo is supplied from the limited reserves of yolk by resorption through the yolk sac and pericardial sac. However, the main stage of development occurs after another mechanism of provision has been established that involves the growth of papillae or villi from the ovarian wall which are referred to as trophonemata by analogy with comparable structures in elasmobranchs. The villi extend into the branchial cavity of the embryo and become closely associated with the gills and may contact the latter via richly vascularised folds in their surface. This association can be interpreted as a potential branchial placenta. In addition to this, the foetus takes in large quantities of embryotroph through the mouth.

The type of gestation is related to the size of the egg and the amount of yolk it contains. Thus, in species with small eggs containing little yolk gestation occurs in the ovarian cavity whereas in those with eggs rich in yolk it occurs in the follicle. Follicular gestation only occurs in members of the Poeciliidae and Anablepidae and, as in the case of ovarian gestation, there are numerous intermediates between species with follicular gestation that are truly ovoviviparous and those which exhibit a pseudoplacental type of viviparity. The whole range of intermediates are represented in the Poeciliidae, in which gestation takes place solely within the follicle. In a range of species there is apparently sufficient yolk stored within the egg to support the entire course of development during gestation, for example the guppy *Lebistes reticulatus* and the swordtail *Xiphophorus helleri*. The embryo is provided with oxygen via a system of portal blood vessels in the wall of the yolk sac, which forms a collar around the embryo's "neck" (Fig. 78C). The wall of the yolk sac forms a close, placenta-like association with the follicle wall that is responsible for the exchange of gases and probably the excretion of nitrogenous wastes but not the supply of nutrients to the embryo. In Poeciliidae, the wall of the follicle is usually richly vascularised and encloses a fluid-filled chamber. In *Gambusia*, the egg has no egg membrane and therefore the embryonic gill filaments float freely within the follicular liquor. The pericardium of *Heterandria* and *Platypoeciliopsis* species forms an organ for the resorption of material. Thus the pericardium and the overlying ectoderm bulge outwards and extend further to cover the embryo like an amnion, thereby forming a pericardial sac (Fig. 78D). The inner layer of this sac is not vascularised and is, thus, comparable to a pseudo-amnion whereas the outer layer, which comes into close contact with the wall of the follicle, possesses a well-developed portal system and can be regarded as a pseudo-chorion. In some of the species in these genera an outpushing of the coelom results in the formation ventrally of an abdominal sac which is similarly in apposition with the follicular epithelium. Both the pericardial and abdominal sacs mediate the exchange of materials between the mother and embryo. The eggs of all these species contain relatively little yolk. An even more specialised structure can be found in species of *Poeciliopsis, Poecilistes* and *Aulophallus*, namely secretory villi which develop from the follicle epithelium which lies against the abdominal sac. Such foeto-maternal associations are referred to as follicular pseudoplacentas. Developments of this type are even more advanced in the Anablepidae. As in the above cases, villi develop from the follicular epithelium but, in addition, resorptive papillae also develop from the ectoderm of the yolk sac to form a follicular pseudoplacenta which becomes func-

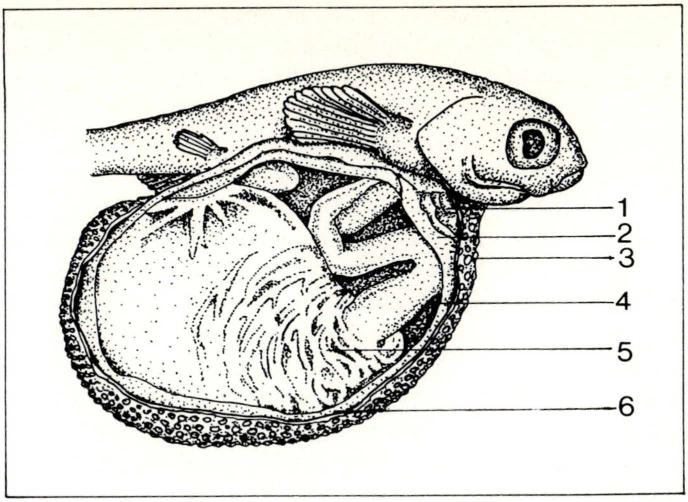

Fig. 79. *Anableps* embryo. The yolk sac has been partly cut away to reveal the organs and cavities within. (After Turner).
1 = heart; 2 = pericardial cavity; 3 = wall of the yolk sac; 4 = coelom; 5 = intestine; 6 = fissure of the pericardial cavity

tional after the reserves of yolk have been exhausted. Rostrally the wall of the yolk sac is underlain by a diverticulum of the pericardium whereas ventrally it is underlain by an extension of the coelomic abdominal sac. The papillae, or bulbi, of the yolk sac are provided with numerous capillaries from the portal system supplying the yolk sac or that supplying the abdominal sac, as is the extension of the pericardial sac. Besides these specialisations, the intestine in *Anableps* species is much enlarged and the yolk sac greatly distended (Fig. 79). These specialisations are involved in the uptake of follicular embryotroph which has been taken in through the mouth and digested in the intestine. These specialisations regress shortly before birth.

To summarise, all the mechanisms by which the embryos of viviparous elasmobranchs are sustained during gestation also occur in viviparous teleosts, although in a specialised form as a result of the constraints of development occurring within the ovary. The development of follicular and branchial pseudoplacentas is a particular feature of bony fish and does not occur in any other group of vertebrates. It must be borne in mind however that data concerning the physiological aspects of the phenomena described above are extremely fragmentary and that some of the assertions in the literature are of a speculative nature.

Evidence relating to the possible hormonal regulation of gestation in teleosts is sparse. Gestation in *Mollienesia latipinna* is apparently not interrupted by hypophysectomy, which in one species of *Gambusia* results in an increased embryo mortality, although this was probably due to the electrolyte balance having been disrupted. The gonadotropin blocker methallibure causes atrophy of the ovarian tissues in *Cymatogaster aggregata* accompanied by a decrease in the growth rate and an increase in the mortality of the embryos. It has been suggested that prolactin may play a role in gestation and that isotocin is involved in the process of birth (see p. 179), although there is no direct proof of this. The role of the corpora lutea (i.e. the CLA and CLO; see p. 247) is similarly unresolved. Species with intrafollicular fertilisation do not possess CLO. One possible function of the CLA is in preventing ovulation of mature follicles or of follicles containing an embryo. In *Cymatogaster aggregata* the activation of the secretory epithelium of the ovary appears to be linked to the formation of CLA. Furthermore, additional CLA are formed during the course of gestation. These structures are thougth to have the effect of reducing the contractility of the ovarian wall.

Finally, the phenomenon of superfoetation must be dealt with briefly. In some Poeciliid species several cohorts of embryos develop in the ovary at the same time with new eggs maturing faster than the existing embryos progress through gestation and leave the ovary. The new eggs are fertilised by spermatozoa that have lain immobilised for several months in sperm chambers of the ovarian epithelium which serve to sustain the spermatozoa. A particularly impressive instance of superfoetation is that exhibited by *Heterandria formosa*, in which living sperm are still present in the ovary 10 months after copulation. Up to nine brood cohorts can develop simultaneously and are eventually given birth to at 10-day intervals.

7.3.3 Viviparity in Amphibians

Viviparity rarely occurs in amphibians. The only known viviparous species of anuran is the toad *Nectophrynoides occidentalis*. Nothing is known in this case concerning the mode of fertilisation, which is presumably internal. The CLO that form persist throughout gestation, which lasts for 9 months. Up to the seventh month no new follicles develop in the ovary nor do any of the fertilised eggs degenerate since the number of corpora lutea usually agrees with the number of embryos. The eggs contain very little yolk and once it has been exhausted the embryos or larvae absorb an embryotroph secreted from the wall of the uterus. The larvae have no gills but it is noteworthy that the head, mouth and arms are formed very early in development. The CLO in the ovary produce progesterone in increasing amounts as gestation proceeds. This hormone seems to be able to affect the development of the embryos, causing it to slow down. This is seen as being of selective advantage since the young are therefore not born during the summer months but at a more favourable time of the year.

Two species of viviparous urodele are known, the Italian cave salamander *Hydromantes genei* and the alpine salamander *Salamandra atra*, although no details are known concerning gestation in the former species. Fertilisation is internal in *Salamandra atra* and is achieved by the transfer of a spermatophore. Gestation is a lengthy process lasting 3–4 years (Fig. 80). A single embryo usually develops in each oviduct even though several eggs are ovulated, and probably also fertilised, resulting in the formation of a corresponding number of CLO within the ovary. However, only one of the several eggs in each limb of the uterus develops further, that being the one nearest the cloaca: the others degenerate. The physiological processes underlying this phenomenon are unknown. The early development of the embryo is supported by the reserves of yolk in the egg. Thereafter the wall of the uterus secretes an embryotroph enriched with leucocytes and material resulting from the degeneration of other eggs. The embryos are provided for in this way during the first 2 years of gestation, the em-

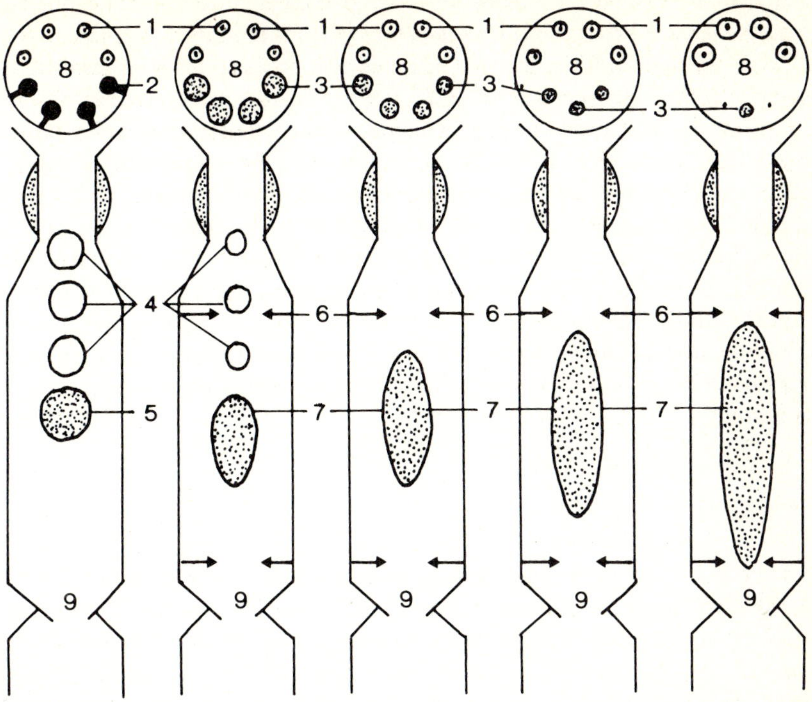

Fertilisation	Year 1	Year 2	Year 3	Year 4
Yolk	Material from degenerating eggs and embryotroph		Embryotroph	

Fig. 80. Gestation in *Salamandra atra*. (After Browning).
1 = previtellogenetic oocyte; 2 = empty follicle; 3 = postovulatory corpus luteum; 4 = undeveloped eggs; 5 = early embryo; 6 = secretion of embryotroph (*arrows*); 7 = embryo; 8 = ovary; 9 = uterus

bryotroph being the sole source of nutrients after the materials from the eggs have been exhausted. The embryotroph contains lymph, tissue debris from the uterine epithelium and blood cells resulting from haemorrhages in the uterus. The larvae possess long, filiform external gills which function as resorptive organs. After about 4 years the larvae are born, having already undergone metamorphosis. During gestation the CLO decrease steadily in size and after the third year in number also so that almost none remain by the time of birth. This regression is countervailed by the formation of new follicles (Fig. 80). The functions of the corpora lutea have not been demonstrated conclusively but are probably the reduction in the contractility of the uterus, promotion of the production of embryotroph, and the inhibition of the development of new eggs. It has yet to be shown that the corpora lutea contain progesterone. It is notable that by the appropriate experimental manipulation *Salamandra atra* can be made to give birth when the young are still larvae and, contrarily, the spotted salamander (*Salamandra salamandra salamandra*), which is described as being ovoviviparous or more precisely larviparous, can be made viviparous so that it gives birth to fully developed young. In the case of the latter species the same number of CLO are formed in the ovary as eggs that are ovulated. Nevertheless, ovariectomy performed during the first 3 months of gestation does not lead to the young being aborted. Similar experiments have not been performed on the alpine salamander.

In general, one may conclude that amphibians exhibit a relatively unspecialised form of viviparity. Placental types are not in evidence and embryotroph is the sole proven means by which the young are nourished, if one interprets the oophagy which occurs in the alpine salamander as a type of embryotroph. Viviparity was undoubtedly of great selective advantage to amphibians with respect to the invasion of relatively arid habitats. In some cases viviparity has evidently been superseded by other types of adaptation.

7.3.4 Viviparity in Reptiles

Reptiles typically have large eggs containing large amounts of yolk which, in contrast to those of amphibians, are primarily "designed" to be laid on land. This is largely a consequence of the evolution of two membranes which enclose the embryo, the amnion and allantois (see p. 240): as a rule these membranes are surrounded by either a soft or a hard calcerous egg membrane. The difficulties of distinguishing between oviparity and ovoviviparity in reptiles have already been dealt with (see p. 238). The evolution of viviparity in this class of vertebrate has taken a single path leading to the development of a placenta which necessarily involved a reduction in the extent of the egg membranes. Some reptiles have a yolk sac placenta (Fig. 74A), as do elasmobranchs, whereas others possess an allantoic placenta (Fig. 74B) and represent the first vertebrates in the course of evolution to possess such a structure. There are, in fact, a number of Australian lizards and snakes which possess both types of placenta (e.g. *Chalcides tridactylus*).

Placental viviparity occurs in the following families: Gekkonidae, Xantudidae, Scincidae (a total of 17 species), Lacertidae, Anguidae, Colubridae, Elapidae, Hydrophiidae and Viperidae. A number of lines of argument concerning the ecology of viviparous reptiles indicate that placentation represents a selective advantage in extremely dry habitats or in regions where it is warm for only a short period. This view is corroborated by physiological evidence which will be discussed elsewhere.

The omphaloplacenta of reptiles has been extensively studied in *Chalcides tridactylus, Hoplodactylus maculatus* and a number of Australian lizards, and in all these cases its structure is essentially the same as the yolk sac placenta of *Mustelus laevis*. The yolk sac is covered in a layer of chorionic ectoderm which is thicker in the region where it overlies the uterine epithelium and also in other, irregularly defined regions beyond this, but it is not usually thickened over the remaining area of the yolk sac. In some cases the uterine epithelium and the omphalochorionic ectoderm are thrown into interdigitating folds whereas in others (species of *Egernias* and *Tiliqua*

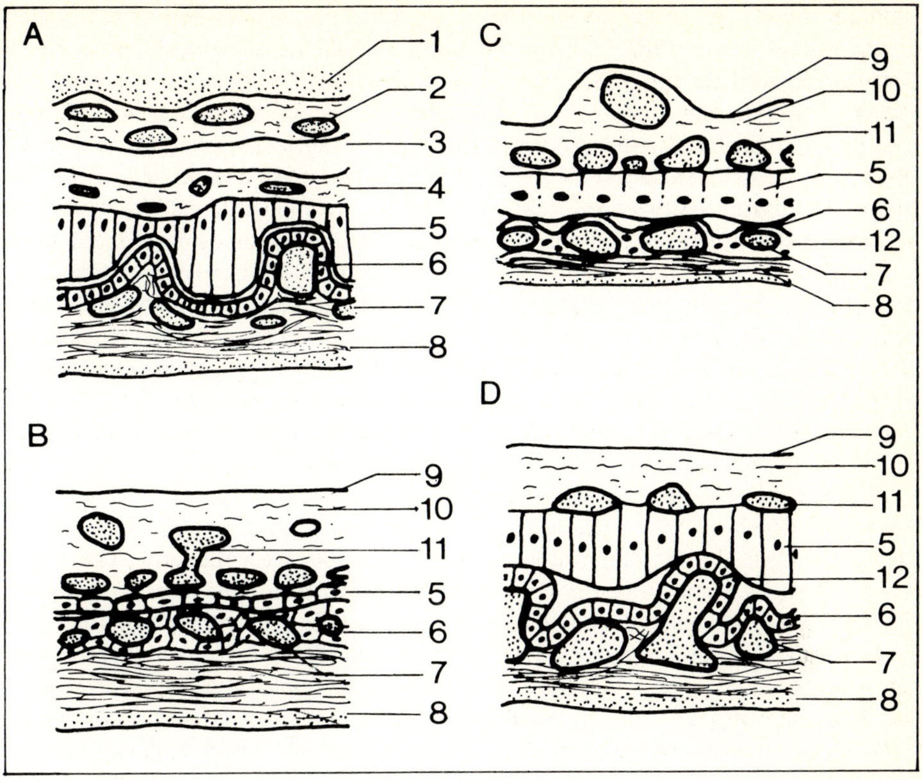

Fig. 81A–D. Histology of the omphaloplacenta (**A**) and the 3 types of allantoic placenta (**B–D**) of reptiles. (After Weekes).
1 = yolk; 2 = yolk sac mesoderm containing blood capillaries; 3 = extraembryonic coelom; 4 = omphalochorionic mesoderm containing capillaries; 5 = chorionic ectoderm; 6 = uterine epithelium; 7 = uterine capillary; 8 = uterine musculature; 9 = allantoendoderm; 10 = allantochorionic mesenchyme; 11 = capillary of the allantois; 12 = uterine fold

for example) only the former is folded. In *Hoplodactylus maculatus* both sides are folded but the folds do not interdigitate. All these types of placentation are primitive, there being no invasion of the foetal tissues into the wall of the uterus (Fig. 81A).

The allantoic placenta of reptiles similarly has a relatively simple structure. One can distinguish 3 main types of allantoic placenta according to their basic structure. In the first type, which occurs in *Chalcides ocellatus* and the snake *Demisonia superba* for example, the capillaries of the allantois lie in close proximity to the chorionic ectoderm and in some cases penetrate this layer so that there is only a thin layer of cytoplasm between the surface of the capillaries and the chorionic epithelium. The latter closely opposes the uterine epithelium, which is similarly invaded by capillaries that extend close to its surface. This configuration greatly reduces the effective barrier to diffusion between the foetal and maternal circulation (Fig. 81B). The second type of allantoic placenta (Fig. 81C) is characterised by the fact that the wall of the uterus and its epithelium are extremely thin. The uterine capillaries thus appear to stand out from the surface of the overlying epithelium, which forms secretory areas in the valleys between the capillaries. Within these valleys lie the large cells of the allantochorionic ectoderm which occasionally give rise to cytoplasmic processes that penetrate the uterine epithelium. In some cases the chorionic ectoderm appears to be a syncytium. Such placentas occur in various species of *Lygosoma*. Placentas of the third type, which also occur in species of *Lygosoma*, have a more or less elliptical shape. The uterine epithelium lies directly over the main longitudinal blood vessels of the uterus that stand out as folds that more or less fit into embayments in the chorionic epithelium. The uterine epithelium is composed of a single layer of ciliated cuboidal cells. The cells of the chorionic ectoderm are also ciliated but large and columnar in form (Fig. 81D). Thus the allantoic placenta of reptiles, like the omphaloplacenta, is basically not different from the omphaloplacenta of sharks except for the fact that the egg membranes are broken down at an early stage of embryonic development. They represent an epithelio-chorial type of placenta which does not involve the intimate association of maternal and foetal elements.

Since those species of reptiles that exhibit the placental form of viviparity also have eggs which contain a relatively large amount of reserve material, one may ask what functions do the omphaloplacenta and allantoic placenta fulfil. The former type reportedly serves primarily to provide the foetus with water, apparently by simple diffusion: proteins in the placental region of the yolk sac have been found to take up water. The water is evidently mainly required for the mobilisation of the reserves of material in the yolk. In addition to this the omphaloplacenta may be involved in the exchange of gases, though conclusive evidence of this is lacking. The principle physiological role of the allantoic placenta likewise appears not to be the provision of nutrients. Rather, it is possible that those of type I and II have exclusively a respiratory function, although it has been found that small amounts of ions and amino acids pass from the mother to the foetus and that excretory products pass in the reverse direction. All of the species that possess these types of placenta in fact have a large enough supply of yolk to sustain the full course of development. A more general but nevertheless important function of the placenta is that of anchoring the foetus within the uterus. The placenta of *Lacerta vivipara* is only poorly developed and it plays such a small role in providing the foetus with nutrients that the foetus can also develop in the absence of a foeto-maternal connection. On the other hand, species which develop a type III placenta have smaller reserves of yolk in their eggs and it is probable that in such cases the placenta does indeed provide the foetus with essential nutrients.

Gestation in reptiles is controlled by hormones. The initial evidence for this came from the fact that when gestation occurs in only one limb of the uterus, the other limb undergoes the same sort of changes as does the pregnant limb. Furthermore, the uterine structures associated with the formation of an allantoic placenta develop prior to placentation of the embryo. However, the involvement of the corpora lutea in this context does not appear to be the same in all viviparous reptiles. In many instances CLA are formed but whether they are of physiological significance remains unsubstantiated. The contrary appears to be the case with respect to CLO. Thus 3 different groups of viviparous species can be recognised. Firstly, those represented by *Lacerta vivipara* and *Anguis fragilis* in which the CLO degenerate immediately after the initial stages of gestation and secondly a group, which includes the Scincidae, where this only occurs during the second half of gestation. In species of the third group the CLO persist throughout gestation, and for up to 6 weeks after birth in the case of the snake *Thamnophis sirtalis*. The CLO of the second and third group can be compared to the corpora lutea graviditatis of mammals. In some of the species that have been investigated ovariectomy performed during the early stages of gestation leads to the death or resorption of the embryos but has no deleterious effects if performed toward the end of gestation. Hypophysectomy yields analogous results. Contradictory evidence exists with respect to the ability of injected progesterone to maintain gestation in animals hypophysectomised during the early stages of gestation. The application of a neurohypophyseal extract has no effect during the first two thirds of the gestation period but leads to premature birth when applied during the last third of the gestation period, indicating that neurohypophyseal hormones may be involved in the phenomenon of birth in reptiles. In *Lacerta vivipara*, where there is only a very loose connection between the foetus and the mother, extirpation of either the corpora lutea or the hypophysis disrupts gestation but has no influence on the development of the embryo. Extracts from the corpora lutea of gravid, viviparous snakes have the same effect as progesterone on the uterus of rats and rabbits treated with oestrodiol. Furthermore, the serum of such reptiles has been found to contain progesterone in a concentration which increases as gestation proceeds. From the above lines of evidence one may conclude that the mechanisms of hormonal control which regulate gestation in viviparous species of primitive amniotes, at least those operating during the last third of gestation, are the antecedents of those controlling gestation in mammals.

CHAPTER 8

Care of the Young Within the Body: Mammals

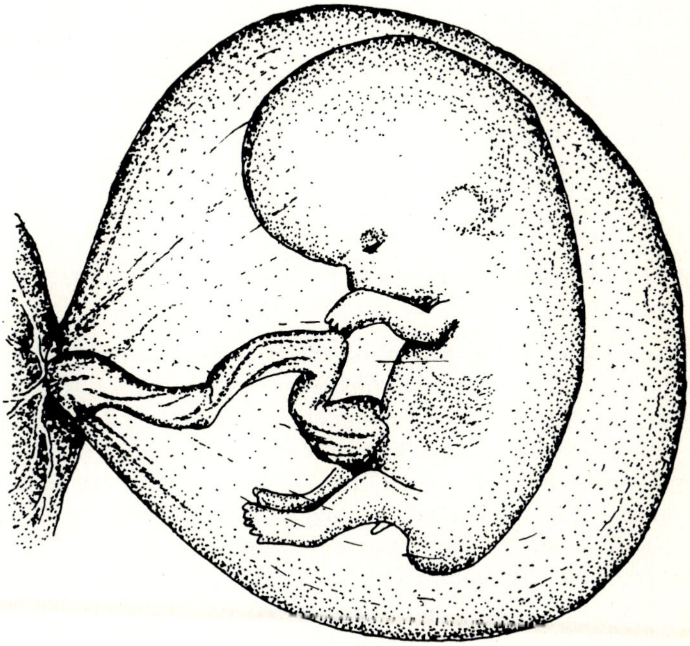

Human foetus at an age of about ten weeks (4 cm long)

The structural interrelationships between the mother and embryo that have developed in submammalians in association with the phenomenon of viviparity, as expounded in Chap. 7, foreshadow the evolution of mammals. These interrelationships facilitate, to a greater or lesser degree, the supply of nutrients to the embryo and the removal of waste products. It was previously stated however that the term placenta does not refer to a distinct morphological structure but rather to a physiological unit, it being irrelevant exactly which part of the mother's body and the embryo's body come into contact. Two types of placenta have proved themselves particularly "successful" in the course of vertebrate phylogenesis, namely the omphalogenic placenta (yolk sac placenta) and the allantogenic placenta (allantoic placenta). In each case the association is between extraembryonic elements developed by the embryo and the uterus of the mother. Both these types of placenta occur within the Theria. Thus most Marsupialia have a yolk sac placenta, as has already been mentioned, whereas with few exceptions (e.g. the shrew *Orycteropus*) Eutheria have an allantoic placenta. This chapter deals firstly with the ways in which the blastocyst makes contact with the uterus and the classification of placentas into different types. This is followed by a description of the most important forms of placentation in Marsupialia and Eutheria. The chapter concludes with an account of the physiology of the mammalian placenta.

8.1 Types of Implantation and the Classification of Placentas

The development of the embryo proceeds as it passes down the tuba uterina (see Chap. 4, p. 149) and by the time it finally reaches the uterus it is at the blastocyst or late morula stage. The uterus of mammals takes one of several forms, namely a uterus duplex, uterus bipartitus, uterus bicornis or uterus simplex (see p. 67 and Fig. 17). The uterus, and its horns if present, is made up of three layers: an outer covering of peritoneum called the perimetrium, a middle layer referred to as the myometrium of greater or lesser thickness and composed of smooth muscle, and an inner epithelial lining called the endometrium. The latter is underlain with glandular and connective tissue that abuts the myometrium. Together they form the mucosa of the uterus. The mucosa varies in thickness not only from species to species but that of a single species can undergo pronounced changes in morphology during the different phases of the oestrus cycle. These changes affect the vascularisation of the mucosa and the development of uterine glands, which are tubular invaginations of the uterine epithelium. Investing these elements of the mucosa is a connective tissue, or stroma. The cyclical changes in the form and function of the endometrium are particularly pronounced in the case of the human menstrual cycle which has already been described (see p. 192). The uterus is attached to the wall of the body cavity by a peritoneal duplicature called the mesometrium (Fig. 82A).

Having arrived in the uterus and shed the zona pellucida the blastocyst makes contact with the uterine epithelium in a process known as implantation. In the simplest case the embryo remains within the lumen of the uterus and is enclosed on all sides by the uterine epithelium. This is referred to as central implantation (Fig. 82A) and occurs in Ungulata, Carnivora, one family of rodents (Leporidae), the Lemuroidea,

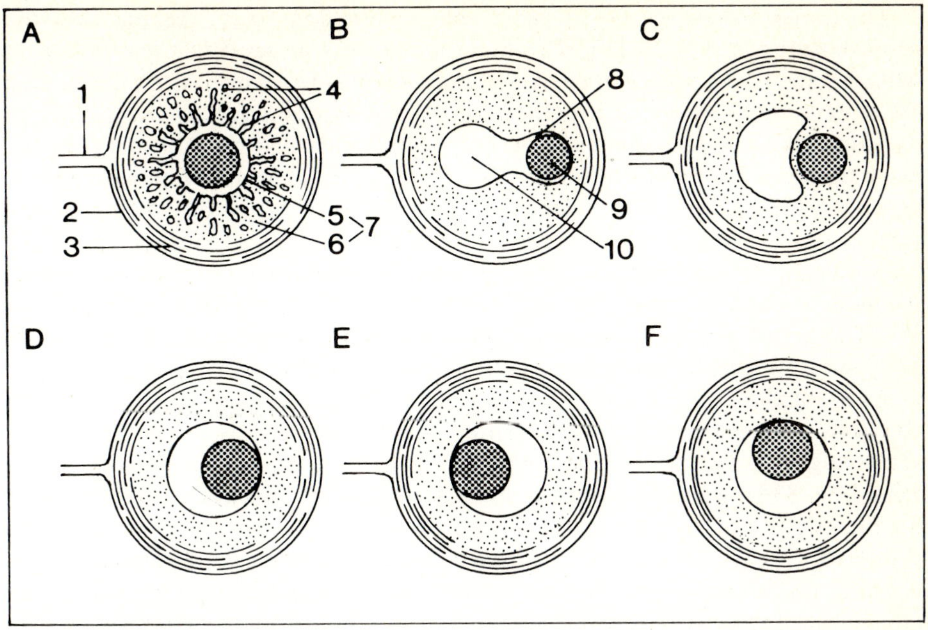

Fig. 82A–F. Diagrammatic representation of the structure of the mammalian uterus (**A**) and of the different types of implantation. **A** Central. **B** Eccentric. **C** Interstitial. **D** Antimesometrial. **E** Mesometrial. **F** Orthomesometrial.
1 = mesometrium; 2 = perimetrium; 3 = myometrium; 4 = uterine gland; 5 = endometrium; 6 = underlying tissue; 7 = mucosa; 8 = diverticulum of the uterus; 9 = blastocyst; 10 = lumen of the uterus

Tarsoidea, Platyrhina, Catarrhina, a few Insectivora and Endendata as well as some Vespertilinoidea. It can be considered the most primitive mode of placentation among mammals. Eccentric implantation is when the blastocyst becomes lodged in a diverticulum of the uterus (Fig. 82B) and this is the case in numerous rodents. Interstitial implantation involves the blastocyst sinking beneath the surface of the endometrium (Fig. 82C). This type of implantation occurs in guinea-pigs (*Cavia cobaya*), a few Chiroptera, the hedgehog (*Erinaceus europaeus*), chimpanzees (*Pan troglodytes*) and man.

The sites of implantation into the wall of the uterus bicornis, in relation to the insertion of the mesometrium, is different in different species. Implantation is referred to as antimesometrial when the blastocyst makes contact with the aspect of the uterine epithelium opposite the mesometrium, as occurs in most rodents and insectivores (Fig. 82D). Implantation into the mesometrial aspect of the uterus is referred to as mesometrial (Fig. 82E) and this occurs in members of the insectivore families Macroscelididae and Megachiroptera (fruit bats). In the case of orthomesometrial implantation the blastocyst attaches first to the dorsal and then to the ventral aspect of the uterus, i.e. lateral to the insertion of the mesometrium (e.g. Tenrecidae, Tupaja). Implantation is referred to as being either dorsal or ventral in the case of the uterus simplex of apes and man.

The names of the first three types of implantation mentioned – central, eccentric and interstitial – only specify the way in which the blastocyst first makes contact with the uterus and not the type of placenta subsequently formed which may not be the same in different species with the same mode of implantation. Four criteria serve to differentiate between the different types of placenta that may develop, the first being the shape of the placenta. Thus the placenta may extend over only part of the surface of the embryo or, in the most extreme case, cover very nearly the entire chorion, in which case it is referred to as a placenta diffusa (Fig. 83A). This type of placenta can be found in pigs, the hippopotamus and camels, among others. In contrast to this type are those that cover only a limited area of the chorion and which are referred to as restricted placentas. An intermediate between the diffuse and the restricted type of placenta is represented by the placenta multiplex which is also referred to as a cotyledonary placenta. In this case there are numerous placental areas distributed evenly over the surface of the chorion that fit into convex or concave uterine thickenings called caruncles. The cotyledons and caruncles together form the placentome (Fig. 83B). Ruminants possess this type of placenta. The next step in the trend for the placenta to become restricted in its coverage of the chorion is represented by the placenta zonaria which occurs in many Carnivora. In this case the placenta forms a ring or belt around the embryonic vesicle (Fig. 83C). If the ring is not continuous around the embryonic vesicle, as occurs in the case of the raccoon *Procyon lotor* for example, the placenta is referred to as being discontinuous (Fig. 83D). A further simplification is seen in the case of the placenta bidiscoidalis of apes where there are two discrete placental areas (Fig. 83E). The final stage of simplification is represented by the placenta discoidalis which is restricted to a single disc-shaped region (Fig. 83F). An example of this type of placenta is that of man.

The second criterion by which different types of placenta can be differentiated from one another involves the functional aspects of this organ. The essential characteristic of the mammalian placenta is that an exchange of material takes place between two circulatory systems which are in close contact but which are not confluent with each other. The maternal tissues of the placenta receive nutrients from and give waste products up to blood vessels of the maternal circulation whereas the embryonic tissues of the placenta are supplied by either the arteria and vena vitellina or the arteria and vena umbilicalis according to whether the placenta develops from the yolk sac or the allantois respectively. The exchange of material between the mother and the embryo (or foetus) occurs in part by diffusion. However, according to the laws of diffusion, the total amount of material that can be exchanged per unit time depends, among other things, on the area over which diffusion takes place. One way in which this limitation could be offset is, for example, by the formation of villi within the small, discrete placental areas, thereby increasing the effective surface area of the embryonic side of the placenta. This is, in fact, the case in many species of mammal but it does not alone account for the increased efficiency of the placenta in relation to its size. The amount of material that can diffuse across the placenta per unit time also depends on a second factor, namely the thickness of the barrier through which the material must diffuse. In the most primitive type of placenta this barrier is made up of 6 layers: the endothelium of the foetal blood vessels, the mesenchyme of the chorion, the chorionic epithelium, the uterine epithelium, the connective tissue of the

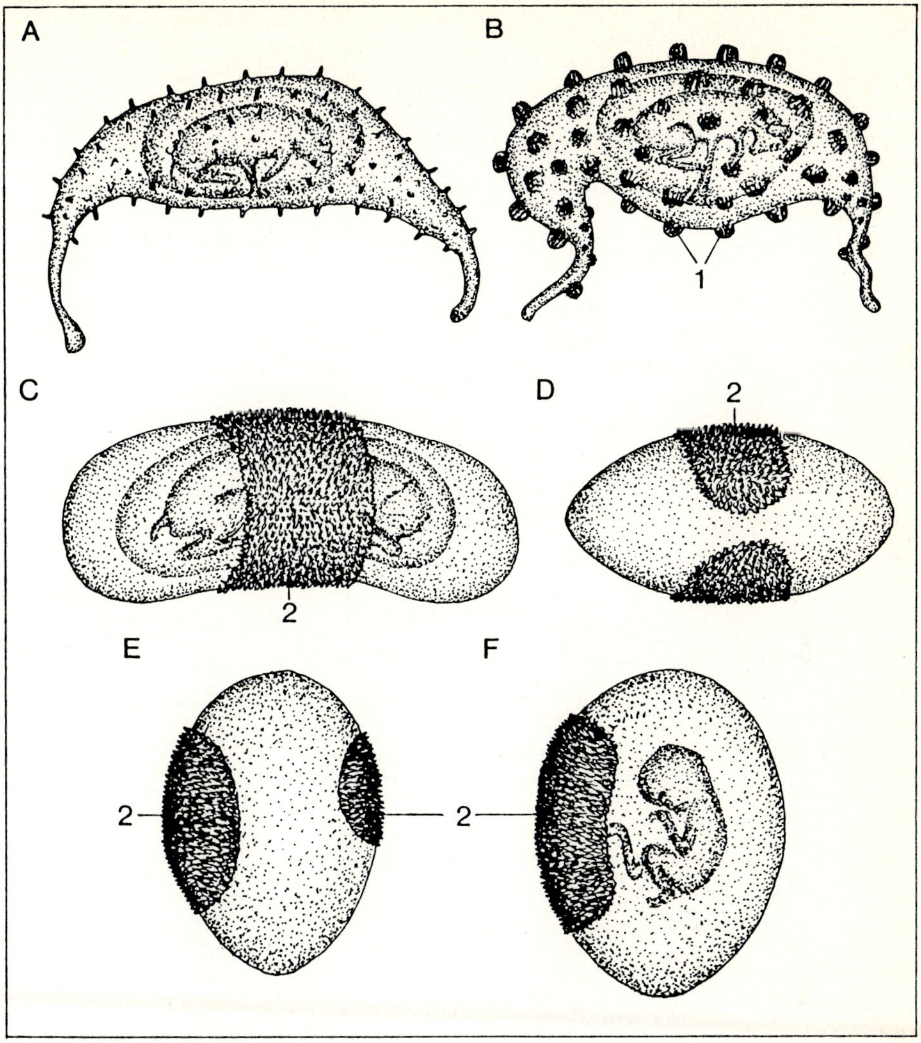

Fig. 83A–F. Gross anatomy of different types of placenta. (After Starck). **A** Placenta diffusa. **B** Placenta multiplex. **C** Placenta zonaria. **D** Discontinuous placenta zonaria. **E** Placenta bidiscoidalis. **F** Placenta discoidalis.
1 = cotyledons; 2 = placenta foetalis

uterine mucosa and lastly the endothelium of the maternal blood vessels. A placenta constructed in this way is called a placenta epithelio-chorialis (Fig. 84A), examples of which occur in the Cetacea, Lemuroidea, Pholidota and a range of Ungulata. In other mammals one can recognise a trend whereby these barriers to diffusion become progressively reduced, beginning with those on the maternal side. The first step in this trend is the breakdown of the uterine epithelium so that the chorion comes into direct contact with the connective tissue of the mucosa. This type of placenta is referred

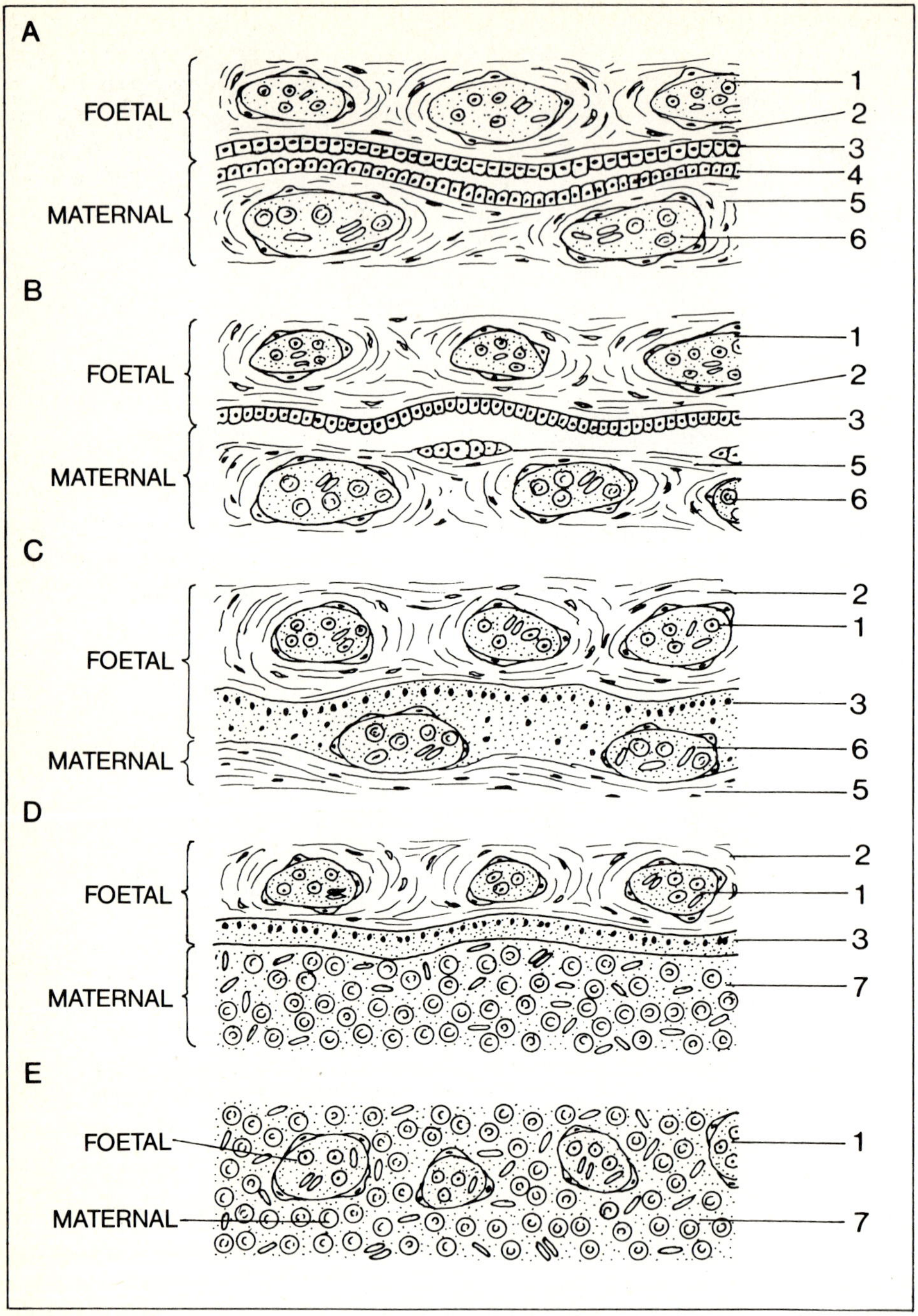

Fig. 84A–E. Histological types of placenta. **A** Placenta epithelio-chorialis. **B** Placenta syndesmo-chorialis. **C** Placenta endothelio-chorialis. **D** Placenta haemo-chorialis. **E** Placenta haemo-endothe-lialis.

1 = foetal blood vessel; 2 = chorionic mesenchyme; 3 = epithelium of the chorion; 4 = uterine epi-thelium; 5 = connective tissue of the uterus; 6 = maternal blood vessel; 7 = maternal blood

to as a placenta syndesmo-chorialis and it occurs in numerous ruminants and the sloth *Bradypus tridactylus* (Fig. 84B). The next step to a placenta endothelio-chorialis involves a reduction in the connective tissue of the mucosa so that the chorion directly opposes the maternal blood vessels (Fig. 84C). Examples of this type of placenta are found among the Carnivora and Chiroptera. The breakdown of the walls of the maternal blood vessels leads to the formation of a placenta haemo-chorialis (Fig. 84D) in which the chorion is partly bathed in maternal blood. Only three of the original six barriers are present in this type of placenta which occurs in elephants, sea cows (Sirenia), rock-rabbits (genus *Hyrax*), armadillos, hare-like rodents, Prosimiae, numerous Insectivora and man. The last step in this progression is represented by the placenta haemo-endothelialis where not only the chorionic epithelium but also the chorionic mesenchyme is lost so that the endothelium of the foetal blood vessels is bathed directly in the maternal blood (Fig. 84E). This rare type of placenta only occurs in a few species of rodent shortly before birth.

The reduction in the number of diffusion barriers undoubtedly increases the efficiency of the placenta but is only one of several factors which determine its efficiency. Thus it is not always the case that a haemo-chorial placenta is more efficient than, for example, an epithelio-chorial placenta. Large ungulates have an epithelio-chorial or a syndesmo-chorial placenta and after a period of gestation that is roughly the same as it is in humans which have a haemo-chorial placenta, give birth to young that are approximately 10–15 times heavier than a new-born child. Furthermore, the above classification of placentas according to histological criteria does not accomodate those which exhibit the characteristics of more than one histological type. Moreover, placentas in which the diffusion barriers break down towards the end of gestation must also go through the preceding steps one after the other starting with the breakdown of the uterine epithelium and, in certain cases, continuing with the breakdown of other layers.

The different histological types of placenta are not restricted in their occurrence to any particular phylogenetic group, rather, each occurs in a wide variety of mammals, as can be seen from the examples given. For example, the seemingly primitive epithelio-chorial type occurs in whales and large ungulates which are, in evolutionary terms, relatively modern forms whereas the "progressive" haemo-chorial type occurs in many insectivores which represent the most primitive form of Eutheria. Two variants of the haemo-chorial placenta exist, the labyrinth placenta and the villous placenta, or placenta olliformis. In the first the chorion, or trophoblast, forms a labyrinth of lacunae which are irrigated by the maternal blood (Fig. 85A). This type of placenta, which also exists in an endothelio-chorial version, occurs in Carnivora. In the case of the placenta olliformis a bowl-shaped cavity filled with maternal blood is formed within the trophoblast into which the chorionic villi project. Specialised attachment villi anchor the foetal tissues of the placenta to the maternal tissues (Fig. 85B).

The third criterion by which different placentas can be classified relates to what happens to the maternal elements at the time of birth. This depends essentially on the closeness of the contact established between the foetal and maternal tissues. When the placental contact is relatively loose and superficial the endometrium sustains little damage during birth, such placentas being referred to as adeciduate. In contrast, the

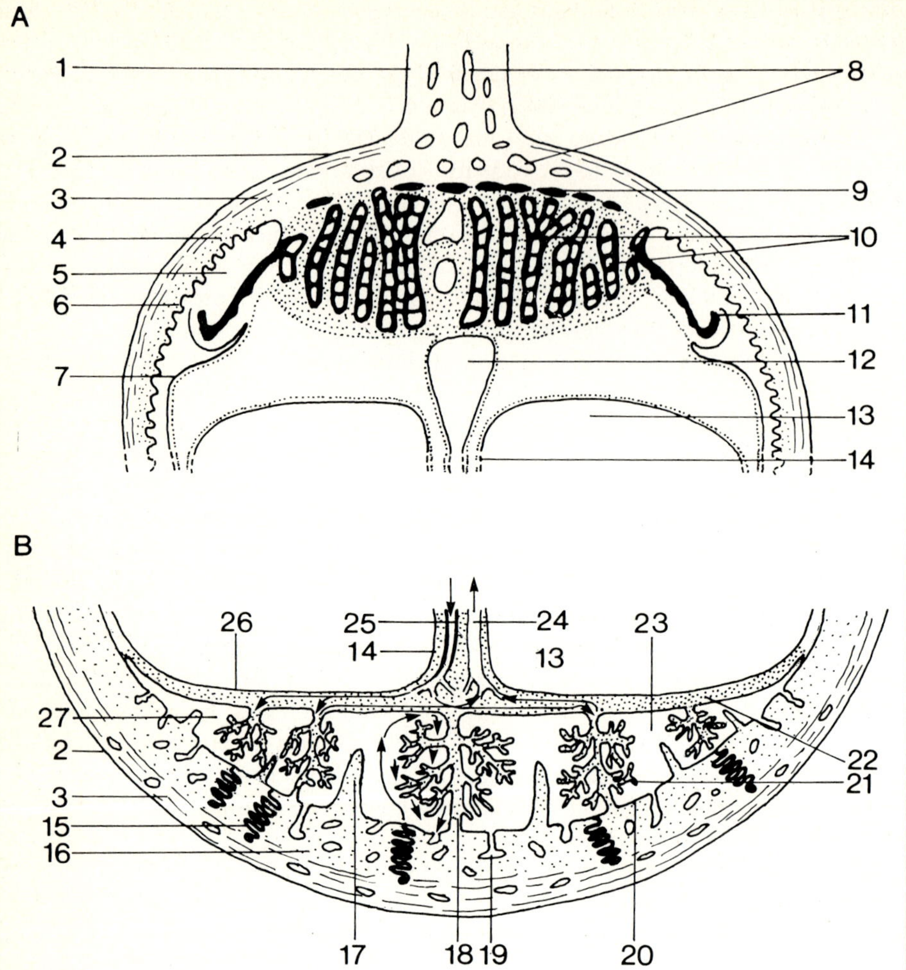

Fig. 85A,B. Diagram of the labyrinth placenta of the rabbit (**A**) and the placenta olliformis of man (**B**). (After Starck).
1 = mesometrium; 2 = perimetrium; 3 = myometrium; 4 = mucosa; 5 = lumen of the uterus; 6 = uterine epithelium; 7 = inner wall of the yolk sac (the outer wall degenerates together with the trophoblast); 8 = maternal blood vessel; 9 = portion of the trophoblast; 10 = labyrinth; 11 = margin of the remaining wall of the trophoblast; 12 = allantois; 13 = amniotic cavity; 14 = umbilicus; 15 = maternal spiral artery; 16 = decidua (basalis); 17 = placental septum; 18 = attachment villus; 19 = maternal vein; 20 = basal trophoblast; 21 = chorionic villus; 22 = chorionic plate; 23 = subchorial space; 24 = vena umbilicalis; 25 = arteria umbilicalis; 26 = epithelium of the amnion; 27 = marginal sinus. *Arrows* indicate the direction of blood flow

foetal part of the placenta penetrates into the mucosa to a greater or lesser extent in the case of an invasive placenta. In placentas of this type a specialised layer is formed within the mucosa called the membrana decidua which is expelled at birth thereby lacerating the uterus to some extent. Placentas of this type are referred to as deciduate and are "thrown off" after birth, whereas those that remain in the uterus after birth and are resorbed are referred to as contradeciduate (e.g. the placenta of the mole). Instances of the latter are rare.

The degree of complexity of the contact between the uterus and chorion varies considerably but the placenta is, at least functionally, always made up of a foetal placenta foetalis and a maternal placenta materna.

Finally, there is a fourth criterion whereby placentas can be divided into two groups, respectively those in which the maternal blood vessels remain intact during birth and those in which these vessels rupture at birth. These groups are referred to as semiplacentae and placenta verae, respectively.

8.2 Transport of the Embryo

After the unfertilised oocyte has reached the tuba uterina through the ostium abdominale it is transported to the site of fertilisation, most probably by the beating of the cilia borne on the epithelium lining the upper regions of the tuba uterina. The early development of the embryo up to the morula or blastocyst stage takes place within the tuba uterina. During this time the embryo is gradually transported towards the uterus by peristaltic movements of the smooth muscles of the tubal wall. In many cases the zona pellucida evidently serves to protect the blastomeres from being damaged by the contractions of the tuba and thereby preserves the structural integrity of the embryo. The time required to transport the embryo into the uterus varies from species to species, taking up to a week in some Carnivora, 2–4 days in many other mammals but only 1 day in some marsupials. This delay between the beginning of embryonic development in the tuba uterina and the arrival of the embryo in the lumen of the uterus is of considerable importance since during oestrus the milieu within the uterus, though optimal for the survival of spermatozoa, is highly toxic to the embryo. It takes at least 1 day for the conditions within the uterus to change so as to be suitable for the survival of the embryo.

When the embryo reaches the lumen of the uterus there is a time lag before it implants that may be of short or long duration depending on the species. In species that bear several young the embryos are then distributed within the uterus: for example, in species possessing a uterus duplex or a uterus bipartitus the embryos are usually lodged in the limb of the uterus on the same side as the ovary that produced the eggs from which they arose. The embryos are initially distributed randomly along the lenght of the uterus by contractions that pass slowly over the myometrial musculature. When the blastocysts reach a certain size each is capable of deforming the wall of the uterus, thereby eliciting waves of contraction that pass more rapidly over the musculature. The fields of contraction elicited by each blastocyst interact in such a way that blastocysts lying next to each other end up separated from one another by a more or less constant distance. As the blastocysts continue to grow their movement becomes progressively slower until a critical size is reached when they become virtually immovable. At this stage the myometrium between the blastocysts contracts more strongly, thereby constraining them in its folds. The above account of what has been called internal migration stems from investigations on the rabbit but one can assume that comparable mechanisms also operate in other species. In species possessing a uterus bicornis the embryos usually stem from only one of the ovaries and, accordingly, only enter the corresponding horn of the uterus. The contractions of the uterine wall nevertheless distribute the embryos over the lining of the entire uterus. Internal migration occasionally occurs in species with a duplex uterus. In general, the embryo does not appear to play an active role in this process: the muscular reaction of the uterus can, for example, be elicited by pieces of non-embryonic tissue.

8.3 The Process of Implantation

Following the migratory phase the embryo, or more precisely the trophoblast, comes into contact with the endometrium of the uterus. This necessitates the shedding of the zona pellucida, which occurs either in the lower regions of the tuba uterina or within the uterus. Successful implantation also requires that the mucous membrane of the uterus is in a suitable state to receive the embryo. This state is established by changes that occur prior to gestation which have been described in an earlier chapter with respect to the human uterus (see p. 192). Implantation takes place in specialised regions which are structurally distinct from the rest of the endometrium. The insectivorous elephant jerboa (*Elephantulus rozeti*) represents an extreme case in point since only a single, cone-shaped implantation area called the menstrual polyp is formed. The uterine mucosa of humans is divided into areas that develop independently of each other and which are each supplied with blood from a so-called spiral artery. In this case it is not uncommon for implantation to occur in the ventral part of the uterus, though it normally takes place in the dorsal part. By the time the mucosa is "ready" for the embryo to implant it consists of 4 main layers: a superficial epithelium underlain by a thin layer of connective tissue; beneath this is a layer made up of the necks of the uterine glands and extensive regions of surrounding stroma; the third layer, the stratum spongiosum, contains the portions of the uterine glands where the secretion is produced – this layer and the second layer are also referred to collectively as the stratum functionalis; the fourth and final layer of the mucosa, the stratum basalis, is made up of the terminal portions of the uterine glands and connective tissue.

Another change that may occur after the embryo has reached the uterus but prior to its implantation is that the trophoblast triggers the formation of deciduoma. This effect can however be induced simply by mechanical stimulation of the endometrium in the absence of a trophoblast.

The nature of the interactions which take place when the trophoblast contacts the uterine epithelium determine to a large extent which type of placenta will subsequently develop. As previously mentioned, two types of placenta can be recognised according to the arrangement of the maternal blood vessels: semiplacentae, which arise in those cases where the trophoblast does not sink into the endometrium, and placentae verae where the trophoblast becomes embeded to a greater or lesser extent within the endometrium. As a general rule blastocysts that result in the formation of the former type of placenta take a long time to establish contact with the endometrium whereas those which embed build up an intimate association with the maternal organism relatively quickly. In Ungulata implantation occurs relatively late, although the blastocyst grows relatively quickly before it attaches to the uterus. The overall secretory activity of the mucosa is heightened in such cases, probably as a result of it being stimulated by the blastocyst. There is no evidence to show that the uterine secretion contains growth factors which specifically affect the embryo, nor do these appear necessary for the growth of the embryo. The degree to which the blastocyst invades the uterine tissues is reflected in the histology of the placenta eventually formed. Thus an epithelio-chorial or syndesmo-chorial placenta is formed in those cases where implantation is relatively superficial. In contrast, a greater degree of invasion leads to the formation of an endothelio-chorial, haemo-chorial or haemo-endothelial placenta, the last two representing the most invasive types of implantation.

The horse and pig are examples of species in which implantation is relatively superficial and leads to the formation of placentae diffusae epithelio-chorialis. In the case of the horse, removal of the zona pellucida results in the blastocyst swelling considerably, due to the uptake of liquid, to reach a diameter of about 5 cm. This is thought to result from the active uptake of dissolved substances drawing water into the blastocyst by osmosis. At this stage the embryo has yet to make contact with the uterine epithelium and is nourished by an embryotroph, or more precisely a histiotroph, i.e. a mixture of uterine secretion and cell debris. At about the end of the third week of gestation specialised groups of cells appear in the trophoblast which promote the uptake of histiotroph and effect a temporary attachment of the trophblast to the uterine epithelium. It is not until the tenth week that chorionic villi begin to develop and extend into folds of the endometrium. The attachment is fully developed by about the fourteenth week. Implantation in the pig follows a similar course as in the horse but does not take quite as long. The first small areas of contact begin to develop on the thirteenth day of gestation. One week later the attachment is fully developed. The blastocysts of the pig are particularly large, reaching about 1 m in length, but are nevertheless nourished by an embryotroph up to the end of implantation.

Implantation in the cow and sheep involves a similarly close association between the blastocyst and uterus. These species, and other ruminants, have a multiplex placenta of either the epithelio-chorial or syndesmo-chorial type, or one which is of a type intermediate between these two. The blastocyst of both species is elongated and supplied with nutrients via a histiotroph. The development of the placentome, i.e. the insertion of the chorionic villi into the caruncles, takes 3 weeks in the sheep but 5 weeks in the cow.

Mice and men can be taken as examples of animals with invasive implantation. That of the mouse is of the eccentric antimesometrial type. Four days after fertilisation the blastocyst, surrounded by the zona pellucida, reaches the uterus which secretes a lytic enzyme from the endometrium that breaks down the zona pellucida. Immediately following this the mucosa responds to the trophoblast by becoming oedematous as a result of the capillaries in the region of contact becoming more permeable (Fig. 86A). Except for the lumen of the implantation diverticulum the uterus then collapses in on itself so that the embryo becomes completely enveloped by the

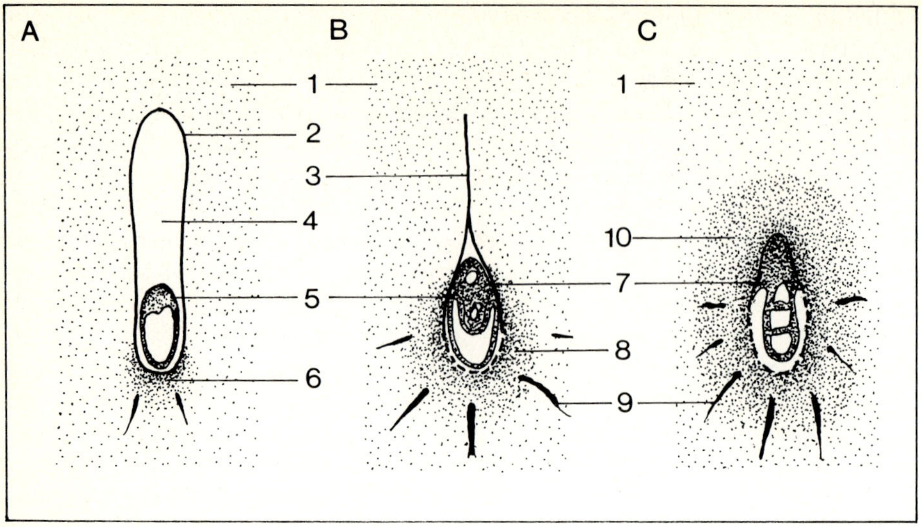

Fig. 86A–C. The process of implantation in Muridae.
1 = uterine stroma; 2 = uterine epithelium; 3 = primary lumen of the uterus (collapsed); 4 = primary lumen of the uterus; 5 = blastocyst; 6 = preimplantation reaction; 7 = carrier; 8 = initial formation of the decidua; 9 = uterine gland; 10 = membrane decidua

endometrium, as is the case in interstitial implantation except that the blastocyst does not lie in the stroma but is simply surrounded by the uterine epithelium (Fig. 86B). Subsequently, the breakdown of uterine tissue gives rise to a secondary lumen diametrically opposite the original lumen. The trophoblast and the uterine epithelium then give rise to microvilli which interdigitate so as to leave only a narrow gap between them of 25–100 nm. However, the microvilli disappear again soon thereafter and the gap between the trophoblast and epithelium becomes smaller. A significant change then takes place in the structure of the trophoblast in that its periphery becomes a syncytium which is supplemented with cells that grow into this layer from the underlying cellular layer. The trophoblast thus has two components, the syncytiotrophoblast and the cytotrophoblast, this being a feature of invasive embryos in general. The carrier region of the embryo then embeds into the uterine epithelium (see p. 152) at the site at which the implantation diverticulum formerly joined the primary lumen of the uterus. This is brought about by proteolytic enzymes released by the trophoblast causing the uterine epithelium to degenerate so that the trophoblast sinks further and further into the stroma of the endometrium. Degenerating cells become incorporated into the syncytiotrophoblast. The uterine tissue reacts to the invasion of the embryo by mesenchymatous fibroblasts metamorphosing into large polygonal decidua cells which rapidly form a thick layer around the blastocyst (Fig. 86C).

Implantation in man is of the interstitial type and although similar to that just described for the mouse differs from it in three respects: contact between the blastocyst and uterine epithelium is established somewhat later, the entire blastocyst becomes embedded within the endometrium, and the decidua reaction occurs more slowly. By the end of the first week of pregnancy the blastocyst, already devoid of its zona pel-

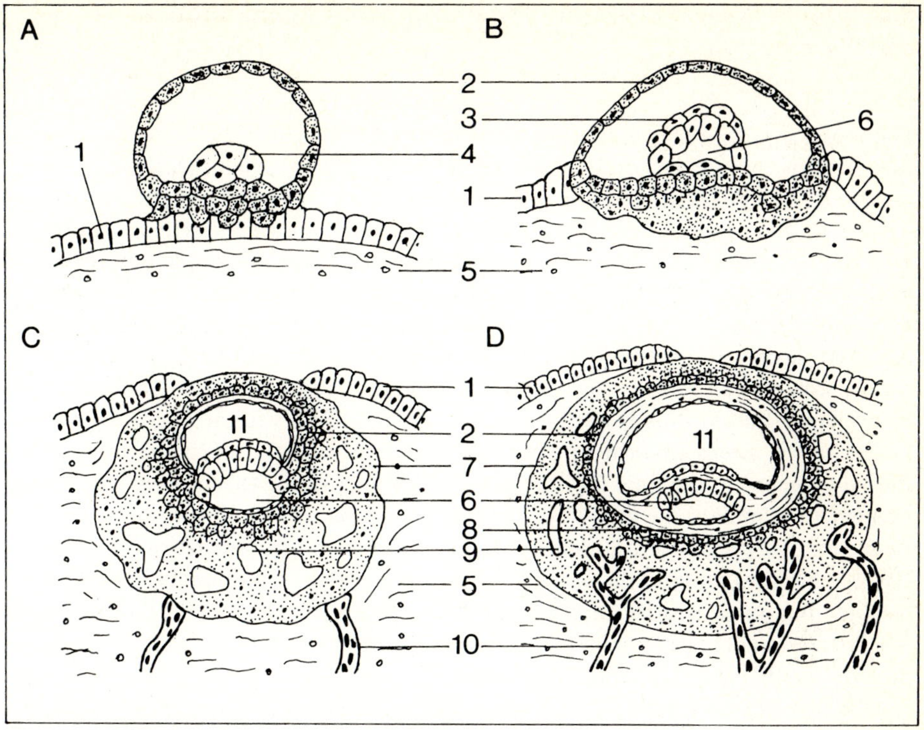

Fig. 87A–D. The process of implantation in man.
1 = uterine epithelium; 2 = cytotrophoblast; 3 = endoderm; 4 = embryonic knob (inner cell mass); 5 = uterine stroma; 6 = amniotic cavity; 7 = syncytiotrophoblast; 8 = extraembryonic mesenchyme; 9 = lacunae; 10 = maternal blood vessels; 11 = exocoel

lucida, has reached the lumen of the uterus. Upon contacting the uterus the blastocyst is orientated so that the embryonic knob lies over the uterine epithelium, the syncytiotrophoblast becoming particularly well developed in the region of contact (Fig. 87A,B). The human embryo is not sustained by a histiotroph but, rather, quickly penetrates into the uterine epithelium and embeds itself in the stroma of the mucosa, after which the uterine epithelium eventually grows back over the site of implantation. Growth of the syncytiotrophoblast results in it enveloping maternal blood vessels so as to form a labyrinth (Fig. 87C,D).

Thus invasive implantation has three main phases. The first, called apposition, includes the orientation of the embryonic knob towards the uterine epithelium and, in the mouse for example, the interdigitation of the trophoblastic and epithelial microvilli. This is followed by adhesion which involves the breakdown of the microvilli in the region of contact and the subsequent drawing together of the blastocyst and uterine epithelium by some as yet unknown mechanism. The third and last phase is the invasion or embedding of the blastocyst into the uterine mucosa.

8.4 Immunological Problems Associated with Implantation and Placentation

The initial impression is that, aside from those cases in which the uterus supplies the enzymes for the dissolution of the zona pellucida, it is the trophoblast which directs the events of implantation. However, from the above account it will be evident that some embryos penetrate into the maternal tissues whereas others, such as that of the pig, do not. Surprisingly however, if a pig embryo is taken and implanted elsewhere, in the urinary duct for example, it is in fact found to possess the ability to penetrate deep into the tissue. This implies that the uterus of the pig has some way of protecting itself against being penetrated by the trophoblast. In light of this, implantation can be seen to be a process of action and reaction between foetal and maternal elements. Nevertheless, this is a subject about which little is yet known. The blastocyst of the rabbit swells prior to implantation and concurrently a protein, uteroglobin, can be detected in the fluid within the lumen of the uterus. This protein was initially thought to stimulate the growth of the trophoblast but is now known to be a progesterone-binding protein and that its synthesis depends on the presence of progesterone. Uteroglobin also inhibits the activity of proteases and can presumably therefore oppose the histolytic activity of the trophoblast. Whether or not such a mechanism exists in other mammals is uncertain. Uteroglobin appears to be either not present or present in only very small amounts in man.

One important aspect of implantation and placentation that has not yet been mentioned is that the embryo is, with respect to the mother, a foreign organism by virtue of the complement of paternal genes it contains and to which it gives expression. When a tissue that is neither from the maternal organism nor the embryo but which may be an isolated piece of embryonic tissue finds its way into a part of the body of the mother other than the uterus, that part usually makes an immunological response that can lead to its being rejected. One may therefore ask how it is that the uterus tolerates an invading blastocyst but this question is still far from receiving a satisfactory answer. It is nevertheless manifestly true that immunologically speaking the uterus represents a safe site where the embryo is protected from the defensive reactions of the mother, which have been shown by transplant experiments to be activated, indicating that the embryonic antigens are in fact effective. It has been found that after implantation the trophoblast is encapsulated in a fibrinous layer composed of an acid mucopolysaccharide which supposedly protects the embryo. Thus when intact mouse embryos are injected into a conspecific mouse of a different strain there is no immune reaction but if the fibrinous layer has previously been removed by treatment with neuramidase, the host animal responds strongly by producing antibodies. However, this may be only part of a more complex mechanism. Gravid mice possess an antibody in the blood against the embryo but this presumably has no deleterious effect on the embryo because the thick layer of decidua cells surrounding it acts as a kind of immunological barrier. This view is supported by the finding that embryos with this protective layer of decidua cells are not rejected when transplanted into females of another strain whereas embryos lacking this layer are rejected. It has recently been found that within only a few hours of fertilisation the embryo produces an "early pregnancy factor" which protects the embryo by suppressing the initial response of the maternal immune system. This factor is not dialysable and is inactivated by a temperature of 72°C but not by a temperature of 56°C. To date it has been found in man, the cow and sheep. Its formation is induced only by the living embryo and not by spermatozoa or oocytes. Another phenomenon of note is that there is no inflammatory reaction associated with the invasive phase of implantation. This may be due to the fact that placentas contain large amounts of histaminase which degrades histamine, the tissue hormone responsible for the inflammatory reaction.

Based on the observations presented above one can hypothesise that the embryo of vertebrates is protected from the maternal immune system as follows. In the first phase following fertilisation the maternal immune reaction is "buffered" by the early pregnancy factor. During the ensuing phase the fibrinous layer protects the blastocyst until it is encapsulated by the decidua. During the latter process the histaminase in the placenta prevents the inflammatory response.

Frequent reference continues to be made in the literature to the close similarity between the "behaviour" of the embryo during implantation and placentation and that of an invasive malignant tumour. In fact, a number of points of correspondence exist. Firstly, cell division in the trophoblast occurs automatically, requiring no stimulus for its initiation. Secondly, the embryo, as do many tumours, invades the host tissue locally by producing a lytic enzyme which infiltrates the tissue. Furthermore, during invasion the cells of the trophoblast adhere more readily to the uterine tissue than to each other. Thirdly, the uterine tissue tolerates the immunologically foreign tissues of the trophoblast, the presence of which, moreover, fails to cause an inflammatory reaction. Fourthly, the maternal circulatory system adapts to the presence of the embryo, by incorporating the embryo into the circulation and providing for its needs, in much the same way as it responds to a tumour. Finally, there are a number of specific points of biochemical similarity. On the other hand, the points of difference, although being fewer in number, may be of greater significance. Thus the growth of the embryo is limited in its extent and duration, the embryo is eventually expelled from the body after a certain period of time, and it does not produce any true metastases (although giant trophoblast cells do find their way into the maternal circulation). This notwithstanding, the similarities between the invasive form of implantation and placentation and invasive tumours are indeed remarkable and it is therefore not surprising that in recent times groups involved in tumour research have begun using the placenta as an "experimental animal".

8.5 Placentation in Marsupialia

Marsupials have a relatively short gestation and give birth to young which can be regarded as still in the foetal stages of development. Correspondingly, the mechanisms of placentation are usually relatively simple and the organ of exchange between the mother and foetus is small in size. The placenta is therefore not expelled after birth but resorbed, placentation in such cases being referred to as contradeciduate. It does not follow however that the placenta is of a simple histological type. Haemo-chorial placentas occur in some species, though they are admitedly rare. As has already been stated in Chap. 4, the majority of marsupials have a yolk sac placenta whereas a small minority have an allantoic placenta. The latter can be found in bandicoots of the genus *Perameles*, the koala *Phascolarctos cinereus* and wombats (genus *Vombatus*).

The simplest type of omphalogenic placenta occurs in opossums of the genus *Didelphis*. The extraembryonic mesenchyme underlies only about half of the ectoderm of the trophoblast thereby forming a cup-like structure. The blood vessels of the yolk sac develop within this mesenchyme and are limited at the periphery of the cup by a ring vessel called the sinus terminalis. The yolk sac is voluminous and fills the greater part of the embryo whereas the exocoel is small in volume. The amnion lies in an embayment beneath the yolk sac and over most of the head region is not underlain with mesenchyme, this part being referred to as the proamnion. The part of the yolk sac covered with mesenchyme, which is therefore vascularised, functions as a simple placenta. During cleavage the needs of the developing embryo are met from the small reserves of yolk (see p. 149) whereas later, when the blastocyst has reached the uterus, it is bathed in an embryotroph secreted by the uterine glands. The uterine epithelium is not disrupted by the implantation of the blastocyst, it merely becomes oedematous. At this stage the embryotroph is made up of the secretion of the uterine glands and an exudate of lymph largely free of cells. Instead of villi the trophoblast develops specialised nurse cells that resorb nutrients. Although an allantois is formed, it does not extend to the chorion. Gestation lasts about 12 days in opossums. The basic structure of the embryo is as shown in Fig. 88A.

The yolk sac placenta of the dasyure *Dasyurus viverrinus* has a more complex structure. Less than half of the ectoderm of the trophoblast is underlain by extraembryonic mesenchyme, which is limited by a sinus terminalis. The latter layer is divided into two structurally distinct regions: a disc-shaped region at one pole of the embryo that directly borders the exocoel and, encircling this, a ring-shaped region underlain by the yolk sac which extends as far as the sinus terminalis. The yolk sac, which lines most of the embryonic vesicle, is thus similarly divided into two regions: the ring-shaped region covered by mesenchyme (and which is therefore vascularised) that extends to the sinus terminalis and the remaining region of the yolk sac which is not covered by mesenchyme and is therefore avascular. This latter region is in turn differentiated into a ring-shaped placental region limited on one side by the sinus terminalis and an unspecialised region beyond this (Fig. 88B). The vascularised region of the yolk sac is also thought to function as a placenta involved in the exchange of gases, although its contribution is of limited physiological significance. The avascular placental belt comes into contact with the uterine epithelium, and trophoblast cells in this region become enlarged and their cell membranes disrupted. The syncytium which arises in this way from phagocytotic nurse cells invades the uterine epithelium and incorporates into itself maternal blood vessels. Such a composite structure, composed of uterine elements and a trophoblastic syncytium, is referred to as a diploplasma. Subsequently, the yolk sac endoderm also becomes a syncytium in which sinuses arise that, towards the end of gestation, are irrigated via the maternal blood vessels with blood from which nutrients are resorbed by the endodermal syncytium of the yolk sac. The histological classification of this type of placenta represents a step "beyond" even a haemo-chorial type since there are no foetal blood vessels and the maternal

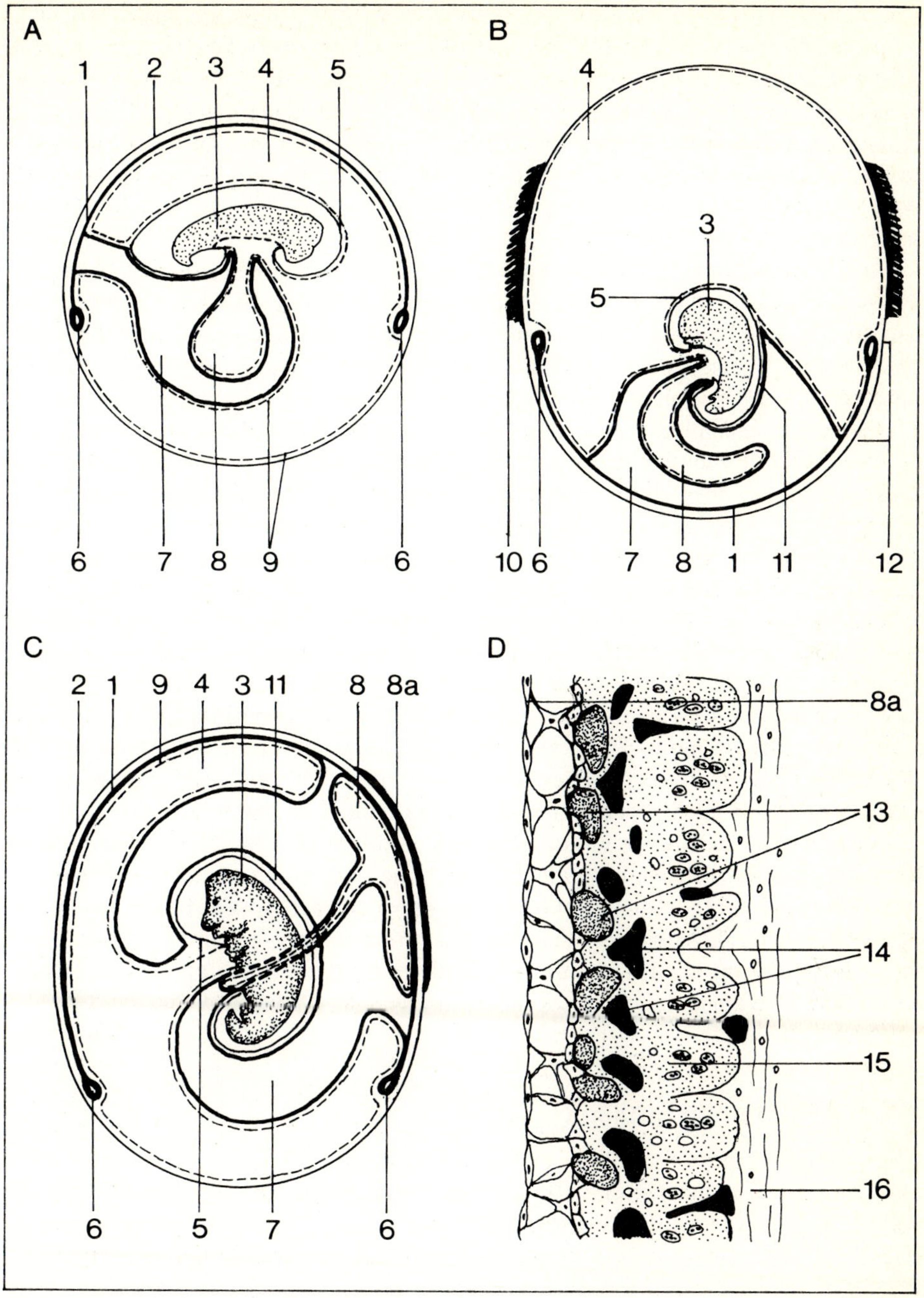

Fig. 88A–D. Diagrams of placentation in Marsupialia. **A** *Didelphis.* (After Minot). **B** *Dasyurus.* (After Hill). **C** *Perameles.* (After Hill). **D** Histology of the placenta of *Perameles.* (After Flynn). 1 = extraembryonic mesenchyme; 2 = extraembryonic ectoderm; 3 = embryo; 4 = lumen of the yolk sac; 5 = proamnion; 6 = sinus terminalis; 7 = exocoel; 8 = allantois; 8a = allantoic endoderm; 9 = endoderm of the yolk sac; 10 = avascular placental region; 11 = amnion; 12 = vascularised placental region; 13 = foetal blood vessels; 14 = maternal blood vessels; 15 = diploplasma; 16 = uterine stroma

blood comes into direct contact with the endoderm. The vascularised part of the placenta has a considerably simpler structure in which the trophoblast and the uterine epithelium merely lie in close proximity to one another and the network of maternal blood vessels is somewhat more extensive than in the avascular region: i.e. it is a placenta epithelio-chorialis. During the course of development an allantois is also formed which is not only vascularised but also comes into contact with the chorion for a short period of time after which however it withdraws again and the blood vessels degenerate.

Finally, in bandicoots of the genus *Perameles* a yolk sac placenta occurs side by side with an allantoic placenta. Prior to implantation the maternal blood vessels present in the endometrium become more extensive and are supplemented by other blood vessels. Sinuses form within the stroma that are filled with lymph and at the same time the uterine glands become more active. The uterine epithelium loses its cellular structure becoming a syncytium which has a lobular form at the border to the mucosa. The cell nuclei migrate from the periphery into the centre of these lobules to form nests of nuclei. After becoming attached to the uterus the trophoblast differentiates into a peripheral syncytiotrophoblast and an inner cytotrophoblast. The former layer invades the maternal syncytium at several points which it incorporates bit by bit to form a diploplasma that engulfs the maternal blood vessels lying in the troughs between the lobules of the maternal syncytium. During this process the cytotrophoblast gradually becomes smaller so that the invading maternal blood vessels eventually come into close contact with the foetal capillaries. The placenta thus formed is a placenta endothelio-endothelialis (Fig. 88D), a histological type which does not exist among the Eutheria. The events described above only occur in the region over which the allantois makes contact with the chorion. The avascular portion of the yolk sac is bathed in embryotroph whereas the vascularised placental region comes into contact with the uterine epithelium and is more richly vascularised here than in the region of the allantoic placenta. The yolk sac placenta presumably functions as an accessory placenta, although it is not known whether it remains functional up until the time of birth. After birth the placental structures and the remaining portion of the allantois are resorbed. The structure of the embryo of *Perameles* and the arrangement of the two placental regions is shown in Fig. 88C.

8.6 Placentation in Eutheria

The main organ of placentation in Eutheria is the allanto-chorion but a temporary yolk sac placenta is formed in some cases, such as in the horse. Placentation in this group of animals takes many different forms but it lies beyond the scope of this book to consider every one of them. Therefore, only the most important types will be dealt with in the following account.

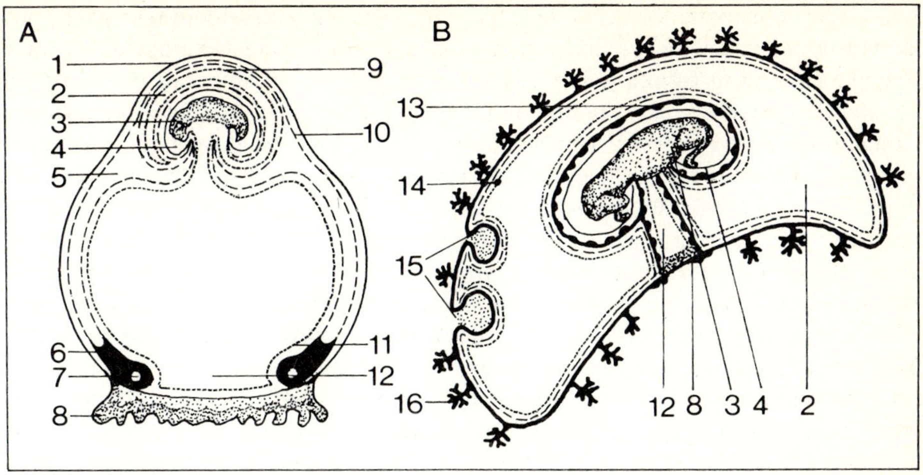

Fig. 89A,B. Diagram of the yolk sac placenta (**A**) and the allantogenic placenta diffusa (**B**) of the horse. (After Bonnet).
1 = trophoblast; 2 = allantois; 3 = embryo or foetus; 4 = amnion; 5 = exocoel; 6 = extraembryonic mesenchyme; 7 = sinus terminalis; 8 = placental region of the yolk sac; 9 = allantoic endoderm; 10 = chorionic mesoderm; 11 = endoderm of the yolk sac; 12 = lumen of the yolk sac; 13 = vascularised mesoderm of the amnion; 14 = allanto-chorion; 15 = hippomanes; 16 = chorionic villus. The trophoblast and chorionic mesoderm are drawn as a *single thick line* in **B**

8.6.1 The Semiplacenta of the Pig and Horse

The domestic pig (*Sus scrofa domesticus*) appears to have a very primitive mode of placentation. The embryonic vesicle has an elongated shape (see Fig. 83A) and during the course of development the allantois rapidly expands to fill the space between the chorion and amnion in all but the pointed ends of the vesicle. In doing this it displaces the originally large and well-vascularised yolk sac which establishes contact with the chorion for a short time but then atrophies and finally disappears altogether. However, it is not known whether this contact constitutes a yolk sac placenta. The blastocyst implants centrally, thereby completely filling the uterus. The surface of the chorion develops folds and villi that fit into depressions in the endometrium. These points of contact are distributed fairly evenly over the surface of the chorion, i.e. a placenta diffusa is formed. In a few places the villi do not contact the uterine epithelium but project into cavities (areolae) filled with uterine secretion which serves to nourish the embryo. In contrast, the embryo is supplied from the maternal blood (haemotrophy) at the sites where the villi contact the uterus. The uterine epithelium remains cellular. The placenta of the pig is of the epithelio-chorial type, although the foetal capillaries at the tips of the villi are only covered by a thin border of cytoplasm so that it represents a limiting case with respect to its histological classification. The placenta can also be classified as adeciduate since no membrana decidua is formed.

The placenta of the horse (*Equus caballus*) is also adeciduate. The horse also has a primitive mode of placentation whereby an omphalogenic and an allantogenic placenta are formed one after the other. By about the 28th day of development a large yolk sac has formed which, below the level of the sinus terminalis, comes into close contact with the chorion to form a yolk sac placenta (Fig. 89A). At this stage the allantois has already grown over the amnion and it continues to expand rapidly thereby displacing most of the yolk sac. However, the allantois does not atrophy but persists in the region overlying the rudiment of the yolk sac placenta, the so-called umbilical vesicle region (Fig. 89B). By about the seventh week of gestation the allanto-chorion is fully developed and gives rise to branched chorionic villi that are distributed more or less evenly over the surface of the elongated embryonic vesicle, being absent only over its pointed ends. The villi extend into correspondingly shaped crypts in the endometrium and thereby anchor the centrally implanted embryo. As is the case in the pig, the uterine epithelium and the trophoblast remain cellular and give rise to a placenta epithelio-chorialis. At birth the chorionic villi are released from the crypts without disruption of the uterine epithelium. The so-called hippomanes are a notable feature of the horse's embryo. These are cavities filled with the secretion of the uterine glands (a histiotroph) which represent invaginations in regions of the allanto-chorion free of villi. The hippomanes are subsequently pinched off into the lumen of the allantois to form free-floating bodies with a diameter of up to 10 cm. Comparable phenomena occur in other ungulates. The amnion of the horse is remarkable in two respects. Firstly, the overlying mesoderm is richly vascularised and secondly the amniotic ectoderm forms thickenings which are referred to as amniotic nipples.

8.6.2 The Semiplacenta of Ruminants

Ruminants have a placenta multiplex, this being where the chorionic villi form tufts or cotyledons from the surface of an otherwise relatively smooth chorion (see Fig. 83B). The cotyledons fit into caruncles that develop in the wall of the uterus prior to implantation and evidently induce the formation of the chorionic villi. Each cotyledon and its corresponding caruncle constitute a placentome. The number of placentomes formed varies not only between different species but also between individuals of the same species. For example, the sheep (*Ovis aries*) has 60–100 placentomes (Fig. 90A), deer (Cervidae) only 4–12, whereas the giraff (*Giraffa camelopardalis*) has approximately 180. The internal structure of a placentome of the cow (*Bos taurus*) is illustrated in Fig. 90B. In this case the placentome is convex and resembles a press-stud with the maternal caruncle locked into the bowl of the cotyledon. The latter gives rise to foetal villi whose branches interdigitate with those of the maternal villi arising from the caruncle. In the cow the placentomes are epithelio-chorial whereas in deer this is the case only at the periphery of the placentome since in the central region the chorionic villi penetrate some way into the stroma of the uterus forming a syndesmo-chorial placentome.

There are various forms of placentome. In the goat they are constructed the other way round to those of the cow with a bowl-shaped caruncle into which fits a convex cotyledon. The maternal and foetal villi nevertheless interdigitate in essentially the same way.

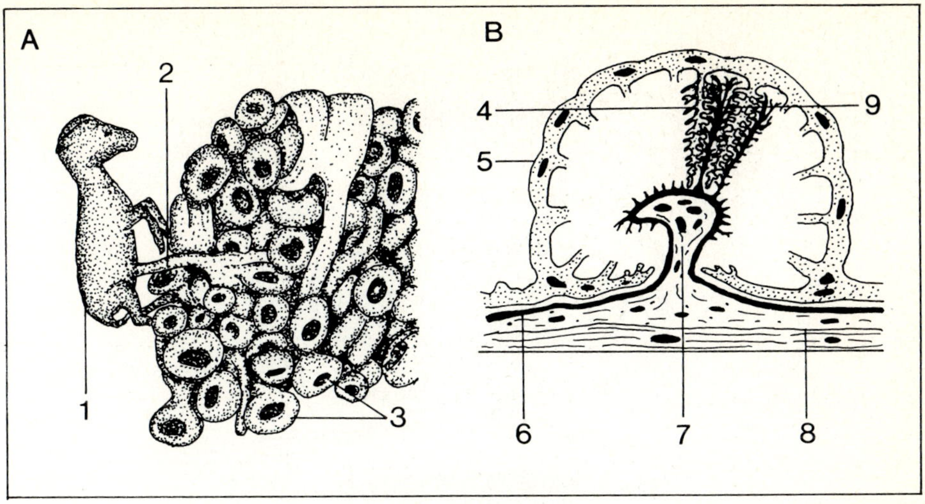

Fig. 90A,B. Placentomes of the sheep revealed by removing the amnion (**A**) and a convex placentome of the cow shown diagrammatically (**B**). (After Starck).
1 = foetus; 2 = umbilicus; 3 = placentome; 4 = foetal villus; 5 = cotyledon; 6 = uterine epithelium; 7 = caruncles; 8 = wall of the uterus; 9 = maternal villus

8.6.3 The Placenta Vera of Carnivores

Implantation in carnivora is invasive and central. The trophoblast gives rise to cellular processes that penetrate into the uterine epithelium, some occluding the openings of uterine glands whereas others grow into the mucosa. The response of the mucosa to the invading trophoblast involves the extension of the network of maternal blood vessels, the differentiation of decidua cells and the breakdown of the cell membranes of the uterine epithelium. More often than not the latter is interpreted in the literature as a process of degeneration and accordingly the structure it gives rise to cannot therefore be regarded as a syncytium. The protoplasmic assemblage including cell nuclei and their remains is therefore called a symplasma. From a wider point of view, however, not only can the superficial cells of the uterine epithelium become symplasmic but also the epithelial cells of the uterine glands. Moreover, the connective tissue cells of the stroma have the same ability. Examples of symplasmas originating from each of these tissues can be found in the placentas of carnivores. As the trophoblast invades the uterus it disrupts the superficial epithelium, the glandular epithelium and some of the decidua cells that have already developed. The maternal capillaries, however, remain intact and become enveloped by the trophoblast. Thus, in a relatively short time an endothelio-chorial labyrinth is formed in which the maternal blood vessels are surrounded by a thin layer of cytotrophoblast. The placenta of dogs and cats has been studied extensively and since the mode of placentation is similar in both these species it will be described here as a representative example for carnivores in general. By the time the blastocyst is ready to attach to the uterus it contains a relatively large yolk sac. The embryonic part of the blastocyst initially remains exposed

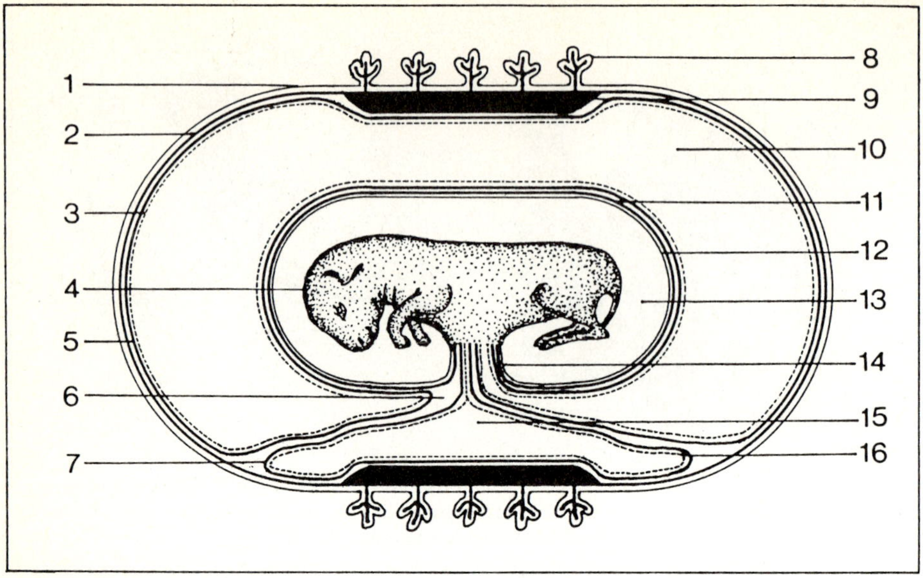

Fig. 91. Diagram of the relationship between the embryonic membranes and the placenta of the cat prior to regression of the yolk sac. (After Amoroso).
1 = chorion; 2 = chorionic mesoderm; 3 = allantoic endoderm; 4 = foetus; 5 = allantoic mesoderm; 6 = exocoel; 7 = mesoderm of the yolk sac; 8 = chorionic villus of the belt placenta; 9 = placental region; 10 = allantois; 11 = amniotic mesoderm; 12 = amnion; 13 = amniotic cavity; 14 = umbilicus; 15 = yolk sac; 16 = endoderm of the yolk sac

to the milieu within the uterus, until it is enclosed by the amniotic folds, so that it is the opposite pole of the blastocyst which first makes contact with the uterine epithelium (i.e. that underlain by the yolk sac). The yolk sac then comes into contact with the overlying trophoblast to give rise to an avascular omphalopleura composed of two layers. Extraembryonic mesoderm subsequently grows in between these two layers and gives rise to blood vessels, although in this case, as in other carnivores, a sinus terminalis is not formed. This vascularised omphalopleura represents a yolk sac placenta which is largely responsible for meeting the needs of the embryo during implantation. The opposing aspect of the yolk sac which underlies the embryo is displaced by the growing embryo, becoming inverted upon itself to some extent: this process is known as incomplete inversion. As development proceeds, namely between the 21st and 24th day of gestation in the cat, the exocoel between the chorion and the yolk sac becomes more extensive. Although this results in the splitting of the yolk sac from the placental region, this layer persists in cats and dogs until birth and evidently plays an important role as a tissue in which blood cells are formed and as an organ in which nutrient fluids destined for the embryo are enriched.

At an earlier stage of development the amnion becomes covered by the allantois but before the allantogenic placenta becomes functional the needs of the embryo are provided for by an accessory chorionic placenta that takes over from the yolk sac pla-

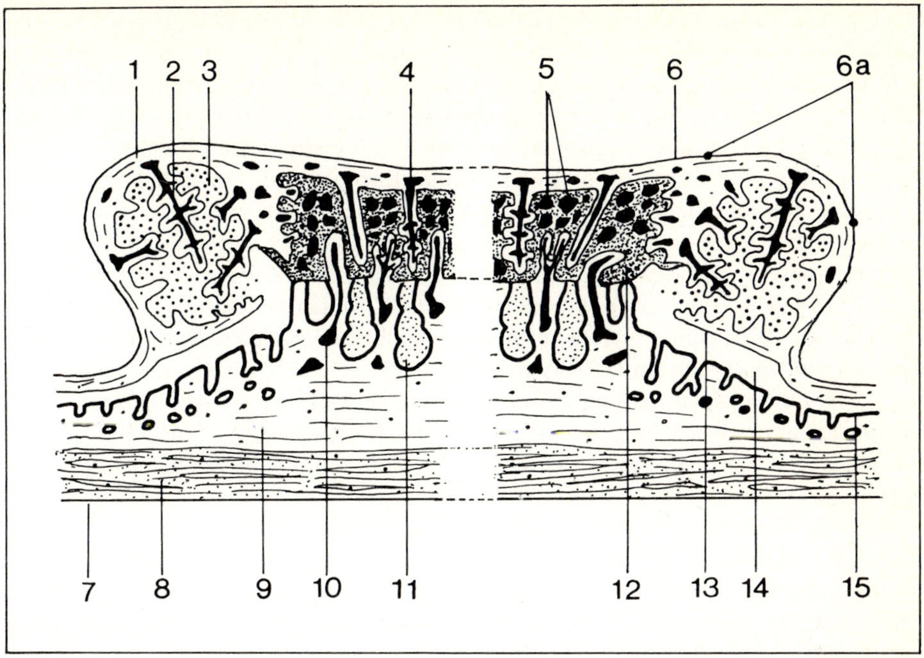

Fig. 92. Diagram of the belt placenta of the dog. (After Grosser and Mossmann).
1 = allanto-chorion; 2 = foetal villus; 3 = maternal blood; 4 = foetal blood vessel; 5 = maternal blood vessel; 6 = endoderm of the allantois; 6a = marginal haematome; 7 = perimetrium; 8 = myometrium; 9 = uterine mucosa; 10 = afferent maternal blood vessel; 11 = uterine gland; 12 = syncytiotrophoblast; 13 = cytotrophoblast; 14 = lumen of the uterus; 15 = uterine epithelium

centa. The chorionic placenta is situated opposite the yolk sac where the exocoel borders the avascular region of the chorion. Up until the time the allantoic placenta begins to form the blastocyst has a spherical shape but thereafter it becomes more elongated. At about the end of the third week of gestation the vascularised allantoic mesoderm makes contact with the region of the chorion lying opposite the yolk sac at the antimesometrial pole of the blastocyst. The trophoblast overlying this region then proliferates more rapidly, evidently due to the better blood supply it now receives, and gives rise to unbranched primary villi that penetrate deep into the mucosa of the endometrium thereby incorporating increasingly more of the symplasma but leaving the maternal blood vessels intact. Vascularised mesenchyme then grows into the villi, which subsequently branch forming secondary and tertiary villi. Most of the trophoblast covering the villi becomes a syncytiotrophoblast with only the tips of the villi remaining cellular. All that finally remains of the maternal tissues are blood vessels that extend through the sponge-like structure of the trophoblast to form an endothelio-chorial labyrinth. The allantoic placenta is fully developed by the fourth week of gestation. It is a placenta zonaria that forms a belt around the middle of the elongated embryonic vesicle (see Fig. 83C). The umbilical cord arises from the mesometrial aspect of the embryo, i.e. on the same side as the yolk sac. In the case of the dog and cat the marginal zones on each side of the labyrinth region of the placenta, where it adjoins the aplacental regions of the chorion, contain so-called marginal haematomes (Fig. 91). In these regions the chorion forms villi which are bathed directly in "blood" that has extravasated out of the maternal blood vessels. The effect of this is to lift the chorion off the uterine epithelium so as to form pockets into which the villi project (Fig. 91). However, this configuration is not comparable to a haemo-chorial placenta where the embryo is nourished by haemotrophy from blood flowing through capillaries. Rather, the blood in the marginal haematomes represents a histiotroph. The marginal haematomes of the dog are coloured green by a product of haemoglobin breakdown (uteroverdin) and are therefore also called the green margins. Those of the cat however have a brown colour. The villi, which are bathed in the extravasated blood, has a cellular surface epithelium which evidently takes up the histiotroph, mainly by phagocytosis. In the labyrinth region of the placenta there are smaller, central haematomes which may be linked to the marginal haematomes by bridges: those of the dog are referred to as the green pockets. The existence of other marginal or central haematomes, or both, is a characteristic feature of the placentas of carnivores. For example, the placenta of the raccoon (*Procyon lotor*) has no marginal haematomes whereas that of the ferret (*Mustela putorius*) is divided by a large central haematome situated antimesometrially. Outside the region of the placenta of carnivores are so-called rosettes, areas of cells corresponding to the areolae. The embryonic membranes and the relationship between the allantois and yolk sac and the placenta of the cat are shown in Fig. 92.

8.6.4 The Placenta Vera of Lagomorphs and Rodents

Among the Lagomorpha, placentation has been investigated most thoroughly in the rabbit which will be taken here as an example of an animal with a deciduate, haemochorial labyrinth placenta. At a stage prior to the arrival of a blastocyst in the uterus the surface of the endometrium is thrown into symmetrical pairs of longitudinal folds, respectively the large placental folds lying mesometrially, the inverted periplacental folds, which are lateral to the first, and the small obplacental folds that lie antimesometrially. The blastocyst first makes contact with the wall of the uterus in the region of the obplacental folds, following which these folds shrink so as to form a depression within which the blastocyst is lodged. Subsequently, the underlying uterine epithelium changes into a symplasma, the region of contact between this and the blastocyst being referred to as the obplacenta. This contact is maintained for only a short time and, after being broken, the blastocyst attaches to the large placental folds on the mesometrial aspect of the endometrium which are also referred to as the placental cushions. However, before this happens an omphalogenic obplacenta is formed by the yolk sac coming to lie directly against the region of the trophoblast, which has become somewhat thicker and which gives rise to processes that extend into the uterine symplasma. The disassociation of the obplacenta involves the detachment of the trophoblast from both the overlying symplasma and the underlying yolk sac. The reattachment of the blastocyst to the placental cushions occurs in a rather special way at a stage prior to the development of the amniotic folds. It begins with the formation of a so-called ectoplacental horseshoe which is a thickened region of the syncytiotrophoblast that forms a wall around the embryonic disc. This wall raises the trophoblast up off the inner surface of the uterus and provides space for the development of the amniotic folds (Fig. 93A–D). This space is known as the sulcus intercotyledonarius. The subsequent development of an allantogenic labyrinth placenta proceeds from the ectoplacental horseshoe. Thus the uterine epithelium underlying this region becomes symplastic and reduced to a thin layer between the trophoblast and the already developed membrana decidua, and finally disappears altogether. This brings the trophoblast into direct contact with the connective tissue of the mucosa. By about the tenth day of gestation the allantois has grown to meet the region of the chorion underlying the amnion, which by this stage completely encloses the embryo (Fig. 93E). The trophoblast gradually penetrates further into and incorporates more and more of the tissue of the uterus. When the advancing syncytiotrophoblast meets the maternal blood vessels it rapidly envelops them and, by about the tenth day of gestation, eventually breaks down the wall of the vessels to give rise to a lamellar labyrinth composed of a system of lacunae, formed by the syncytiotrophoblast, that is embedded in foetal mesenchyme (Fig. 93F). Since the labyrinth is irrigated with maternal blood, this is an example of a haemo-chorial placenta.

As mentioned above, the breakdown of the preplacenta entails the separation of the trophoblast from the underlying endoderm of the yolk sac, whereby the yolk sac becomes continuous with the lumen of the uterus (Fig. 93C). Subsequently, the amniotic cavity, which completely surrounds the embryo by this stage, sinks into the remaining yolk sac and begins to gradually rotate within it until the region of the amnion covering the dorsal aspect of the embryo is underlain by the wall of the yolk sac (Fig. 93D,E). The embryo thus rotates 180° from its original orientation and the arrangement of the yolk sac wall is exactly reversed in that the endoderm now lies outside the extraembryonic mesoderm (Fig. 93F). This process is known as yolk sac inversion or germ layer inversion. The stalk of the yolk sac remains intact during this process and in continuity with the lumen of the uterus so that the embryo is connected to the wall of the uterus only in the region of the allantoic placenta. The remains of the trophoblast form a narrow flange around the placenta whose free margin is limited by the sinus terminalis. The rest of the embryonic vesicle is covered by the inverted endoderm of the yolk sac which, beyond the sinus terminalis, still bears the remnants of the former duplicature (Fig. 93F).

Placentation in the mouse (*Mus musculus*) and rat (*Rattus rattus*) has been studied in considerable detail. The former species will serve here as a representative example of placentation in Rodentia. It has already been mentioned that, in contrast to the rabbit, implantation in the mouse takes place eccentrically in a diverticulum of the uterus which subsequently closes so that the blastocyst is surrounded on all sides by uterine tissue, including a membrana decidua, as is the case after interstitial implantation (see p. 261 and Fig. 82). A decidua basalis develops in the carrier region at the site of placentation whereas the decidua capsularis encapsulates the blastocyst within a so-called egg chamber. The carrier, which is also referred to as the ectoplacental cone, penetrates into the uterine mucosa, expands into a disc, and forms the haemochorial labyrinth. The way in which the amnion, exocoel and the allantois are formed within the carrier has already been described in Chap. 4 (see p. 152). However, the allantois of the mouse has no lumen and is represented only by a strand of vascularised

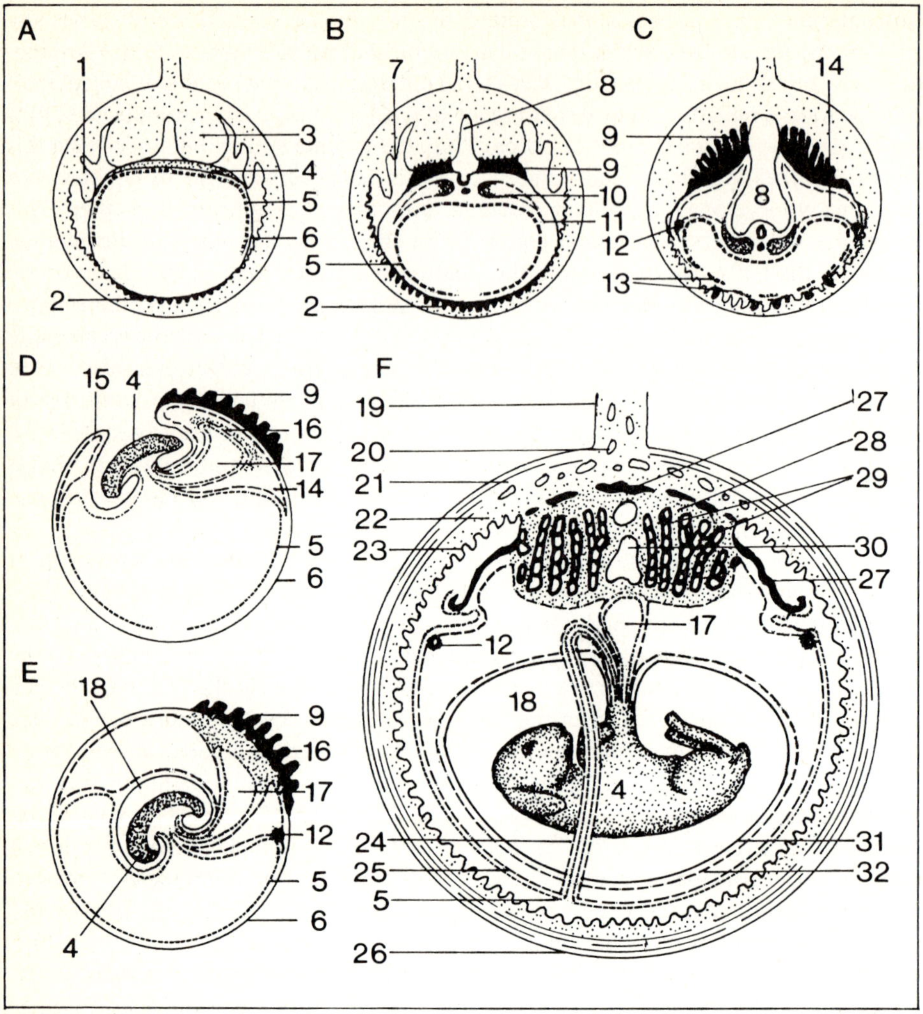

Fig. 93A–F. Series of diagrams showing placentation in the rabbit (**A**, **C**, **F** after Amoroso; **B**, **D**, **E** after Starck). **A** Attachment of the blastocyst in the region of the obplacenta. **B** Formation of the ectoplacental horseshoe. **C** Dissolution of the obplacenta (**A–C** are cross-sections). **D**,**E** Sagittal longitudinal sections showing the formation of the amnion and allantois placenta. **F** Fully formed labyrinth placenta (foetus drawn rotated 90° about the umbilicus).
1 = periplacental fold; 2 = region of the obplacenta; 3 = placental cushion; 4 = embryonic region of the blastocyst, becoming later the embryo and then the foetus; 5 = yolk sac endoderm; 6 = trophoblast; 7 = lumen of the uterus; 8 = sulcus intercotyledonarius; 9 = ectoplacental horseshoe; 10 = embryonic mesoderm; 11 = extraembryonic mesoderm; 12 = sinus terminalis; 13 = remains of the trophoblast and endoderm of the yolk sac in the region of the obplacenta; 14 = exocoel; 15 = amniotic fold; 16 = allantoic mesoderm; 17 = allantois; 18 = amniotic cavity; 19 = mesometrium; 20 = maternal blood vessel; 21 = myometrium; 22 = mucosa of the uterus; 23 = uterine epithelium; 24 = stalk of the yolk sac; 25 = mesoderm of the inverted yolk sac; 26 = perimetrium; 27 = remains of the trophoblast; 28 = mesoderm in the placental region; 29 = labyrinth; 30 = remnants of the sulcus intercotyledonarius; 31 = amnion; 32 = amniotic mesoderm

mesenchyme which carries the foetal blood vessels to the placenta. The yolk sac undergoes inversion as in the rabbit but in this case the process is considerably more pronounced. There is one further structure which remains to be mentioned, namely Reichert's membrane. This is formed on the seventh day of pregnancy between the distal ectoderm of the trophoblast and the yolk sac and subsequently expands beneath the trophoblast. It is a thin homogeneous, acellular layer which, following the dissolution of the trophoblast, comes into close proximity with the tissues of the uterus, to which it is apparently anchored by giant trophoblast cells, each of which contains a single nucleus. Reichert's membrane is apparently the basal membrane of the trophoblast. The membrane disappears again later when the outer wall of the yolk sac degenerates. In general, placentation in the mouse and rabbit are very similar, specific differences being the occurrence of giant trophoblast cells in the mouse, the development in this species of a discrete decidua capsularis, and the more pronounced inversion of the germ layers. In addition, the region of the yolk sac endoderm peripheral to the placenta of the mouse gives rise to outgrowths resembling villi that project into the uterine fluid which contains the uterine secretion. These outgrowths, which are provided with foetal blood vessels, extend into the periphery of the placenta where they follow the course of the maternal blood vessels towards the labyrinth: such structures are referred to as an endodermal sinus. The structure of the placenta of the mouse is shown in Fig. 94A.

In the guinea-pig (*Cavia cobaya*) the inversion of the yolk sac is even more pronounced than it is in the mouse. Placentation in the guinea-pig begins on about the seventh day of pregnancy after the blastocyst has crept out of the zona pellucida, which it does by amoeboid movement. The blastocyst then implants into the antimesometrial aspect of the uterus wall, implantation being of the interstitial type. After having embedded into the uterus it becomes encapsulated by a decidua capsularis. In contrast to the mouse and rabbit, the portion of the trophoblast lying outside the placental region degenerates rapidly whereas the outer wall of the yolk sac does not even develop. Rather, the yolk sac endoderm builds a layer of ectoplacental endoderm around the carrier which remains uninterrupted until penetrated by root-like processes arising from the trophoblast that extend into the tissue of the uterus. A spongy haemo-chorial ectoplacenta is formed initially which is later provided with foetal blood vessels from a central cone of mesenchyme. The allantois of *Cavia*, like that of the mouse, has no lumen. As the haemo-chorial labyrinth develops the base of the hollow cone of mesenchyme extends over a greater area of the uterus wall and gives rise to mesenchymatous lamellae that project into the developing placenta. The structure so formed is referred to as a subplacenta and is responsible for the resorption of histiotroph. On the other hand, within the true placenta the foetal blood vessels and mesenchyme displace the labyrinth formed by the syncytiotrophoblast in such a way that a lobular structure is produced. The placenta is covered by a layer of yolk sac endoderm which is separated from the syncytiotrophoblast by a layer of giant trophoblast cells. Towards the end of gestation the decidua capsularis breaks up and the extraplacental region of the yolk sac endoderm comes into contact with the wall of the uterus. Additionally, the subplacenta gradually dedifferentiates and the placenta rises up from the wall of the uterus on a stalk. The latter development is interpreted as being of selective advantage in that it reduces the area of the uterus la-

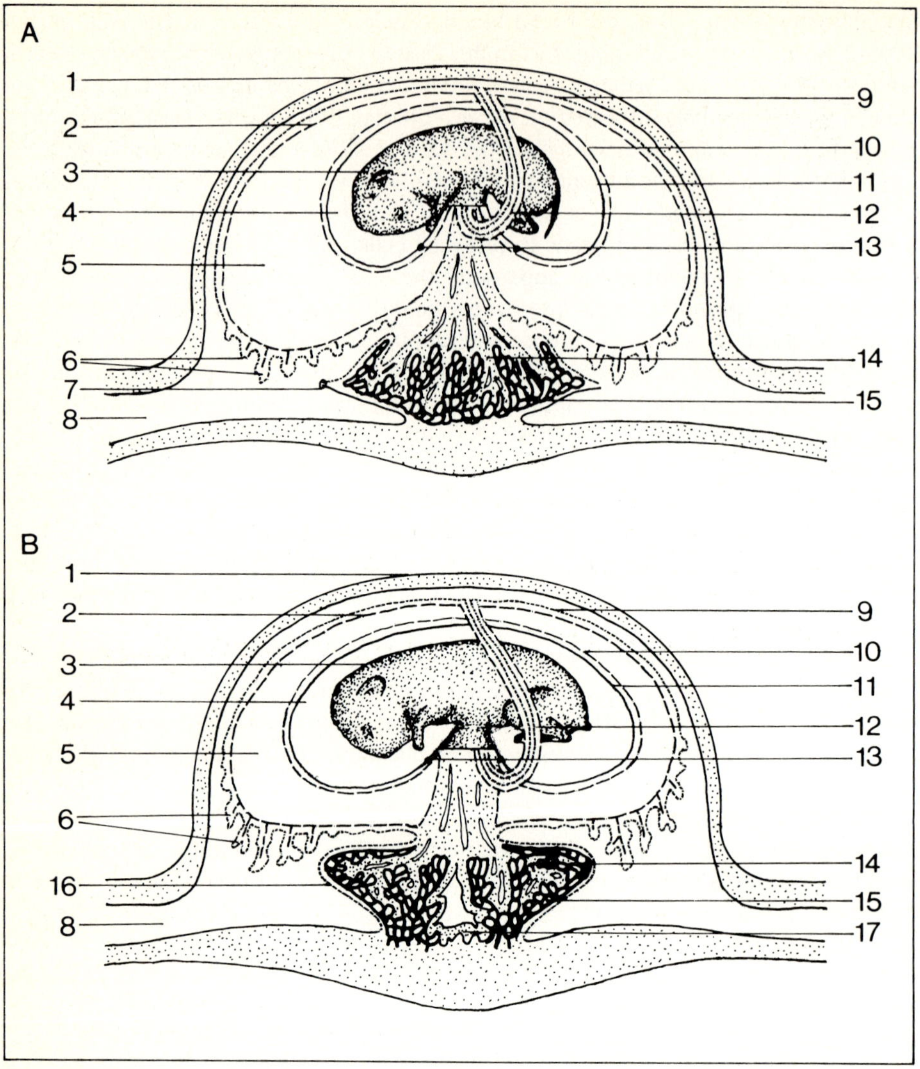

Fig. 94A,B. Diagrams of the placenta of the mouse (**A**) and the guinea-pig (**B**). (After Mossmann). 1 = wall of the uterus; 2 = mesoderm of the yolk sac; 3 = foetus; 4 = amniotic cavity; 5 = exocoel; 6 = endodermal villi; 7 = remains of the outer wall of the yolk sac; 8 = lumen of the uterus; 9 = endoderm of the yolk sac; 10 = amniotic mesoderm; 11 = amnion; 12 = stalk of the yolk sac; 13 = umbilicus; 14 = mesenchyme containing foetal blood vessels; 15 = labyrinth containing maternal blood; 16 = yolk sac endoderm overlying the placenta; 17 = subplacenta. In both cases the placenta is illustrated at a stage after the decidua capsularis has broken down

cerated by the detachment of the placenta at birth. As in the mouse, the placental region of the yolk sac endoderm gives rise to vascularised villi, although in this case they do not contact the placenta and therefore do not form an endodermal sinus. The structure of the placenta of *Cavia* after the breakdown of the decidua capsularis is illustrated in Fig. 94B.

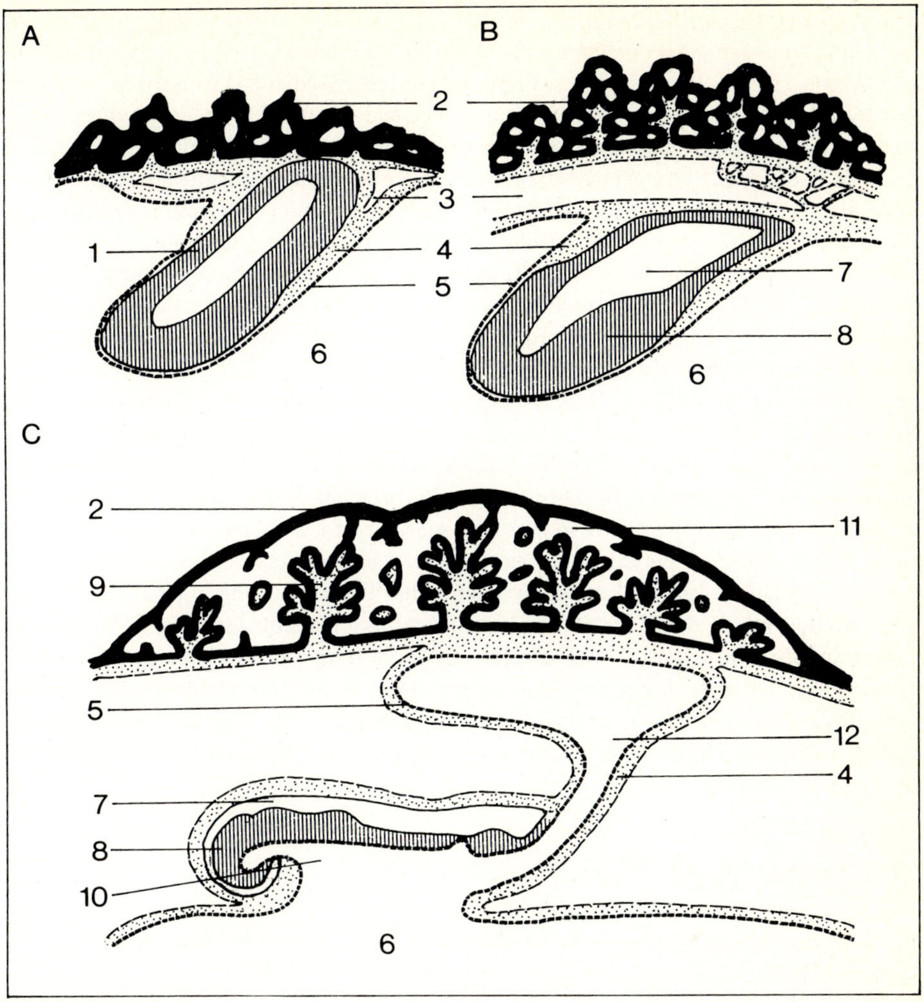

Fig. 95A–C. Diagrams of different stages in the placentation of *Cynocephalus*. (After de Lange). **A** Stage in which the embryo is attached by the primary body stalk. **B** Stage at which the amnion separates from the placental region. **C** The placenta after the amnion has formed.
1 = embryocyst, or early amnion; 2 = trophoblast in the placental region; 3 = exocoel; 4 = extra-embryonic mesenchyme; 5 = endoderm of the yolk sac; 6 = lumen of the yolk sac; 7 = amniotic cavity; 8 = embryo; 9 = villus with a core of mesenchyme; 10 = archenteron; 11 = intervillous space, 12 = allantois

8.6.5 The Placenta Vera of Dermoptera

The placenta of Dermoptera occupies an intermediate position between the labyrinth type of placenta and the discoidal or villous placenta characteristic of higher primates. Indeed, the systematic position of the Dermoptera is a subject of contention since they exhibit a combination of primitive features more typical of insectivores and other features which are characteristic of primates. The Dermoptera include members of the genus *Cynocephalus* (= *Galeopithecus*) which are represented by only two south-east Asian species. These flying mammals have large folds of skin between the fore and hind limb which serve as a parachute. The blastocyst of *Cynocephalus* contains an embryocyst which is connected to the prospective placental region by a so-called primary body stalk. The glands of the uterine epithelium within the uterine attachment site hypertrophy and the connective tissue of the mucosa becomes swollen. Upon attachment the trophoblast rapidly proliferates over this site, penetrates into the tissue of the uterus and forms an ectoplacenta with a labyrinth containing maternal blood (Fig. 95A). Extraembryonic mesenchyme subsequently grows into the labyrinth and, at the same time, the region of the placenta so that for a short time there is no connection between the embryo and the placenta (Fig. 95B). Development proceeds with the growth of an allantois which establishes a secondary connection to the placenta (Fig. 95C). The placenta continues to develop in the meantime and when fully developed represents a composite made up of elements of a labyrinth placenta and elements of a villous placenta. Thus foetal villi, provided with a core of mesenchyme, are formed which project into sponge-like spaces filled with blood. Those villi adjacent to the tissues of the mucosa make contact with, and thereby attach the embryo to, the trophoblast layer, of which little remains. The part of the placenta containing the maternal blood therefore resembles the hominid placenta (see Fig. 85B) whereas the presence of an allantois with a relatively large lumen can be regarded as a primitive feature. These features therefore reflect the intermediate status of the Dermoptera. There is a noticable reduction in the allantois, as is the case in lagomorphs and rodents, whereas a body stalk containing the rudiment of the allantois is present, this being a characteristic feature of higher primates.

8.6.6 The Placenta of Primates

Prosimian primates of the family Lemuridae have an epithelio-chorial placenta diffusa whose form corresponds to that of the pig (see p. 277). Early in the development of these animals the umbilical blood vessels give rise to four branches that extend over the placental region of the chorion and thus the allantois has four lobes (Fig. 96A).

In the case of the tarsier or spectral lemur (*Tarsius tarsius*) the extraembryonic mesenchyme is formed before the primitive streak has begun to develop, a feature which is also characteristic of higher primates. Implantation occurs centrally and the amnion is formed from amniotic folds and not by entypy, which is a feature of development that has been lost secondarily. The trophoblast proliferates over the site of its attachment to the mesometrial aspect of the uterus and forms an ectoplacenta which, however, does not invade the tissues of the uterus. At the same time as the mesenchyme grows into the placenta the trophoblast gradually transforms into a syncytium and sinuses develop within it which fill with blood from the modified uterine mucosa underlying the placenta. The continued growth of mesenchyme and of foetal blood vessels into the placental region eventually gives rise to a spongy, reticular structure that is invested by a thin layer of syncytiotrophoblast and which constitutes a discoidal, haemo-chorial labyrinth placenta. A body stalk composed of mesenchyme

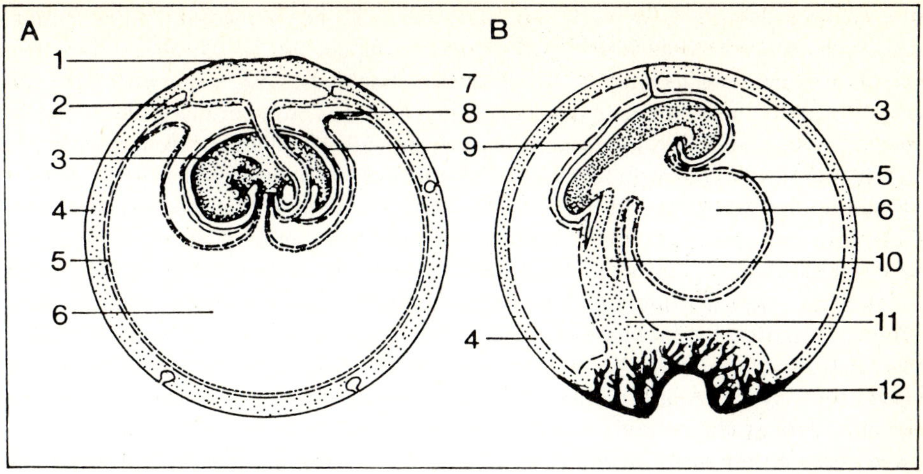

Fig. 96A,B. Diagram of the placenta of the primates *Loris tardigradus* (**A**, after Hill et al.) and *Tarsius tarsius* (**B**, after Hubrecht).
1 = region of the allanto-chorion; 2 = accessory lobe of the allantois; 3 = embryo, or foetus; 4 = mesoderm of the chorion; 5 = endoderm of the yolk sac; 6 = lumen of the yolk sac; 7 = main lobe of the allantois; 8 = exocoel; 9 = amnion; 10 = allantois; 11 = body stalk; 12 = placenta

is formed at the caudal end of the primitive streak relatively early in the course of development and makes contact with the anlage of the placenta. The stalk is surrounded by the exocoel and, in *Tarsius*, still contains a relatively large allantoic cavity and, lying parallel to this, an extension of the amniotic cavity. The body stalk of *Tarsius* has therefore been considered to prove that the mesenchymatous anlage of the allantois is homologous to the body stalk of higher primates. The structure of the embryo of *Tarsius* and the configuration of its placenta are shown schematically in Fig. 96B.

Three stages of placentation, which can be regarded as evolutionary stages, can be identified in apes (Simiae). The first, the Platyrrhinian stage, is represented by the broad-nosed or New World apes (Platyrrhina). In this case implantation is central, the blastocyst being initially attached to the uterus by its embryonic pole and then subsequently by the opposite pole. Correspondingly, two disc-shaped placentas are formed, i.e. a placenta bidiscoidalis. The amnion is formed by the process of entypy which gives rise to a cavitation amnion. The trophoblast proliferates over the two areas of placentation and differentiates rapidly into a superficial syncytiotrophoblast and an inner cytotrophoblast. In this case the trophoblast is invasive and penetrates into the stroma of the uterus thereby engulfing maternal blood vessels so that a system of blood-filled lacunae is formed prior to the ingrowth of mesenchyme. When the latter eventually invades the trophoblast a network of interconnected villi and shelf-like structures are formed which are covered by a layer of trophoblast that progressively changes into a syncytium. The syncytium connecting the shelves and villi then degenerates giving rise to large blood-filled sinuses into which some of the villi project. Thus, in this case, the form of the placenta is intermediate between that of a placenta olliformis and a labyrinth placenta, and resembles that of *Cynocephalus* which has already been described. There are no attachment villi in the placenta of Platyrrhina, rather, the placenta is anchored to the maternal tissues by the periphery of the trophoblast which remains intact to a large extent. The body stalk only connects the embryo, or foetus, with the main placenta which develops at the primary site of attachment. The second, accessory placenta is provided with blood vessels from the first.

The placenta of Catarrhina (Old World monkeys and apes) is basically very similar to that of Platyrrhina but differs from the latter in a number of respects. The trophoblast does not grow as fast but, because it invades the uterine tissue more rapidly, it nevertheless comes to invest the maternal blood vessels just as quickly. In Platyrrhina a relatively large trophoblast labyrinth is formed into which there is a secondary growth of mesenchyme whereas in Catarrhina, villi develop which contain mesenchyme from the outset. The growth of these villi behind the primary trophoblast pushes this layer forwards so that intervillous sinuses filled with maternal blood are formed without the secondary breakdown of tissue. Nevertheless, connections between the villi do occur but in the majority of cases this is probably due to the villi becoming "stuck" to one another secondarily. The only connection established by the villi which can be regarded as probably being primary is that to the basal boundary layer (i.e. on the uterine side of the placenta). The body stalk and the blood supply to the accessory placenta is formed in the same way as in Platyrrhina.

Placentation in anthropoid apes (superfamily Hominoidea) will be dealt with here by taking man as an example since the process of implantation and of development up to the stage at which lacunae form in the trophoblast have already been described (see pp. 153 and 270 respectively). The next stage is characterised by the development of primary villi from the trophoblast between the lacunae (Fgi. 97A). At about the same time the endometrium surrounding the embryo degenerates thereby forming a penetration zone composed of an amorphous mass of fibres and a so-called defense zone characterised by the presence of maternal leucocytes. As the trophoblast continues to sink into the wall of the uterus it disrupts not only the maternal blood vessels but also the uterine glands, whose lumena may contain blood at this stage. Furthermore, those maternal blood vessels in the vicinity of the embryo which are intact become more extensive and the decidua continues to develop. The trophoblast and particularly the cytotrophoblast then begins to grow more rapidly, the latter becoming syncytial. In the course of this process the number of villi is increased by the formation of secondary villi which contain a core of mesenchyme, though they are still avascular. The secondary villi then branch and the space formed between them is then referred to as the intervillous space (Fig. 97B). As the villi continue to grow they push the unbroken layer of cytotrophoblast (the basal trophoblast) in front of them progressively deeper into the uterine mucosa: the basal trophoblast therefore represents the peripheral wall of the intervillous space. This layer eventually forms septa between the villi which extend towards, but fall short of, the chorionic plate on the embryonic side of the placenta thereby leaving a subchorial cleft (Fig. 97C). The villi, which are by this stage vascularised and have many branches, are referred to as tertiary villi. In general, villi that possess a core of mesenchyme are called chorionic villi. In the human placenta, as distinct from those of Platyrrhina and Catarrhina, villi are formed over the entire surface of the trophoblast, although they are most numerous in the region of placentation. In this region the tertiary villi consist of a

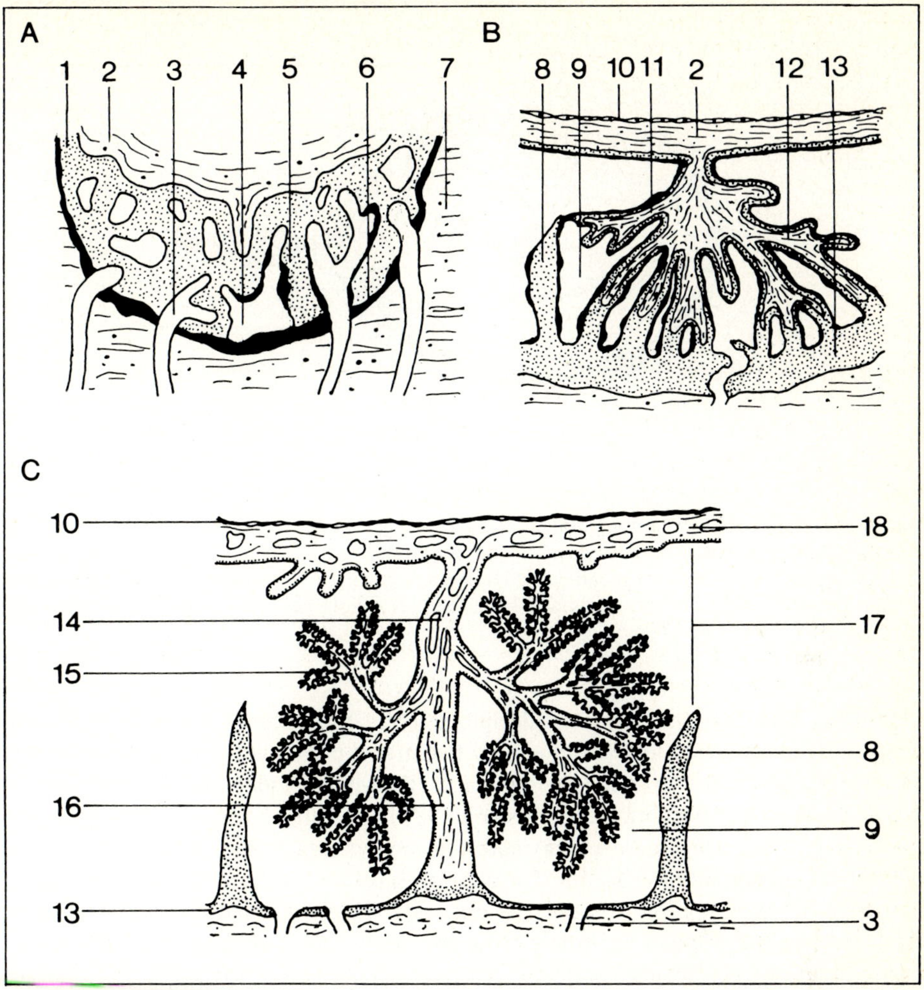

Fig. 97A–C. Diagrams of the human placenta at three stages of development characterised by primary (**A**), secondary (**B**) and tertiary villi (**C**), respectively. (After Starck).
1 = cytotrophoblast; 2 = extraembryonic mesenchyme; 3 = maternal blood vessels; 4 = initial formation of the intervillous space; 5 = primary villus; 6 = syncytiotrophoblast; 7 = uterine tissue; 8 = septum; 9 = intervillous space; 10 = amniotic epithelium; 11 = mesenchymatous core of an avascular villus; 12 = secondary villus; 13 = basal trophoblast; 14 = mesenchymatous core of a vascularised villus; 15 = tertiary villus; 16 = attachment villus; 17 = subchorial cleft; 18 = chorionic plate

core of mesenchyme covered by an inner cytotrophoblastic layer of Langerhans' cells and, outside this, a syncytiotrophoblastic layer which bears numerous microvilli. The embryo grows considerably in size as development proceeds causing the wall of the uterus in which it is embedded to distend into the lumen of the uterus. When fully developed, the membrana decidua consists of a decidua basalis lying directly beneath the placenta and a decidua capsularis which encloses the embryo within the wall of

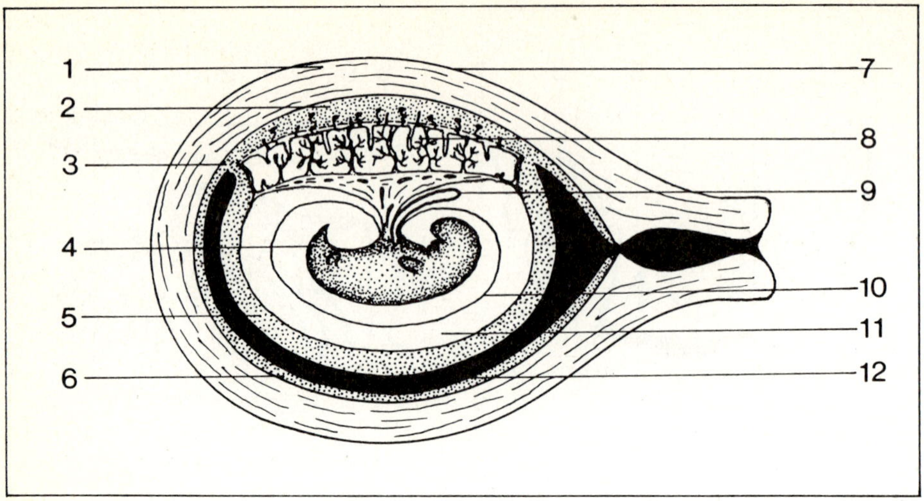

Fig. 98. Diagram of a sagittal longitudinal section through the gravid human uterus to illustrate the configuration of the membrana decidua. (After Starck).
1 = myometrium; 2 = decidua basalis; 3 = decidua marginalis; 4 = foetus; 5 = decidua capsularis; 6 = decidua parietalis; 7 = perimetrium; 8 = placenta; 9 = yolk sac; 10 = amnion; 11 = exocoel; 12 = lumen of the uterus

the uterus. The development of the decidua also encompasses the region of the endometrium lying opposite and peripheral to the site of implantation, this region of the decidua being referred to as the decidua parietalis, which borders the other regions of the decidua as the decidua marginalis (Fig. 98). The villi which initially develop in the region overlain by the decidua capsularis are eventually lost so that villi persist only in the region of the placenta. The chorion is referred to as a chorion frondosum if it bears villi and a chorion laeve if it does not. The continued growth of the embryo distends the decidua capsularis more and more until it finally contacts and becomes joined to the decidua parietalis, whereby the lumen of the uterus is obliterated. The human placenta reaches its maximum thickness of between 18–21 mm in the fourth month of pregnancy and thereafter only increases in area. Although it grows more slowly after the seventh month it evidently still continues to grow up to the time of birth. Between the fifth and sixth month the cells of Langerhans disappear from the surface of the villi, which are then covered only by a layer of syncytiotrophoblast.

The structure of the placenta when almost fully developed is illustrated in Fig. 85B. Each highly branched villus projects from the chorionic plate into a chamber or sinus, of which there are some 20–30, that are partially separated from each other by septa. The villi and the chambers bordered by the septa are somewhat inappropriately referred to as cotyledons and caruncles respectively that together form placentomes, based on a loose analogy to the placenta of ruminants. The villi consist of a main stem which bears first, second and third order branches. Attachment villi attach the chorionic plate to the basal trophoblast (Fig. 97C). The total volume of the villi in the mature placenta is so great that the intervillous space comprises only about

25% of the entire volume of the placenta. The effective surface area of the villi totals some 14 m^2 which corresponds to roughly ten times the surface area of the body of a grown man or woman.

The basal trophoblast, also referred to as the basal plate, is composed of cells of the cytotrophoblast lying adjacent to the decidua basalis. Current views are divided as to whether the basal trophoblast is the sole element of which the septa are composed or whether maternal elements are also present.

The maternal blood enters each placental chamber via a spiral artery which extends through the basal trophoblast. Since there is a pressure difference of about 60–80 mm Hg between the arteries and the intervillous space, the maternal blood squirts out like a fountain against the chorionic plate and flows slowly over the branches of the villi to the basal trophoblast where it is collected again by the funnel-shaped openings of the maternal veins. The intervillous sinus at the periphery of the placenta is largely free of villi but to what extent it serves as a drain for blood from the central region is not known. The exchange of materials takes place across the network of foetal capillaries which are only present at the tips of the villi. There is a perivascular capillary network associated with the larger blood vessels of the villi which probably supplies the villi themselves.

8.6.7 The Physiology of the Placenta

To recapitulate briefly, the placenta is an organ responsible for supplying the embryo or foetus with materials and removing waste products, an organ in which the maternal circulation is separated from the foetus by at least one barrier. The different ways in which the placenta is organised structurally can be ordered in a series representing on the one hand a progressive reduction in the physical barriers between the maternal and foetal blood and, on the other, a progressive increase in the area over which the exchange between the maternal and foetal blood takes place. It has been noted that these trends result in a more efficient placenta. At the same time however, according to the physical laws of diffusion the histology of the placenta, and of homologous structures derived from it such as the placenta olliformis, does not provide an infallible criterion for assessing the overall efficiency of the organ, as was clearly demonstrated by the comparison between the simple placenta diffusa of the horse and the highly specialised placenta of humans (see p. 265). Only a few of the substances that pass to the embryo or foetus do so by simple diffusion, oxygen being a case in point. Other passive transport mechanisms which operate include facilitated diffusion and diapedesial filtration. These are supplemented by active, energy-consuming mechanisms which are either enzymatic processes or are manifest on a larger scale as pinocytosis. Such mechanisms play a key role in the complex processes whereby materials are transported across the placenta. All endogenous substances, with the exception of the respiratory gases, are transferred against an electrochemical gradient. The mechanisms of active transport are highly specific and in many cases discriminate between different substances belonging to the same chemical or biochemical family. In man for example, more gamma globulin than albumin is transported into the foetus, and only low concentrations of HCG (see p. 180) can be detected in the foetal blood although it is produced in large quantities by the trophoblast. Moreover, there is a differential transport of globulins to the foetus, the alpha and beta immunoglobulins being transported in greater amounts than others, and differences also exist in the selectivity of the mechanisms for different alpha globulins. The processes of active transport across the placenta are often considerably more complex than comparable processes associated with cell membranes since in the former case the material may be transported across several cells and, furthermore, may be modified in the process. The metabolic energy necessary for the active transport of materials is provided exclusively by the placenta itself.

The placenta is sometimes regarded in the older literature as constituting a protective barrier that withholds "bad" substances from the foetus and provides for the foetus by letting "good" substances through. While there is more than a germ of truth to this description, it reduces the functional complexity of the placenta to a level at which it acts as a mere filter. Indeed, the fact that specific mechanisms of active transport operate within the placenta, in addition to mechanisms of passive transport, is sufficient to show that it is more than a passive filter. Rather, it plays an active and selective role in the transport of substances. In attempting to define the full range of functions performed by the placenta it is important to recognise that the placenta is an organ, like the liver and kidneys, that possess the full complement of metabolic machinary. It has, for example, its own biosynthetic capacity and can undertake the molecular modification of exogenous and endogenous substances. Thus the placenta has its own distinctive "metabolic identity". The principle functions of this metabolically active organ are the transfer of respiratory gases from the maternal blood to the foetus, and vice versa, the transfer of metabolites and nutrients, the formation of a barrier which inhibits infection, the provision of the foetus with maternal immunoglobulins, the biosynthesis of hormones, and the storage of certain substances, not to mention the maintenance of its own growth and basal metabolism. Moreover, these functions do not remain qualitatively and quantitatively the same but continually change during the different "ages" the placenta goes through in the course of its short existence.

The placenta therefore represents an intermediary between the supportive activities of the maternal organism and the embryo or foetus which is not only responsible for the transfer of substances but which can itself intervene in the processes of transfer and modify them metabolically. It is appropriate here to recall that the embryo can be provided for in two different ways, namely, either via an embryotroph or histiotroph or via a haemotroph. In those cases where there is a substantial placental barrier, i.e. in epithelio-chorial and syndesmo-chorial placentas, both these processes play a central role whereas haemotrophy takes on an increasingly important role in the case where the blastocyst invades the uterine tissues, this being reflected primarily in the progressive breakdown of the barriers between the maternal and foetal blood. Such changes can also be observed during the course of the development of epithelio-chorial and syndesmo-chorial types of placentas, as is the case in the pig for example. Thus the connective tissue layers become thinner and there is a growth of capillaries between the epithelial cells. The thickness of the placental barrier in such cases is nevertheless still substantial – 20 to 100 μm.

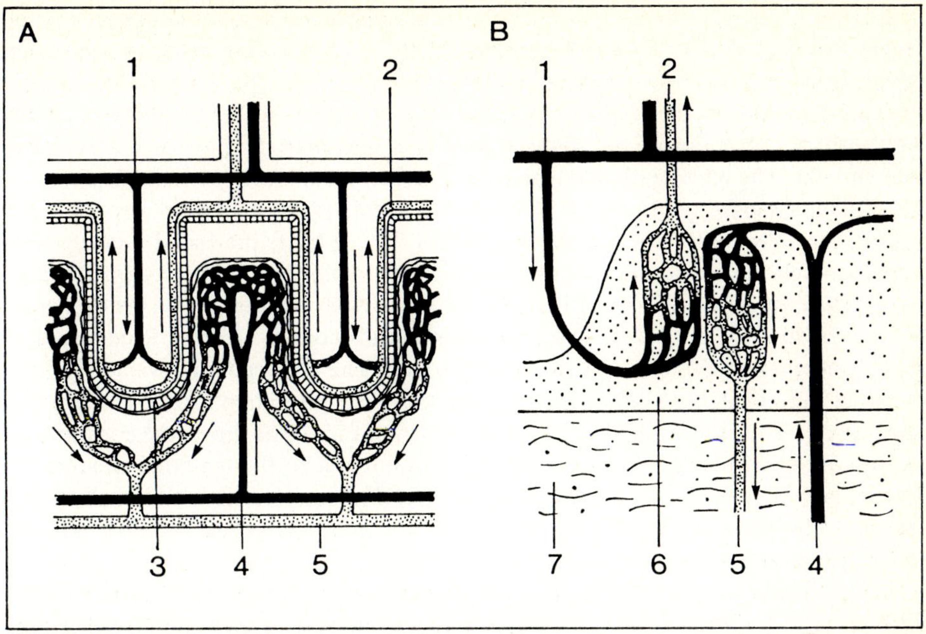

Fig. 99A,B. Diagram showing that the direction of blood flow in the placenta of ruminants **(A)** and in the labyrinth placenta **(B)** agrees with Mossmann's rule. (After Barron and Starck, respectively).
1 = foetal artery; 2 = foetal vein; 3 = foetal villus; 4 = maternal artery; 5 = maternal vein; 6 = trophoblast; 7 = uterine tissues. The network of maternal and foetal vessels in the exchange region have been drawn side by side in **B** for clarity though in fact they intermesh so that individual foetal capillaries lie parallel to portions of the labyrinth

The placenta is the respiratory organ of the foetus, the foetal lung so to speak. Oxygen is delivered by the maternal blood and diffuses across the placental barrier. If the maternal and foetal blood in the region of exchange flow parallel to each other in the same direction then there is, at best, an equalisation of the concentration of material in the two vascular compartments. If however the maternal and foetal blood flow in opposing directions then a great deal more than half of the material is exchanged from each compartment to the other. The greater efficiency of exchange systems which operate according to such a counter-current principle is exemplified by the process of gas exchange in the gills of bony fish. It is therefore not surprising that this principle of exchange is also embodied in the organisation of the placenta. Mossmann's rule states that the foetal blood flows in the opposite direction to the maternal blood in labyrinth placentas. Thus the maternal current flows from the foetal to the maternal side of the maternal circulation whereas the foetal current flows from the maternal to the foetal side of the foetal circulation. This means that maternal blood vessesls extend to the inner border of the trophoblast then reverse their direction to convey the blood back via the labyrinth to the maternal side, whereas the reverse is the case with respect to the foetal blood vessels (Fig. 99B). This configuration is also evident in the placenta of ruminants where the foetal vessels convey blood to the tips of the villi and thence beneath the periphery of the villus to its base whereas the maternal vessels extend to the tips of the corresponding maternal villi from where blood returns via a capillary bed (Fig. 99A). It is thought that partial countercurrents may also be generated in the placenta olliformis of higher primates as the blood which spurts out of the spiral arteries hits the chorionic plate and flows back over the villi. The blood within the maternal arteria uterina of man contains 14.5 vol. % oxygen at a partial pressure of 96 Torr and is 97% saturated with oxygen. This blood then flows into the intervillous sinuses. In contrast, the foetal blood which flows through the arteria umbilicalis into the villi contains 5.5 vol. % oxygen at a partial pressure of only 15 Torr and is 25% saturated with oxygen. After the exchange of gases in the villi the blood in the vena umbilicalis contains 13 vol. % oxygen at a partial pressure of 30 Torr and is 60% saturated. The corresponding values for the maternal blood fall to 8.3 vol. %, 35 Torr, and 55% saturation after exchange has taken place. This is illustrated in Fig. 100A whereas Fig. 100B shows the corresponding process of carbon dioxide exchange from the foetal to the maternal circulation. The uptake of oxygen by the foetus of sheep, cattle and primates is increased by the fact that the foetus possesses a specialised haemoglobin F which has a greater affinity for oxygen than the pigment in the maternal blood.

Water is conveyed across the placental barrier from mother to foetus primarily by differences in hydrostatic and osmotic pressure. In those animals which have been investigated that do not possess a haemo-endothelial placenta, the content of protein, and of high molecular weight proteins in particular, in the foetal blood is lower than it is in the maternal blood. The result is that the osmotic pressure is roughly the same on both sides of the placental barrier. For example, in the sheep it is 20–25 mm Hg, and in this case the blood pressure in the capillaries is also about the same on both sides, namely 10–15 mm Hg. The transfer of water to the foetus in this case presumably occurs at the base of the foetal villi whereas the reverse process occurs at the tips of the villi. It has been shown that in animals of small size the rate at which water is transferred to the foetus is much higher than the rate at which it is required by the foetus.

The pinciple minerals transferred across the placenta include sodium, potassium, calcium, iron, chloride and phosphate. The permeability of the placental barrier to sodium and potassium is different, nevertheless their concentration in the maternal and foetal blood is about the same. The rate at which minerals are transported across the placenta is lower in those types of placenta in which the physical barrier is more substantial. Experiments have shown that during the last phase of pregnancy more sodium and chloride passes from the mother to the foetus than the foetus actually needs. This finding is taken to indicate that both of these ions can pass freely across the placental barrier in both directions but does not exclude the possibility that they are also actively transported. The amount of sodium transported across the placenta per unit time is greater in those animals in which the placental barrier is relatively thin than in those in which it is more substantial. For example, in the pig the rate is 0.03 mg g^{-1} h^{-1} whereas in the cat it is 0.8 mg g^{-1} h^{-1} and in the rabbit 0.3 mg g^{-1} h^{-1}. The concentration of calcium and of phosphate in the foetal blood is higher than in the maternal blood, this evidently being the result of an active transport mechanism. The rate at which these two ions are transported increases greatly during

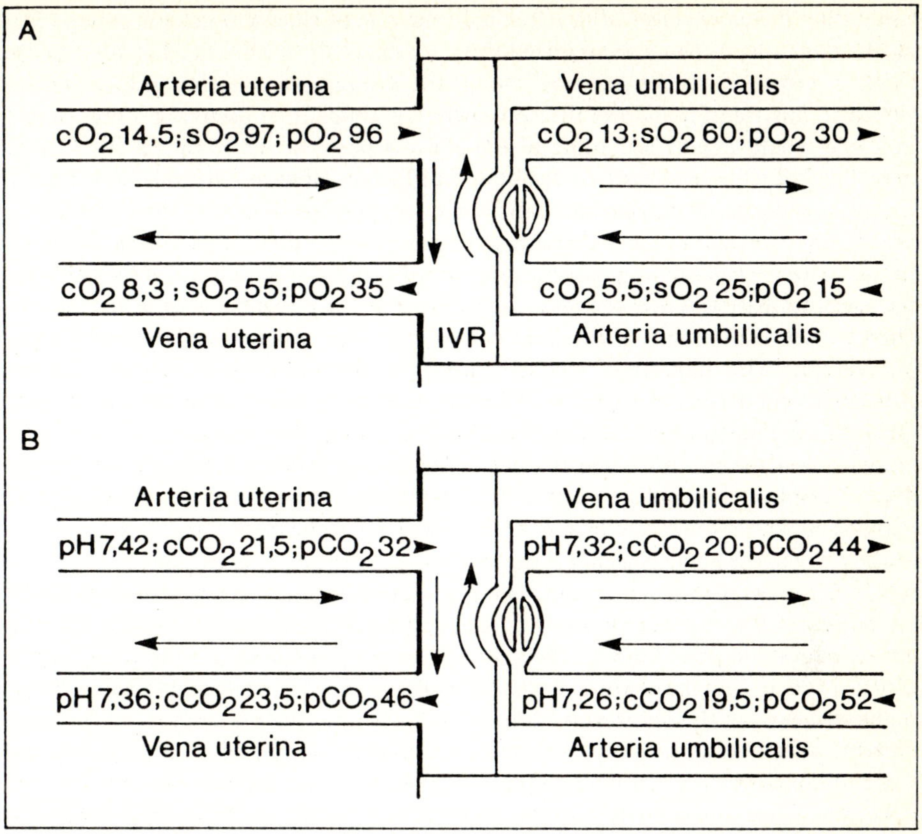

Fig. 100A,B. Schematic representation of oxygen (**A**) and carbon dioxide exchange (**B**) in the human placenta.
IVS = intervillous space; c = content (volume %); p = partial pressure in Torr; s = percentage degree of saturation

the formation of the skeleton in the foetus, their concentration in the blood thereby being kept relatively constant. The iron required by the foetus is supplied by embryotrophy and haemotrophy. The haematomes within the placenta of carnivores are sites at which iron is taken up through the trophoblast from the embryotroph.

Vitamins are probably actively transported into the foetus. Thus, in the pig for example, the level of vitamin B in the foetal blood is higher than it is in the maternal blood.

The transport of nutrients from the maternal blood into the foetus and the metabolism of the placenta, primarily that of man, will be dealt with together in order to present a coherent picture of the different energy sources associated with the processes of active transport.

Glucose is the most important component of the intermediate metabolism of the placenta and also the main source of energy. Large amounts of glucose are taken up by the placenta, the rate of uptake exceeding even that of the liver. While this is the case during the early stages of development, the rate of glucose uptake decreases during the later stages. The placenta itself uses an average of about a third of the glucose which it receives from the maternal blood, the rest being transferred to the embryo. Although this transfer follows the concentration gradient, it occurs at a higher rate than one would expect according to the existing physical and chemical conditions, indicating that this is a case of facilitated diffusion. However, little is known concerning the details of this mechanism. The concentration gradient between the maternal and foetal sides of the placenta is evidently dependent upon the thickness of the placenta is evidently dependent upon the thickness of the placental barrier since the placental barrier since the gradient is less steep in placentas with a thin barrier than in those with a more substantial barrier, as is the case in ungulates and the pig for example. The transfer of glucose appears to depend in some way on oxygen since under conditions of oxygen deficit the level of glucose in the maternal blood, but not in the foetal blood, is found to rise. Conditions of oxygen deficit also lead to an increase in the rate at which the placenta takes up glucose, this being the Pasteur effect. The physiological basis of this effect remains obscure. The pathways of glucose metabolism within the placenta appear to differ from those in the liver. Thus phosphorylation of glucose in the placenta is mediated by an unspecific hexokinase, and perhaps by an additional mechanism as well. The placenta possesses all the necessary enzymes and cofactors associated with the two main pathways of glucose metabolism, namely glycolysis, involving the Embden-Meyerhof pathway in conjunction with the Krebs or citric acid cycle, and the pentose phosphate pathway. The principle metabolic pathway in the placenta is glycolysis and, due to its high rate, the placenta produces large amounts of lactate. In one way or another the metabolism of glucose provides the energy and precursors for the metabolism of proteins, pyrimidines and fats. The type of carbohydrate metabolism which occurs depends on the stage the embryo has reached in its development. Thus the pentose phosphate pathway is used almost exclusively in the blastocyst but gradually diminishes in importance after implantation when glycolysis becomes the dominant pathway. The lactate produced by the Embden-Meyerhof pathway is selectively transported to the foetus, although theoretically it could diffuse through the placental barrier in either direction. Lactate is believed to represent an important metabolic substrate in the foetus since 25 % of all metabolism is oxidative. The lactate remaining in the placenta is probably oxidised where it is by the lactate dehydrogenase present in this organ. Theoretically, the products of oxidation could be used for the synthesis of amino acids or proteins. Since the placenta is supplied with sufficient glucose from the maternal blood it does not actually "need" any glycogen reserves, yet these are nevertheless present in the vicinity of the larger foetal blood vessels, though not in those regions where exchange takes place. The foetal liver, in contrast, contains a considerable amount of glycogen. The physiological significance of the glycogen reserves in the placenta is unknown but it has been suggested that they may be used by the blood vessels or be involved in the formation of aromatic compounds in the course of steroid biosynthesis in the placenta. The vast majority of the total amount of glucose in the placenta stems from the maternal blood and only a very small proportion is produced by the placenta itself. Thus, in the early stages of development the placenta has the ability to synthesise glucose from pyruvate, but this ability is lost as development proceeds. Relatively high levels of fructose are present in the foetus of ungulates but not of man or the rat. The transport of fructose across the placenta is slower than that of glucose. In the sheep fructose is synthesised in the placenta. Fructose presumably represents a reserve of energy for the foetus that can be rapidly mobilised.

The placenta has a higher rate of protein metabolism than any other tissue of the body, even the liver. This is primarily a consequence of the growth of the placenta which, in man, surpasses that of the foetus during the first half of pregnancy: the situation is however reversed during the second half. In the tenth week of pregnancy when the placenta weighs about 50 g, it produces about 1.5 g of protein per day. When almost fully developed and weighing some 450 g the rate of protein biosynthesis is approximately 7.5 g of protein per day. Proteins are synthesised either from amino acids or from metabolites of carbohydrate metabolism, the latter process being reflected by the fact that the rates of protein and carbohydrate metabolism vary in a parallel manner during the course of development. One factor that limits the rate at which proteins are synthesised is the availability of pentoses, which limit the synthesis of nucleic acids and thus of protein. Most of the amino acids necessary for the synthesis of foetal and placental proteins stem from the maternal blood and only a small proportion are synthesised by the placenta itself or made available by the degradation of plasma proteins. The level of amino acids in the foetal blood is higher than that in the maternal blood, indicating that there is a mechanism for their active transport across the placenta. However, this process appears to take place in three steps. The first is the active and, moreover, selective transport of amino acids from the maternal blood into the placenta. The second step involves the accumulation of amino acids in the placenta which are subsequently used for the synthesis of proteins to either support the growth of the placenta or to be used in some other metabolic pathway. The transport of amino acids into the foetal blood represents the last step and may occur by simple diffusion. Thus the active energy-dependent step is the uptake of amino acids, as is the case in muscle for example. The rate of this selective uptake depends on the chemical properties of the amino acid, the energy required being provided mainly by glycolysis. Certain maternal proteins, mainly immunoglobulins, are transported directly to the foetus. The gamma globulins in particular, which neither the placenta nor the foetus is capable of synthesising, are provided exclusively by the maternal organism. The gamma globulins are selectively transported across the placenta, as indicated, for example, by the fact that immunoglobulin G is transported more readily than immunoglobulin M. The former, which has a molecular weight of about 160,000 Dalton, belongs to the 7S globulin fraction that includes some 90% of the maternal antibodies. In Carnivora and Rodentia the blood of the new-born has roughly the same globulin content as the maternal blood whereas in those animals in which the placental barrier is relatively thick, such as the cow and pig, the foetus receives few if any maternal antibodies before it is born. In such cases, antibodies are passed on in the first milk (or colostrum) the mother gives the newborn young during its first days of life. The thickness of the placental barrier in different species similarly influences the transport of other plasma proteins. Experiments have shown that protein synthesis in the placenta is influenced by certain hormones. It is, for example, stimulated by oestradiol, although this is probably not a direct effect but the result of more energy being made available for synthesis due to the promotion of transhydrogenation by oestradiol. Proteins are probably degraded within the placenta by the "conventional" mechanisms operating in other tissues. The placenta also possesses enzymes, transaminases and deaminases, that degrade amino acids. It is unlikely that the placenta has any reserves of protein destined for the foetus. It is a curious yet unexplained fact that the placenta evidently has "priority" over the foetus since the placenta maintains an almost constant size in conditions of protein deficiency whereas the foetus becomes smaller.

The placenta contains little fat in comparison to other tissues. During pregnancy in man, the overall level of lipid in the maternal blood is raised, only the level of one particular phospholipid being lower. On the other hand, the concentration of free fatty acids, free cholesterol, of phospholipid, lecithin, lysolecithin and sphingomyelin in the tissue of the placenta are all higher than in the maternal blood whereas that of cholesterol esters and of triglycerides is lower. The placenta is thought to play an active role in the transfer of lipids to the foetus and in their biosynthesis and degradation. For example, the placenta is capable of transporting fatty acids to the foetus, the rate of transport varying between different species: in the rat and sheep the rate is low compared with that in the guinea-pig, rabbit and in primates. However, the transport of free fatty acids, which may stem either from the placenta itself or the maternal blood, is evidently limited in some way. In man, fatty acids are transferred across the placenta to the foetus down the gradient in their concentration. On the other hand, it has been shown that palmitate can be transported in the opposite direction in the mature placenta maintained under experimental conditions by perfusion. Similarly, free fatty acids are transported from the foetus to the mother in humans and in animals commonly studied in the laboratory. Furthermore, this process is selective in that different fatty acids are transported at different rates. The major source of fatty acids destined for the foetus are the triglycerides in the maternal blood. Fatty acid synthesis within the placenta provides only about 2% of the foetus' total requirement in the case of the rat. The human placenta is capable of forming free fatty acids by esterification of triglycerides which it either synthesises itself or obtains from the maternal blood. Esterification depends on the synthesis of glycerophosphate, the necessary energy coming from adenosine triphosphate (ATP) produced by the process of glycolysis. In general, triglycerides do not appear to pass into the foetus but are first degraded into fatty acids. Despite the fact that the human placenta can synthesise triglycerides rapidly, their concentration is low, indicating that these substances undergo a rapid turnover. It is notable that about 75% of the lipids in the placenta are phospholipids, most of which are probably synthesised by the placenta itself. Their function within the placenta is unknown but they may be involved in the transport of amino acids to the foetus. Cholesterol is not only transported into the placenta but also synthesised there. The placenta is able to synthesise cholesterol esters out of cholesterol and free fatty acids. Cholesterol plays a key role in the synthesis of steroids in the placenta.

8.7 The Duration of Gestation

The placenta is a transitory organ of the body that develops during the course of pregnancy, fulfils its function, and is usually expelled as a whole at birth, its life-time being the period between implantation and birth. The duration of gestation in different species of mammals, disregarding for the moment those with delayed implantation, differs considerably. A general rule of thumb for the Eutheria is that the bigger the animal, the longer is the period of gestation, although there is no direct correlation between the weight of the body and the duration of gestation. For example, elephants (Elephantidae) have the longest gestation of any mammal (22 months) whereas that of the great baleen whales (Mystacoceti) lasts only about a year although just the tongue of the blue whale (*Balaenoptera musculus*), with a weight of more than 3000 kg, is as heavy as a cow elephant. Another example which illustrates the limited validity of the above rule of thumb is the comparison between the polar bear (*Thalarctos maritimus*) and man. The former has a body weight of 600–700 kg, which is about ten times as heavy as a decidedly portly woman, yet has a gestation period of from 6–7 months whereas that of man lasts about 9 months.

One can, nevertheless, classify mammals into broad groups according to the duration of gestation. Thus Marsupialia have the shortest period of gestation of all mammals, as one might expect from their nature. For example, the dasyure (*Dasyurus viverrinus*) gives birth after about 11 days. Eutheria with a period of gestation of up to 1 month include the hedgehog (*Talpa europaea*: 30 days), the rat (*Rattus rattus*: approximately 21 days) and the rabbit (*Oryctolagas cuniculus*: 30 days).

The second group of animals, in which gestation takes up to 4 months, includes domestic cats (9 weeks), dogs (*Canis familiaris*: about 9 weeks), the wolf (*Canis lupus*: about 9 weeks) and the puma (*Panthera concolor*: 15 weeks). The third group includes those animals with a period of gestation of from 5–22 months, for example the alpine ibex (*Capra ibex ibex*: 5 months), the sheep (*Ovis aries*: 5–6 months), the tiger (*Panthera tigris*: about 5 1/2 months), the chamois (*Rupicapra rupicapra*: 5–6 months), the fallow deer (*Dama dama*: 8 months), the musk-ox (*Ovibos moschatus*: 9 months), the lama (*Lama huanachos glama*: 10–11 months), the bactrian camel (*Camelus bactrianus*: 13 months) and the giraffe (*Giraffa camelopardalis*: 14 1/2 months).

As mentioned, this classification should not be given undue significance since within each group there are instances which deviate considerably from the relationship between the duration of gestation and body size, body weight or body volume. For example, gestation in the guinea-pig (*Cavia cobaya*) takes 62 days whereas in the physically larger rabbit it takes 30 days. Again, the lion (*Panthera leo*) carries its young for about 4 months, which is a shorter gestation than in the sheep. The main reason for these variations relates to the different stage of development at which the young are born, the consequences of which are considered in the next chapter.

CHAPTER 9

Care of the Young Outside the Body

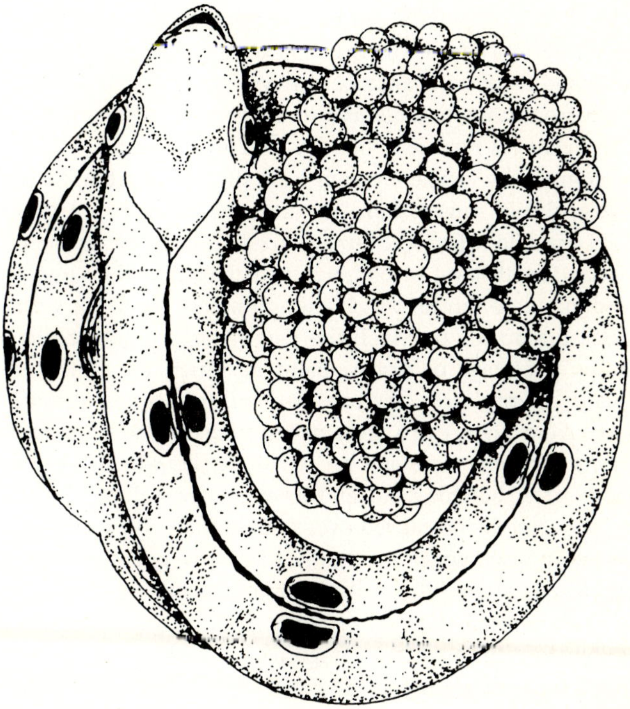

The gunnel *(Pholis gunnellus)* guarding its spawn

The methods of caring for the young within the body dealt with in the last two chapters represent the most specialised means of increasing the chances of the offspring's survival. In the overwhelming majority of cases this is a passive phenomenon, that is, the parent animal carries the young in or on its own body and sustains them in some way from its own metabolism but does not engage in any specific behavioural activity. In contrast to this, such behaviour is the defining feature of the way the young are cared for once they are outside of the body. For example, the animal may build a nest in which to lay the hatch or young; it may guard the brood and in certain circumstances defend it; the animal must find more food to feed not only itself but also the young, and in many cases it familiarises the young animals with certain environmental situations. The subject of this chapter is thus a diverse phenomenon encompassing many different levels of complexity.

Four basic levels of complexity can be distinguished. In the simplest case the eggs are laid and fertilised in a place which in some way gives the developing embryos or newly hatched brood a better chance of survival because it is hidden from nest robbers or protected from unfavourable abiotic factors or, alternatively, set in a favourable spot. These measures represent the full extent of parental care in this case. The second level of complexity is where one or both of the parent animals guard, bood or take care of the brood in some other way until the young hatch out. The third level of complexity is reached when the young are fed, guarded or cared for after they have hatched. The fourth level of complexity is basically the same as the third but there is a preceding phase during which the young are cared for within the body. In the last two cases the degree to which the parents are committed to caring for their young depends on the stage of development at which the young hatch out or are born. Accordingly, among birds one can distinguish between altricial species where the young emerge from the egg in a more or less helpless condition and must be cared for by the parents for some time and precocial species where the young are able to stand immediately after hatching out and have a more well-developed plumage. This distinction can, in principle, be carried over to mammals to give the following classification: nest-bound infants, which are born at a relatively early stage of development with their eyes closed; active infants that come into the world possessing hair and a well-developed body and are capable of some form of locomotion; and suckling infants which fall between the first two catagories.

Nest building frequently occurs at the beginning of the reproductive period in species of vertebrate that do not care for the young inside their body, and is evidently sexually motivated in many cases (see Chap. 6). But whatever motivates the animal to build a nest, the nest always serves as the site at which the parents care for their young. The activities associated with the latter behaviour follow on from the act of spawning or from the fertilised eggs being laid, and take place in or around the nest. In the simplest case the eggs are merely guarded and defended against nest robbers. Commonly, however, the parents take more care of their young. Some teleosts, for example, clean the eggs by mouthing them, remove unfertilised eggs or those covered with fungus, and may provide the developing embryos with fresh, oxygenated water by fanning them (see p. 219). Birds, on the other hand, brood their eggs by sitting on them. Care of the young after they have hatched out or have been born can involve them being guarded, hidden, led, fed and in some cases tought certain behaviour patterns by the parent animals. The diversity of ways in which vertebrates care for their young is so great that it is impossible here to deal comprehensively with all the multifarious aspects in detail. The following account will therefore deal mainly with nest building and comparable phenomena, the most important types of behaviour associated with caring for the young in the nest, and the feeding of the young.

Vertebrates which nest either make the entire nest themselves or make use of some existing cavity such as a rock crevice, empty mussel shell, a cavity in a tree or even a tin can as a nest which may or may not then be furnished with other materials brought into the nest. Between these two basic types of nest are many intermediate types. Thus nests which are built can be differentiated according to the type of material used to construct them, that may be produced wholly or partly by the animal itself. An extreme example of this is where the eggs or brood are carried about on the body of the parent, which therefore serves as a sort of living nest. In general, the term nest can be very loosely defined as a delimited area in which the fertilised eggs or young are deposited and where the brood go through at least part of their development.

9.1 Pisces

A range of teleosts do not, strictly speaking, care for their young but simply lay their eggs in some favourable place, an example being the perch (*Perca fluviatilis*). Females of this species lay long strings of eggs which stick together to form a network that is draped over water-plants; this is thought to ensure that the developing embryos are well supplied with oxygen. Salmonids go one step further and bury their spawn by beating out a shallow trench in the river bed which they subsequently fill in again after the eggs have been laid by performing sinuous movements with their body. This can then be regarded as the simplest type of nest built by fish. It is usually made by the male. The nest may be formed in sand, gravel or in muddy ground by the fish either beating with its tail or digging with its mouth and often involves the excavation of remarkably large amounts of material. Nest of this type are built by members of the Petromyzontia, Holostei, Teleostei and the Dipnoi. An example of the simplest form of this type of nest, a round shallow trench in sand or gravel, is that formed by sun perch (Centrarchidae). In this group of fish it is the males which fan, clean and guard the eggs as they lie in the nest. The nest of the cyprinid *Semotilus atromaculatus* is constructed in a more complex fashion. It is built at a site of running water and has an elongated oval shape. The side facing the current is marked by a wall of gravel that shields the site from the current whereas the other end is made of sand (Fig. 101A).

According to an old report the Australian catfish *Arius australis* likewise builds a nest out of gravel and stones, but not by excavation. This fish spawns over a layer of gravel and small stones which it has previously collected and deposited over a site that has a diameter of about 50 cm. The male then covers the fertilised eggs with several layers of larger stones that he gathers in with his tail from around the nest. The result is that the nest is encircled by a ditch approximately 1 m in diameter.

A different type of nest is built by the dwarf catfish *Ictalurus nebulosus*. In this case both parents pull up or bite off plants over an area of ground about 1.2 m in dia-

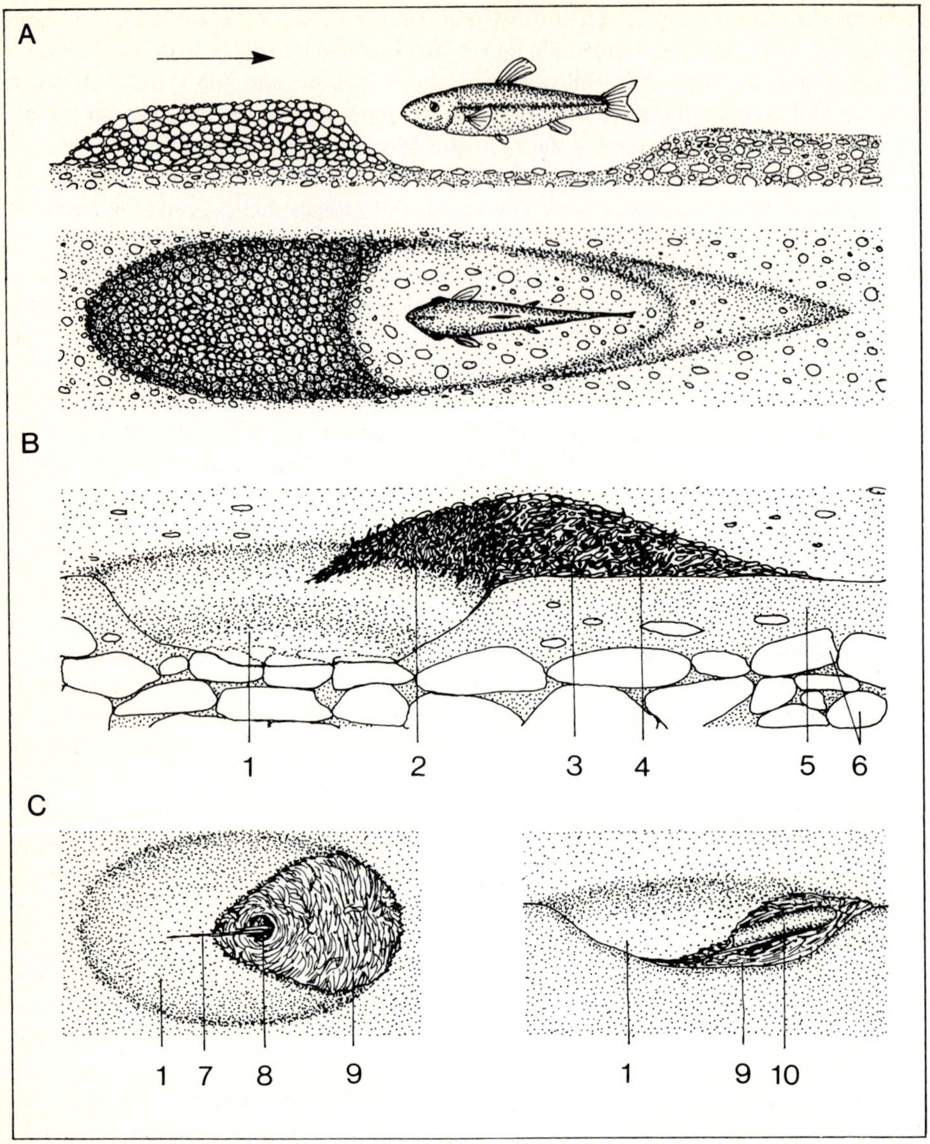

Fig. 101A–C. The nests of teleosts: I. **A** The nest of *Semotilus atromaculatus* seen in longitudinal section (*upper figure*) and from above (*lower figure*). *Arrow* indicates the direction of the current. (After Reighard). The nest of *Crenilabrus quinquemaculatus* (**B**) is formed from a pit and a mound of algae. (After Soljan). That of *Gasterosteus aculeatus* (**C**) consists of a pit partly furnished with plant material. (After Wunder).
1 = pit; 2 = steep inner wall built of algae; 3 = pieces of *Cystoseira* which surround the eggs; 4 = covering stones; 5 = sand; 6 = stones; 7 = large pieces of plant material; 8 = opening of the nest; 9 = small pieces of plant material (algae); 10 = cavity of the nest

meter and build them up into a surrounding wall several centimeters thick. The eggs and the brood within the nest are then attended to for some time.

Male mudfish (*Amia calva*) also graze plants from the area where they dig their nest, in which they subsequently guard and fan the developing young for some months. It has been reported that in dry periods, when the spawning pools of this species frequently dry up, the male sprays water over the eggs with his tail. After hatching out the young attach themselves by a sucker-like organ to the plants forming the rim of the nest.

South American lungfish of the genus *Lepidosiren* and African species of the genus *Protopterus* dig pits or tunnels in the muddy bottom in which they deposit about 5000 eggs. The male is responsible for taking care of the developing embryos and supplying them with oxygenated water by performing sinuous fanning motions with his body.

The eel *Monopterus javanensis* reportedly builds a tubular, U-shaped nest in which the male produces a foamy mass that froths from one of the openings, though the reason for this is unknown.

Several species use plant material to build a rather different nest from the enwalled constructions mentioned above. Thus the Mediterranian wrasse *Crenilabrus ocellatus* builds a cup-shaped nest, mostly out of green *Cladophora* algae, that is reminiscent of a bird's nest. The male gathers this material from an area within a 10 m radius of the nest, the nest being built upon the contrastingly coloured strands of brown *Cystoseira* algae. The male subsequently spawns together with one female, covers the eggs with algae, and goes on to spawn with other females until the wall of the nest is filled with several hundred eggs. After spawning the male is responsible for fanning the eggs and cleaning the nest. The living algae of which the nest is made are thought to improve the supply of oxygen to the developing eggs. The nest of the related species *Crenilabrus quinquemaculatus* is designed somewhat differently. It consists of a pit in the sand fenced on one side by a crescent-shaped mound made up of pieces of *Cystoseira*. The inner face of the mound is steep but the outer face slopes gradually into the sand. The mound is covered, amongst other things, with small stones and fragments of mussel shell (Fig. 101B). The nest is built by the male who then spawns with several females. In each case the eggs are deposited on the inner face of the mound and then covered with algae. The eggs are subsequently fanned and attended to by the male.

The sticklebacks (family Gasterosteidae) also build a nest of plant material but in this case it is held together by a sticky secretion from the kidneys of the male fish. The marine stickleback (*Spinachia spinachia*) builds a round nest, out of pieces of algae and other material, above the sea bed in, for example, fonds of seaweed. The three-spined stickleback (*Gasterosteus aculeatus*) first beats out a depression in the substratum, within one side of which it builds a nest about the size of a walnut out of algae and other pieces of plant material. The cavity of the nest has a single opening (Fig. 101C) through which the female swims to lay her eggs, though she exits by breaking down the wall opposite the opening. The male spawns with several females and afterwards fans and guards the eggs, and he continues to do this after the young have hatched out. Young that leave the nest are often taken up by the male in his mouth and brought back to the nest. The four-spined stickleback (*Apeltes quadraeus*)

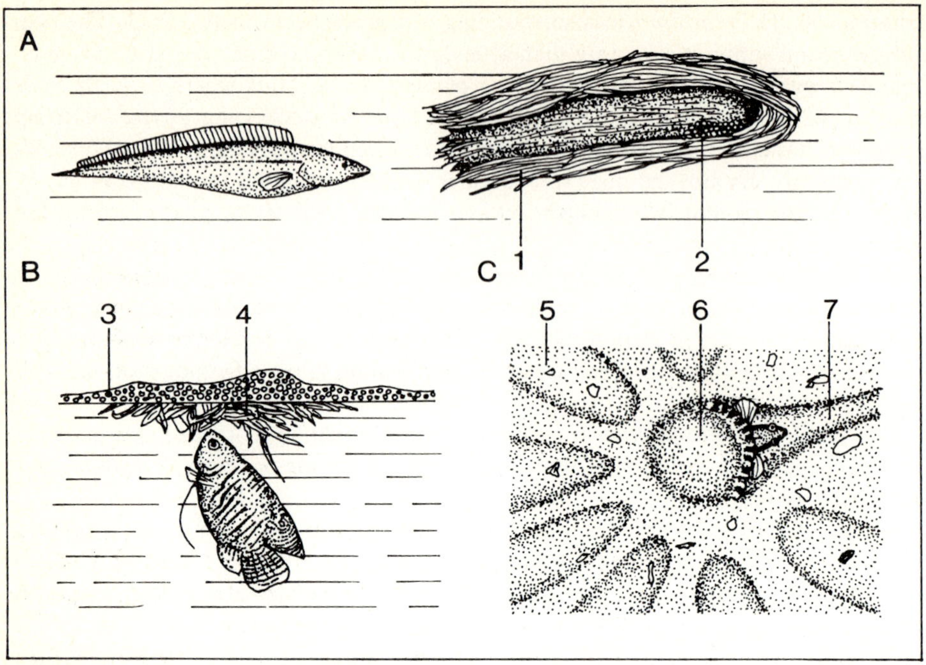

Fig. 102A–C. The nests of teleosts: II. **A** Floating nest of *Gymnarchus niloticus*. (After Daget). **B** Nest of foam and plant material built by *Colisa lalia*. (After Forselius). **C** Nest of *Gobius minutus* formed under a mussel shell covered with sand. (After Guitel).
1 = plant material; 2 = eggs; 3 = bubbles of mucus; 4 = small pieces of plant material; 5 = furrows in the sand; 6 = mussel shell covered with sand; 7 = entrance groove

ventilates the nest in a rather special way. The male puts his head in one of the two nest openings and creates a current of fresh water over the brood by sucking water in through his mouth.

The gymnarchid *Gymnarchus niloticus* builds a floating nest from plant material, the nest being about 50–60 cm long and 20–30 cm wide (Fig. 102A). The cavity within it holds some 1000 eggs but how they are spawned and how the nest is built remains obscure. Anabantid species, which occur in south-east Asia and in Africa, also build a floating nest. These animals have an accessory respiratory organ which enables them to breath atmospheric air. When building the nest the male takes air into his mouth and envelops it in a secretion produced by numerous, single-celled glands in the mucus membrane of the mouth. The air bubble is then spat out and it rises to the surface where the bubbles collect to form a floating nest. Many species add pieces of plant material to the nest structure, one such being *Colisa lalia* (Fig. 102B). The male of this species spawns with only a single female, spawning taking place beneath the nest. The fertilised eggs usually sink but they are taken into the mouth by the male, though sometimes by either the male or the female, and blown out again into the nest. The male then produces more bubbles of mucus so that the eggs become sandwiched between two layers of bubbles. Thereafter the male guards the eggs until the young hatch out. It is thought that the bubbles, by reflecting and refracting light, protect the eggs from the deleterious effects of strong sunlight, which include overheating. Similar nests composed of bubbles and plant material are constructed by catfish of the genera *Callichthys* and *Hoplosternum*.

Female bull-head fish of the species *Cottus gobio* lay their eggs on the underside of a stone where they are then guarded by the male. The Blenniidae make use of a rock crevice or a hole made by a boring mussel as a place to nest and spawn. The male first cleans the walls of the nest with his mouth or fins and then spawns successively with several females that deposit their eggs in a layer covering the wall of the nest. The eggs are then guarded and fanned by the male until the young hatch out. The marine goby *Gobius minutus* uses the empty shell of a cockle or a venus mussel (or, more precisely, one half of the shell) as a nest. If the shell does not already lie with its inside facing downwards the male turns it over and having done this digs out the sand beneath the shell with his mouth and fins. He then bulldozers sand over the shell using his pectoral fins and tail to give rise to a number of radially arranged furrows in the ground around the nest (Fig. 102C). Leading up to the single opening of the nest is an entrance furrow made firm with a secretion of the skin. The eggs are laid against and stick securely to the inner surface of the shell. The gunnel (*Pholis gunnellus*) also spawns in an empty mussel shell and the male subsequently guards the egg cluster. If a suitable mussel shell cannot be found, however, he coils his body around the eggs (see title picture to this chapter).

Nilsson's glass goby (*Crystallogobius nilssonii*) lays its eggs in the unoccupied tubes of polychaete worms where they are guarded by the male.

The South American characine (*Copella arnoldi*) has a particularly remarkable way of caring for its young. The nest is a leaf hanging over the surface of water. The male and female fish leap out of the water and hang for a moment with their belly pressed against the leaf and in this moment they spawn (see the picture on the cover of this book). Spawning is complete after the fish have made several leaps and thereafter the male tends to the developing young by remaining beneath the nest and from time to time by splashing water over the eggs with his tail.

In a few cases the eggs are laid on the male's or the female's body, which can thus be regarded as a kind of nest. For example, male fish of the teleost species *Kurtus gulliveri* from New Guinea have a bony process on their forehead from which hang two clusters of eggs by a central ligament (Fig. 103A). In syngnathids of the genus *Entellurus* and *Nerophis* the female sticks her eggs to the belly of the male who then carries them around until they hatch. This type of parental care can be regarded as leading to the more specialised mechanisms that occur in this family described in Chap. 7. Male sea dragons of the genus *Phyllopteryx*, which also belong to the Syngnathidae, carry the eggs in a groove on the ventral side of the tail.

The South American catfish *Aspredo cotylophorus* similarly carries its eggs on the body, although in this case it is the female that cares for the young. The eggs are borne on the tips of numerous stalk-like processes which grow from the belly of the female (Fig. 103B). In the closely related syngnathid genus *Solenostoma* (the tube mouthed fishes) the female's pelvic fins are partially fused to the body forming an open marsupium into which project branched appendages that carry the eggs at their tips (Fig. 103C). This arrangement can be regarded as bordering on that where the young are cared for within the body.

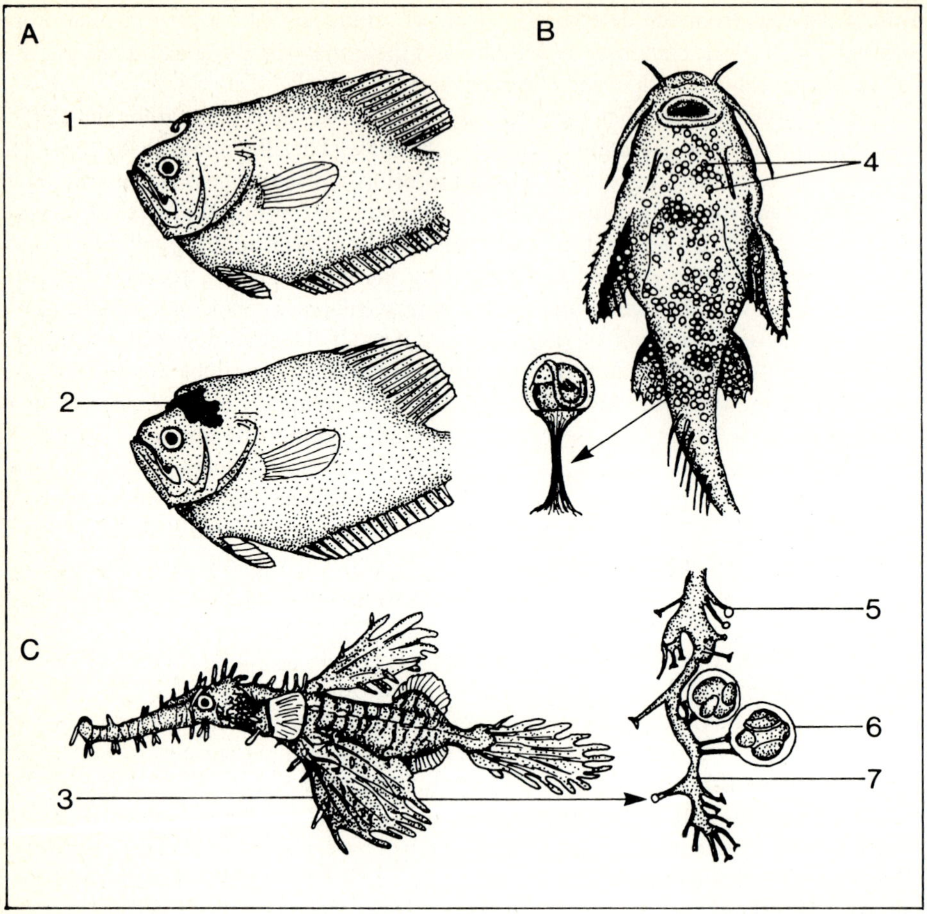

Fig. 103A–C. Teleosts which care for their young on their own bodies. **A** *Kurtus gulliveri. Upper figure*: male without eggs; *lower figure*: male with eggs. (After Guitel). **B** Female of an *Aspredo* sp. with eggs hanging from stalks (detailed) on the underside of the fish. (After Wyman). **C** Female of the genus *Solenostoma* with marsupium and cutaneous appendages (detailed *right*) on which the eggs are carried. (After Willey).
1 = process arising from the forehead; 2 = hanging cluster of eggs; 3 = marsupium; 4 = eggs; 5 = branch bearing an egg; 6 = embryo within an egg; 7 = cutaneous appendage

In no case within the Pisces are the young known to be actively fed by the parent fish. However, in all species of South American cichlids of the genus *Symphysodon* the young are fed passively with a nutrient secretion produced by the parents during the phase when they care for their young. The epidermis becomes thicker and there is a rapid increase in the number of mucus-producing goblet cells, which result in the body becoming covered in a thick layer of mucus and proliferating epidermal cells. This layer is eaten by the young fish, for which it represents an essential dietary supplement. These changes in the skin can be induced experimentally by prolactin (see p. 219). The parent fish relieve each other from feeding the young by performing a ceremonial jerk of the fins.

9.2 Amphibia

In teleosts, building the nest and caring for the young is usually the sole responsibility of the male whereas in Amphibia it is frequently the female that takes care of the eggs. In all three subclasses of amphibians it is generally the case that the eggs are attended to and that only in a few cases is parental care extended to the larvae once they have hatched out.

Among Gymnophiona that do not give birth to live young are 4 species in which the female, having laid her eggs in a moist place, has been observed to coil her body around the eggs (*Siphonops paulensis, S. annulatus, Idiocranium russeli* and *Ichthyopsis glutinosus*). This behaviour can also be observed in, for example, salamanders (Urodela) which have no lungs (Plethodontidae). These animals deposit their eggs on land in a damp spot, such as under a stone or a piece of wood. In most cases the female cares for the young by curling herself around the eggs, thereby keeping them moist with her skin. It is also possible that the cutaneous secretion of the female acts as a fungicide and a bactericide. Other species reported to take care of their young in this way include *Desmognathus fuscus, Aneides lugubris* and *Plethodon cinereus*. In the latter species the female also defends the eggs against robbers by pretending to bite the intruder. The giant aquatic salamanders of the family Cryptobranchidae similarly guard the eggs until they hatch, although in this case it is the male that performs this role. Both the male and female eat their own eggs but enough usually survive to hatch out as larvae since there are more eggs than the male is able to eat and, moreover, the female drives the male away from the clutch of eggs. Mud puppies of the genus *Necturus*, which belong to the Proteidae, also live in water which is where they lay their eggs. From time to time the female shakes her head vigorously over the eggs, thereby moving them about with her large external gills, an action which evidently improves the supply of oxygen to the eggs. A comparable behaviour that has the same effect is exhibited by the cave salamander (*Proteus anguinus*). This species also guards the eggs: the male either does this alone or together with other females with which he has spawned, if more than one. In addition, the parent animals fan fresh water over the eggs with their tails. The three-toed amphiuma (*Amphiuma tridactylum*) lay their eggs in a hollow in the bottom of muddy pools. These pools usually dry up before the larvae hatch out, in which case the female guarding the nest coils herself around the eggs so as to prevent them touching the ground.

Urodeles seldom build a nest. One of the few species that does is the long-toed salamander (*Ambystoma opacum*). In this species the female forms a simple pit on dry land in which she lies in such a way as to maintain body contact with the eggs. This contact is believed to help keep the eggs moist. One by one the female rolls the eggs over the ground, thereby covering them with earth that prevents them from sticking to each other. The circular nest area is always free of fungi, even when their spores cover the surrounding ground. This is thought to be because the cutaneous secretion produced by the female acts as a fungicide. The larvae hatch out after heavy falls of rain fill the nest with water. Drought may result in this development being delayed until the succeeding year. Members of the genus *Aneides* (salamanders lacking lungs), *Plethodon*, and species of the genus *Hemidactylina* reportedly also guard the larvae after they have hatched out.

The ways in which anurans care for their young are similarly diverse. In the simplest case only one of the parents remains with the eggs, this being the male in the case of the amphicoelid frog *Leiopelma hochstetteri*, for example. These animals dig a hole some 30 cm deep with entrances that lie above the water level. The eggs are deposited in the damp region of the entrance to the nest, partly in and partly out of the water. It has been reported that males of the species *Pyxidephalus adspersus*, a large African frog, guards his swarm of tadpoles and will even defend them against animals that drink at "his" water hole, although it is questionable whether this behaviour is in fact a specific aspect of parental care. Many anurans make their nest partly in and partly out of the water although among Anura in general there are far fewer species which nest in water than species which nest on land. A striking example of the latter is provided by certain South American species of frog, for example, the white-lipped frog *Leptodactylus ocellatus* which build floating nests of foam. During spawning the male rides on the back of the female who secretes a thick proteinaceous fluid from her cloaca that is whisked into a stiff foamy mass with the hind legs. This mass containing the fertilised eggs floats on the surface of the water by virtue of the large amount of air it contains. The two partners float in the middle of their nest of foam, leaving a hole in it when they swim off. The nest is about 30 cm acorss and 5 cm deep. When the tadpoles hatch out the foam becomes fluid and the nest dissolves.

The crater nest of the South American frog *Hyla faber*, also called the blacksmith frog because of the metallic sound of its call, represents an intermediate type between those built on land and those built in water. The nest resembles a circular atoll and is built by the male out of small balls of mud he fashions with his front feet. The nest has a diameter of about 30 cm and an enclosing embankment about 8–10 cm high (Fig. 104B). After completing the nest the male attracts a female with his distinctive call and they spawn together in the nest and subsequently depart. The form of the nest ensures that enough water remains in the crater to cover the eggs when the level of the surrounding water falls. When they hatch out the tadpoles have long, branched, feather-like gills which stick to the surface of the water. This no doubt represents an adaptation to the limited amount of oxygen contained in the small pool of water within the nest.

A range of species from the families Microhylidae, Bufonidae, Dendrobatinae and many Leptodactylidae lay their eggs in some damp place on land. A characteristic of all these species is that embryonic development is greatly accelerated with the result that metamorphosis may take place within the egg membrane. The African frog *Leptopelis karissimbensis* builds a nest in a hollow in the ground which it lines with plant material before laying its eggs in the nest (Fig. 104A). Other species, such as those of the genus *Phyllomedusa* which includes the maki frogs, lay their eggs on leaves hanging above water. The animals spawn in a previously formed "leaf-bag" whose edges are held together by the sticky eggs (see title picture to Chap. 1). In the case of two other species, *P. callidryas* and *P. danicolor*, the male calls to the female and on them coming together mounts the female and they both enter the water. The male fills his bladder and climbs onto a leaf where the female lays her eggs and he fertilises them. In doing this the male empties his bladder over the eggs causing the gelatinous egg membranes to swell, a process essential for the development of the embryos. The same procedure is repeated several times to produce a number of separate clutches of eggs.

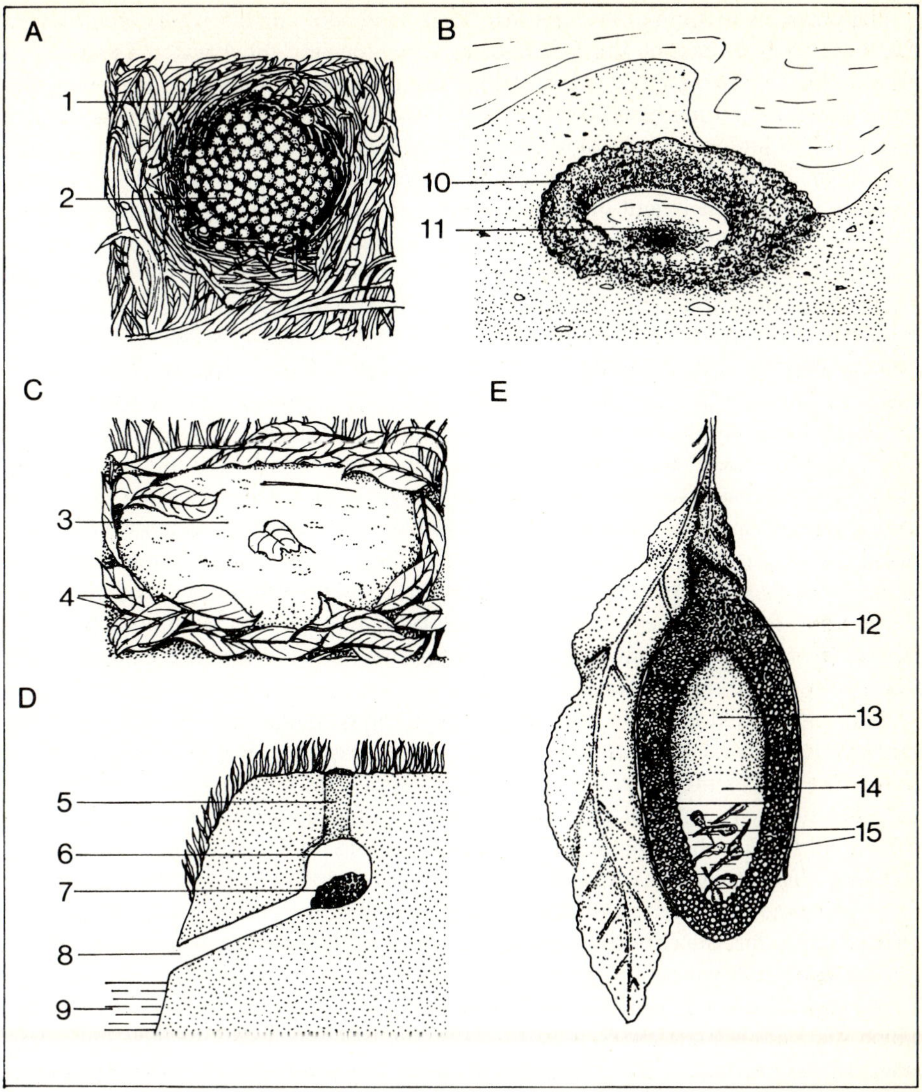

Fig. 104A–E. The nest of amphibians. **A** Terrestrial nest of the African frog *Leptopelis karissimbensis*. (After Mertens). **B** Mud nest of a tree frog, possibly related to *Hyla faber*. (After Mertens). **C** Foam nest of the giant American frog *Leptodactylus pentadactylus*. (After Mertens). **D** Diagram of the nest of *Rhacophorus schlegeli*. (After Brandes and Schoenichen). **E** Nest of leaves and foam made by the flying frog *Polypedates reinwardtii* cut open to reveal its structure. (After Siedlecki).

1 = plant material; 2 = eggs; 3 = foam; 4 = leaves; 5 = entrance to the nest (filled in); 6 = nest chamber; 7 = eggs; 8 = exit tunnel of the nest; 9 = water; 10 = surrounding embankment; 11 = water containing the eggs; 12 = capsule of foam containing degenerating eggs; 13 = cavity of the nest; 14 = fluid from the eggs; 15 = larvae

Many species of the family Leptodactylidae and some of the family Rhacophoridae form nests of foam on the ground or on leaves hanging above water. They do this in essentially the same way as other species form foam nests on water. The female releases a fluid from her cloaca which she whips into a foam with her hind legs. The eggs are then laid and at the same time another fluid is released which, after the eggs have been fertilised, is also whipped into a foam that envelops the eggs. The outer layer of foam hardens and presumably serves to protect the eggs from desiccation. After the tadpoles hatch out the inner mass of foam turns into a fluid in which the tadpoles live. Metamorphosis may take place inside the nest or in neighbouring bodies of water which the tadpoles find their way into after it has rained. The terrestrial foam nest of the giant American tropical frog *Leptodactylus pentadactylus* is illustrated in Fig. 104C. The Javanese flying frog *Polypedates reinwardtii* makes a foam nest in the leaves of a suitable plant. The female, with the male mounted on her back, climbs up the plant and together they produce a mass of foam containing the fertilised eggs in the manner described above, although in this case they both whip the foam up with their hind legs. The male then departs leaving the female to shape the foam mass into an ovoid structure with her hind legs, which she then uses to press leaves against the sticky foam so that they enclose the nest and serve as its means of suspension. As development proceeds the foam in the core of the nest turns into a fluid, forming a cavity partly filled with liquid wherein the tadpoles live until rain softens the outer foam layer of the mass and washes the young animals into a larger body of water (Fig. 104E).

Anura also build nests on the ground, most simply by digging a pit and laying the eggs within it, as do the Australian leptodactylids *Pseudophryne australis* and *P. bibroni*. The nest is then guarded by the adult frogs until rainfall fills it and the tadpoles hatch out. The nest of the Japanese aogaeru *Rhacophorus schlegeli* is more elaborate (Fig. 104D). The female, with the male mounted on her back, begins by digging a vertical shaft in the bank of a river, the bottom of which is broadened into a chamber 6–9 cm across. The female then makes the wall of the chamber smooth with her belly and also fills the entrance shaft with earth. During spawning the female evidently secretes a fluid with the eggs which she beats into a foam with her hind legs. She then digs a tunnel sloping downwards and opening close to the surface of the water which she and her mate use to leave the nest. The foam enclosing the developing eggs gradually becomes fluid and finally the tadpoles hatch out and slip down the tunnel into the water.

Some anuran species make use of prefabricated nest sites in which to spawn, such as a cavity beneath a stone or some other object. A small number of tree frogs (Hylidae) spawn in water that collects in the concavities of plants, bromeliacia funnels for example, or in an empty bee hive or within a hollow tree trunk.

In certain anuran species the parents take care of their young by carrying them about on their body. For example, the fertilised eggs of the European midwife toad (*Alytes obstreticans*) are carried about on the hind legs of the male. Spawning in this case takes place on land by the male mounting the female and beating her with his hind legs, inducing her to form a receptacle for the eggs with her hind legs. From 20–60 oocytes are laid and immediately fertilised by the male. This done he shifts forwards on the back of the female and stretches his hind legs through the mass of eggs

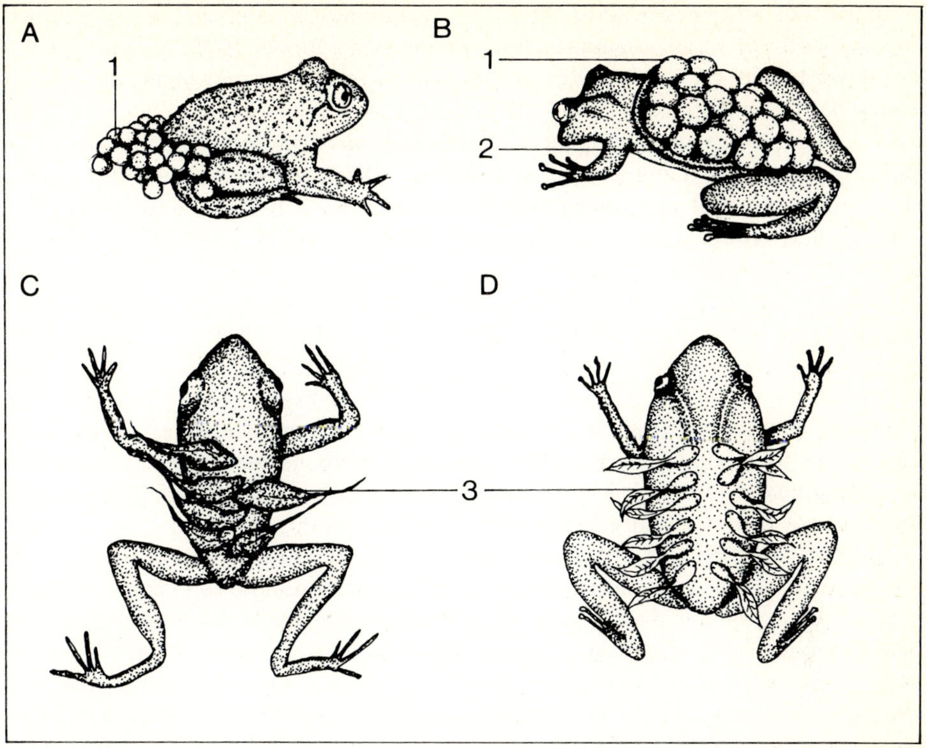

Fig. 105A–D. Amphibia which care for their young on their bodies. **A** Male midwife frog (*Alytes obstreticans*) with eggs clustered around his hind quarters. (After Mertens). **B** Female *Flectonotus goeldei* with eggs on her back. (After Mertens). **C** Male *Sooglossus seychellensis* with tadpoles on his back. (After Brauer). **D** Female *Hylodes lineatus* with tadpoles on her back. (After Wiedersheim)

so that they form a ruff around his thighs. The male then separates from the female and carries the eggs around with him until the tadpoles hatch out (Fig. 105A). During this period, which lasts about four weeks, the eggs are kept moist by the dew but occasionally the male may also take a quick swim. Finally, however, he spends a longer time in the water and the tadpoles eventually hatch from the egg membrane and swim off. Comparable phenomena occur within the Hylidae. Thus females of the genus *Fritziana* have two folds of skin on their back, one on each side, between which the eggs are deposited. Females of the genus *Cryptobatrachus, Stefania* and *Hemiphractus* have round indentations in their backs in which the eggs are seated whereas female frogs of the species *Flectonotus goeldei* have a fold of skin on the back that forms the rim of a large depression where the eggs are carried (Fig. 105B). Other species of frog carry their eggs against their abdomen, the male of the Ceylonese webbed frog *Rhacophorus reticulatus* for example. In all the species mentioned above the tadpoles become independent after they have hatched out.

Several species of the family Dendrobatidae from Central and South America carry the tadpoles on their backs. These species lay their eggs on land where they are usually guarded by the male until the tadpoles hatch out. The tadpoles then crawl onto his back and become stuck there by a secretion of the skin and are thus carried to some body of water, usually that in a bromeliacian funnel. In the case of *Hylodes lineatus* the tadpoles are thought to go through the greater part of their development on the back of their mother (Fig. 105B).

The male of the pelobatid species *Sooglossus seychellensis* from the Seychelles similarly carries the tadpoles on its back (Fig. 105C), but in this case they remain there until fully metamorphosed, which is interpreted as being an adaptation to its living in an arid environment.

9.3 Reptilia

The ways in which reptiles care for their young are not anywhere near as diverse or complex as those exhibited by teleosts and amphibians. In most species it is limited to concealment of the eggs in a favourable place and is solely the task of the female in all but one case in which the male participates in building a nest.

Chelonia lay their eggs either in a hole in the ground or in a mass of vegetable remains. All those species which build the former type of nest do so in much the same way. At first, a hole is dug which is slightly larger than the female's body. She then uses her hind legs to dig an egg chamber in the bottom of the hole into which the eggs are deposited. While laying the eggs the female urinates over them so that they remain moist when subsequently covered with earth or sand. After the egg chamber has been sealed the hole is also filled in and the surface levelled off with the shell. The female then leaves the nest site.

Many species of lizard conceal their eggs in a similar way. Some use cavities under stones, dead plants or other objects whereas others dig a nest in the ground, or build one. A sunny or a shaded spot is chosen for the nest according to the local climate. Females of the species *Varanus bengalensis*, for example, use their front legs to dig a cup-shaped chamber which, after the eggs have been laid, it carefully seals again. Around this nest the female then digs a number of "decoy" nests that do not contain any eggs but which may serve to distract potential egg robbers from the real nest. The chameleon *Chamaeleo dilepis* digs a tunnel in the ground big enough to accomodate the female's body. After the eggs have been laid the tunnel is filled in and its entrance concealed with dry grass and small twigs. As mentioned, other species of lizard merely hide their eggs. The behaviour of *Varanus niloticus* and *V. varius* is, in contrast, quite "refined" – they deposit their eggs in large termite nests. They do this in the rainy season when the otherwise very hard walls of the nest are soft enough for the animal to break through the wall easily and deposit its eggs inside. The termites then repair the breach in the wall, and the steady temperature within the nest and its continuous ventilation provide the eggs with excellent conditions in which to develop.

The only lizard to take more care of their young than just hiding the eggs somewhere are species of the North American skinks of the genus *Eumeces* and glass lizards of the genus *Ophiosaurus*. The latter coil themselves around their clutch of eggs and also defend them. A similar behaviour is shown by the skink species *Eumeces laticeps* and *E. fasciatus*. Moreover, upon returning to the nest after a short excursion the female licks and turns the eggs or occasionally takes them into her mouth and moves them to some other place. A more specialised form of parental care is exhibited by *E. obsoletus*. The female guards the eggs, occasionally licking and turning them, and when they are ready to hatch she presses against them with her head, body and legs, thereby stimulating the young within, who respond by making movements which eventually free them of the egg shell. Once they have hatched out the female continues to care for the young by guarding them, by picking them up in her mouth and setting them down in a new bedding site, and by licking them clean around the region of the cloaca.

Nothing is known about the ways in which many species of snake build their nests or care for their young: indeed, only a relatively few species are thought to build a true nest. Observations made on Indian cobras (*Naja naja*) kept in zoos constitute the only available evidence that the male participates in the care of the young. According to these observations the male, together with the female, digs a ditch in the ground with his head and body that is large enough to hold both of them. The eggs are then laid and the female guards them until the young hatch out, although when she leaves the nest, to drink for example, her part is apparently taken over by the male. The only snake that builds a nest out of plant material is the king cobra (*Ophiophagus hannah*). Again, this is known from observations on captive animals. The female uses her body to form a mound out of sand and plant material, in the top of which she makes an egg chamber where the eggs are deposited. The female then plugs the chamber with plant material and coils herself up on the top of her nest. The male of the species is not involved. He is, in fact, usually driven away from the nest by the female.

Species of the genus *Python* (*P. moratus* for example) brood their eggs after they have been laid by the female coiling her body around them several times and capping the top with her head. Snakes of this group are thermogenic and the heat produced by the body can raise the temperature of the eggs by several degrees centigrade above the ambient temperature. The female ceases to care for her young once they have hatched out.

Crocodiles are one of the few groups of reptiles that take more care of their young. The American crocodile (*Crocodylus acutus*) and the Nile crocodile (*Crocodylus niloticus*), specifically the females thereof, dig a nest some 60 cm deep in sand or earth. The nest always lies in a shady spot, near trees for example, not more than 200 m from the water. The eggs are laid in this hole, covered with sand or earth, and subsequently watched over by the female. The cover of the hole frequently dries to a crust. Prior to the young hatching out of the eggs they emit sounds which bring the mother to the nest where she removes the top layer of nest material. She continues to guard her brood after their emergence and reportedly may lead them into the water like a duck leading her ducklings.

The American alligator (*Alligator mississipiensis*), the swamp crocodile (*Crocodylus porosus*) and some species of cayman build nests out of plant material and mud. The female of the first mentioned species collects plant material together to make a mound about 1.8 m across and 90 cm high. She crawls to the top of this mound and makes a crater-like depression in it with her hind legs that she subsequently fills in again with mud and plant material. She then makes a new depression to receive the eggs, it too being subsequently filled with mud and plant material which she transports to the nest in her mouth. The nest is compacted into a cone shape by the female crawling over it and, this done, she guards the nest either remaining close by or lying on top of it. The eggs are incubated by heat produced in the nest from the fermentation of the plant material. When the young are ready to hatch out they call to their mother who responds by removing the top layer of the nest so as to clear an exit for the emerging young. The young alligators remain in still bodies of water under the protection of their mother for a year or more.

9.4 Aves

The complexity of the ways in which birds care for their young greatly exceeds that of all those described so far. Although there are species which do not exhibit any special nest-building behaviour, simply laying their eggs on the ground, this class of animals includes species that build the most complex nests of any vertebrate, Moreover, with few exceptions, birds brood their eggs: in other words, the heat necessary for the development of the young within the eggs is provided by the parent animal which, as a homoeotherm, has a constant body temperature of between 39°C and 44°C – it does this by sitting on the eggs. The transfer of heat to the eggs is facilitated by the development of an especially well vascularised region on the ventral side of the body, the brood patch, which has no feathers (see p. 181). The plumage, and in many cases the lining of the nest as well, serve to further insulate the eggs from the environment. Although the eggs may be brooded by both the male and female bird, it is usually the latter that spends most time brooding. In the case of monogamous species both partners normally brood the eggs and, accordingly, both develop a brood patch. However, even in this case the female spends a longer time sitting on the eggs than the male. In most polygamous species the eggs are brooded only by the female, the male having no brood patch. In a few species, however, it is the males that develop a brood patch and brood the eggs; for example, cassowaries (Casuaridae), rhea (Rheidae), the ostrich *Rhynchotus rufescens* and the phalaropes (Phalaropidae). In many cases the parent birds not only sit on the eggs but care for them in other ways, for example the eggs are turned with the beak or sometimes covered over with leaves before the brooding parent leaves the nest. Different species brood for anywhere between 8 days and 10 weeks. Thus songbirds (Oscines) brood for from 10 to 16 days, ducks (Anatinae) for about 3 weeks and ostriches (Struthionidae) for 8 weeks. The amount of time required before the brood hatch out is evidently correlated with the size of the bird or the size of the eggs. The males of many species participate in looking after the young, if they are not already permanently or temporarily engaged in brooding the eggs. For example, male birds of prey (Accipetres), gulls (Laridae), storks (Ciconiidae) and male swans (Cygneae) defend the nest whereas the male partner of monogamous species (e.g. the blue tit *Palus coeruleus*) feeds the brooding female.

Nearly all birds take care of their young once they have hatched in greater measure than they do their eggs. During their first weeks of life the young of altricial species need to be kept warm and therefore the brooding parent continues to warm them with its body for a least 2 weeks. In addition, most birds feed their young until they have fully fledged and, in many cases, continue to do so even after the young are able to leave the nest. This task is initially performed by the parent bird which is not occupied with brooding the hatchlings, usually the males, but later on both parents gather food for their young.

Beyond those mentioned above, birds display other patterns of behaviour which are less directly to do with parental care but which are nevertheless of importance in this respect. For example, the parent birds may distract potential predators away from the nest by performing some particularly "eye-catching" behaviour. The parents may call to their chicks (the clucking of hens for instance), lead their young (as ducks do), or hide them under their bodies.

The nest is usually built by the female, though she is frequently helped in this by the male who gathers material for the nest. In some species there is a roughly equal division of labour, as is the case with the sand martin (*Riparia riparia*) and the long-tailed titmouse (*Aegithalos caudatus*). There are also a few species in which the male builds his own nest alongside the nest containing the brood.

The nests of those few species that do not brood their eggs closely resemble those of reptiles. Thus the maleo (*Megacephalon maleo*), which lives on Celebes, either lays its eggs individually or lays them together in a ditch it shovels out of the sand and subsequently fills in again; the female does not tend the nest further. In this case the sun provides the warmth the eggs require for their development. After hatching out, the young birds are already at an advanced stage of development and immediately run from the nest and fend for themselves.

The nest of the South Australian Mallee fowl (*Leipoa ocellata*), which also belongs to the mound fowl (Megapodiidae), corresponds to that of *Alligator mississipiensis* in both its form and function. The nest is built in winter (May to August) by both the male and female, usually in a clearing surrounded by bushes. At first the birds rake the sand to make a hollow banked by a surrounding wall of sand. They then brush leaves and twigs into the hollow with their wings until a mound about 60 cm high is formed. The nest is then left in this state for 4–5 months during which time the leaves etc. become soaked through with rain and begin to decay. Shortly before the laying period the hen returns and rakes out an egg chamber in the middle of the mound that is about 50 cm across and 40–60 cm deep and has a smooth wall. The empty chamber is provided with a roof of twigs overlain with rotting vegetable matter and the whole nest is then covered with sand forming a structure up to 3 m across and about 1 m high. Every 3 days the female carefully opens the egg chamber, lays a single egg within so that its air chamber (which is at the less pointed end) is uppermost, and seals the nest again. Laying is complete after 25–30 days, by which time there are between 6 and 20 eggs deposited in several tiers. The subsequent control of incubation is undertaken by the male. At this stage the nest is effectively a large compost heap in which heat is produced by fermentation, but it is not at all easy to see how this process can be regulated so that the conditions within the nest are optimal for the development of the embryos. The de facto temperature in the egg chamber is 33 ± 1°C, although the ambient temperature varies from +44°C during the day to −8°C at night. This constancy is the result of the male modifying the nest and the fact that he can sense the absolute temperature of the nest, which he does by periodically pushing his beak into the nest: at the same time he checks on the condition of the eggs. According to whether the nest temperature is too low or too high, material is either removed or added to the nest so as to correct the temperature imbalance. Furthermore, the nest is covered with deadwood when it rains. Because of its ability to determine the absolute temperature of the nest, *Leipoa ocellata* is also called the thermometer bird. The parent birds take no further care of their young once they have hatched.

In the case of the brush or wattle turkey (*Alectura lathami*), which is closely related to *L. ocellata*, the male rakes together a heap of plant material about 1.5 m high in which the female lays her eggs one by one. As in the case of *Leipoa*, the nest is subsequently guarded and its temperature regulated, but the male looks after the young for about 3 days after they hatch out by helping them out of the egg chamber, to which he returns them every evening. These birds use the same nest for a number of years. Sometimes several males build a communal nest which has a separate egg chamber for each female and may reach a diameter of 20 m and a height of 3 m – without doubt the largest nest built by any bird.

Species that brood their young in a nest occur within a wide range of latitudes. Some of these species make use of heat provided by the sun. The Egyptian plover (*Aegyptius pluvialis*), for example, buries its eggs in the warm sand and only broods them during the night.

In some cases making a nest is simply a process of laying the eggs in an existing hollow in the ground, or one the bird grubs out with its body. Simplest of all, the eggs may be laid directly on the flat earth. Such primitive nests are sometimes surrounded or meagerly furnished with stones, feathers or pieces of plant material: they are characteristic of many species of gull (Laridae), tern (Sternidae), the pewit (*Vanellus vanellus*) and some gallinaceous birds (Rasores). The goatsucker (*Caprimulgus europaeus*) simply lays its eggs on the ground.

Other species of bird deposit their eggs in some existing hollow or cavity, be it natural or man-made. The tropic birds (*Phaeton* sp.) lay a single egg in a rock crevice whereas the shell-duck (*Bucephala clangula*) nests in a tree or a hollow in the ground.

The aptly named miner birds dig their own underground nests: examples include the diving petrels (Pelecanoidae), the sand martin (*Riparia riparia*), the penguin *Spheniscus magellanicus*, the puffin (*Fratercula arcica*), the European kingfisher (*Alcedo atthis*) and the European bee-eater (*Merops apiaster*). This type of nest is usually built on steep slopes, the banks of rivers, lakes etc. by the bird digging a tunnel that rises, falls or extends horizontally to a more expansive nesting cavity (Fig. 106A). This cavity is sometimes furnished with pieces of vegetation or feathers, for example, to cushion the eggs. In the case of the bee-eater the bedding for the eggs is made up of fragments of chitin regurgitated by the parent birds. Birds that construct this type of nest include those which make their nest in trees, albeit they usually do this by expanding an existing cavity within the tree. In most cases the inside wall is made smooth and the bird fashions its own entrance to the nest, as do the European woodpeckers (Picidae).

Seen as a series of nests of increasingly more complex structure which therefore demand more of the bird in their construction, the next step is represented by those which are pits or hollows in the ground that are provided with some form of lining. The nest of the lark *Otocoris alpestris praticola* is one such. It is laid down in a hol-

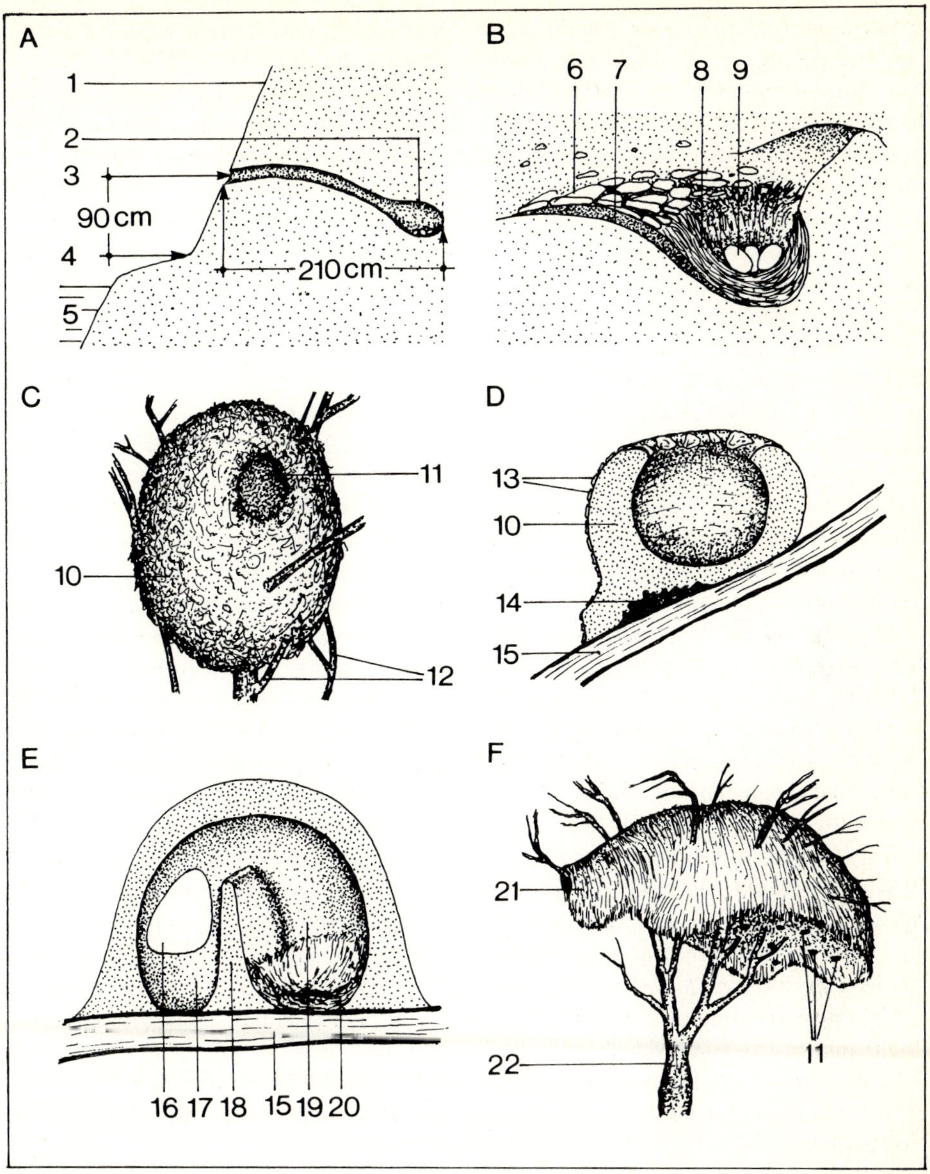

Fig. 106A–F. Birds' nests. **A** Underground nest of the kingfisher *Ceryle torquata*. (After Young). **B** Longitudinal section through the nest of *Otocoris alpestris praticola*. (After Pickwell). **C** Nest of *Aegithalos caudatus*. (After Smolik). **D** Section through the nest of *Archilochus colubris*. (After Herrick). **E** Section through the nest of *Furnarius rufus*. **F** Communal nest of *Philetarius socius*. (After Smolik).

1 = steep slope; 2 = nesting chamber; 3 = entrance tunnel; 4 = river bank; 5 = water; 6 = stone pavement; 7 = excavated earth; 8 = plant material lining the nest; 9 = eggs; 10 = substance of the nest; 11 = entrance to the nest; 12 = twigs; 13 = camouflage material; 14 = saliva; 15 = branch; 16 = nest entrance; 17 = entrance chamber; 18 = partition; 19 = brooding chamber; 20 = nest lining of feathers and cotton wool; 21 = communal nest; 22 = tree

low to one side of a small mound and furnished with a thick layer of plant material and a stone pavement on the side opposite the adjacent mound (Fig. 106B).

A large number of birds make their nests in trees or bushes. The simplest form of nest consists of a platform with a shallow depression in it built at a fork in the supporting branch out of small branches and twigs. Ring doves (*Columba palumbus*), the turtle dove (*Streptopelia tutur*) as well as storks (Ciconiidae), herons (Ardeidae) and eagles (genus *Aguila*) construct such nests. Some birds of prey build nests of a similar type but of considerable depth and, in the case of the magpie (*Pica pica*), the nest is even provided with a roof of twigs. Other species improve upon this basic type of nest by lining it with thin or small pieces of plant material, fine root fibrils or animal hair (e.g. the bullfinch *Pyrrhule pyrrhula*). The great crested grebe (*Podiceps cristatus*) builds a floating nest of this type.

Nests woven out of dry plant material typically have a relatively deep well in them which may be lined or sealed with additional material. Crows and ravens (Corvidae) and many birds of prey (Accipetres) construct such nests. The blackbird (*Turdus merula*) seals the inner surface of the nest with moist earth or mud whereas the song-thrush (*Turdus philomelos*) may also use cow dung. Nests which are truly woven are also made of materials such as thin blades of grass or animal hairs woven into the fabric of the nest. Such nests may be open or closed. The goldfinch (*Carduelis carduelis*) and chaffinch (*Fringilla coelebs*) build an open nest of this type. The latter species uses mainly the woolly hairs of animals woven together with less flexible hairs to form the bulk of the nest. The rim of the nest often has feathers woven into it whereas the outer surface is camouflaged with moss, twine and small pieces of bark. The long-tailed titmouse (*Aegithalos caudatus*), on the other hand, constructs a nest which is closed, but for an entrance in its side, and lined mostly with the feathers of other birds (Fig. 106C). The category of closed nests includes the sack-like nest of the penduline titmouse (*Remiz pendulinus*) and the often structurally complex hanging nests of the true weaver birds (Ploceidae). In general woven nests are held together by twigs or blades of grass as, for example, is the nest of the golden oriole (*Oriolus oriolus*) and the reed warbler (*Acrocephalus scirpaceus*). Hanging nests are commonly ballasted with earth or stones to make them more stable.

Nests woven out of extremely fine materials such as animal hairs, fine pieces of plant material or spiders' webs are referred to as felt nests: many species of humming bird build nests of this type. The outer surface is frequently camouflaged with moss, twine and pieces of bark, among other things, and the nest as a whole is affixed to the branches supporting it by a sticky type of saliva (Fig. 106D).

Other species of bird use loam, clay or earth in the construction of their nests. A well-known example is the bowl-shaped nest of the house martin (*Delichon urbica*) which is built of earth and saliva on the walls of houses and which possesses a slit-shaped opening. The nest of the swallow *Hirundo rustica* resembles a fixture for holding soap. It is built on the roof of a building and is made of clay, earth, fragments of wood and straw and saliva. The similarly shaped nests of the swiftlets (*Collocalia fuciphage*) are composed only of saliva and are considered a delicacy in south-east Asia when made into soup. The related species of Javanese tree swift (*Dendrochelidon clecho*) sticks twine and small pieces of bark to thin branches with saliva to form a bowl-shaped nest 3–4 cm across and about 1 cm deep with paper-thin walls which holds only a single egg. The walls are so thin that the bird must sit on a branch while brooding its egg. In relation to the size of the bird's body (it is about 18 cm long), this is one of the smallest of all bird's nests. Flamingos (Phoenicopteridae) build a nest of mud shaped like a truncated cone some 40 cm high, on which the bird sits in a shallow depression. The earthen nest of the South American oven bird *Furnarius rufus* has a considerably more complicated form. Both the male and female fashion balls of moist clay into a dome-shaped structure which rests on a horizontal branch. It is partitioned by a dividing wall into two rooms, a large brood chamber and an entrance chamber accessed through a roughly semicircular hole (Fig. 106E). The brood chamber is furnished with cotton-wool and feathers.

Clay or earth is also used by species that build enclosed nests in trees in order to reduce the size of the entrance hole. For example, the nuthatch (*Sitta europaea*) uses small lumps of clay and saliva to this end. The hornbills (Bucerotidae) exhibit a more extreme example of this habit. Again, the entrance to the nest is reduced in size by building up the wall with various materials but then the female enters the nest and the remaining opening is closed (in part with the bird's droppings) except for a thin slit. Depending on the species, this is done by either the male, both the male and female, or mainly the female. During the entire period of brooding the incarcerated female is provided with food by the male, with the result that she frequently starves and dies shortly after brooding is over.

The last type of nest to be considered here is the sewn nest made by the Indian tailor bird *Orthotomus sutorius*. This bird seeks out a large leaf and several thread-like pieces of plant material which it covers with its sticky saliva. It then bores holes with its beak in one edge of the leaf, through which it pulls about two-thirds of the length of the threads. The hardening of the saliva secures the threads to the leaf. The other edge of the leaf is then sewn to the first in the same manner, forming a bag in which the actual nest is built out of finely textured material.

Some birds use their bodies as a nest, as do certain other animals already mentioned. Thus the female emperor penguin (*Aptenodytes forsteri*) lays a single egg which she covers with a feathered fold of skin, the so-called brood fold: the egg does not lie on the ground but on her feet. The male also has a brood fold and only a few hours after the egg has been laid he takes over the brooding of the egg in a handing-over ceremony. While the female returns to the sea the male broods the egg under his brood fold for a total of 63 days, during which time he does not feed. Shortly after the young penguin hatches out the female returns and takes it into her brood fold and the male takes to the sea. King penguins (*Aptenodytes patagonicus*) brood in the same way except that the male and female regularly change places.

Many species of bird brood in colonies, which means that there is a congregation of nests in one place and that the individual members of the colony brood at about the same time. The social weaver (*Philetarius socius*), in contrast, builds a so-called colonial nest under a common roof. Each pair of birds build their own nest with its own entrance but the nests lie so close to one another that they fuse to become a single mass. Every year new nests are built beneath the old ones so that the roof steadily grows in thickness. The nests are built of grass (Fig. 106F). Up to a thousand pairs of birds may brood in one such colonial nest.

A number of species of bird are brood parasites. Thus, other than a few tropical species that nest in constructions built by insects (whether occupied or not), some birds which normally build a nest themselves nevertheless tend to lay their eggs in the nest of another species of bird. Sparrows, for example, sometimes use swallows' nests and certain birds of prey occasionally occupy the nests of Corvidae. These, and a range of other examples, have in common the fact that one species of bird usurps the nest of another in which to raise its young.

The South American black-headed duck *Heteronetta atricapilla*, on the other hand, is a true brood parasite which lays its eggs in the nests of other species of duck and even in that of herons and birds of prey. It is doubtful, however, that the young reach the stage of hatching out in all these cases. The rosy-billed pochard *Metopiana peposaca* is evidently the specific host of the black-headed duck.

The definitive exponents of brood parasitism among birds are the cuckoos (Cuculidae). This family includes some two hundred species of which eighty are brood parasites. One can conveniently reconstruct the probable evolution of this life-style as a series of steps towards full parasitism. Thus not all females of the common ani (*Crotophaga ani*) build their own nests, rather, those that do not brood in communal nests which, however, are very disordered in their arrangement. Such nests are occupied by up to six females but there is frequently insufficient room for one or the other of the birds to lay all of her eggs, in which case the remaining eggs are either laid in a neighbouring communal nest or in the nest of another species of cuckoo. Having done this she returns to brood the eggs in her "own" nest. The guira cuckoo (*Guira guira*) also builds communal nests but these are so loosely constructed that the eggs frequently fall through. With this restriction on laying sites, the females often fly about as if in panic to find somewhere to lay their eggs, finally setting them down on branches or the ground but also in the nests of other birds. The female yellow-billed cuckoo (*Coccyzus americanus*) builds its own nest, but sometimes uses the unoccupied nests of other birds, occasionally laying its eggs in the occupied nest of a mocking bird (*Mimus polyglottus*). The European cuckoo (*Cuculus canorus*) and many other species of cuckoo do not build a nest but usurp that of another, host species, usually a species of songbird much smaller than themselves. The young females are thought to seek out the same host species as the one that raised them. The female cuckoo then watches the chosen nest until she can move in unobserved (often some time in the afternoon) to lay her own egg within. Alternatively, she may manouvre the egg into the nest with her beak. In most cases one of the eggs of the host bird is removed from the nest. A single female can parasitise about 20 nests in a period of between 35 and 46 days. The eggs of the cuckoo are similar in shape and colour to those of the host species, possibly because the female has previously been fed with food gathered by that species, the same species in whose nest she preferentially chooses to lay her own eggs. It is nevertheless not uncommon that the host notices the interpolation of a foreign egg and either removes it or leaves the nest. If this does not happen and the egg is brooded along with the host's own eggs, then it is the young cuckoo which hatches out first and after about 10 h begins to throw all the other eggs out of the nest, a characteristic and innate behaviour pattern of the cuckoo, until it alone remains in the nest.

Comparable behaviour is exhibited by orioles (Icteridae) and honey guides (Indicatoridae). All the species and families of brood parasites mentioned above have a common type of mating, namely, polyandry. This type of reproduction is thought to be linked in some way with the evolution of brood parasitism. On the other hand, some weaver birds (Ploceidae) are also brood parasites, although they are evidently polygynous.

It will be recalled that most birds which feed their young have beak markings on the inside of the upper half of their beaks. As a rule, the parents either feed their young directly with food they have gathered or with partially digested food. However, the food the young receive is usually not the same as that which the adult birds eat. For example, the young of species which are purely herbivorous are commonly fed insects, worms, snails etc. at first and only later is their diet supplemented with plant material. In many cases predigested food is regurgitated from the stomach or crop and passed to the young with the beak, as do finches when they change over from feeding their young insects to feeding them grain. Numerous species of parrot feed their young predigested nuts and grains, whereas woodpeckers and swallows provide their young with insects. Herons and storks offer their young predigested fish and amphibians. Species of gull disgorge food onto the ground in front of their brood. Common to all these cases is the fact that the food is gathered by the parent birds and also predigested by them in certain cases.

Pigeons (Gyrantes) of both sexes produce a secretion from their crop, so-called crop milk (see p. 181), with which the young are fed for 9 or 10 days. The milk is a pasty substance produced by the proliferation of the crop epithelium under the influence of prolactin.

9.5 Mammalia

In all mammals, except egg-laying monotremata, the period of brooding the eggs is replaced by gestation – birth marking the beginning of the phase during which the parent animals actively care for their offspring. The pre-eminent characteristic of this phase is that, without exception, the young are nourished for a certain length of time with milk produced by the mother's mammary glands. The building of a nest is of less significance in the case of mammals for although many species build often complicated underground works, as does the mole *Talpa europaea* for example, these are not nests in the sense of the definition employed here that a nest is a place where the young are taken care of while they develop. Such constructions as are fabricated by mammals are better regarded as living quarters of one kind or another which also serve as a place where the young are normally born and raised. Many species, such as mice and rats, build a simple "nest" out of some suitable material but even this does not comply fully with the definition used here and is better referred to as a storage nest. The large fabrications of beavers (Castoridae) and the muskrat (*Ondatra zibethica*) are also not built solely for the purpose of raising the young, although this is one purpose to which they are put. Tree squirrels (Sciuridae), which include the European squirrel *Sciurus vulgaris*, build a storage nest of fairly complex structure for

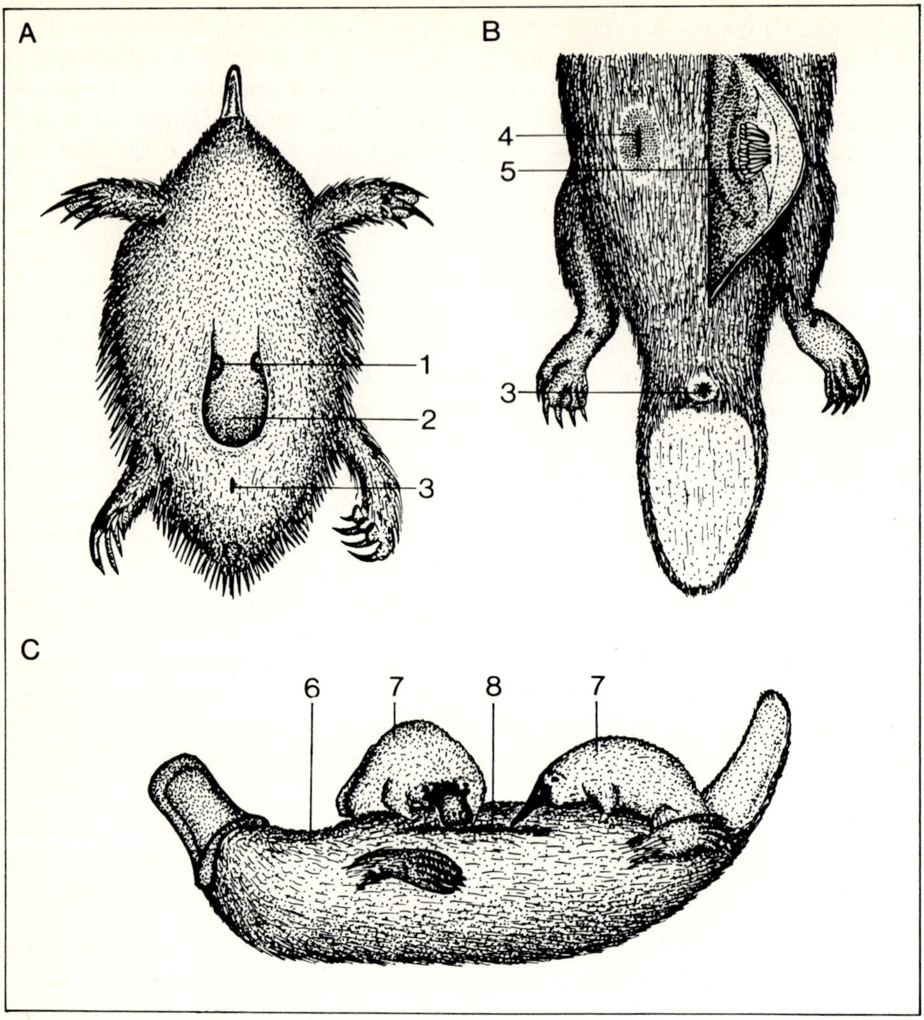

Fig. 107. A Ventral view of a female *Echidna aculeata*. (After Haacke). Ventral view of a female *Ornithorhynchus paradoxus* (**B**) with the left side of the body opened to show the milk glands. (After Klaatsch). **C** Young duckbill platypuses (*O. paradoxus*) suckling. (After Hartig).
1 = milk fields (areolae); 2 = marsupium; 3 = cloaca; 4 = naked region of the milk field; 5 = milk glands; 6 = the mother animal; 7 = the young animal; 8 = abdominal groove

their young sables and next to this a sleeping nest and play nest. Many mammals and in particular those able to stand or walk immediately after being born (e.g. horses) do not make any sort of nest or establish any permanent or semipermanent living quarter. On the other hand, the pouch of marsupials serves as a nest for their young. The different types of behaviour associated with the ways in which mammals care for their offspring, over and above those associated with feeding the young, are highly specialised and complex and range from preening to the guidance or instruction of the young (see Chap. 6).

The monotreme *Ornithorhynchus paradoxus*, commonly known as the duck-billed platypus, normally lives in pairs in an underground burrow. However, prior to laying her eggs the female digs a separate brood hole and furnishes it with wet leaves and grass. She usually lays two eggs that she broods for about 2 weeks. When the young hatch out they are held in a sort of pouch formed by the female holding her flat tail up under her abdomen. The young hatch out at a stage of development corresponding to that of a foetus having fairly well-developed fore legs but very poorly developed hind legs. Olfaction is very probably the only sensory modality of the hatchling that is functional at this stage. Instead of mammary glands with teats, monotremes have two milk fields, (or areolae) located ventrally on either side of the mid-line in the abdominal region. Each is a naked area of skin where the underlying milk glands open individually to the surface (Fig. 107B). The young do not suckle but must work the milk out with their mouths (beaks). In the case of *Ornithorhynchus* the mother lays on her back and by contracting the abdominal musculature forms a groove medially into which milk flows from the areolae and from which the milk is sipped up by the young animals (Fig. 107C). Platypuses tend their young for about 20 weeks.

Spiny anteaters (Echidnidae) do not build nests. Rather, the female possesses a marsupium in the abdominal region formed by a fold of skin (Fig. 107A) into which she lays her two eggs directly by tucking her tail between her legs. The eggs are secured in this position by a secretion from the cloaca. After the young have hatched out they lick up milk released from the areolae, which in these animals are inside the marsupium. The young remain for about 10 days in the marsupium and are then set down in a shallow depression in the ground, evidently because by this time their spines have begun to grow, where they continue to be provided with milk for several weeks.

Most marsupials possess a marsupium, as their name denotes, formed by a fold of skin (Fig. 108A) as is the case in monotremes, although it is very probably not homologous to that of Echidnidae. The marsupium is a muscular organ that encloses the mammary glands. The pelvic girdle of both marsupials and monotremes has a pair of so-called marsupial bones (prepubis) which were previously believed to provide support for the pouch (and its contents), but this view is not endorsed by current opinion. The marsupium has a very different structure in different species of marsupials. Primitive species, such as the little water opossum (*Lutcolina crassicaudata*), have no pouch at all. The young of this species hang by their mouths from their mother's teats, their bodies nestling in the hair covering the abdomen. They subsequently climb onto the mother's back, anchoring themselves there with their prehensile tails, and are thus carried about by the mother. The marsupium of the opossum *Philander marsupialis* consists simply of two thin, longitudinal folds of skin adjacent to the teats. The two species just mentioned belong to the family Didelphidae (opossums) that includes the North American opossum *Didelphis virginiana* which has a well-developed pouch provided with 13 teats. Moreover, the opening of the pouch can be closed by a sphincter muscle or directed forwards, backwards and downwards to some extent. The marsupium of kangaroos (macropodidae) opens cranially whereas that of the Tasmanian devil (*Sarcophilus harrisii*), the koala (*Phascolarctos cinereus*) and the bandicoot (*Perameles bougainvillei*), among others, opens caudally.

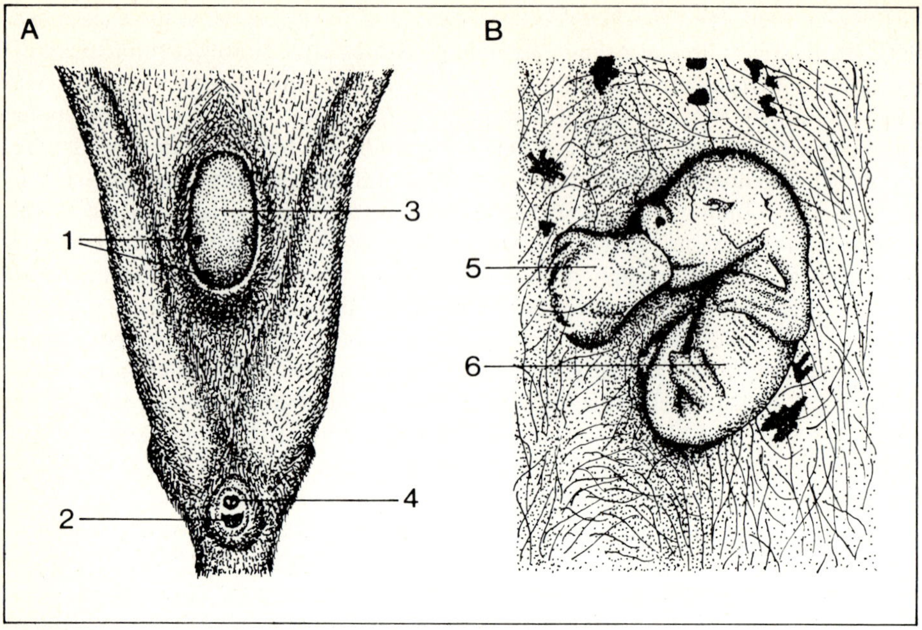

Fig. 108. A Abdomen of a female marsupial (*Thylacinus cynocephalus*) with marsupium. (After Carlsson). **B** Young kangaroo (*Megaleia rufa*) on the teat. (After Griffiths).
1 = teats; 2 = anus; 3 = marsupium; 4 = orifice of the sinus urogenitalis; 5 = teat; 6 = young animal

In those marsupials which do not have a pouch the teats of the mammary glands are variable in number and distributed over the entire abdomen whereas in those with a marsupium the teats are only found within the confines of the pouch. The number of teats possessed by species of the latter group is also variable, the significance of this being that it limits the number of young which can be suckled at any one time. A few species give birth to more young than they have teats, with the consequence that those for which there is no teat die.

Marsupials are born at a foetal stage of developement, as are monotremes, and may weigh as little as 5 mg. The young nevertheless have well-developed forelegs with which they crawl into the marsupium. In some cases the mother lays down a trail of saliva from the orifice of her vagina to the pouch which the young follows and, if it strays off course, the mother helps it on its way with her mouth and paws. Once inside the marsupium the young animal locates a teat and grasps it in its mouth, whereupon the teat swells and provisionally secures the animal to its mother. Subsequently, the lips of the new-born grow together at the edges of the mouth, thereby providing a stronger attachment (Fig. 108B). The young animal cannot suckle, however. Rather, the milk is ejected from the mammary gland by the same mechanism as that described in a previous chapter (see p. 202). The animal breaths through its large nostrils and one might suppose that it would easily choke since it cannot influence the flow of milk. However, this problem does not arise because the larynx is displaced into the choanae so that air passes through the trachea in front of the oesophagus,

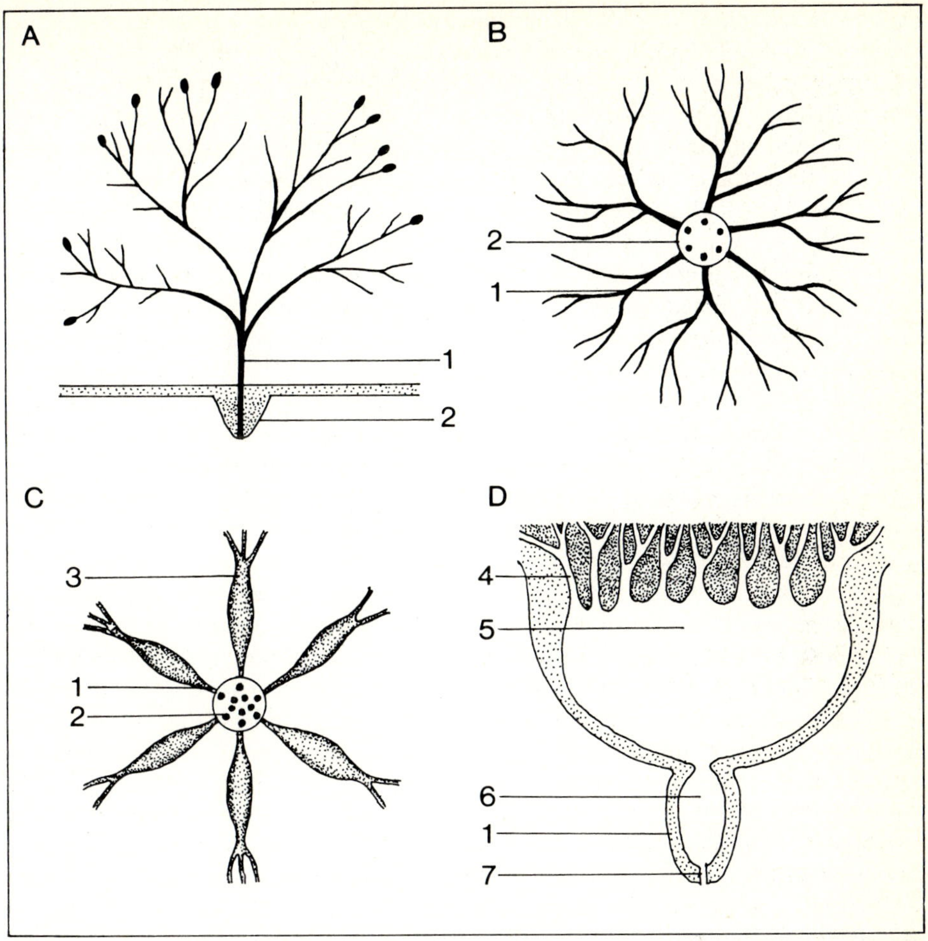

Fig. 109A–D. Diagrams showing the configuration of milk ducts in the mammary glands of different mammals. **A** Rat. **B** Rabbit. **C** Man. **D** Ruminant. (After Cowie).
1 = galactophore; 2 = teat; 3 = milk sinus; 4 = main duct; 5 = gland cistern; 6 = teat cistern; 7 = orifice of the teat cistern

thereby permitting uninterrupted breathing. After a certain length of time, which varies from species to species, the young animal frees itself from the teat, for only short periods at first but eventually for good. The larynx subsequently falls back into its normal position.

The young of Eutheria are born at a considerably more advanced stage of development than the young of marsupials, to the extent that in a few cases (e.g. the guinea-pig *Cavia cobaya*) they have no need of the mother's milk if sufficient other food is available. In most cases, however, milk is an essential foodstuff, although the length of time the young suckle varies from species to species. As in marsupials, the number of young is related to the number of mammary glands the animal possesses.

The microscopic anatomy of the mammary glands and salient facts concerning the composition of milk have been presented in Chap. 5 (see p. 199). The glands which actually produce the milk are laid down during embryonic development as so-called milk ridges that run ventrolaterally from the armpit to the groin. Symmetrical pairs of milk glands subsequently develop along the length of the ridges, each gland opening to the surface in a teat via a number of ducts. More precisely, each duct may have a separate opening which together form a nipple or the ducts may unit into a single, common collecting duct that comes to the surface as a teat. Mammary glands are also present in males but are usually not functional. With regard to the number of mammary glands, the guinea-pig, sheep, goats and man have but a single pair whereas pigs have 7–9 pairs. In some species the glands are fused to form an udder, as for example in sheep, goats and in the cow, which has 2 pairs of mammary glands. The position of the glands is variable: in bats, elephants, apes and man they are situated in the thoracic region, in whales in the abdominal region, in ruminants they are situated in the region of the groin whereas in pigs, mice and rabbits they extend over the entire ventral aspect of the body. Those of the coypu or nutria *Myocastor coypus* are located laterally on either side of the back. The form of the mammary glands in also variable. They may protrude or hang from the body, as in man and those species with an udder for example, or they may have a low profile, as is the case in rats and mice. As mentioned above, the ducts of the mammary gland may unite to form a common duct, the galactophore (Fig. 109A), or converge to form several galactophores, each of which has a separate orifice (Fig. 109B). A variation of the latter arrangement involves the expansion of the galactophores into sinuses wherein the milk is stored (Fig. 109C). In the case of the udder of ruminants the large milk ducts merge into a common gland cistern which is continuous with a further chamber, the so-called teat cistern, from which the milk is expelled (Fig. 109D).

It has already been pointed out that the composition of the milk differs from one species of mammal to the next (see p. 200) and also varies with the nutritional condition of the animal as well as during the course of lactation. For example, colostrum has a very different composition to the mother's normal milk. Colostrum is produced a few days before and after the time of birth and in most cases contains more protein, sodium and chloride than the normal milk but less potassium and lactose. Both colostrum and normal milk are not only a source of nutrients; colostrum especially is an important source of antibodies (see p. 299). This is of particular significance in those mammals in which the placental barrier is relatively thick (e.g. marsupials and ungulates). This more or less prevents any antibodies from reaching the foetus, which is incapable of producing them itself. Although the antibodies are not actually absorbed by the gut of the new-born animal in most cases, they are nevertheless presumed to protect against infection of the gut. The foetus is afforded the same protection in those species in which maternal immunoglobulins are transported to the foetus via the placenta. The immunoglobulins in colostrum and normal milk are produced by plasma cells in the mammary glands. Colostrum and milk also contain other bactericides, such as lysozyme, lactoferrin and the lactoperoxidase system.

References

General

Johnson MH, Everett BL (1984) Essential reproduction, 2nd edn. Blackwell Oxford

Meisenheimer J (1921) Geschlecht und Geschlechter im Tierreiche, vol 1. Fischer, Jena

Nalbandov AV (1976) Reproductive physiology of mammals and birds. Freeman, San Francisco

Parkes AS (ed) (1956) Marshall's physiology of reproduction, 4 vols. Longman and Green, London

Parkes AS (1976) Patterns of sexuality and reproduction. Oxford University Press, London

Rowlands JW, Perry JS (eds) (1969) Comparative biology of reproduction in mammals. J Reprod Fertil Suppl 6

van Tienhoven A (1968) Reproductive physiology of vertebrates. Saunders, Philadelphia

References to Chapter 1

Books

Armstrong CN, Marshal AF (eds) (1964) Intersexuality in vertebrates including man. Academic Press, London

Bresch C, Hausmann R (1972) Klassische und Molekulare Genetik, 3rd edn. Springer, Berlin Heidelberg New York

Bullough WS (1951) Vertebrate sexual cycles. Wiley, New York

Enders AC (ed) (1963) Delayed implantation. University of Chicago Press, Chicago

Lack D (1968) Ecological adaptations for breeding in birds. Methuen, London

McKeown BA (1983) Fish migration. Groom and Helm, Beckenham

Money J, Ehrhardt AA (1975) Männlich-Weiblich. Die Entstehung der Geschlechtsunterschiede. Rowohlt, Hamburg

Ortavant R, Pelletier J, Ravault JP (eds) (1981) Photoperiodism and reproduction in vertebrates. INRA Colloquium 6, INRA, Paris

Reinboth R (ed) 1974) Symposium on intersexuality in the animal kingdom. Springer, Berlin Heidelberg New York

Sadleir RMFS (1969) The ecology of reproduction in wild and domestic mammals. Methuen, London

Reviews

Amoroso ES, Marshall FHA (1960) External factors in sexual periodicity. In: Parkes AS (ed) Marshall's physiology of reproduction, vol I (Pt 2). Longmans, London, pp 707–831

Baker JR (1938) The relation between latitude and breeding season in birds. Proc Zool Soc Lond 108: 557

Baker JR (1947) The season in a tropical rain forest. Part B: Lizards (*Emoia*). J Linn Soc Lond Zool 41: 243–247

Baker JR, Marshall AJ, Harrison TH (1940) The seasons in a tropical rain forest (New Hebrides). Part 5: birds (*Pachycephalus*). J Linn Soc Lond Zool 41: 50–70

Ballinger RE (1978) Variation in and evolution of clutch and litter size. In: Jones RE (ed) The vertebrate ovary. Plenum, New York, pp 789–826
Beatty RA (1967) Parthenogenesis in vertebrates. In: Metz A, Monroy A (eds) Fertilization. Academic Press, New York, pp 413–440
Beatty RA (1970) The genetics of the mammalian gamete. Biol Rev 45: 73–119
Benoit J (1950) Reproduction – charactères sexueles et hormones determinisme du cycle sexuel saisonnier. In: Grassé PP (ed) Traité de zoologie, T. XV. Masson, Paris, pp 384–478
Bertin L (1958) Sexualité et fécondation. In: Grassé PP (ed) Traité de zoologie, T. XIII, F. II. Masson, Paris, pp 1791–1812
Brown JA Jr (1972) The "clocks" timing biological rhythms. Am Sci 60: 756–766
Chan STH, Yeung WSB (1983) Sex control and sex reversal in fish under natural conditions. In: Hoar WS, Randall DJ, Donaldson EM (eds) Fish physiology, vol 9B. Academic Press, New York, pp 171–222
Clarke JR (1981) Physiological problems of seasonal breeding in eutherian mammals. In: Finn CA (ed) Oxford reviews of reproductive biology. Clarendon, Oxford, pp 244–312
Crump MI (1974) Reproductive strategies in a tropical anuran community. University of Kansas Museum of Natural History. Misc Publ No 61
Cuellar O (1966) Oviductal anatomy and sperm storage structures in lizards. J Morphol 119: 7–20
Daniel JC Jr (1970) Dormant embryos of mammals. Bioscience 20: 411–415
Davis DE (1955) Breeding biology of birds. In: Wolfson A (ed) Recent studies in avian biology. University of Illinois Press, Urbana, pp 264–308
Duvall D, Guilletta LJ, Jones RE (1982) Environmental control of reptilian reproductive cycles. In: Gans C (ed) Biology of reptilia, vol 13. Academic Press, London, pp 201–232
Emlen ST, Oring LW (1977) Ecology, sexual selections, and the evolution of mating systems. Science 197: 215–223
Farner DS (1964) The photoperiodic control of reproductive cycles in birds. Am Sci 52: 137–156
Farner DS, Follett BK (1979) Reproductive periodicity in birds. In: Barrington EJW (ed) Hormones and evolution. Academic Press, London, pp 829–872
Fitch H (1970) Reproductive cycles of lizards and snakes. University of Kansas Museum of Natural History, Misc Publ 52, 1–247
Follett BK, Davies T (1975) Photoperiodicity and the neuroendocrine control of reproduction in birds. In: Peakes M (ed) Avian physiology. Symp Zool Soc Lond, vol 35. Academic Press, London, pp 199–224
Fontaine M (1954) Du déterminisme physiologique des migrations. Biol Rev 29: 390–418
Fox W (1956) Seminal receptacles in snakes. Anat Rec 124: 519–539
Frisch RE (1978) Population, food intake and fertility. Science 199: 22–30
Frisch RE (1982) Malnutrition and fertility. Science 215: 1272–1273
Gauthreaux SA (1982) The ecology and evolution of avian migration systems. In: Farner DS, King JR, Parkes KC (eds) Avian biology, vol 6. Academic Press, New York, pp 93–168
Gibson RN (1978) Lunar and tidal rhythms in fish. In: Thorpe J (ed) Rhythmic activity of fishes. Academic Press, New York
Gorman GC, Licht P (1974) Seasonality in ovarian cycles among tropical lizards. Ecology 55: 360–369
Grassé PP (1970) La parthénogenése. In: Grassé PP (ed) Traité de zoologie, T. XIV, F. III. Masson, Paris, pp 972–978
Griffin DR (1970) Migrations and homing of bats. In: Winsatt WA (ed) Biology of bats, vol 1. Academic Press, New York, pp 233–264
Guibé J (1970) La reproduction. In: Grassé PP (ed) Traité de zoologie, T. XIV, F. III. Masson, Paris, pp. 859–892
Gwinner E (1977) Cirannual rhythms in bird migration. Annu Rev Ecol Syst 8: 381–405
Hildemann WH, Walford RL (1963) Annual fishes – promising species as biological control agents. J Trop Med Hyg 66: 163–166
Howarth B Jr (1974) Sperm storage: as a function of the female reproductive tract. In: Johnson AD, Foley CW (eds) The oviduct and its functions. Academic Press, New York, pp 237–270

Hurlbert SH (1970) The post-larval migration of the red-spotted newt, *Notophthalmus viridescens* (Rafinesque). Copeia 515–528

Immelmann K (1971) Ecological aspects of periodic reproduction. In: Farner DS, King JR, Parkes KC (eds) Avian biology, vol 1. Academic Press, New York, pp 342–392

Jenni DA (1974) The evolution of polyandry in birds. Am Zool 14: 129–144

Jones RE (1978) Ovarian cycles in nonmammalian vertebrates. In: Jones RE (ed) The vertebrate ovary. Plenum, New York, pp 731–762

Jørgensen CB (1973a) Mechanisms regulating ovarian function in amphibians (toads). In: Peters H (ed) The development and maturation of the ovary and its function. Excerpta Medica, Amsterdam, pp 133–151

Karsch FJ (1980) Seasonal reproduction: a saga of reversible fertility. Physiologist 23: 29–38

Keast JA, Marshall AJ (1954) The influence of drought and rainfall on reproduction in Australian desert birds. Proc Zool Soc Lond 124: 494–499

Kleiman DG (1977) Monogamy in mammals. Q Rev Biol 52: 39–69

Lam TJ (1983) Environmental influences on gonadal activity fish. In: Hoar WS, Randall DJ, Donaldson EM (eds) Fish physiology, vol 96. Academic Press, New York, pp 65–116

Leggett WC (1977) The ecology of fish migrations. Annu Rev Ecol Syst 8: 285 308

Licht P (1966) Reproduction in lizards: influence of temperature on photoperiodism in testicular recrudescence. Science 154: 1668–1670

Licht P (1972) Environmental physiology of reptilian breeding cycles: role of temperature. Gen Comp Endocrinol Suppl 3: 447–487

Lincoln GA, Short RV (1980) Seasonal breeding: nature's contraceptive. Recent Prog Horm Res 36: 1–52

Lofts B (1975) Environmental control of reproduction. In: Peaker M (ed) Avian physiology. Symp Zool Soc Lond, vol 35. Academic Press, London, pp 177–198

Lofts B, Murton RK (1968) Photoperiodic and physiological adaptations regulating avian breeding cycles and their ecological significance. J Zool Soc London 155: 327–394

Marshall AJ (1960) Reproduction. In: Marshall AJ (ed) Biol Comp Physiol Birds, vol 2, Academic Press, New York, pp 169–214

Marshall AJ (1960) Breeding seasons and migration. In: Marshall AJ (ed) Biol Comp Physiol Birds, vol 2. Academic Press, New York, pp 307–340

Marshall FAH (1956) The breeding season. In: Parkes AS (ed) Marshall's physiology of reproduction, vol 1. Longman and Green, London, pp 1–44

Maslin TP (1971) Parthenogenesis in reptiles. Am Zool 11: 361–380

Mathey R (1969) Les chromosomes et l'évolution des mammiferes. In: Grassé PP (ed) Traité de zoologie, T. XVI, F. VI. Masson, Paris, pp 858–911

Mayaud N (1950) Biologie de la reproduction. In: Grassé PP (ed) Traité de zoologie, T. XV. Masson, Paris, pp 539–653

Meas GW, Bertlesen E, Cohen DM (1964) Reproduction among deep sea fishes. Deep-Sea Res 11: 569–596

Moreau RE (1950) The breeding season in African birds. I. Land birds. Ibis 92: 223–267

Murton RK (1975) Ecological adaption in avian reproductive physiology. In: Peaker M (ed) Avian physiology. Symp Zool Soc Lond, vol 35. Academic Press, London, pp 149–176

Negus NC (1972) Environmental factors and reproductive processes in mammalian populations. In: Verlardo NT, Kasprow BA (eds) Biology of reproduction, basic and clinical studies. 3rd Pan American Congress of Anatomy, New Orleans

Orians GH (1969) On the evolution of mating systems in birds and mammals. Am Nat 103: 589–603

Oring LE (1982) Avian mating systems. In: Farner DS, King Jr, Parkes KC (eds) Avian biology, vol 6. Academic Press, New York, pp 1–92

Reinboth R (1970) Intersexuality in fishes. Mem Soc Endocrinol 18: 515–541

Rowan W (1958) Light and seaonal reproduction in animals. Biol Rev Camb Philos Soc 13: 374–402

Salthe SN, Duellman WE (1973) Quantitative constraints associated with reproductive mode in anurans. In: Evolutionary biology of the anurans. University of Missouri Press, Columbia, pp 229–249

Short RV (1979) Sex determination and differentiation. Br Med Bull 35: 121–127

Taylor MH, Leach GH, DiMichele L (1979) Lunar spawning cycle in the mummichog, *Fundulus heteroclitus* (Pisces: Cyprinodontidae). Copeia 291–297

Tienhoven A van, Planck RJ (1973) The effect of light on avian reproductive activity. In: Handbook of physiology, sect 7. Endocrinology, female reproductive tract (pt 1), vol 2. American Physiological Society, Washington, pp 79–107

Uzzell T (1970) Meiotic mechanisms of naturally occurring unisexual vertebrates. Am Nat 109: 433–445

Vlaming VL de (1974) Environmental and endocrine control of teleost reproduction. In: Schreck CB (ed) Control of sex in fishes. Sea Grant Extension Division, Virginia Polytechnic Institute and State University, Blacksburg, VA, pp 13–83

Witschi E (1961) Sex and secondary sexual characters. In: Marshall AJ (ed) Biology and comparative physiology of birds, vol 2. Academic Press, New York, pp 115–168

Yamamoto T (1969) Sex differentiation. In: Hoar WS, Randall DJ (eds) Fish physiology, vol 3. Academic Press, New York, pp 117–177

References to Chapter 2

Books

Burger HG, Kresler DM de (1981) The testis. Raven, New York

Jones RE (ed) (1978) The vertebrate ovary. Comparative biology and evolution. Plenum, New York

Peters H, McNatty (1980) The ovary. Granada, London

Starck D (1982) Vergleichende Anatomie der Wirbeltiere auf evolutionsbiologischer Grundlage, vol 3. Springer, Berlin Heidelberg New York

Reviews

Benoit J (1950) Organes urogénitales. In: Grassé PP (ed) Traité de zoologie, T. XV. Masson, Paris, pp 341–378

Bremer JL (1916) The interrelations of the mesonephros, kidney and placenta in different classes of mammals. Am J Anat 19: 179–209

Broek HJP van den (1933) Gonaden und Ausführgänge. In: Bolk L, Göppert E, Kallius E, Lubosch W (eds) Handbuch der vergleichenden Anatomie der Wirbeltiere, vol 6. Urban und Schwarzenberg, Berlin, pp 1–154

Broek HJP van den, Van Oordt GJ, Hirsch GC (1938) Urogenitalsystem. In: Bolk L, Göppert E, Kallius E, Lubosch W (eds) Handbuch der vergleichenden Anatomie der Wirbeltiere, vol 5. Urban und Schwarzenberg, Berlin, pp 683–854

Gerhardt U (1933) Kloake und Begattungsorgane. In: Bolk L, Göppert E, Kallius E, Lubosch W (eds) Handbuch der vergleichenden Anatomie der Wirbeltiere, vol 6. Urban und Schwarzenberg, Berlin, pp 267–349

Guibé J (1970) L'appareil urogénital. In: Grassé PP (ed) Traité de zoologie, T. XIV, F. III. Masson, Paris, pp 801–829

Guraya SS (1978) Maturation of the follicular wall in nonmammalian vertebrates. In: Jones RE (ed) The vertebrate ovary. Plenum, New York, pp 261–330

Hardisty MW (1971) Gonadogenesis, sex differentiation and gametogenesis. In: Hardisty MW, Potter IC (eds) The biology of lampreys, vol 1. Academic Press, New York, pp 295–359

Hardisty MW (1978) Primordial germ cells and the vertebrate germ line. In: Jones RE (ed) The vertebrate ovary. Plenum, New York

Haseltine FP, Ohno S (1981) Mechansims of gonadal differentiation. Science 211: 1272–1278

Jones RE (1978) Evolution of the vertebrate ovary: an overview. In: Jones RE (ed) The vertebrate ovary. Plenum, New York, pp 827–840

Lehmann R (1968) Die Nieren der Wirbeltiere. Naturwiss Rundsch 21: 139–144
Lombard-des-Gouttes MN (1982) Dévelopment de l'appareil urogénital. In: Grassé PP (ed) Traité de zoologie, T. XVI, F. VII. Masson, Paris, pp 993–1090
Merchant-Larios (1978) Ovarian differentiation. In: Jones RE (ed) The vertebrate ovary. Plenum, New York, pp 47–82
Nagahama Y (1983) The functional morphology of teleost gonads. In: Hoar WS, Randall DJ, Donaldson EM (eds) Fish physiology, vol 9A. Academic Press, New York, pp 223–276
Ottow B (1955) Biologische Anatomie der Genitalorgane und der Fortpflanzung der Säugetiere. Fischer, Jena
Peters H (1970) Migration of gonocytes into the mammalian gonad and their differentiation. Philos Trans R Soc Lond B Biol Sci 259: 91–101
Prasad MRN (1974) Männliche Geschlechtsorgane, Säugetiere. In: Helmcke JG, Starck D, Wermuth H (eds) Handbuch Zoologie, vol 8, Lfg 51. De Gruyter, Berlin, pp 1–50
Strauss F (1966) Weibliche Geschlechtsorgane (Säugetiere). In: Helmcke JG, Starck P, Wermuth H (eds) Handbuch Zoologie, vol 8, Lfg 36, Lfg 40. De Gruyter, Berlin 1964/1966, pp 1–96, 97–202
Raynaud A (1969) Les organes génitales des mammifères. In Grassé PP (ed) Traité de zoologie, T. XVI, F. VI. Masson, Paris, pp 152–636
Wachtel SS, Koo GC, Boyse EA (1975) Evolutionary conservation of H-Y (male) antigen. Nature (Lond) 254: 270–272
Wachtel SS, Ohno S, Koo GC, Boyse EA (1975) Possible role for H-Y antigen in the primary determination of sex. Nature (Lond) 257: 235–236
Witschi E (1967) Biochemistry of sex differentiation in vertebrate embryos. In: Weber R (ed) The biochemistry of animal development, vol 2. Academic Press, New York, pp 193–225

Original Works

Boyd MMM (1940) The structure of the ovary and the formation of the corpus luteum in *Hoplodactylus maculatus* Gray. Q J Microsc Sci 82: 337–376
Fischel A (1930) Über die Entwicklung der Keimdrüsen der Menschen. Z Anat Entwicklungsgesch 92: 34–72
Hardisty MW (1965a) Sex differentiation and gonadogenesis in lampreys. I. The ammocoete gonads of the brook lamprey, *Lampetra planeri*. J Zool (Lond) 146: 305–345
Hardisty MW (1965b) Sex differentiation and gonadogenesis in lampreys. II. The ammocoete gonads of the landlocked sea lamprey, *Petromyzon marinus*. J Zool (Lond) 146: 346–387
Swift CH (1915) Origin of the definitive sex-cells in the female chick and their relation to the primordial germ-cells. Am J Anat 18: 441–470
Saidapur SK (1978) Follicular atresia in the ovaries of nonmammalian vertebrates. Int Rev Cytol 54: 225–244
Tribe M, Brambell FWR (1932) The origin and migration of the primordial germ-cells of *Sphenodon punctatus*. Q J Microsc Sci 75: 251–282

References to Chapter 3

Books

Afzelius BA (ed) (1975) The functional anatomy of the spermatozoon. Pergamon, Oxford
Austin CR, Short RV (ed) (1972) Reproduction in mammals. Book 1. Germ cells and fertilization, 2nd edn. Cambridge University Press, Cambridge
Bacetti B (1970) Comparative spermatology. Academic Press, New York
Bacetti B, Afzelius BA (1976) The biology of the sperm cell. Karger, Basel
Holstein AF, Horstmann E, Schirren C (1970) Morphological aspects of andrology. Grosse, Berlin

Holstein AF, Rosen-Runge EC (1981) Atlas of human spermatogenesis. Grosse, Berlin
Kunz W, Schäfer U (1978) Oogenese und spermatogenese. Fischer, Stuttgart
Raven CP (1961) Oogenesis: the storage of developmental information. Pergamon, Oxford
Rosen-Runge EC (1977) The process of spermatogenesis in animals. Cambridge University Press, Cambridge

Reviews

Anderson E (1974) Comparative aspects of the ultrastructure of the female gamete. In: Bourne GH, Danieli JF (eds) Int Rev Cytol Suppl 4. Academic Press, New York, pp 1–70
Baker TG, O WS (1976) Development of the ovary and oogenesis. Clin Obstet Gynecol 3: 3–26
Beatty RA (1970) The genetics of the mammalian gamete. Biol Rev 45: 73–119
Billard R, Jalabert B, Breton B (1972) Les cellules de sertoli des poissons teléosteens. I. Etude ultrastructurale. Chem Biol Anim Biochem Biophys 12: 19–32
Erickson RP (1973) Haploid gene expression versus meiotic drive: the relevance of intercellular bridges during spermatogenesis. Nature New Biol 243: 210–211
Fawcett DW (1975) Ultrastructure and function of the sertoli cell. In: Hamilton DW, Greep RO (ed) Handbook of physiology, vol 5, sect 7. Male reproductive system. American Physiological Society, Washington DC, pp 21–55
Follett BK, Redshaw MR (1974) The physiology of vitellogenesis. In: Lofts B (ed) Physiology of the amphibia, vol 2. Academic Press, New York, pp 219–298
Franchi LL, Mandl AM, Zuckerman S (1962) The development of the ovary and the process of oogenesis. In: Zuckerman S (ed) The ovary, vol 1. Academic Press, New York, pp 1–88
Franklin LE (1970) Fertilization and the role of the acrosomal region in non-mammals. Biol Reprod Suppl 2: 159–176
Franzen A (1970) Phylogenetic aspects of the morphology of spermatozoa and spermiogenesis. In: Baccetti B (ed) Comparative spermatology. Academic Press, New York, pp 29–46
Goetz FW (1983) Hormonal control of oocyte final maturation and ovulation in fishes. In: Hoar WS, Randall DJ, Donaldson EM (eds) Fish physiology, vol 9B. Academic Press, New York, pp 117–170
Gondos B (1978) Oogonia and oocytes in mammals. In: Jones RE (ed) The vertebrate ovary. Plenum, New York, pp 83–120
Gorbman A (1983) Reproduction in cyclostome fishes and its regulation. In: Hoar WS, Randall DJ, Donaldson EM (ed) Fish physiology, vol 9A. Academic Press, New York, pp 1–30
Grier HE (1981) Cellular organization of the testis and spermatogenesis in fishes. Am Zool 21: 345–357
Jones RE (1978) Control of follicular selection. In: Jones RE (ed) The vertebrate ovary. Plenum, New York, pp 763–788
Lake PE (1975) Gamete production and the fertile period with particular reference to domesticated birds. In: Peaker M (ed) Avian physiology. Symp Zool Soc Lond, vol 35. Academic Press, London, pp 225–244
Lofts B (1968) Patterns of testicular activity. In: Barrington EJW, Barker–Jørgensen C (eds) Perspectives in endocrinology. Academic Press, London, pp 239–304
Ng BT, Idler DR (1983) Yolk formation and differentiation in teleost fishes. In: Hoar WS, Randall DJ, Donaldson EM (eds) Fish physiology. Academic Press, New York, pp 373–404
Nørrevang A (1968) Electron microscopic morphology of oogenesis. Int Rev Cytol 23: 113–186
Simkiss K, Taylor TG (1971) Shell formation. In: Bell DJ, Freeman BM (eds) Physiology and biochemistry of the domestic fowl, vol 55. Academic Press, New York
Smith LD (1975) Molecular events during oocyte maturation. In: Weber R (ed) Biochemistry of animal development, vol 3. Academic Press, San Fransisco, pp 1–46
Stoss J (1983) Fish gamete preservation and spermatozoan physiology. In: Hoar WS, Randall DJ, Donaldson EM (eds) Fish physiology, vol 9B. Academic Press, New York, pp 305–350
Thibault G (1969a) La spermatogenése chez les mammiféres. In: Grassé PP (ed) Traité de zoologie, T. XVI, F. VI. Masson, Paris, pp 800–858

Tokarz RR (1978) Oogonial proliferation, oogenesis, and folliculogenesis in nonmammalian vertebrates. In: Jones RE (ed) The vertebrate ovary. Plenum, New York, pp 145–180

Tsafriri A (1978) Oocyte maturation in mammals. In: Jones RE (ed) The vertebrate ovary. Plenum, New York, pp 409–442

Wahli W, Dawid IB, Ryffel GU, Weber R (1981) Vitellogenesis and the vitellogenin gene family. Science 212: 298–304

Wallace RA, Bergink EW (1974) Amphibian vitellogenin: properties, hormonal regulation of hepatic synthesis and ovarian uptake, and conversion to yolk proteins. Am Zool 14: 1159–1175

Wartenberg H (1974) Spermatogenese-Oogenese: ein cyto-morphologischer Vergleich. Verh Anat Ges 68: 63–92

Wasserman WJ, Smith DL (1978) Oocyte maturation in nonmammalian vertebrates. In: Jones RE (ed) The vertebrate ovary. Plenum, New York, pp 443–468

Wischnitzer S (1976) The lampbrush chromosomes: their morphology and physiological importance. Endeavour (Oxf) 124: 27–31

Original Works

Anand Kumar TC (1974) Oogenesis in adult prosimian primates. Contrib Primatol 3: 82–96

Guraya SS (1962) The structure and function of the so-called yolk-nucleus in the oogenesis of birds. Q J Microsc Sci 103: 411–415

Guraya SS (1963a) Histochemical studies on the yolk nucleus in the oogenesis of Indian reptiles. Anat Rec 146: 17–21

Jones RE, Fitzgerald KI, Duvall D (1978) Quantitative analysis of the ovarian cycle of the lizard *Lepidodactylus lugubris*. Gen Comp Endocrinol 35: 70–76

Marza VD, Marza EV (1936) The formation of the hen's egg. Parts I–IV. Q J Microscr Sci 78: 133–189

Munson JP (1904) Researches on the oogenesis of the tortoise *Clemmys marmorata*. Am J Anat 3: 311–347

Wartenberg H (1962) Elektronenmikroskopische und histochemische Studien über die Oogenese der Amphibieneizelle. Z Zellforsch Mikrosk Anat 58: 427–486

Zanandrea G (1957) Neoteny in a lamprey. Nature (Lond) 179: 925–926

References to Chapter 4

Books

Austin CR (1965) Fertilization. Prentice Hall, Englewood Cliffs, New York

Austin CR (1968) The ultrastructure of fertilization. Holt, New York

Austin CR, Short RV (1972) Reproduction in mammals. I. Germ cells and fertilization. Cambridge University Press, Cambridge

Austin CR, Short RV (ed) 1972) Reproduction in mammals, book 2. Embryonic and fetal development, 2nd edn. Cambridge University Press, Cambridge

Balinsky B (1981) An introduction to embryology, 5th edn. Saunders, Philadelphia

Berril NJ (1971) Developmental biology. McGraw-Hill, New York

Goodrick ES (1958) Studies on the structure and development of vertebrates. Dover, New York

Hadek R (1969) Mammalian fertilisation. An atlas of ultrastructure. Academic Press, New York

Hamilton NC, Boyd JP, Mossman HW (1947) Human embryology. Heffer, Cambridge

Kerr JG (1919) Textbook of embryology. II. Vertebrata with the exception of mammalia. Macmillan, London

Lillie FR (1919) Problems of fertilization. University of Chicago Press, Chicago

Metz CB, Monroy A (1967) Fertilization. Academic Press, New York

Mogishi KS, Hatez ESE (1972) Biology of mammalian fertilisation and implantation. Thomas Springfield

Monroy A (1965) Chemistry and physiology of fertilization. Holt, Rinehart and Winston, New York
Needham J (1966) Biochemistry and morphogenesis. Cambridge University Press, London
Nelsen OE (1953) Comparative embryology of the vertebrates. Blakiston, New York
Pflugfelder O (1962) Lehrbuch der Entwicklungsgeschichte und Entwicklungsphysiologie der Tiere. Fischer, Jena
Sannders JW (1970) Patterns and principles of animal development. The Macmillan biology series. In: Giles NH, Torrey JG (eds) Macmillan, London
Siewing R (1969) Lehrbuch der vergleichenden Entwicklungsgeschichte der Tiere. Parey, Hamburg
Starck D (1959) Ontogenie und Entwicklungsphysiologie der Säugetiere. De Gruyter, Berlin
Starck D (1975) Embryologie, 3rd edn. Thieme, Stuttgart
Torrey TW (1971) Morphogenesis of the vertebrates, 3rd edn. Wiley, New York
Tyler A, Borstel RC von, Metz CB (1957) The beginnings of the embryonic development. American Association for the Advancement of Science, Washington
Weber R (ed) (1965) The biochemistry of animal development. Academic Press, New York

Reviews

Bellairs R (1960) Development of birds. In: Marshall AJ (ed) Biol Comp Physiol Birds, vol 1. Academic Press, New York, pp 127–189
Blaxter JHS (1969) Development: eggs and larvae. In: Hoar WS, Randall DJ (eds) Fish physiology, vol 3. Academic Press, New York, pp 178–252
Boyd JD, Hamilton WJ (1956) Cleavage, early development and implantation of the egg. In: Parkes AD (ed) Marshall's physiology of reproduction, vol 2. Longsmans and Green, London, pp 1–26
Flynn TT, Hill JP (1939) The development of the monotremata. IV. Growth of the ovarian ovum, maturation, fertilization and early cleavage. Proc Zool Soc Lond 24: 495–582
Gomot L, Lucarz-Bietry A (1982) Dévelopment embryonaire des marsupiales. In: Grassé PP (ed) Traité de zoologie, T. XVI, F. VII. Masson, Paris, pp 34–83
Gurdon JB (1977) Egg cytoplasm and gene control in development. Proc R Soc Lond B Biol Sci 198: 211–247
Mulncard JG, Pasteels JJ (1982) Le dévelopment des mammifères euthériens du stade indivis a la jeune-neurula. In: Grassé PP (ed) Traité de zoologie, T. XVI, F. VII. Masson, Paris, pp 84–164
Oppenheimer J (1970) Cells and organizers. Am Zool 10: 45–88
Pasteels J (1950) Dévelopment embryonaire. In: Grassé PP (ed) Traité de zoologie, T. XV. Masson, Paris, pp 479–520
Pasteels J (1958) Dévelopment embryonaire. In: Grassé PP (ed) Traité de zoologie, T. XIII, F. II. Masson, Paris, pp 1685–1754
Pasteels JJ (1970) Dévelopment embryonaire. In: Grassé PP (ed) Traité de zoologie, T. XIV, F. III. Masson, Paris, pp 893–971
Thibault C (1969) La fécondation chez les mammifères. In: Grassé PP (ed) Traité de zoologie, T. XVI, F. VI. Masson, Paris, pp 912–966
Wourms JP (1977) Reproduction and development in chondrichthyan fishes. Am Zool 17: 379–410
Yamamoto T (1961) Physiology of fertilization in fish eggs, Int Rev Cytol 12: 361–405

References to Chapter 5

Books

Austin CR, Short RV (eds) (1984) Reproduction in mammals, vol 3. Hormonal control of reproduction, 2nd edn. Cambridge University Press, Cambridge

Bentley PJ (1982) Comparative vertebrate endocrinology, 2nd edn. Cambridge University Press, Cambridge
Cowie AT, Forsyth IA, Hart IC (1980) Hormonal control of lactation. Springer, Berlin Heidelberg New York
Ensor DM (1978) Comparative endocrinology of prolactin. Chapman and Hall, London
Griffith M (1978) The biologiy of the Monotremes. Academic Press, New York
Gorbman A, Bern HA (1962) A textbook of comparative endocrinology. Wiley, New York
Idler DR (ed) (1972) Steroids in nonmammalian vertebrates. Academic Press, New York
Norman RL (ed) (1983) Neuroendocrine aspects of reproduction. Academic Press, New York
Reinboth R (1980) Vergleichende Endokrinologie. Thieme, Stuttgart
Schmidt GH (1971) Biology of lactation. Freeman, San Fransisco
Turner CD (1966) General endocrinology. Saunders, Philadelphia
Vandenberg FG (ed) (1983) Pheromones and reproduction in mammals. Academic Press, New York
Wolstenhome GEW, Knight J (eds) (1970) The pineal gland. Livingstone, London
Wurtman RJ, Axelrod J, Kelly DE (1968) The pineal. Academic Press, New York
Yen SSC, Jaffe RB (ed) (1978) Reproductive endocrinology. Saunders, Philadelphia

Reviews

Bahl OP (1973) Chemistry of human chorionic gonadotropin. In: Li CH (ed) Hormonal proteins and peptides, vol 1. Academic Press, New York, pp 171–199
Behrman HR (1979) Prostaglandins in hypothalamo-pituitary and ovarian function. Annu Rev Physiol 41: 685–700
Bjersing L (1978) Maturation, morphology and endocrine function of the follicular wall in mammals. In: Jones RE (ed) The vertebrate ovary. Plenum, New York, pp 181–214
Blüm V (1977) Prolaktin: Phylogenetische Aspekte. Gynäkologe 10: 51–61
Callard IP, Chan SWC, Potts MA (1972a) The control of the reptilian gonad. Am Zool 12: 273–287
Callard IP, Doolittle J, Banks WL, Chan SWC (1972b) Recent studies on the control of the reptilian ovarian cycle. Gen Comp Endocrinol Suppl 3: 65–75
Chieffi G (1967) The reproductive system of elasmobranchs: developmental and endocrinological aspects. In: Gilbert PW, Mathewson F, Rall DP (eds) Sharks, skates and rays. Johns Hopkins University Press, Baltimore, pp 553–580
Davies IJ, Ryan KJ (1971) Comparative endocrinology of gestation. Vitam Horm 30: 223–279
Dodd JM (1972) Ovarian control in cyclostomes and elasmobranchs. Am Zool 12: 325–339
Dodd JM (1975) The hormones of sex and reproduction and their effects in fish and lower chordates: twenty years old. In: Barrington EJW (ed) Trends in comparative endocrinology. Am Zool Suppl 1 15: 137–171
Dodd JM, Evennett PJ, Goddard CK (1960) Reproductive endocrinology in cyclostomes and elasmobranchs. Symp Zool Soc Lond 1: 77–103
Donaldson EM (1973) Reproductive endocrinology of fishes. Am Zool 13: 909–927
Dorner G (1979) Hormones and sexual differentiation of the brain. In: Sex, hormones and behaviour. Ciba Found Symp 62. Excerpta Medica, Amsterdam, pp 81–101
Doty RL (1976) Reproductive endocrine influences upon human nasal chemoreception: a review. In: Doty RL (ed) Mammalian olfaction, reproductive processes and behavior. Academic Press, New York, pp 295–321
Follett BK, Davies DT (1979) The endocrine control of ovulation in birds. In: Hawk HW (ed) Barc Symp 3. Animal reproduction. Halsted, New York, pp 323–344
Follett BK, Robinson JE (1980) Photoperiod and gonadotropin secretion in birds. Prog Reprod Biol 5: 39–51
Fostier A, Halabert B, Billard R, Breton B, Zohar Y (1983) The gonadal steroids. In: Hoar WS, Randall DJ, Donaldson EM (eds) Fish physiology, vol 9A. Academic Press, New York, pp 277–372

Gallien I (1959) Endocrine basis for reproductive adaptations in amphibia. In: Gorbman A (ed) Comparative endocrinology. Wiley, New York, pp 479–487

Goldman BD (1983) The physiology of melatonin in mammals. Pineal Res Rev 1: 145–182

Goodman RL, Karsch FJ (1981) The hypothalamic pulse generator: a key determinant of reproductive cycles in sheep. In: Follett BK, Follett DF, Wright J (eds) Biological clocks and seasonal reproductive cycles. Colston Papers No 32. Bristol, 223 p

Green JD (1951) The comparative anatomy of the hypophysis with special references to its blood supply and innervation. Am J Anat 88: 225–312

Hearn JP (1977) Pituitary function in marsupial reproduction. In: Stonehouse B, Gilmore D (eds) The biology of masurpials. Macmillan, London, pp 337–344

Idler DR, Ng TB (1983) Teleost gonadotropins: biochemistry and function. In: Hoar WS, Randall DJ, Donaldson EM (eds) Fish physiology, vol 9A. Academic Press, London, pp 187–222

Josso N, Picard JY, Tran D (1977) The antimullerian hormone. Recent Prog Horm Res 33: 117–163

Jost A (1979) Basic sexual trends in the development of vertebrates. In: Sex hormones and behaviour. Ciba Found Symp 62. Excerpta Medica, Amsterdam, pp 5–18

Knobil E (1980) The neuroendocrine control of the menstrual cycle. Recent Prog Horm Res 36: 53–88

Lance V, Callard JP (1978) Hormonal control of ovarian steroidogenesis in nonmammalian vertebrates. In: Jones RE (ed) The vertebrate ovary. Plenum, New York

Larsen LO (1978) Hormonal endocrinology. Elsevier North-Holland, Amsterdam, pp 105–108

Licht P (1979) Reproductive endocrinology of reptiles and amphibians: gonadotropins. Annu Rev Physiol 41: 337–351

Licht P (1983) Evolutionary divergence in the structure and function of pituitary gonadotropins of tetrapod vertebrates. Am Zool 23: 675–683

Licht P, Papkoff H, Farmer SW, Muller CH, Tsui HW, Crews D (1977) Evolution of gonadotropin structure and function. Recent Prog Horm Res 33: 169–248

MacLusky NJ, Naftolin F (1981) Sexual differentiation of the central nervous system. Science 211: 1294–1303

McNatty KP (1978) Follicular fluid. In: Jones RE (ed) The vertebrate ovary. Plenum, New York, pp 215–260

Means AR, Fakunding JL, Hukkins C, Tindall DJ, Vitale R (1976) Follicle-stimulating hormone, the Sertoli cell and spermatogenesis. Recent Prog Horm Res 32: 477–527

Naaktgeboren C, Slijper EJ (1970) Biologie der Geburt. Parey, Hamburg

Nathanielsz PW (1978) Endocrine mechanisms of parturition. Annu Rev Physiol 40: 411–445

Neumann F, Elger W, Steinbeck H, Gräf KJ (1975) The role of androgen in sexual differentiation of mammals. In: Reinboth R (ed) Intersexuality in the animal Kingdom. Springer, Berlin Heidelberg New York, pp 407–421

Nicoll CS (1980) Ontogeny and evolution of prolactin's function. Fed Proc 39: 2563–2566

Oksche A (1978) Evolution, differentiation and organization of hypothalamic systems controlling reproduction. In: Scott DE et al. (eds) Brain-endocrine interaction. III. Neural hormones and reproduction. Karger, Basel

Ozon R (1972) Androgens in fishes, amphibians, reptiles and birds. In: Idler DR (ed) Steroids in nonmammalian vertebrates. Academic Press, New York, pp 329–389

Perry JS (1971) The ovarian cycle of mammals. University reviews in biology, No 13. Oliver and Boyd, Edinburgh

Peter RE (1983) Evolution of neurohormonal regulation of reproduction in lower vertebrates. Am Zool 23: 685–695

Peter RE (1983) The brain and neurohormones in teleost reproduction. In: Hoar WS, Randall DJ, Donaldson EM (eds) Fish physiology, vol 9A. Academic Press, New York, pp 97–163

Peter RE, Crim LW (1979) Reproductive endocrinology of fishes: gonadal cycles and gonadotropin in teleosts. Annu Rev Physiol 41: 323–335

Peters H (1978) Folliculogenesis in mammals. In: Jones RE (ed) The vertebrate ovary. Plenum, New York, pp 121–144

Pfeiffer W (1974) Pheromones in fish and amphibians. In: Birch MC (ed) Pheromones. Frontiers of biology, vol 32. North-Holland, Amsterdam, pp 269–296

Porter DG (1979) Relaxin: old hormone, new prospect. In: Finn CA (ed) Oxford reviews of reproductive biology, vol 1. Clarendon, Oxford, pp 1–57
Ralph CL (1975) The pineal complex: a retrospective view. Am Zool Suppl 1 15: 105–116
Redshaw MR (1972) The hormonal control of the amphibian ovary. Am Zool 12: 289–306
Reinboth R (1970) Intersexuality in fishes. In: Benson GK, Phillips JG (eds) Hormones and the environment. Mem Soc Endocrinol 18: 515–543
Reinboth R (1972) Hormonal control of the teleost ovary. Am Zool 12: 307–324
Reiter RJ (1980) The pineal and its hormones in the control of reproduction in mammals. Endocr Rev 1: 109–131
Reiter RJ (1980) The pineal gland: a regulator of regulators. Prog Psychobiol Physiol Psychol 9: 323–355
Richards JS (1978) Hormonal control of follicular growth and maturation in mammals. In: Jones RE (ed) The vertebrate ovary. Plenum, New York, pp 331–560
Schuetz AW (1974) Role of hormones in oocyte maturation. Biol Reprod 10: 150–178
Sharman GB (1980) Reproductive physiology of marsupials. Science 167: 1221–1228
Sharpe RM (1982) The hormonal regulation of the leydig cell. In: Finn CA (ed) Oxford reviews of reproductive biology, vol 4. Clarendon, Oxford, pp 241–317
Steinberger E (1971) Hormonal control of mammalian spermatogenesis. Physiol Rev 51: 22
Toran-Allerand CD (1978) Gonadal hormones and brain development: cellular aspects of sexual differentiation. Am Zool 18: 553–565
Tyndale-Biscoe CH, Evans SM (1980) Pituitary-ovarian interactions in marsupials. In: Schmidt-Nielsen K, Bolis L, Taylor CR (eds) Comparative physiology: primitive mammals. Cambridge University Press, Cambridge, pp 259–268
Wilson JD, George FW, Griffin JE (1981) The hormonal control of sexual development. Science 211: 1278–1284
Wingstrand K (1966) Comparative anatomy and evolution of the hypophysis. In: Harris GW, Donovan BT (eds) The pituitary gland, vol 1. University of California Press, Berkeley, pp 58–146
Yoshinaga K (1978) Cyclic hormone secretion by the mammalian ovary. In: Jones RE (ed) The vertebrate ovary. Plenum, New York, pp 691–730

References to Chapter 6

Books

Balthazart J, Pröve E, Giles R (1983) Hormones and behaviour in higher vertebrates. Springer, Berlin Heidelberg New York
Bastock M (1967) Courtship: a zoological study. Heinemann, London
Broadbent DE (1961) Behaviour. Eyre and Spottiswoode, London
Brown JL (1975) The evolution of behavior. Norton, New York
Dethier VG, Stellar E (1970) Animal behavior, 3rd edn. Prentice-Hall, Englewood Cliffs, New York
Deutsch JA (1960) The structural basis of behavior. Cambridge University Press, Cambridge
Dimond SJ (1970) The social behaviour of animals. Batsford, London
Ehrman L, Parsons PA (1976) The genetics of behavior. Sinauer, Sunderland, MA; Freeman, Reading
Hinde RA (1973) Das Verhalten der Tiere, vol 1,2. Suhrkamp, Frankfurt
Konorski J (1948) Conditioned reflexes and neuron organization. Cambridge University Press, Cambridge
Levine S (ed) (1972) Hormones and behaviour. Academic Press New York
Lorenz KZ (1966a) Evolution and modification of behaviour. Methuen, London
Manning A (1979) Verhaltensforschung. Springer, Berlin Heidelberg New York
Thorpe WH (1963) Learning and instinct in animals, 2nd edn. Methuen, London
Tinbergen N (1966) Instinktlehre, 4th edn. Parey, Berlin

Reviews

Adkins EK (1978) Sex steroids and the differentiation of avian reproductive behavior. Am Zool 18: 501–509

Adler NT (1974) The behavioral control of reproductive physiology. In: Montagna W (ed) Reproductive behaviour. Plenum, New York, pp 259–286

Baerends GP (1976) The functional organization of behaviour. Anim Behav 24: 726–738

Baggerman B (1968) Hormonal control of reproductive and parental behaviour in fishes. In: Barrington EJW, Barker-Jorgensen C (ed) Perspectives in endocrinology. Academic Press, London, pp 357–404

Blair WF (1958) Mating call in the speciation of anuran amphibians. Am Nat 92: 27–51

Blest AD (1961) The concept of ritualization. In: Current problems in animal behaviour. Cambridge University Press, Cambridge, pp 102–124

Blüm V (1974) Die Rolle des Prolaktins bei der Cichlidenbrutpflege. Untersuchungen zum Prinzip hormonaler Verhaltenssteuerung. Fortschr Zool 22: 155–166

Crews D (1978) Endocrine control of reptilian reproductive behavior. In: Beyer C (ed) Endocrine control of sexual behavior. Raven, New York

Crook JH (1965) The adaptive significance of avian social organizations. Symp Zool Soc Lond 14: 181–218

Fiedler K (1974) Hormonale Kontrolle des Verhaltens bei Fischen. Fortschr Zool 22: 268–309

Guiton P (1962) The development of sexual responses in the domestic fowl, in relation to the concept of imprinting. Symp Zool Soc Lond 8: 227–234

Harlow HF (1958) The evolution of learning. In: Roe A, Simpson GG (eds) Behavior and evolution. Yale University Press, New Haven, pp 269–290

Herbert J (1972) Behavioural patterns. In: Austin CR, Short RV (eds) Reproduction in mammals. IV. Reproductive patterns. Cambridge University Press, Cambridge, pp 34–68

Kelley DB (1978) Neuroenatomical correlates of hormone sensitive behaviours in frogs and birds. Am Zool 18: 477–488

Kelley DB (1981) Social signals – an overview. Am Zool 21: 111–116

Krebs JR, May RM (1983) The evolutionary basis of behaviour. Nature (Lond) 306: 533–534

Lehrman DS (1961) Hormonal regulation of parental behavior in birds and infrahuman mammals. In: Young WC (ed) Sex and internal secretions, 3rd edn. Bailliere, Tindal and Cox, London, pp 1268–1382

Lehrman DS (1964) The reproductive behavior of ring doves. Sci Am 211: 48–54

Liley NR (1969) Hormones and reproductive behaviour in fishes. In: Hoar WS, Randall DJ (eds) Fish physiology, vol 3. Academic Press, New York, pp 73–116

Liley NR, Stacey NE (1983) Hormones, pheromones and reproductive behaviour in fish. In: Hoar WS, Randall DJ, Donaldson EM (eds) Fish physiology, vol 9B. Academic Press, New York, pp 1–64

Lorenz KZ (1958) The evolution of behavior. Sci Am 199(6): 67–78

Manning A (1966) Sexual behaviour. Symp R Entomol Soc Lond 3: 59–68

Manning A (1971) Evolution of behavior. In: McGaugh J (ed) Psychobiology: biological bases of behavior. Academic Press, New York, pp 1–52

Morris D (1956) The feather postures of birds and the problem of the origin of social signals. Behaviour 9: 75–113

Michael RP (1966) Action of hormones on the cat brain. In: Gorski RA, Whalen RE (eds) The brain of gonadal function, vol 3. University of California Press, Berkeley, pp 81–98

Shillito Walser E (1977) Maternal behaviour in mammals. In: Peaker M (ed) Comparative aspects of lactation. Symp Zool Soc Lond 41: 313–332

Ward IL (1974) Sexual behavior differentiation: prenatal hormonal and environmental control. In: Friedman RC, Richart RM, van de Wiele RL (eds) Sex differences in behavior. Wiley, New York, pp 3–17

Wood-Gush DGM, Gilbert AB (1975) The physiological basis of a behaviour pattern in the domestic hen. In: Peaker M (ed) Avian physiology. Symp Zool Soc Lond 35: 261–276

Original Works

Abbott DH, Hearn JP (1979) The effects of neonatal exposure to testosterone on the development of behaviour in female marmoset monkeys. In: Sex, hormones, and behaviour. Ciba Found Symp 62. Excerpta Medica, Amsterdam, pp 299–316

Baerends GP, Brouwer R, Waterbolk HTJ (1955) Ethological studies on *Lebistes reticulatus* (Peters). I. An analysis of the male courtship pattern. Behaviour 8: 250–334

Beach FA (1942) Analysis of the stimuli adequate to elicit mating behavior in the sexually inexperienced male rat. J Comp Psychol 33: 163–207

Blüm V, Fiedler K (1965) Hormonal control of reproductive behavior in some cichlid fish. Gen Comp Endocrinol 5: 186–196

Braddock JC, Braddock ZI (1959) Development of nesting behaviour in the Siamese fighting fish *Betta splendens*. Anim Behav 7: 222–232

Buntin JD (1979) Prolactin release in parent ring doves after brief exposure to their young. J Endocrinol 82: 127–130

Cheng M-F (1973a) Effect of ovariectomy on the reproductive behavior of female ring doves (*Streptopelia risoria*). J Comp Physiol Psychol 83: 221–233

Cheng M-F (1973b) Effect of estrogen on the behavior of ovariectomized ring doves (*Streptopelia risoria*). J Comp Physiol Psychol 83: 234–239

Cheng M-F (1977) Role of gonadotrophin releasing hormones in the reproductive behaviour of female ring doves (*Streptopelia risoria*). J Endocrinol 74: 37–45

Hale EB (1966) Visual stimuli and reproductive behavior in bulls. J Anim Sci Suppl 25: 36–44

Hamilton WD (1964) The genetical evolution of social behaviour, I and II. J Theor Biol 7: 1–52

Harris GW, Michael RP (1964) The activation of sexual behaviour by hypothalamic implants of oestrogen. J Physiol Lond 171: 275–301

Hutchison JB (1970) Influence of gonadal hormones on the hypothalamic integration of courtship behaviour in the Barbary dove. J Reprod Fertil Suppl 11: 15–41

Hutchison JB (1976) Hypothalamic mechansims of sexual behaviour with special reference to birds. Adv Study Behav 6: 159–200

Hutchison RE (1975) Effects of ovarian steroids and prolactin on the sequential development of nesting behaviour in female budgerigars. J Endocrinol 67: 29–39

Iersel van JJA (1953) An analysis of the parental behaviour of the male three-spined stickleback. Behaviour Suppl 3: 1–159

Komisaruk BR (1967) Effects of local brain implants of progesterone on reproductive behaviour in ring doves. J Comp Physiol Psychol 64: 219–224

Komisaruk BR, Adler NT, Hutchison J (1972) Genital sensory field: enlargement by estrogen treatment in female rats. Science 178: 1295–1298

Lehrman DS (1953) The physiological basis of parental feeding behaviour in the ring dove (*Streptopelia risoria*). Behaviour 7: 241–286

Lincoln GA, Youngson RW, Short RV (1970) The social and sexual behaviour of the red deer stag. J Reprod Fertil Suppl 11: 71–103

Lorenz K, Tinbergen N (1938) Taxis und Instinkthandlung in der Eirollbewegung der Graugans. I. Z Tierpsychol 2: 1–29

Michael RP, Scott PP (1964) The activation of sexual behaviour in cats by the subcutaneous administration of oestrogen. J Physiol (Lond) 171: 254–274

Rubin RT, Reinisch JM, Haskett RF (1981) Postnatal gonadal steroid effects on human behavior. Science 211: 1318–1324

Schein MW, Hale EB (1959) The effect of early social experience on male sexual behviour of androgen-injected turkeys. Anim Behav 7: 189–200

Sevenster P (1961) A causal analysis of a displacement activity (fanning in *Gasterosteus aculeatus* L.). Behaviour Suppl 9: 1–70

Watson A (1970) Territorial and reproductive behavior of red grouse. J Reprod Fertil Suppl 11: 3–14

References to Chapter 7

Books

Bellairs A (1970) The life of reptiles. II. The universe natural history series. Universe Books, New York
Breder CM Jr, Rosen DE (1966) Modes of reproduction in fishes. Natural History, Garden City, New York
Gilbert PW, Mathewson RF, Rall DP (ed) (1962) Sharks, skates and rays. Hopkins Baltimore, MD
Mertens R (1960) The world of amphibians and reptiles. McGraw-Hill, New York
Tyler MJ (ed) (1983) The gastric brooding frog. Groom and Helm, London

Reviews

Amoroso EC (1956) Placentation. In: Parkes AS (ed) Marshall's physiology of reproduction, vol 2. Longmans and Green, London, pp 127–311
Amoroso EC (196) Viviparity in fishes. Symp Zool Soc Lond 1: 153–181
Amoroso EC, Heap AB, Renfree MB (1979) Hormones and the evolution of viviparity. In: Barrington EJW (ed) Hormones and evolution. Academic Press, New York, pp 925–989
Bauchot R (1965) La placentation chez les reptiles. Ann Biol IV 940: 547–575
Bertin L (1952) Oviparité, ovoviviparité, viviparité. Bull Soc Zool Fr 77: 84–88
Bertin L (1958) Viviparité des téléostéens. In: Grassé PP (ed) Traité de zoologie, T. XIII, F. II. Masson, Paris, pp 1755–1790
Dodd JM (1983) Reproduction in cartilaginous fishes (Chondrichthyes). In Hoar WS, Randall DJ, Donaldson EM (eds) Fish physiology, vol 9A. Academic Press, New York, pp 31–96
Fitch HS (1970) Reproductive cycles in lizards and snakes. University of Kansas Publications Museum of Natural History, vol 11, pp 63–326
Hoar WS (1969) Reproduction. In: Hoar WS, Randall DJ (eds) Fish physiology, vol 3. Academic Press, New York, pp 1–72
Lofts B (1974) Reproduction. In: Lofts B (ed) Physiology of the amphibia. Academic Press, New York, pp 107–218
Matthews LH (1955) The evolution of the viviparity in vertebrates. Mem Soc Endocrinol 4: 129–148
Neill WT (1964) Viviparity in snakes: some ecological and zoogeographic considerations. Am Nat 98: 35–55
Packard GC, Tracy CR, Roth JJ (1977) The physiological ecology of reptilian eggs and embryos, and the evolution of viviparity within the class reptilia. Biol Rev 52: 71–105
Parker HW (1956) Viviparous caecilians and amphibian phylogeny. Nature (Lond) 178: 250–252
Salthe SN, Mecham JS (1974) Reproductive and courtship patterns. In: Lofts B (ed) Physiology of the amphibia, vol 2. Academic Press, New York, pp 309–521
Turner CL (1937) Reproductive cycles and superfetation in poeciliid fishes. Biol Bull 72: 145–164
Turner CL (1940) Viviparity in teleost fishes. Sci Monthly (Lond) 65: 508–518
Wourms JP (1977) Chondrichthyan reproduction. Am Zool 2: 379–410
Wunder W (1931) Brutpflege und Nestbau bei Fischen. Ergeb Biol 7: 118–192
Wunder W (1932) Nestbau und Brutpflege bei Amphibien. Ergeb Biol 8: 180–222
Xavier F (1977) An exceptional reproductive strategy in Anura, *Nectophrynoides occidentalis* Angel (Bufonidae), and adaptation to terrestrial life by viviparity. In: Hecht MK et al. (eds) Major patterns of vertebrate evolution. Plenum, New York, pp 545–552
Yaron Z (1972) Endocrine aspects of gestation in viviparous snakes. Gen Comp Endocrinol Suppl 3: 663–673

Original Works

Corbin CJ, Ingram GJ, Tyler MJ (1974) Gastric brooding: unique form of parental care in an Australian frog. Science 186: 946–947

Drewry GE, Jones KL (1976) A new ovoviviparous frog, *Eleutherodactylus jasperi* (Amphibia, Anura, Leptodactylidae), from Puerto Rico. J Herpetol 10: 161–165

Smith CC, Rand CS, Schaeffer B, Atz JW (1975) *Latimeria*, the living Coelacanth, is ovoviviparous. Science 190: 1105-1106

Wake MH (1978) The reproductive biology of *Eleutherodactylus jasperi* (Amphibia, Anura, Leptodactylidae), with comments on the evolution of live bearing systems. J Herpetol 12: 121–133

Xavier F (1974) La pseudogestation chez *Nectophrynoides occidentalis* Angel. Gen Comp Endocrinol 22: 98–115

Xavier F, Ozon R (1971) Recherches sur l'activité endocrine de l'ovaire de *Nectophrynoides occidentalis* Angel (amphibien anoure vivipare). II Synthèse in vitro de stéroids. Gen Comp Endocrinol 16: 30–40

References to Chapter 8

Books

Anderson JM (1972) Nature's transplant: the transplantation immunology of viviparity. Butterworths, London

Austin CR, Short RV (1972) Reproduction in mammals. Book 2. Embryonic and fetal development. Cambridge University Press, Cambridge

Bartels H (1970) Prenatal respiration. North Holland, Amsterdam

Beaconsfield P, Billee C (eds) (1979) Placenta. A neglected experimental animal. Pergamon, Oxford

Boyd JD, Hamilton WJ (1970) The human placenta. Heffer, Cambridge

Brambell FWR (1970) The transmission of passive immunity from mother to young. North Holland, Amsterdam

Grosser O (1909) Vergleichende Anatomie und Entwicklungsgeschichte der Placenta. Braumüller, Wien

Grosser O (1927) Frühentwicklung, Eihautbildung und Placentation des Menschen und der Säugetiere. Bergmann, München

Longo LD, Bartels H (ed) (1972) Respiratory gas exchange and blood flow in the placenta. Dhew, Bethesda

Ramsey EM (1975) The placenta of laboratory animals and man. Holt, Rinehart and Winston, Washington

Snoeck J (1958) Le placenta humain. Aspects morphologiques et fonctionnels. Masson, Paris

Villee CA (1960) The placenta and fetal membranes. Williams and Wilkins, New York

Reviews

Aitken RJ, Beaconsfield R, Ginsburg J (1979) Origin and formation of the placenta. In: Beaconsfield P, Villee L (eds) Placenta. A neglected experimental animal. Pergamon, Oxford, pp 152–163

Amoroso EC (1956) Placentation. In: Parkes AS (ed) Marshall's physiology of reproduction, vol 2. Longmans and Green, London, pp 127–311

Amoroso EC (1959) The biology of the placenta. 5th conference on gestation. In: Villee CA (ed) Gestation. Josiah Macy Jr Found, New York

Assali NS, Dilts PV, Plentl AA, Kirschbaum TH, Gross SJ (1968) Physiology of the placenta. In: Assali N (ed) Biology of gestation, vol 1. Academic Press, New York, pp 185–289

Bartels H (1964) Comparative physiology of oxygen transport in mammals. Lancet 9: 599–604
Beaconsfield P, Ginsburg J (1979) Carbohydrate, fat and protein metabolism in the placenta: a clinician's review. In: Beaconsfield P, Villee L (eds) Placenta, a neglected experimental animal. Pergamon, Oxford, pp 34–58
Billington WD (1976) The immunobiology of trophoblast. In: Scott JS, Jones WR (eds) Immunology of human reproduction. Academic Press, London
Canivence R, Mayer G (1982) La nidation de l'oeuf et ses modalités. In: Grassé PP (ed) Traité de zoologie, T. XVI, F. VI. Masson, Paris, pp 967–998
Dawes GS (1960) Foetal blood gas homeostasis. In: Wolstenholme GW, O'Connor H (eds) Foetal autonomy. Ciba Found Symp Churchill, London
Edwards RG (1977) Early human development: from the oocyte to implantation. In: Philipp EE, Barnes J, Nenton M (eds) Sci Found obstetrics and gynecology, Ch XV. Heinemann, London
Huggelt ASG, Hammond J (1956) Physiology of the placenta. In: Parkes AS (ed) Marshall's physiology of reproduction, vol 2. Longmans and Green, London, pp 312–397
Kelly WA, Eckstein P (1972) Implantation, development of the fetus and fetal membranes. In: Cole HH, Cupps PT (ed) Reproduction in domestic animals, chapter 14, 2nd edn. Academic Press, New York, pp 385–413
Klopper A (1980) The new placental proteins. Placenta 1: 77–89
Lemtis H (1970) Physiologie der Plazenta. Fortschr Geburtshilfe Gynäkol 41: 1–52
Luckett WP (1974) Comparative development and evolution of the placenta in primates. In: Kuhn H (ed) Contributions to primatology, vol 3. Karger, Basel, pp 142–234
Luckett WP (1977) Ontogeny of amniote fetal membranes and their application to phylogeny. In: Hecht MK et al. (eds) Major patterns of vertebrate evolution. Plenum, New York, pp 439–516
Mossman HW (1937) Comparative morphogenesis of the fetal membranes and accessory uterine structures. Contrib Embryol Carneg Inst 26: 129–246
Mossman HW (1965) The principle interchange vessels of the chorioallantoic placenta of mammals. In: De Haaen, Ursprung (eds) Organogenesis. Holt, Rinehart and Winston. New York, pp 771–786
Panigel M (1982) Les annexes embryonaires et le placenta des mammifères. In: Grassé PP (ed) Traité de zoologie, T. XVI, F. VII. Masson, Paris, pp 215–296
Patten BM (1964) The extra-embryonic membranes of mammals and the relations of the embryo to the uterus. In: Foundations of embryology, chapter 15, 2nd edn. McGraw-Hill, New York, pp 302–335
Starck D (1959) Ontogenie und Entwicklungsphysiologie der Säugetiere. In: Kükenthal W (ed) Handbuch der Zoologie, 8–9 (7). De Gruyter, Berlin, pp 1–276
Wimsatt WA (1962) Some aspects of the comparative anatomy of the mammalian placenta. Am J Obstet Gynecol 84: 1568–1594
Wimsatt WA (1975) Some comparative aspects of implantation. Biol Reprod 12: 1–40

Original Works

Amoroso EC, Perry JS (1964) The foetal membranes and placenta of the african elephant (*Loxodonta africana*). Philos Trans R Soc Lond B Biol Sci 248: 1–34
Amoroso ES, Hancock NA, Kellas L (1958) The foetal membranes and placenta of the hippopotamus (*Hippopotamus ampibius* L.). Proc Zool Soc Lond 130: 437–447
Andresen A (1922) Über die Semiplacenta multiplex des *Cervus rufus* Cuv. Z Anat Entwicklungsgesch 65: 544–569
Andresen A (1927) Die Plazentome der Wiederkäuer. Morph Jahrb 57: 410–485
Barcroft J, Barron DH (1946) Observations upon the form and relations of the maternal and fetal vessels in the placenta of the sheep. Anat Rec 94: 569
Björkman N, Wimsatt WA (1968) The allantoic placenta of the vampire bat (*Desmodus rotundus murinus*): a reinterpretation of its structure based on electron microscopic observations. Anat Rec 162: 83–98

Boyd JD, Hamilton WJ (1976) Development and structure of the human placenta from the end of the 3rd month of gestation. J Obstet Gynaecol Br Commonw 74: 161–226

Bridgman J (1948a) Morphological study of the development of placenta of the rat. I. Development of the placenta of the white rat. J Morphol 83: 61–85

Bridgman J (1948b) II. An histological and cytochemical study of the development of the chorioallantoic placenta of the white rat. J Morphol 93: 195–224

Carter AM, Gothlin J, Olin T (1971) An angiographic study of the structure and function of the uterine and maternal placental vasculature in the rabbit. J Reprod Fertil 25: 201–210

Davidoff M (1973) Guinea-pig placenta: fine structure and development. Acta Anat 86 Suppl 61: 23–47

Drieux H, Thiery G (1949) Placentation chez les mammifères domestiques. Placenta des euqidés. Rec Med Vet (Alfort) 125: 197–214

Drieux H, Thiery G (1951) Placenta des bovidés. Rec Med Vet (Alfort) 127: 5–25

Drieux H, Thiery G (1952) Placenta des ovidés. Rec Med Vet (Alfort) 128: 5–18

Duval M (1889) Le placenta des rongeurs: le placenta du lapin. J Anat (Paris) 25: 309–342, 573–627

Duval M (1890) Le placenta des rongeurs: le placenta du lapin. J Anat (Paris) 26: 1–48, 273–344, 551–592

Enders AC (1960a) Development and structure of the villous haemochorial placenta of the nine-banded armadillo (*Dasypus novemcinctus*). J Anat 94: 34–35

Hamilton WJ, Boyd JD (1960) Development of the human placenta in the first three months of gestation. J Anat 94: 297–328

Harrison RJ, Young BA (1966) Functional characteristics of the pinniped placenta. Symp Zool Soc Lond 15: 47–67

Hill P, Young M (1973) Net placental transfer of free amino acids against varying concentrations (guinea-pig). J Physiol (Lond) 235: 409–422

Hitzig WH (1949) Über die Entwicklung und Ausbildung des Chorions der Placenta zonaria bei Katze, Hund und Fuchs. Z Anat Entwicklungsgesch 143: 25–42

Laga E, Driscoll S, Murno H (1973a) Quantitative studies of human placenta. I. Morphometry. Biol Neonat 23: 231–260

Laga E, Driscoll S, Murno H (1973b) II. Biochemical characteristics. Biol Neonat 23: 260–285

Miller RK, Bernot WO (1975) Mechanisms of transport across the placenta, an in vitro approach. Life Sci 16: 7–30

Novy MJ (1972) The placenta and its circulation in primates. Primate News 10: 6–11

Panigel M (1970) Structure et ultrastructure comparée de la membrane placentaire chez certains primates non humains: *Galago demidovii, Erythrocebus patas, Macaca mulatta, Macaca irus (fascicularis)* et *Papio cynocephalus*. Bull Assoc Anat 145: 319–337

Reneau DD, Gullbeau E, Cameron JM (1974) Theoretical analysis of the dynamics of oxygen transport and exchange in the placental-fetal system. Microvasc Res 8: 346–361

Schauder W (1944) Der gravide Uterus und die Placenta des Tapirs mit Vergleich von Uterus und Placenta des Schweines und Pferdes. Morph Jahrb 89: 407–456

Stieve H (1948) Der Bau der Primatenplacenta. Anat Anz 96: 299–329

Strauss F (1942) Vergleichende Beurteilung der Placentation bei den Insektivoren. Rev Suisse Zool 49: 269–282

Stump CW, Robins JP, Garde ML (1960) The development of the embryo and membranes of the humpback whale *Megaptera nodosa* (Bonnaterre). Aust J Mar Freshwater Res 11: 365–386

Wimsatt W (1974) Morphogenesis of the fetal membranes and placenta of the black bear, *Ursus americanus*. Am J Anat 140: 471–496

References to Chapter 9

Books

Breder CM Jr, Rosen DE (1966) Modes of reproduction in fishes. Natural History, Garden City
Duellman WE (1970) The hylid frogs of Middle America. Monograph of the Museum of Natural History. The University of Kansas, No 1. Lawrence, KS
Farner DS (ed) (1972) Breeding biology of birds. Proc Symp Breeding behavior and reproductive physiology of birds. Natl Acad Sci, Washington DC
Griffiths M (1978) The biology of the monotremes. Academic Press, New York
Mertens R (1960) The world of amphibians and reptiles. McGraw-Hill, New York
Noble GK (1954) The biology of the amphibia. Dover, New York
Peaker M (1977) Comparative aspects of lactation. Zool Soc Lond Academic Press, London
Peaker M, Vernon RG, Knight CH (1983) Physiological strategies in lactation. Symp Zool Soc Lond, No 51. Academic Press, London
Porter KR (1972) Herpetology. Saunders, Philadelphia
Skutch AF (1976) Parent birds and their young. University of Texas Press, Austin
Stebbins RC (1951) Amphibians of western North America. University California Press, Berkeley
Stonehouse B, Gilmore D (eds) (1977) The biology of marsupials. Macmillan, London
Tyndale-Biscoe H (1973) Life of marsupials. Arnold, London
Viel JL (ed) (1973) Evolutionary biology of the anura. University Missouri Press, Columbia, MO

Reviews

Bertin L (1958) Nidification. In: Grassé PP (ed) Traité de zoologie, T. XIII, F. II. Masson, Paris, pp 1653–1684
Chadwik A (1977) Comparison of milk-like secretions found in nonmammals. In: Peaker M (ed) Comparative aspects of lactation. Symp Zool Soc Lond 41: 341–358
Drent R (1973) The natural history of incubation. In: Farner DS (ed) Breeding biology of birds. Natl Acad Sci Washington, DC, pp 262–311
Drent R (1975) Incubation. In: Farner DS, King IR, Parkes KC (eds) Avian biology, vol 5. Academic Press, London, pp 333–420
Guibé (1970) La reproduction. In: Grassé PP (ed) Traité de zoologie, T. XIV, F. III. Masson, Paris, pp 859–892
Jeness R (1974) The composition of milk. In: Larson BL, Smith VR (eds) Lactation, vol 3. Academic Press, New York, pp 3–107
Le Maho Y (1977) The emperor penguin: a strategy to live and breed in the cold. Am Sci 65: 680–693
Lutz B (1947) Trends towards non-aquatic and direct development in frogs. Copeia 242–252
Mayoud N (1950) Biologie de la reproduction. In: Grassé PP (ed) Traité de zoologie, T. XV. Masson, Paris, pp 539–653
Pond CM (1977) The significance of lactation in the evolution of mammals. Evolution 31: 177–199
Raynaud A (1969) Mamelles. In: Grassé PP (ed) Traité de zoologie, T. XVI, F. VI. Masson, Paris, pp 1–147
Rose W (1956) Parental care in batrachians. Afr Wildl 10: 257
Saltke SN, Mecham JS (1974) Reproductive and courtship patterns. In: Lofts B (ed) Physiology of the amphibia, vol 2. Academic Press, New York, pp 309–521
Wunder W (1931) Brutpflege und Nestbau bei Fischen. Ergeb Biol 7: 118–192
Wunder W (1932) Nestbau und Brutpflege bei Amphibien. Ergeb Biol 8: 180–221
Wunder W (1934) Nestbau und Brutpflege bei Reptilien. Ergeb Biol 10: 1–36
Wunder W (1937) Brutpflege und Nestbau bei Säugetieren. Ergeb Biol 14: 280–348

Original Works

Ashby E (1929) Notes on the unique methods of nidification of the australian mallee-fowl (*Leipoa ocellata*) with original data supplied by B.W. Leake. R.A.O.U. Auk Lancaster, vol 46. Lancaster, pp 294–305

Hunt WH (1976) Maternal behavior in the Morelet's crocodile, *Crocodylus moreleti*. Copeia 763–764

Hutchinson VH, Dowling HG, Vinegar A (1966) Thermoregulation in a broding female indian python: *Python molurus bivittatus*. Science 151: 3711, 694–696

Noble GK (1935) The brooding habits of the blood python and other snakes. Copeia Ann Arbor 1: 1–3

Noble GK, Masson ER (1935) Experiments on the brooding habits of the lizard *Eumeces* and *Ophisaurus*. Am Mus Novit NY No 619

List of Figure Sources

Cover Figure: after a photograph taken by Dipl. Biol. E Stretzke
Preface Figure: after Tyler MJ (1983) The gastric brooding frog. Groom and Helm, London

Chapter Figures

Chap. 1: after Budgett JS (1899) Notes on the Batrachians of the Paraguayan Chaco. Quart J Microsc Sci 42

Chap. 3: after Zeller. In: Meisenheimer J (1921) Geschlecht und Geschlechter im Tierreich, Jena

Chap. 4: after Ginsburg and Detlav. In: Grassé PP (ed) (1958) Traité de zoologie, T. XIII, F. II. Masson, Paris

Chap. 6: after Gudger. In: Grassé PP (ed) (1958) Traité de zoologie, T. XIII, F. II. Masson, Paris

Chap. 7: after Turner. In: Grassé PP (ed) (1958) Traité de zoologie, T. XIII, F. II. Masson, Paris

Chap. 9: after Le Danois. In: Grassé PP (ed) (1958) Traité de zoologie, T. XIII, F. II. Masson, Paris

Text Figures

Fig. 5 A: after Lagler KF, Bardach JE, Miller RR, May Passino DR (1977) Ichthyology. Wiley, New York

B: after Buchner. In: Czihak G, Langer H, Ziegler H (eds) (1981) Biologie, 3rd edn. Springer, Berlin Heidelberg New York

C: after Noble GK (1954) The biology of the Amphibia. Dover, New York

Fig. 9: after Balinsky BI (1975) An introduction to embryology. Saunders, Philadelphia

Fig. 14: after Turner. In: Romer AS, Parsons TS (1983) Vergleichende Anatomie der Wirbeltiere. Parey, Hamburg

Fig. 18: after Leigh-Sharpe WH (1920) The comparative morphology of the secondary sex characters of Elasmobranch fishes. J Morphol 34: 245–265

Fig. 19: after Wibault-Isebree Moens. In: Bolk L, Göppert E, Kallius E, Lubosch W (eds) (1933) Handb d vergl Anat d Wirbeltiere, vol 6. Urban und Schwarzenberg, Berlin

Fig. 20: after Boas. In: Bolk L, Göppert E. Kallius E, Lubosch W (eds) (1933) Handb d vergl Anat d Wirbeltiere, vol 6. Urban und Schwarzenberg, Berlin

Fig. 21 A,B: after Boas. In: Bolk L, Göppert E. Kallius E, Lubosch W (eds) (1933) Handb d vergl Anat d Wirbeltiere, vol 6. Urban und Schwarzenberg, Berlin

C: after von den Broek. In: Bolk L, Göppert E, Kallius E, Lubosch W (eds) (1933) Handb d vergl Anat d Wirbeltiere, vol 6 Urban und Schwarzenberg, Berlin

Fig. 23 A: after Billard R, Jalabert B, Breton B (1972) Les cellules de Sertoli des poissons teleosteens. I Etude ultrastructurale. Ann Biol anim Bioch Biophys 12: 19–32

B: assembled from Figs. from: Nicander L (1967) An electron microscopical study of cell contacts in the semiferous tubules of some ammals. Z Zellforsch mikr Anat 4: 375–395

Dym M, Fawcett DW (1970) The blood-testis barrier in the rat and the physiological compartimentation of the semiferous epithelium. Biol Reprod 3: 308–326

Fig. 24: after Holstein AF, Rosen-Runge EC (1981) Atlas of human spermatogenesis. Grosse, Berlin

Fig. 29 A: after Götting KJ (1967) Der Follikel und die peripheren Strukturen der Oocyten der Teleosteer und Amphibien. Z Zellforsch mikr Anat 79: 481–491

C: after Balinski BI (1975) An introduction to embryology. Saunders, Philadelphia

D: after Borcea. In: Bolk L, Göppert E, Kallius E, Lubosch W (eds) (1933) Handb d vergl Anat d Wirbeltiere, vol 6. Urban und Schwarzenberg, Berlin

E: after Lillie. In: Balinsky BI (1975) An introduction to embryology. Saunders, Philadelphia

Fig. 30: after Colwin and Colwin. In: Torrey W (1971) Morphogenesis of the vertebrates. Wiley, New York

Fig. 31: after Austin. In: Austin CR, Short RV (eds) (1972) Reproduction in mammals, vol 1. Germ cells and fertilisation. Cambridge University Press, Cambridge

Fig. 34: after Balinski BI (1975) An introduction to embryology. Saunders, Philadelphia

Fig. 39: after Portmann A (1969) Einführung in die vergleichende Morphologie der Wirbeltiere. Schwabe & Co, Basel Stuttgart

Fig. 42 B,C: after Dalq and Pasteels. In: Schwartz V (1973) Entwicklungsgeschichte der Tiere. DTV and Thieme, Stuttgart

D,E: after Wilson. In: Siewing R (1969) Lehrbuch der vergleichenden Entwicklungsgeschichte der Tiere. Parey, Hamburg

Fig. 43 B,C,E,F: after Will. In: Schwartz R (1969) Entwicklungsgeschichte der Tiere. DTV and Thieme, Stuttgart

Fig. 44 B–G: after Nelsen, Hamburger and Hamilton. In: Schwartz V (1973) Entwicklungsgeschichte der Tiere. DTV and Thieme, Stuttgart

Fig. 45: after Weissenberg. In: Schwartz V (1978) Entwicklungsgeschichte der Tiere. DTV and Thieme, Stuttgart

Figs. 46–49: after Grosser, Hertig, Rock and Starck. In: Schwartz V (1978) Entwicklungsgeschichte der Tiere. DTV and Thieme, Stuttgart

Fig. 50: after Pflugfelder O (1962) Lehrbuch der Entwicklungsgeschichte und Entwicklungsphysiologie der Tiere. Fischer, Jena

Fig. 62: after Hansel W, Echternkamp SE (1972) Control of ovarian function in domestic animals. Amer Zool 12: 225–243

Fig. 63 B: after Kalra PS, Krulich L, Quijada M, Kalra SP, Fawcett CP, McCann SM (1971) Feedback effects of gonadal steroids on gonadotropins and prolactin in the rat. In: James VHT, Martini L (eds) Hormonal steroids. Excerpta Medica, Amsterdam, pp 708–715

combined with Bentley PJ (1982) Comparative vertebrate endocrinology. Cambridge University Press, Cambridge London

Fig. 66: after Neumann F, Elger W, Steinbeck H, Gräf KJ (1975) The role of androgens in sexual differentiation of mammals. In: Reinboth R (ed) Intersexuality in the animal kingdom. Springer, Berlin Heidelberg New York

Fig. 68: after Tinbergen N (1966) Instinktlehre. Vergleichende Erforschung angeborenen Verhaltens. Parey, Hamburg

Fig. 72 A: after Wiedersheim R (1900) Brutpflege bei niederen Wirbeltieren. Biol Zbl 20
B: after Huot A (1902) Recherches sur les poissons lophobranches. Ann des Sci nat VIII, s. Zool
C: after Brandes G, Schoenichen W (1901) Die Brutpflege der schwanzlosen Batrachier. Abh naturforsch Ges 22, Halle
D: after Sarasin P, Sarasin S (1887–1890) Ergebnisse naturwiss. Forschungen auf Ceylon. Wiesbaden

Fig. 73: after Wiedersheim R (1900) Brutpflege bei niederen Wirbeltieren. Biol Zbl 20

Fig. 75: after Ranzi and Needham. In: Hoar WS, Randall DJ (eds) (1969) Fish physiology, vol 3. Academic Press, New York

Fig. 76 A: after Wood-Mason and Alcock. In: Grassé PP (ed) (1958) Traité de zoologie, T. XIII, F, II. Masson, Paris
B: after Leuckart FS (1836) Untersuchungen über die äußeren Kiemen der Embryonen von Rochen und Haien. Verlag Stuttgart

Fig. 77: after Ranzi S (1934) Le basi fisio morfologische dello sviluppo embrionale dei Selaci. Parti II e III. Publ Staz Zool Napoli 13: 331–437

Fig. 78 A,B: after Turner CL (1933) Viviparity superimposed upon ovo-viviparity in the Goodeidae, a family of cyprinodont teleost fishes of the mexican plateau. J Morphol 55: 207–251
Turner CL (1940) Pericardial sac, trophotaeniae, and alimentary tract in embryos of goodeid fishes. J Morphol 67: 291–297
C,D: after Turner CL (1940) Pseudamnion, pseudochorion, and follicular pseudoplacenta in poeciliid fishes. J Morphol 67: 59–89

Fig. 79: after Turner CL (1940) Pseudamnion, pseudochorion, and follicular pseudoplacenta in poeciliid fishes. J Morphol 67: 59–89

Fig. 80: after Browning HC (1973) The evolutionary history of the corpus luteum. Biol Reproduction 8: 128–155

Fig. 81 A: after Weekes HC (1927) A note on reproductive phenomena in some lizards. Proc Linn Soc NSW 52: 25–32
B–D: after Weekes HC (1935) A review of placentation among reptiles, with particular regard to the function and evolution of the placenta. Proc Soc Zool Soc (London) 2: 625–645

Fig. 83: after Starck D (1975) Embryologie. Thieme, Stuttgart

Fig. 85: after Starck D (1975) Embryologie. Thieme, Stuttgart

Fig. 88 A: after Minot. In: Parkes AS (ed) (1952) Marshall's physiology of reproduction, vol II. Longmans and Green, London
B,C: after Hill. In: Parkes AS (ed) (1952) Marshall's physiology of reproduction, vol II. Longmans and Green, London
D: after Flynn. In: Parkes AS (ed) (1952) Marshall's physiology of reproduction, vol II. Longmans and Green, London

Fig. 89: after Bonnet. In: Siewing R (1969) Lehrbuch der vergleichenden Entwicklungsgeschichte der Tiere. Parey, Hamburg

Fig. 90: after Starck D (1975) Embryologie. Thieme, Stuttgart

Fig. 91: after Grosser and Mossmann. In: Starck D (1975) Embryologie. Thieme, Stuttgart

Fig. 92: after Amoroso. In: Parkes AS (ed) (1952) Marshall's physiology of reproduction, vol II. Longmans and Green, London

Fig. 93 A,C,F: after Amoroso. In: Starck D (1975) Embryologie, Thieme, Stuttgart
B,C,E: after Starck. In: Siewing R (1969) Lehrbuch der vergleichenden Entwicklungsgeschichte der Tiere. Parey, Hamburg

Fig. 94: after Mossmann HW (1973) Comparative morphogenesis of the fetal membranes and accessory uterine structures. Contrib Embryol Carneg Inst 26: 129–246

Fig. 95: after de Lange. In: Bolk L, Göppert E, Kallius E, Lubosch W (eds) (1933) Handb d vergl Anat d Wirbeltiere, vol 6. Urban und Schwarzenberg, Berlin

Fig. 96 A: after Hill. In: Starck D (1975) Embryologie. Thieme, Stuttgart
B: after Hubrecht. In: Bolk L, Göppert E, Kallius E, Lubosch W (eds) (1933) Handb d vergl Anat d Wirbeltiere, vol 6. Urban und Schwarzenberg, Berlin

Fig. 97: after Starck D (1975) Embryologie. Thieme, Stuttgart

Fig. 98: after Starck D (1975) Embryologie. Thieme, Stuttgart

Fig. 99 A: after Barron. In: Parkes AS (ed) (1952) Marshall's physiology of reproduction, vol II. Longmans and Green, London
B: after Starck D (1975) Embryologie. Thieme, Stuttgart

Fig. 101 A: after Reighart JE (1908) Methods of studying the habits of fishes, with an account of the horned dace (*Semotilus atromaculatus*). Proc 4th Int Fish Congr 2: 1111–1136
B: after Soljan T (1930) Brutpflege und Nestbau bei *Crenilabrus quinquemaculatus* RISSO, einem adriatischen Lippfisch. Z Morphol u Ökol Tiere 20
C: after Wunder W (1930) Experimentelle Untersuchungen an Stichlingen (Kämpfen, Nestbau, Laichen, Brutpflege). Z Morphol u Ökol Tiere 16

Fig. 102 A: after Daget J (1952) La nidification de *Gymnarchus niloticus*. Aquarium et Poissons 2: 16–18
B: after Forselius S (1957) Studies of anabantid fishes, I–III. Zoologiska Bidrag fran Uppsala 32: 93–599
C: after Guitel F (1892) Observations sur les moeurs du *Gobius minutus*. Arch Zool Exp Gen 6: 423–467

Fig. 103 A: after Guitel F (1913) L'appareil fixateure de l'oeuf de *Kurtus gulliveri*. Arch Zool Exp Gen 52: 1–11
B: after Wyman J (1859) On some unusual modes of gestation in batrachians and fishes. Am J Sc 27: 5–13
C: after Willey A (1910) The occurrence of *Solenostoma* off the coast of Ceylon. Spolia Zeylanica 6: 102–107

Fig. 104 A–C: after Mertens R (1960) The world of amphibians and reptiles. McGraw-Hill, New York
D: after Brandes G, Schoenichen W (1901) Die Brutpflege der schwanzlosen Batrachier. Abh Naturforsch Ges 22, Halle
E: after Siedlecki M (1909) Zur Kenntnis des javanischen Flugfrosches. Biol Zentralbl 29: 704–737

Fig. 105 A,B: after Mertens R (1960) The world of amphibians and reptiles. McGraw-Hill, New York
C: after Brauer. In: Wunder W (1932) Nestbau und Brutpflege bei Amphibien. Erg Biol 8: 180–221
D: after Wiedersheim. In: Wunder W (1932) Nestbau und Brutpflege bei Amphibien. Erg Biol 8: 180–221

Fig. 106 A: after Young. In: Grassé PP (ed) (1950) Traité de zoologie, T. XV. Masson, Paris

B: after Pickwell. In: Grassé PP (ed) (1950) Traité de zoologie, T. XV. Masson, Paris

C,F: after Smolik HW (1968) rororo-Tierlexikon. Bertelsmann Lexikon-Verlag, Gütersloh

D: after Henrick FH (1911) Nest and nest-building in birds. J Anim Behaviour 1: 154–192

Fig. 107 A: after Haacke. In: Wunder W (1937) Brutpflege und Nestbau bei Säugetieren. Erg Biol 14: 280–348

B: after Klaatsch. In: Meisenheimer J (1921) Geschlecht und Geschlechter im Tierreiche. Jena

C: after Hartig. In: Remane A, Storch V, Welsch U (1980) Systematische Zoologie. Fischer, Stuttgart New York

Fig. 108 A: after Carlsson. In: Meisenheimer J (1921) Geschlecht und Geschlechter im Tierreiche. Jena

B: after Griffiths M (1978) The biology of the Monotremes. Academic Press, New York

Fig. 109: after Cowie AT. In: Austin CR, Sort RV (1984) Reproduction in mammals, vol 3. Hormonal control of reproduction. Cambridge University Press, Cambridge London

Glossary

The terms in this glossary are explained primarily with respect to the vertebrates. Definitions extending beyond this are marked with "in general". Further information is available in subjects indicated with →.

Abdominal funnel: Funnel-like extension of the coelomic cavity which may open into the exterior with an "abdominal pore". In Cyclostomata it forms, together with the end portion of the →primary urinary duct, a →sinus urogenitalis. →Genital funnel.

Acrosome: A vesicular cell organelle in front of the nucleus in the vertebrate →spermatozoon. It may contain soluble and membrane-bound lytic enzymes (Hyaluronidase e.g.) necessary for the penetration of secondary and primary →egg membranes. The acrosome lacks in the primitive spermatozoa of most teleosts.

Acrosomal filament: A tubular structure formed by the →subacrosomal material after the release of acrosomal enzymes, which penetrates the →egg membrane and fuses with the →oocyte during the initial process of →fertilisation.

Acrosome reaction: The lysis of egg membranes by enzymes of the →acrosome (mammals) and after this the formation of an →acrosomal filament contacting the oocyte as the initial process of →fertilisation. The second event is investigated in some invertebrate species and it is suggested that it takes place also in vertebrates possessing →subacrosomal material in their →spermatozoa.

Acystic spermatogensis: →Spermatogenesis which takes place between neighbouring →Sertoli cells from the periphery to the centred lumen of the semiferous tubule in amniota.

Adeciduate placenta: A type of →placenta in which the inner epithelium of the →uterus is not heavily damaged during birth, i.e. the maternal-foetal contact is relatively loose and a →membrana decidua is not formed.

Adelphophagy: A subtype of →viviparity. The embryo consumes sibbling embryos which are at an advanced stage of development.

Adenohypophysis: The epithelial part of the →hypophysis. It consists of a cranial pars distalis and a pars intermedia as a rule lying caudally from the former. In tetrapods it involves a third part, the pars infundubularis (pars tuberalis) lying round the →infundibulum.

Agglutination reaction: General term for "sticking together". In reproductive biology: the sticking together of eggs and →spermatozoa.

Allanto-chorion: The →chorion covering the →allantois.

Allantoic placenta: The advanced type of the typical vertebrate →placenta. The region of the →chorion covering the →allantois makes contact with the uterus wall.

Allantois: A diverticulum of the hind gut in Amniota which is overgrown by splanchnic →mesoderm.

Altricial species: A species whose young stay within the nest or under the guardianship of the parents after birth or hatching.

Amnion: Extraembryonic organ in Reptilia, Aves and Mammalia deriving from →ectoderm duplicatures underlain by →mesoderm (amniotic folds) which grow over the embryo and eventually melt together, thus forming an "aquarium" filled with amniotic fluid. The amnion may be interpreted as an adaptation of the pressures of development in a dry environment. Reptiles, birds and mammals are therefore all called Amniota. In mammals in many cases it develops in a different way within an →embryonic knob.

Amniotic cauity: The cavity enclosed by the →amnion wich is filled with the amniotic fluid.

Ampulla ductus deferrentis: A temporary or permanent swelling of the caudal part of the →ductus deferens serving for sperm storage.

Anadromous migration: "Upstream" migration of a species, i.e. from seawater to freshwater.

Analis: The anal fin in fishes.

Androgens: The male sex hormones. In vertebrates these are steroids, i.e. they possess a common typical tetracyclic chemical basic structure. They contain 19 carbon atoms. The main androgen is testosterone.

Animal pole: The part of an egg or the →blastomeres deriving from it which form the nervous system, the sensory organs and the skin of a developing animal. In eggs containing →yolk as a rule it is the yolk-free or yolk-poor part.

Antidiuretic hormone: A →neurohormone stored in the →neurohypophysis of mammals. It is also called vasopressin.

Antimesometrial implantation: Type of →implantation. The →blastocyst contacts the →mucosa of the uterus opposite to the side of the insertion of the →mesometrium in the →uterus bicornis.

Anti-müllerian duct hormone: A substance produced by the foetal →testes which inhibits the development of the →müllerian duct system. It is also called factor X or anti-müllerian hormone (AMH).

Antrum folliculi: The cavity of the mammalian →ovarian follicle in certain stages of development. It is filled with the →liquor folliculi.

Antrum testis: A peripheral cavity upon the →testis of some birds collecting sperm from the →semiferous tubuli.

Apopyle: Cranial opening of the →pterygopodium.

Appendices testis: →Mesonephric rudiment in male amniotes.

Appendices vesiculosae: →Mesonephric rudiment in female amniotes.

Appendiculata: Appendages of the yolk sac stalk of some eleasmobranchs which presumably faciltate the resorption of →embryotroph.

Archenteron: →Primitive gut. This term is also sometimes used for the →gastrocoel.

Area opaca: The ring of →germinal disc tissue contacting the →yolk mass in Sauropsida.

Area pellucida: The somewhat translucent central area of the sauropsidan →germinal disc where it covers the →blastocoel remant or the →subgerminal cavity, or both.

Areolae: →Milk fields.

Arginine vasopressin: A →neurohypophyseal peptide (→neurohormone). The common →antidiuretic hormones in mammals except Suiformes.

Arginine vasotocin: A →neurohypophyseal peptide (→neurohormone) of all submammalians.

Aspartocin: A →neurophypophyseal peptide (→neurohormone) in Elasmobranchii.

Autonomous hormone gland: An →endocrine gland which is not influenced by other hormones. This therm must be used very carefully, because in most cases possible interactions are not yet clar. It is possibly a "hypothetical" term.

Auxocytes: →Primary spermatocytes.

Basal trophoblast: The region of the →trophoblast lying between the →intervillar space and the uterus tissue.

Belt placenta: A →restricted placenta surrounding the chorion either totally or with an interruption like a belt.

Bidders organ: Embryonic ovary lying upon the cranial part of the bufonid →testis.

Blastocoel: The cavity of the →blastula. It is also called primary body cavity.

Blastocyst: The result of a special form of development in mammals other than Monotremata. It resembles a blastula and there are differing opinions whether it is homologous to a blastula or not. It consists of two physiologically distincty parts, the trophoblast (trophoderm, trophoecteroderm) and the embryoblast, which corresponds to an early →germinal disc. The latter is either enclosed within a vesicle whose wall represents the trophoblast or it is incorporated into the trophoblast.

Blastoderm: The wall of the →blastula.

Blastodisc: →Germinal disc.

Blastomeres: The cells which rise in the course of →cleavage.

Blastomeric anarchy: An irregular type of →total cleavage which in vertebrates occurs only in marsurpials.

Blastopore: The orifice of the →primitive gut. In vertebrates (as in all Deuterostomia) it becomes the anus in later stages of development. It is also called primitive mouth and is surrounded by the primitive mouth lips.

Blastopore lips: The embryonic tissue lining the →blastopore.

Blastula: The hollow cell ball which arises from the →morula in the case of →total cleavage. In →discoidal cleavage the late →germinal disc corresponds to the blastula.

Body stalk: The mesenchymatic connection of the embryo together with →amnion and →yolk sac with the →chorion in primates.

Boundary cells: Cells surounding the ampullae or tubuli of the →testis which are thought to be homologous to the →Leydig cells in most fishes and Urodela. For the former, newer results indicate that this opnion may be doubtful.

Bursa ovarii: Ovarian pouch formed by the walls of the →infundubulum around the →ovary of some mammals.

Canalis neurentericus: The connection between the lumina of the →primitive gut and the neural tube.

Canalis urogenitalis: The "male urethra" of mammals. It derives from the junction of the →ductus deferens and the urethra and runs through the →penis. It may be compared with a →sinus urogenitalis.

Capacitation: Processes which establish the conditions for the →acrosome reaction in mammalian →spermatozoa. It is thought that these involve some form of "demasking" of the spermatozoan membrane. Comparable phenomenons have been observed in a few submammalian species.

Carrier: The enlarged pole of the →blastocyst of Muridae whose base encorporates the →embryonic knob. It is also called ectoplacental cone and plays an important role in →placentation.

Caruncle: Convex or concave structure of the →mucosa of the uterus fitting together with a →cotyledon.

Catadromous migration: "Downstream" migration of a species, i.e. from freshwater to seawater.

Cavitation amnion: A special type of →amnion in many mammals which does not derive from →amnictic folds. It is, moreover, the derivative of the cavity of the →embryocyst.

Central Haematome: A region comparable to the →marginal haematome of the →placenta of carnivores which is situated other than at the margins.

Central Implantation: Primitive type of →implantation. The →blastocyst lies within the cavity of the →uterus and is enclosed by the →endometrium.

Chiasma opticum: →Optic chiasma.

Chorda canal: The lumen of a structure homologous to the →archenteron formed by invagination of the →primitive knot of Reptilia. Therefore its →endoderm is referred as primary endoderm in contrast to the →secondary endoderm which is built up prior to it. The chorda canal also occurs in in some birds and some mammals.

Chorda dorsalis: The typical transient embryonic axial sceleton of vertebrates. It is the permament axial sceleton in the Tunicata, Acrania and, within vertebrates, in Cyclostomata. It is also called notochord

Chordamesoderm: The portion of mesoderm by which the →chorda dorsalis is formed.

Chorioallantois: The outer layer of the →allantois lying against the →chorion.

Chorion: Term with twofold meaning: (A) The →primary egg membrane in fishes. (B) The outer layer of the amniote embryonic membranes.

Chorion Frondosum: The villus-bearing portion of the human →chorion.

Chorion laeue: The villus-free region of the human →chorion.

Chorionic plate: The plate which lines the →intervillous space at the side of the →foetus in the →placenta olliformis. It projects the villi into the intervillous space and is vascularised by the vessels of the →umbilical cord from the oposite side.

Cleavage: The initial →mitotic divisions of the →zygote leading to the →blastula in the case of →total cleavage and to a →discoblastula in the case of →discoidal cleavage.

Clitoris: Rudimentary →penis in female mammals

Cloaca: The common termination of the gastrointestinal system and the urogenital system. (Elasmobranchii, Dipnoi, Amphibia and Sauropsida).

Coelom: Secondary body cavity enclosed by mesoderm.

Colostrum: A secretion of the →mammae which is produced a few days before and after the time of birth having a different composition than the normal mother's milk. It is also called pre-milk.

Chontinuous hormone secretion: A hormone is secreted continuously into the bloodstream.

Contradeciduate placenta: A type of →placenta which is not expelled after birth.

Copulation: The behavioural action of transferring the male gametes into the female genital duct system by direct body contact (e.g. pressing together the →cloacae in birds). This term is used in some textbooks also in the sense of →karyogamy.

Copulatory organ: Any structure of the male body which is introduced into the male genital duct system for sperm transfer.

Corona radiata: The "envelope" of the →oocyte within the typical mammalian →ovarian follicle in certain stages of development consisting of →granulosa cells.

Corpus albicans: The rudiments of a degenerated →corpus luteum.

Corpus atreticum: Structure of varying appearance resulting of the degeneration of an unovulated →oocyte in different developmental stages. It is often found in submammalians and sometimes resembles morphologically a →corpus luteum. Therefore it is sometimes called corpus luteum praeovulatorium.

Corpus cavernosum: An erectile structure in the chelonian →penis. This term is also used in some anatomy textbooks for a structure of the mamalian →penis which is named →corpus fibrosum in this book.

Corpusculum renis: Bowman's capsule with the internal glomerulus. It is also called Malpighian corpuscle.

Corpus fibrosum: Erectile structure in the mammalian →penis. Non-erectile tissues with the same name occur in the penises of chelonia and birds. By means of a homogenous terminology thuse of this term in this book deviates from some others.

Corpus luteum: Ovarian structure formed mainly by transformed →granulosa cells (granulosa luteal cells) after the event of →ovulation. It occurs in many species throughout the vertebrate classes. In mammals it produces mainly progesterone which has important functions in the preparation and maintenance of pregnancy. It is also called corpus luteum postovulatorium.

Corpus luteum atreticum: →Corpus atreticum.

Corpus luteum graviditatis: The →corpus luteum present during →gestation in mammals.

Corpus luteum ovulatorium: →Corpus luteum.

Corpus luteum preaovulatorium: →Corpus atreticum.

Corpus luteum postovulatorium: →Corpus luteum.

Corpus spongiosum: Fastening tissue structure surrounding the urinary canal in the →penis of mammals. It seems to be homologous to the →corpus cavernosum of the lower amniotes possessing a →penis.

Cortical granule reaction: The release of the contents of the →cortical granules into the →perivitelline space just after contact with the fertilising →spermatozoon (only in species which possess cortical granules).

Cortical granules: Granules lying at the periphery of the cytoplasm of an →oocyte (the cortex) in many vertebrate and invertebrate species (e. g. frog).

Courtship: The series of behavioural patterns of sexual partners which leads to the release of the →gametes or →copulation.

Cotyledon: A focal placental area in the →Placenta multiplex.

Cumulus oophorus: The "stalk" consisting of →granulosa cells bearing the →oocyte within the typical mammalian →ovarian follicle in certain stages of development.

Cystic spermatogensis: →Spermatogensis which takes place in cysts of →Sertoli-cyst cells containing each a clone of germ cells deriving from a primary →spermatogeonium. This form is typical in Anamnia.

Cytogamy: The fusion of the cytoplasm of the gametes as the first step in →fertilisation.

Cytotrophoblast: The inner layer of the →invasive trophoblast which remains as a tissue consisting of single cells.

Decidua basalis: The →membrana decidua in the region of the →placenta.

Decidua capsularis: The →membrana decidua surrounding the embryo in the non-→placental region.

Decidua marginalis: The →membrana decidua connecting the →decidua basalis with the →decidua capsularis and the →decidua parietalis in the gravid human uterus.

Decidua parietalis: The →membrana decidua surrounding the lumen of the gravid human →uterus.

Deciduate placenta: A type of →placenta which penetrates into the →mucosa of the uterus which results in the formation of a →membrane decidua. This placenta is expelled after birth which results in a certain damage of the →mucosa of the uterus.

Delayed implantation: "Freezing in" the development of a mammalian embryo in an early stage before →implantation into the →uterus. Occurs in species which live in particularly constraining conditions with respect to their reproduction and their survival.

Dermatome: →Somites.

Descensus testiculorum: The temporary or permanent transfer of the →testes from the dorsal wall of the body cavity into a →scrotum hanging from the body. More correct term: descensus testis.

Desquamation: →Menstruation.

Deuterogonium: A cell in *Petromyzon* →spermatogenesis corresponding to the →secondary spermatogonium.

Diadromous migration: Migration of a species from freshwater to seawater and then back to freshwater.

Diencephalon: The interbrain (between-brain) which is situated between the forebrain and the midbrain. It consists of three main parts: dorsally the epithalamus, laterally the thalamus and ventrally the hypothalamus. The latter is an important source of →neurohormones including →releasing and inhibiting hormones influencing the →hypophysis.

Dioestrus phase: The phase following the →metoestrus phase within the shortened reproductive cycle in some mammals (e.g. in the rat).

Diplogenotypic sex determination: The combination of chromosomes determining sex is first present in a definitive form only in the diploid phase.

Diploplasma: A composite structure consisting of uterine elements and →trophoblastic syncytium in some mammalian →placentas.

Discoblastula: The differentiated →germinal disc of most sauropsids. Below the original germinal disc a second cell layer develops which is called hypoblast. The germinal disc covering it is now named epiblast. The space between these corresponds to the →blastocoel. The space between the hypoblast and the undivided yolk-mass corresponds to the →subgerminal cavity. In a modified form the epiblast and the hypoblast also occur in mammalian embryonic development.

Discoidal cleavage: Subtype of →partial cleavage in which a →germinal disc is formed.

Disturbed invagination: →Invagination in an inequally cleaved →gastrula.

Ductuli efferentes: Small canals leading sperm from the →testis ampulae or tubuli out of the male →gonad.

Ductuli epididymidis: The →mesonephric →nephrons used for sperm transport and storage in the males of most vertebrate species.

Ductus deferens: The "free" part of the →primary kidney duct or →Wolffian duct between the →epididymis and its joining with the exterior or other body structures (e.g. the →cloaca). In mammals it is associated with nerves, blood vessels and a well-developed musculature, thus forming the funiculus spermaticus.

Ductus efferens: The direct connection between →testis and →Wolffian duct existing in a few anuran species. This term is sometimes also used for the single connection between testis and →epididymis in some male reptiles.

Ductus epididymidis: The portion of the →primary kidney duct (Wolffian duct) leading through the →epididymis.

Ductus vitello-intestinalis: →Yolk sac stalk.

Early pregnancy factor: A non-dialysable substance produced by the embryo within only a few hours after →fertilisation. It inhibits the initial response of the maternal immune system which is directed against the →blastocyst. It has been found in man, cow and sheep.

Eccentric implantation: Type of →implantation. The →blastocyst in enclosed by a diverticulum of the lumen of the →uterus.

Ectoderm: →Germ layers.

Ectoplacenta: An early →placenta formed by the embryo of many mammals (comparable to the →ectoplacentar cone of the muridae).

Ectoplacental cone: →Carrier.

Ectoplacental horseshoe: A horseshoe-shaped thickened region of the →syncytiotrophoblast which forms a well around the →embryonic disc in lagomorphs.

Egg activation: The initiation of the embryonic development in general. It is a complex process generally induced by the mechanical stimulus of sperm penetration through the →oocyte membrane.

Egg membrane: Any structure surrounding →oocytes or →ova.

Embden-Meyerhof pathway: The transformation of glucose to pyruvate in the course of →glycolysis.

Embryoblast: The portion of the mammalian →blastocyst which forms the embryo sensu stricto.

Embryocyst: A hollow →embryonic knob.

Embryonic knob: A compact →embryoblast enclosed by the →trophoblast.

Embryonic shield: A thickened part of the →germinal disc of teleosts and reptiles.

Embryotroph: Nutritory fluid which is secreted by the uterus wall or the ovary and by which the embryo is provided for. It is also reffered as uterine milk. In many cases it also contains leucocytes and tissue debris and is then called histiotroph.

Emigration: A type of →mesoderm formation. It refers the formation of endodermal ridges on either side of the →primitive gut which become hollowed out secondarily to form the →coelom.

Eminentia mediana: The ventral region of the →hypothalamus lying just cranially to the →infundibulum.

Endocrine gland: A special ductless gland producing →glandular hormones.

Endoderm: →Germ layers.

Endodermal sinus: Villi of the →endoderm of the →yolk sac projecting into the →embryotroph of the →uterus in the mouse for example.

Endomesoderm: Embryonic tissue in teleosts from which →mesoderm and →endoderm arise in a somewhat different manner than in other vertebrates.

Endometrium: The innermost epithelial layer of the mammalian →uterus.

Enterocoel formation: A type of →mesoderm formation. Mesodermal sacs derive from bilateral segmental diverticles of the dorsolateral →primitive gut which are pinched off. They enclose the secondary body cavity or →coelom.

Entypy: The →embryoblast is enclosed by the →trophoblast.

Entovarian cavity: The cavity within an →ovary (well expressed in the amphibian ovary). It is not an →oviduct; the →oocytes are ovulated in direction of the coelomic cavity.

Entovarian oviduct: The cavity within the →ovaries of many teleosts formed by lateral folding-up of the ovary and its attachment to the dorsal wall of the body cavity. It serves a an →oviduct within the ovary and the →oocytes are ovulated into it.

Epiblast: →Discoblastula.

Epiboly: A growth of a cell layer from the →animal pole around →yolk-containing cells or an undivided mass of yolk.

Epididymis: Structure of →mesonephric origin which is closely associated with the →testis. In consists of the →ductuli epididymidis and the →ductus epididymidis in most male vertebrates with →urogenital connection. In mammals the →ductuli efferentes of the testis run directly into the →ductus epididymidis. The epididymidis serves for sperm transport and storage.

Epiphysis: An organ extending from the →epithalamus which represents a rudimentary sense organ for light and contains photoreceptors up to the phylogenetic level of reptiles. It is an →endocrine gland producing the hormone melatonin. Correct term: epiphysis cerebri. It is also called pineal organ.

Epiphysis cerebri: →Epiphysis.

Episodic hormone secretion: →Pulsatile hormone secretion.

Epitesticular duct: A special secondary →spermiduct lying on the testis surface in salmonids.

Epithalamus: The "roof" of the →diencephalon.

Epoophoron: →Meseonephric rudiment in female amniotes.

Equal cleavage: Subtype of →total cleavage. Cleavage of →alecithal eggs which result in →blastomeres of equal sizes.

Exocoel: Short form for →extraembryonic coelom. This term is mostly used for mammalian development.

Extraembryonic coelom: The →coelom within the extraembryonic region of vertebrates with →meroblastic development.

Extraembryonic region: A part in an embryo with →meroblastic development which contains the →yolk mass. It consists of the →yolk sac covered with →mesoderm and →ectoderm and the →yolk sac stalk. In amniotes this term describes the "embryonic membranes" including the →yolk sac, the →allantois and the →amnion covered by the →chorion.

Extratesticular duct: The portion of a secondary →spermiduct lying outside the →testis.

Factor X: →Anti-Müllerian duct hormone.

Female pronucleus: The nucleus resulting from the second maturation division of the →secondary oocyte in the course of →meiosis.

Fertilisation: The fusion of the →gametes including the melting together of the maternal and paternal nuclei which results in a unique mixture of the genetic material.

Fertilisin-antifertilisin reaction: An →agglutination reaction investigated in invertebrate species. The water-soluble and more or less species-specific fertilisins are present in the egg membrane and the anti-fertilisins are bound to the →spermatozoan membrane. Each fertilisin molecule is able to bind more than one anti-fertilisin molecule and vice versa so that an network of fixed →spermatozoa surround the →oocyte. The fertilisin-antifertilisin reaction is comparable with an antigen-antibody reaction. It is discussed as a model also for vertebrates.

Fetus: →Foetus.

Fimbria: Finger-like protrusions of the uppermost part of the →infundibulum of the →Müllerian duct of some vertebrates species lining the →ostium abdominale.

First-order endocrine gland: An →endocrine gland which is directly influenced by hypothalamic →releasing or →inhibiting hormones.

First-order oocyte: →Primary oocyte.

First-order spermatocytes: →Primary spermatocytes.

Foetus: A vertebrate embryo between the end of →organogensis and birth or hatching. It shows already characteristics of the systematic order or family of the corresponding species (e.g. in humans an embryo after the 85th day of development). It is also called fetus.

Follicle stimulating hormone: →Gonadotropins.

Follicular atresia: The degeneration of an →ovarian follicle.

Follicular gestation: Partial or total embryonic development within the →ovarian follicle (e.g. some →ovoviviparous teleosts).

Follicular pseudoplacenta: A →placenta-like assocation of villi of the wall of the →ovarian follicle which make contact with certain embryonic structures in teleost species with →follicular gestation.

Folliculogenesis: The formation of the →ovarian follicle in the course of →oogenesis.

Funiculus spermaticus: →Ductus deferens.

Galactophore: The common duct through which the milk is led out of the →mammae.

Galactopoesis: The continuous production of milk in the mammary gland.

Gametes: The functional germ cells. In vertebrates the female gametes are the immobile eggs (ova) and the male gametes are the mobile spermatozoa.

Gametogenesis: Production and maturation of the →gametes in general.

Gartner's duct: →Mesonephric rudiment in female amniotes.

Gastrocoel: The lumen of the →primitive gut.

Gastrulation: The formation of →endoderm.

Genitalia: External and internal structures of the reproductive system.

Genital funnel: A coelomic funnel in salmonids which opens to the exterior with a genital pore. It serves as a →gonoduct. It is also called Lickteig's funnel.

Genital ridges: →Gonadal anlage.

Genotypic sex determination: Sex determination caused by genetic factors; the sex realisers are genes. It is the rule in vertebrates.

Germinal disc: A disc-shaped patch of separated cells lying upon an undivided yolk-rich part of the embryo in the course of →meroblastic development.

Germinal line theory: In multicellular organisms there are two components, the soma and the germinal tissue. The former is represented by proliferating "agamonts" which increase in their number by mitosis. It is the mortal part of the organism. The germinal tissue is potentially immortal. It consists of "gamonts" and the male and female gametes. This germinal line links the successive generations of a species into a continuous lineage.

Germinal vesicle: The enlarged nucleus of the→primary oocyte.

Germ layers: Tissue components of the embryo which more or less specifically form certain groups of organs in an embryo. There are three germ layers, the ectoderm, the endoderm and the mesoderm. In vertebrates the first forms the epithelial component of the skin and its derivatives, the epithelia of the mouth and cloaca and the nervous system including the retina of the eye and the derivatives of the neural crest. The second gives rise to the gut epithelium and the epithelial layers of the large digestive glands and a considerable portion of the respiratory system. The third forms the organs "between" the first and the second, e.g. musculature, skeleton, vasculary and urogenital system and the lining of the body cavity. It must be pointed out that this system is merely a crude orientation help for the understanding of the developmental processes.

Gestation: The period in which embryonic development takes place within the mother's body in →viviparous and →ovoviviparous species. In mammals also called pregnancy.

Gestagens: Ovarian hormones which are involved in the regulation of →gestation but also in other reproductive processes in submammalians. They are steroids, i.e. they possess a common tetracyclic basic chemical structure. They contain 21 carbon atoms. They are also called progestins. In mammals the main biologically active gestagen is progesterone.

Glandula coagulans: Gland associated with and deriving from the →ductus deferens in some mammals (e.g. rat).

Glandula cowperi: Cowper's gland, associated with and deriving from the →canalis urogenitalis in mammals.

Glandula nidamentatia: Glandular structure linked with the →tuba uterina in female elasmobranchs. It produces alimentary secretions and secondary egg membranes.

Glandula praeputialis: Gland opening into the "preputial pouch" between the →glans penis and the →praeputium in mammals.

Glandula prostatica: The prostate gland associated with and deriving from the →canalis urogenitalis in mammals.

Glandula vesicularis: Gland associated with and deriving from the →ductus deferens in mammals.

Glandotropic hormone: A →glandular hormone which infuences the activity of another hormone gland, i.e. the hormone of a →first-order endocrine gland which acts upon a →second-order endocrine gland.

Glandular hormones: Hormones which are produced in special "endocrine" glands without release ducts. As a rule the hormones are produced and stored intracellularly (exception: thyroid gland). They are eventually released directly into the blood vessels within the glandular tissue.

Glans penis: The tip or the "head" of the amniote →penis.

Glumitocin: A →neurohypophyseal peptide (→neurohormone) in Elasmobranchii.

Glycolysis: The anaerobous transformation of glucose to lactate "used" for the production of chemical energy in animal cells.

Gonadal anlage: Epithelial ridges in the wall of the coelom of a vertebrate embryo just lateral to each side of the dorsal mesentery. They develop to the somatic elements of the gonads in the course of ontogeny. Also named "genital ridges".

Gonadal hormones: All hormones produced by the →gonads.

Gonadoliberin: The →releasing hormone for hypophyseal →gonadotropins.

Gonadotropins: Generally: hormones stimulating the maturation and function of the →gonads. In vertebrates there are, according to the species, one ore two gonadotropins which are produced within the →adenohypophysis. They are glycoproteins. If only one gonadotropin is present (e.g. in most teleosts) there exists no special name. If two gonadotropins are produced (e.g. in birds and mammals) they are called follicle-stimulating hormone (FSH) and luteinising hormone (LH). They are also named gonadotropic hormones.

Gonadotropic hormones: →Gonadotropins.

Gonads: The organs in which the male and female gametes are produced. In vertebrates as a rule the male and female gonads are separated organs: the ovaries (→ovarium) are the female and the testes (→testis) are the male gonads.

Gonium: Cell deriving from the →promordial germ cells by →mitotic cell divison in general.

Gonocyte: A cell deriving from the →primary germ cell which becomes a →gonium at a certain stage of development.

Gonoduct: A duct leading gametes from the gonads out of the body.

Graafian follicle: The typical liquor-filled ripe follicle containing the →oocyte in mammals.

Granulosa: Exact term: Membrana granulosa. The "inner" wall of the vertebrate →ovarian follicle consisting of one or more cell layers. It produces a part of the primary →egg membranes and has endocrine functions.

Grey crescent: A pigmented area of the ranid →oocyte which is formed by movements of peripheral cytoplasmatic layers shortly after penetration of the oocyte membrane by a →spermatozoon.

Hemipenes: The →copulatory organs of squamate reptiles deriving from the dorsocaudal wall of the →cloaca.

Hensens's node: The thickened border of the head end of the →primitive groove which corresponds to the dorsal →blastopore lip.

Hermaphroditism: The realisation of both sexes in one individual. Monoecious species possess →testes as well as →ovaries in their body. It occurs in some teleosts and in early stages of life in a few amphibians. Simultaneous hermaphrotides have functional ovaries and testes at the same time. In consecutive hermaphrodites the male and female gonad are mature one after another.

Hippomanes: Invaginations of the →allanto-chorion filled with →histiotroph wich are pinched off into the lumen of the →allantois in horse embryo.

Histiotroph: An →embryotroph containing cell material.

Holoblastic development: The embryonic development of vertebrates with →total cleavage.

Hologamy: Primitive type of sexuality. Haploid "normal" individuals (NI) under certain circumstances become →gametes (morphologically not different from the NI). Each two of the gametes fuse and their nuclei melt together thus forming a →zygote. This eventually goes through meiosis, which results in haploid NI, which then reproduce themselves by →mitosis.

Holonephros: Segmentated primitive vertebrate kidney type which does not occur in any existent species. Similar forms can be observed during myxinoid development.

Hormone receptors: Proteins located within the cell membrane or within the cytoplasm which are able to bind hormones with a high specifity.

Human chorionic gonatotropin: A →gonadotropin produced within the human →placenta during →gestation.

Human choriosomatomammotropin: →Human placental lactogen.

Human placental lactogen: A →prolactin-like hormone produced by the human →placenta during →gestation. It is also called human choriosomatomammotropin

Hymen: Flap-like constrictions of the virgin →vagina in humans and the horse.

Hypoblast: →Discoblastula.

Hypochordal plate: The connection to the roof of the →primitive gut formed by the →chordamesoderm in amphibian development.

Hypophyseal cleft: A cleft between the pars distalis and the pars intermedia of the →adenohypophysis of many vertebrates which is interpreted as a remnant of the lumen of →Rathke's pouch.

Hypophyseal recess: The part of the third brain verticle within the →hypothalamus reaching into the →neurohypophysis of many vertebrates. Correct anatomical term: recessus hypophyseus.

Hypophysis: A neuroepithelial organ consisting of an epithelial part which represents and important source of →glandular hormones and a →neurohaemal organ. The first is called the adenohypophysis and the latter the neurohypophysis. The hypophysis is also called pituitary gland.

Hypopyle: Caudal opening of the →pterygopodium.

Hypothalamus: The ventral region of the →diencephalon containing important neurohormonal centers.

Implantation: The more or less intimate "fastening" of the →blastocyst at or into the →uterus wall prior to →placentation. There are different types of implantation according to the intensity of the contact with the →mucosa of the uterus and the locus of implantation with respect to the →mesometrium.

Implantation area: An area of the →mucosa of the uterus which is in the morphological and physiological condition to make contact with a →blastocyst resulting in →implantation.

Induced ovulation: →Ovulation is induced by an external stimulus via an →neuroendocrine reflex arc (e.g. rabbit).

Inequal cleavage: Subtype of →total cleavage. Cleavage of →yolk-rich eggs which results in →micromeres and →yolk-containig →macromeres.

Inhibin: A water-soluble substance in the testes of some mammals which specifically inhibits the secretion of →follicle-stimulating hormone.

Inhibiting hormone: A hypothalamic →neurohormone inhibiting the release of a →glandular hormone from the →adenohypophysis.

Infundibulum: Term with twofold meaning. (A) The most ventral part of the →hypothalamus forming the →neurophypophsis in most fishes or the stalk of the →hypophysis in tetrapods. (B) The funnel-like most cranial part of the →Müllerian duct.

Insemination: The transfer of spermatozoa into the female reproductive tract.

Interstitial cell stimulating hormone: →Luteinizing hormone.

Interstitial implantation: Type of →implantation. The →blastocyst invades the →mucosa of the uterus and remains enclosed outside the uterine cavity.

Interstitial tissue. The tissue between the →semiferous tubuli or ampullae of the vertebrate testis containing the →Leydig cells, connective tissue fibers, blood vessels etc.

Intervillar space: The space between the villi within the →placenta oliformis.

Invagination: A type of →gastrulation in the course of which a →primitive gut is formed by invagination of the →blastoderm of the →vegetative pole resulting in a cup-shaped bilayered embryo.

Invasive placenta: A type of →placenta penetrating into the tissue of the →uterus to a greater or lesser extent.

Invasive trophoblast: A →trophoblast which penetrates the →endometrium and invades into the →mucosa of the uterus.

Isolecithal egg: Egg with equal distributed yolk.

Isotocin: A →neurohypophyseal peptide (→neurohormone) in Actinopterygii. It is also called ichthyotocin.

Isthmus: A tapering of the →tube uterina just before its joining with the →uterus in vertebrates with complex →Müllerian ducts.

Karyogamy: The essential part of →fertilisation: the melting together of maternal and paternal gamete nuclei.

Labia majora: Structures of the external female mammalian genitalia which are homologous to the male →scrotum lying laterally to the →labia vulvae.

Labia minora: →Labia vulvae.

Labia vulvae: Together with the →preaputium of the →clitoris the lining of the →vestibulum vaginae.

Labyrinth placenta: An anatomical type of the →placenta haemo-chorialis and →placenta endothelio-chorialis. A part of the →trophoblast contains numerous lacunae which are irrigated by maternal blood.

Lactation: The whole penomenon of milk production in mammals.

Lactogenesis: The initial production of milk in the mammary gland.

Lampbrush chromosomes: Chromosomes with conspicuous side-loops in the →primary oocyte nuclei of some submammalians during the diplotene stage of →meiosis.

Larviparous animals: →Ovoviviparous or →viviparous animals whose young are born in a larval stage of development.

Lateral plates: The two undivided lateroventral →mesoderm plates enclosing the →coelom. They are connected with the →somites by the →somite stalks. They each divide into a visceral layer (lying next to the gut), the splanchnopleure and a somatic layer (facing the →ectoderm), the somatopleure. In later development the lateral plates detach from the somites.

Lateral kidney canal: A canal linked between the →ductuli efferentes and the →nephrons of the mesonephros in the males of many vertebrate species.

Leydig cells: Somatic cells situated within the →interstitial tissue of the testis which produce male sex hormone.

Leydig's gland: A →mesonephric structure in some elasmobranchs which is closely associated with the →epididymis. The secretion of its modified →nephrons possibly activates the spermatozoa.

Lickteig's funnel: →Genital funnel.

Lipovitellin: →Yolk platelets.

Liquor folliculi: The fluid filling the cavity within mammalian →ovarian follicles in certain stages of development.

Luteal phase: The time in a mammalian female reproductive cycle in which a →corpus luteum is formed and actively present.

Luteinising hormone: →Gonadotropins.

Lysine vasopressin: A →neurohypophyseal peptide (→neurohormone). The →antidiuretic hormone of Suiformes.

Macromeres: The yolk-containing →blastomeres of the →vegetal pole in →inequal cleavage.

Male pronucleus: The enlarged →spermatozoan nucleus within the cytoplasm of the →ovum just before →syngamy.

Macrolecithal egg: Egg with a large amount of →yolk. It is also called polylecithal egg.

Mammae: The mammary glands after which the mammals are named.

Marginal haematome: A marginal region of the carnivore →placenta in which villi of the →chorion are bathed directly within maternal extravasated blood which in this case has the function of a histiotroph.

Marsurpium: A brood pouch formed by any structure outside the female genital tract.

Maturation agent: A substance, most likely →progesterone, which stimulates the later maturation of amphibian oocytes.

Meiosis: Cell division with reduction of the two chromosome sets to a single one. It consists of two division steps. the first maturation division which mostly is a reduction division (each daughter cell contains the single chromosome set) and the second maturation division (segregation division) which follows the same course as →mitosis.

Meiotic cell division: →Meiosis.

Melatonin: A hormone produced within the →epiphysis.

Membrana decidua: A special type of tissue which is formed within the →mucosa of the uterus around an embryo or a →placental region. The term means decaying membrane.

Menstruation: The detachment of the uterus epithelium (desquamation) at the end of a →menstrual cycle which involves tearing of special arteries the non-coagulating blood of which carries out the tissue debris. Menstruation occurs solely in primates. It is also called "menses".

Menstrual cycle: A female mammalian reproductive cycle which involves a →menstruation bleeding.

Menstrual polyp: A single, cone-shaphed →implantation area in the uterus of *Elephantulus.*

Merogamy: Type of sexuality. Diploid "normal" individuals (NI, "agamonts") eventually become "gamonts" which undergo meiosis resulting in each four haploid gametes. Each two of these melt together forming a →zygote, which then undergoes mitotic cell divisions, which produce new NI. If the gametes are morphologically equal the merogamy is called isogamous. If two morphologically different gamete types are formed, merogamy is anisogamous.

Meroblastic development: The embryonic development of vertebrates with →discoidal cleavage.

Mesoderm: →Germ layers.

Mesolecithal egg: Egg with a moderate amount of →yolk.

Mesometrial implantation: Type of →implantation. The →blastocyst contacts the →mucosa of the uterus at the side where the →mesometrium of the →uterus bicornis inserts.

Mesometrium: The peritoneal duplicature attching the mammalian →uterus to the wall of the body cavity.

Mesorchium: The duplicature of the peritoneal epithelium connecting the →testes with the dorsal wall of the coelomic cavity.

Mesotocin: A →neurohypophyseal peptide (→neurohormone) in amphibians.

Mesovarium: The duplicature of the peritoneal epithelium attaching the →ovaries to the dorsal wall of the coelomic cavtiy.

Mesonephros: The second "kidney generation" in vertebrates. It is situated behind the →pronephros and is the functional kidney of most fish and amphibians. In many vertebrates in the male sex it plays partially or totally an important role in sperm transport. It is also called primitive kidney.

Metanephros: The third "kidney generation" in vertebrates. It is the functional excretory organ of the Amniota.

Metoestrus phase: The phase just after the →oestrus within the shortened reproductive cycle in some mammals (e.g. in the rat).

Microlecithal egg: Egg containing very little yolk.

Micromeres: The yolk-free or yolk-poor →blastomeres of the →animal pole in →inequal cleavage.

Micropyle: The preformed orifice for the entry of →spermatozoa in the teleost egg membrane, the →chorion. This term is also used in a somewhat different meaning in plant reproduction.

Milk fields: Two naked areas of skin underlain by milk glands in monotremes. They are located ventrally on either side of the mid line of the abdominal region. They are also called "areolae".

Milk sinus: An expanded →galactophore which serves for the storage of milk within the →mammae.

Mitosis: Cell division without reduction of the chromosome sets (in vertebrates in general two). The "normal" cell division of somatic cells.

Mitotic cell division: →Mitosis.

Mixopterygium: →Pterygopodium.

Monocyclic species: Species which have only one reproductive cycle per year.

Monogamy: Basic mating type. One individual mates with only one partner of the opposite sex.

Monospermy: One →spermatozoon penetrates and fertilises the →oocyte (e.g. teleosts and mammals).

Morgagnic hydatids: →Mesonephric rudiment in female amniotes.

Morula: The mulberry-like compact ball of →blastomeres which is the result of total cleavage.

Mossmann's rule: It states that the blood of the →foetus flows in the opposite direction to the maternal blood in the →labyrinth placenta thus establishing a countercurrent principle. It is also applicable to the ruminant →placenta.

Motivation: An underlying drive which results in specific types of behaviour. In this book the term is used in the sense of a specific physiological "readiness" of certain central nervous centers responsible for the generation of specific behaviour patterns in either the →sexual or the →parental phase of the reproductive cycle, for example.

Mouth breeding: A male or a female takes the fertilised or unfertilised eggs into its mouth, where embryonic development partially or totally takes place. It occurs in some teleosts and in only one amphibian (frog) species.

Mucosa of the uterus: The tissue lining the cavity of the uterus. In mammals it is defined to be formed by the →endometrium and the underlying glandular and connective tissue down to the →myometrium.

Mucous membrane of the uterus: →Mucosa of the uterus.

Müllerian ducts: The female →gonoducts in most vertebrates. They occur basically paired, one may be reduced (most Aves). They are duplicates of the →primary kidney ducts (Elasmobranchii, Amphibia) or are "new" structures deriving from peritoneal folds lying lateral to the →gonadal anlage (all other vertebrates). In the different vertebrate groups they are of varying complexity.

Myometrium: The smooth muscular layer of varying thickness of the mammalian →uterus.

Myotome: →Somites.

Negative feedback loop: A cybernetic definition which describes the fact that a measured value (e.g. the hormone level in blood) is fed back to a regulatory instance (e.g. the →hypothalamus), which compares this value with a preset one and then induces a reaction in the opposite direction according to the deviation between the preset and the measured value (i.e. if the measured value is lower it is increased and vice versa).

Neoteny: Sexual maturity and reproduction in the larval phase (e.g. axolotl).

Nephron: The basic unit of the vertebrate excretory system. It consists of a kidney duct and a ciliated funnel which opens into the coelom or a closed "Bowman's capsule". The latter structure is always intimately associated with a ball of tangled blood vessels (glomerulus).

Nephrostome: The opening of the funnel in primitive "open" →nephrons.

Nephrotomes: The embryonic origin of the vertebrate →nephrons. Deriving from the →somite stalks, they are situated between the →somites and the →lateral plates.

Neuroectoderm: The dorsal portion of →ectoderm which forms the neural tube.

Neuroendocrine reflex arc: An interaction between the nervous system and the hormone system. A special stimulus is taken up by a receptive structure and then is led immediately and with high velocity to the →hypothalamus via nerve fibers. There a →neurosecretory neuron is affected to release a →neurohormone which directly or indirectly induces a specific reaction of a target organ (e.g. the →ovulation in the rabbit which is induced by the copulation stimulus in the →vagina).

Neurohaemal organ: The connection of the modified axon endings of →neurosecretory neurons with blood vessels (e.g. in the neurohypophysis).

Neurohormones: Hormones which are produced in special nerve cells (neurosecretory neurons), transported along the axon and stored within a modified axon ending. The latter contacts a blood capillary and the hormone is eventually released into it.

Neurohypophyseal peptides: The →neurohormones stored within the →neurohypophysis.

Neurophypophysis: The neural part of the →hypopysis representing a →neurohaemal organ. As a rule in fishes it is formed by the →infundibulum, in tetrapods this region only represents the hypophyseal stalk, whereas the main part is called the pars nervosa.

Neuropore: The opening of the lumen of the neural tube into the exterior.

Neurosecretory neuron: →Neurohormones.

Neurulation: The formation of a neural tube in the course of acraniate and vertebrate development.

Notochord: →Corda dorsalis.

Obplacenta: The region of contact between the →symplasma of the eptithelium of the →uterus and the →blastocyst during early →placentation in Lagomorpha.

Oestradiol-17β: The main vertebrate →oestrogen.

Oestrogens: The vertebrate female sex hormones. They are steroids, i.e. they possess a common typical tetracyclic basic chemical structure. They contain 18 carbon atoms. The main oestrogen is oestradiol-17β.

Oestrone: An →oestrogen.

Oestrus: The time of "heat" in the reproductive cycle of many female mammals, i.e. the time where the female is sexually active and ready for copulation. The oestrus time as a rule occurs shortly prior to or at the time of →ovulation.

Oestrus cycle: A female mammalian reproductive cycle containing an →oestrus.

Oestrus phase: The time in a female mammalian reproductive cycle in which the →oestrus and the →ovulation take place.

Omphaloplacenta: The primitive type of the typical vertebrate →placenta. The region of the →chorion overlaying the →yolk sac makes contact with the uterus wall. It is also called yolk sac placenta or preplacenta.

Oogamy: Special highly evolved modification of anisogamous →merogamy. One →gamete type is immobile (female gamete, egg) and the other is mobile, mostly bearing locomotion organelles (male gametes, spermatozoa). Oogamy is the sexuality type of all higher organisms, including the vertebrates.

Oocytes: The cells entering and running through →meiosis in the course of egg formation.

Oogenesis: The formation of →oocytes from →oogonia. This term is also used for the whole process of the genesis of eggs from →primordial germ cells.

Oogonial proliferation: The formation of →oogonia from →primordial germ cells by multiple →mitotic cell division. There are two types of ooginial proliferation in vertebrates. In the first it is restricted to the embryonic or larval phase (e.g. Petromyzontia and most mammals) and the second encloses the entire reproductively active period of the life (e.g. most teleosts and Reptilia).

Oogonium: The product of multiple →mitotic divisions of the →primordial germ cells in the course of →oogonial prolifereation. Female →gonium.

Oophagy: A subtype of →viviparity. The embryo is nourished with the substance of other eggs.

Opisthonephros: A term of comparative anatomy referring together →mesonephros and →metanephros.

Optic chiasma: The crossing optic nerves just before their entrance into the →diencephalon. Exact anatomical term: chiasma opticum.

Organogenesis: The development of the body organs within embryonic development excluding their histological differentiation.

Orthomesometrial implantation: Type of →implantation. The →blastocyst inserts laterally (as a rule alternating left and right) to the locus of insertion of the →mesometrium in the →uterus bicornis.

Os clitoridis: A bony structure within the →clitoris, homologous to the →os penis.

Os penis: The "penis bone". An bony structure within the →penis of many mammalian species (e.g. Carnivora and primates other than man).

Ostium abdominale: The opening of the →Müllerian duct into which the →oocytes are ovulated in most female vertebrates.

Ovarian follicle: The fibrous and cellular "envelope" of the →oocytes in the vertebrate ovary. →Granulosa and →theca.

Ovarian gestation: The embryo partially or totally develops within a cavity of the →ovary outside the →ovarian follicles.

Ovarian stroma: The tissue between the →ovarian follicles consisting of connective tissue, steroidogenic cells, blood and lymph vessels etc.

Ovarium: The female gonad (ovary). Vertebrates (exceptions: a few fish species and most birds) as a basic pattern have two ovaries. They are covered with coelomic epithelium and consist of the somatic →ovarian stroma and the spherical somatic follicles containing the →oocytes. There are different types of ovaries within vertebrates concerning the relation of the mass of the single components to each other.

Ovary: →Ovarium.

Oviduct: Any structure leading eggs or →oocytes out of the body.

Oviparity: Fertilized eggs leave the female's body into the environment. It occurs in egg-laying species with internal fertilisation (egg-laying Elasmobranchii, most Urodela, many reptiles and all birds).

Ovoviviparity: The embryo develops partially or totally within the egg membranes in the mother's genital system← (→ovarian follicle, →ovarian cavity, →oviduct) without any supply mechanisms living only from →vitellus (many elasmobranchs, some teleosts, some Urodela and some Reptilia). This definition of the term is used in this book; others are present in literature.

Ovulation: The process of expelling the mature →oocytes out of the rupturing ovarian follicles into the body cavity and the →oviduct.

Ovuliparity: Unfertilized eggs leave the female's body and are fertilized in the environment (Petromyzontia, Teleostei, most Anura, e.g.).

Oxytocin: A →neurohormone stored in the →neurohypophysis of mammals and Holocephala.

Paradidymis: →Mesonephric rudiment in male amniotes.

Parental phase: The time in a reproductive cycle in which the behavioural activities are mainly directed towards the care of the young.

Paroophoron: →Mesonephric rudiment in female amniotes.

Paraphysis: An organ with unknown function just cranially to the →epithalamus which is thought to be a derivative of the dorsal forebrain.

Parietal organ: An intracranial unpaired eye which extends from the →epithalamus and which is situated between the →paraphysis and the →epiphysis. It is present in only a few reptile species. It is also called parietal eye.

Pars distalis: The cranioventral part of the →adenohypophysis.

Pars epigonalis: The caudal portion of the vertebrate →gonadal anlage. In Urodela and Gymnophonia it is, together with the →pars progonalis, the origin of the gonadal fat body. In some elasmobranchs it forms the "epigonal organ", the function of which is unknown. It remains undeveloped in all other vertebrates.

Pars gonalis: The median part of the vertebrate →gonadal anlage which gives rise to the gonads during the course of embryonic development.

Pars infundibularis: The part of the →adenohypophysis of tetrapods lying round the →infundibulum.

Pars intermedia: The part of the →adenohypophysis next to the →neurohypophysis.

Pars progonalis: The cranial portion of the vertebrate →gonadal anlage. It forms a gonadal fat body in anuranas and the same together with the →pars epigonalis in urodela and gymnophiona during the course of gonadal development. It remains undeveloped in all other vertebrates.

Pars ventralis: A part of the →adenohypohysis of elasmobranchs which is situated ventrally to the →pars distalis.

Parthenogenesis: Reproduction from unfertilized eggs, also called "virgin birth". Occurs rarely in vertebrates and primarily in hybrids (a few teleosts, Amphibia and reptiles).

Partial cleavage: Only the →blastomeres which contain little or no →yolk are separated totally in the course of →cleavage. In the yolk-rich part of the →zygote and the embryo there are only divisions of the nuclei and not of the cytoplasm (e.g. elasmobranchs, teleosts, Sauropsida).

Pathologigal polyspermy: Two or three →spermatozoa penetrate the →oocyte membrane in a species with →monospermy. The developing embryos as a rule do not survive by reason of tri- or tetraploidy.

Penis: The erectile →copulatory organ of crocodiles, chelonians, some birds and all mammals deriving from the wall of the caudal section of the →cloaca and from a sex protuberance which results from the paritioning of the →cloaca and the →perineum in mammals.

Penis appositus: A →penis which is partially fastened at the ventral skin so that it is directed cranially (e.g. dog).

Penis pendulus: A →penis which hangs totally free from the body (e.g. man).

Perforatorium: →Subacrosomal material.

Periblast: Syncytial boundary layer of the →yolk mass in vertebrates with →meroblastic development lying just under the germinal disc and which plays an important role in the provision of the embryo with nutrients from the yolk.

Perineum: A structure separating the openings of the gastrointestinal and urogenital systems im mammals other than Monotremata.

Perivitelline space: The space between the →egg membrane and the →oocyte membrane.

Phenotypic sex determination: Factors other than genetical ones determine the sex of an organism.

Pheromone: A substance whose functions in co-ordinating events extend beyond the individual. It is produced by one individual and released into the environment. Then it is perceived by an individual of the same species in which certain responses are induced. The signal pheromones arise a behavioural reaction in the perceiving individual within a short time span. The primer pheromones activate physiological mechanisms, especially hormonal systems, resulting in long-term changes.

Perimetrium: The outer peritoneal layer of the mammalian →uterus.

Phosvitin: →Yolk platelets.

Phoretic spawning: The release of →gametes of both sexes into animals of another species, where the development of the embryos takes place partially or totally.

Physiological polyspermy: Several →spermatozoa penetrate the →oocyte membrane but only the nucleus of one of them fuses with oocyte nucleus the others degenerating in the cytolplasm. This type as a rule is found in vertebrate species with yolk-rich eggs (e.g. Elasmobranchii, Aves).

Pineal antigonadotropin: A substance which is present in the →epiphysis of some mammals e.g. cattle) which induces a decrease of the level of →luteinising hormone in blood.

Pineal organ: →Epiphysis.

Pituitary gland: →Hypophysis.

Placenta: In general: a physiological interdependence between the embryo and the maternal organism without the definition of the structures involved. In elasmobranchs, reptiles and mammals it consists of varying portions of maternal (→uterus) and embryonic (→chorionic) tissue.

Placenta bidiscoidalis: A →restricted placenta consisting of two disc-shaped →placental areas.

Placenta cotyledonaria: →Placenta multiplex.

Placenta diffusa: Anatomical type of →placenta. The placenta covers widely or totally the surface of the embryo.

Placenta discoidalis: A →restricted placenta with the shape of a disc.

Placenta endothelio-chorialis: A histological type of →placenta. The chorion contacts directly the endothelium of capillaries of the →uterus.

Placenta endothelio-endothelialis: A special histological type of →placenta occurring in some marsurpials (bandicoots). Foetal blood vessels directly contact such of the mother.

Placenta epithelio-chorialis: A histological type of →placenta. The →chorion contacts the →endometrium of the →uterus.

Placenta haemo-chorialis: A histological type of →placenta. The →chorion is partially bathed directly within the maternal blood.

Placenta haemo-endothelialis: A histological type of →placenta. Foetal blood vessels are bathed directly within maternal blood after the →chorion and its underlying →mesoderm is broken down.

Placental area: The area of an embryo which contacts the maternal tissue with the function of a →placenta.

Placenta multiplex: Anatomical type of →placenta. Several →placental areas, the cotyledons, are spotted over the surface of the embryo. It is also called placenta cotyledonaria.

Placenta olliformis: Anatomical type of the →placenta haemo-chlorialis. A cavity within the bowl-shaped placental area of the →trophoblast is filled with maternal blood into which chorionic villi project. It is also called villous placenta.

Placenta syndesmo-chorialis: A histological type of→placenta. The →endometrium of the →uterus breaks down and the →chorion comes into direct contact with the connective tissue of the →mucosa of the uterus.

Placentation: The formation of a placenta. The formation of specialised structures whereby a close association or fusion of the embryonic and maternal tissues is established which facilitates the physiological exchange of substances between the maternal blood and the embryo.

Placenta vera: A type of →placenta in which the maternal blood bessels are ruptured during birth.

Placenta zonaria: Anatomical type of →placenta. The →restricted placenta sensu stricto. There are different forms of the placenta in this type.

Placentome: The unit of a →cotyledon and a →caruncle.

Polar body: Rudimentary →secondary oocyte or →ova in vertebrates.

Polyandry: Subtype of →polygamy. A female mates with several males.

Polycyclic species: Species which have more than one reproductive cycle per year.

Polygamy: Basic mating type. One individual mates with more than one partner of the opposite sex.

Polygyny: Subtype of →polygamy. A male mates with several females. This term is sometimes also used for the typical →karyogamy in vertebrates in which no membrane is formed around the →zygote nucleus.

Polylecithal egg: →Macrolecithal egg.

Portal system: A venous capillary system.

Postluteal phase: The time in a mammalian female reproductive cycle in which a formerly active →corpus luteum degenerates.

Postnuptial spermatogenesis: →Spermatogenesis which takes place shortly after the mating phase (e.g. in the majority of poicilothermic vertebrates).

Praeputium: A skin duplicature covering the →glans penis in mammals.

Precocial species: A species whose young leave the nest or the guardianship of the parents soon after birth or hatching.

Pregnancy: →Gestation.

Prenuptial spermatogenesis: →Spermatogensis which takes place shorty before the reproductive period. It is the typical spermatogenesis in homoiothermic vertebrates.

Preplacenta: →Omphaloplacenta.

Prespermatid: →Secondary spermatocyte.

Prespermatogensis: The formation of →gonocytes from primordial germ cells in the testis.

Priapum: →Copulatory organ in some Teleostei deriving from elements of the ventral fins, ribs and the pectoral girdle.

Primary body cavity: →Blastocoel.

Primary egg membrane: Any →egg membrane produced within the ovary.

Primary endoderm: →Chorda canal.

Primary kidney duct: The duct into with the →pronephric and →mesonephric nephrons run. In many male vertebrates it serves as an →urospermiduct or →spermiduct. It is also called Wolffian duct and represents the primary urinary duct.

Primary oocyte: The cell which arises by growth from the last cell of the →oogonial proliferation. The formation of the primary oocyte is the →oogenesis s. str. It is also called first order oocyte.

Primary sex cords. Cords of mesodermal tissue of discussed origin growing within the mesenchyme of the "medulla" of the vertebrate →gonadal anlage. Primordial germ cells migrate into these structures, which give rise to the ampullae or tubuli of the →testis including the →Sertoli-cells.

Primary spermatocytes: The cells which arise from the last →mitotic cell division in the course of →spermatocytogensis. The last step of their development is a growth phase and they therefore also called auxocytes. Synonym: First-order spermatocytes.

Primary spermatogonium: A →gonocyte in a certain developmental stage ready to enter →spermatocytogenesis.

Primary urinary duct: →Primary kidney duct.

Primer pheromone: →Pheromone.

Primitive groove: The groove in the middle of the →primitive streak in the avian →germinal disc or the mammalian →embryoblast. It corresponds to a →blastopore.

Primitive knot: The invaginated →embryonic shield in the repitilian →germinal disc.

Primitive gut: The →endoderm formed in early embryonic development. It is also called archenteron.

Primitive streak: A thickened elongated zone in the middle of the aviam →germinal disc and the mammalian →embryoblast which follows the head-tail line of the embryo.

Primordial germ cells: Special cells of →endodermal or ←mesodermal origin which migrate into the epithelium of the→gonadal anlage of an vertebrate embryo during the course of development. They are the stem cells of the →gametes. They are also called primordial sex cells.

Primordial sex cell: →Primordial germ cell.

Proamnion: An area in front of the head of an amniote embryo which develops into an amniotic fold. →Amnion.

Progesterone: The most important →gestagen in mammals.

Progestins: →Gestagens.

Prolactin: A hormone of the →adenohypophysis in all vertebrates except the Cyclostomata. In mammals and in some other vertebrates it has important functions in parental care. It is also called luteotropic hormone.

Prolactoliberin: A →releasing hormone for hypophyseal →prolactin which is possibly present in some vertebrate species.

Prolactostatin: The →inhibiting hormone for hypophyseal →prolactin.

Pronephros: The first "kidney generation" in vertebrate embryonic development. It persists as a functional excretory organ only in Myxinoidea and a few bony fish.

Prooestrus phase: The phase just prior to the →oestrus within the shortened reproductive cycle of some mammals (e.g. in the rat). In general the time preceding the →oestrus in a mammalian →oestrous cycle.

Proliferative phase: The time in a →menstrual cycle in which the uterine epithelium thickens and becomes vascularised prior to →ovulation.

Prostaglandins: A group of polyunsaturated fatty acids connected with a cyclopentane ring which have several regulatory or modulating functions in vertebrate reproduction. They are not hormones s.str. but belong to a not well defined group of regulatory active substances called parahormones.

Protandry: Type of →consecutive hermaphroditism in which the male phase comes before the female phase.

Protogonium: A cell in *Petromyzon* →spermatogensis corresponding to the →primary spermatogomium.

Protogyny: Type of →consecutive hermaphroditism, also named "proterogyny". The female phase comes prior to the male phase.

Pterygopodium: The →copulatory organ of Elasmobranchii and Holocephala deriving from the ventral fins.

Puberty: The time span from the beginning of the maturation of the gonads to the animal becoming fully capable of reproducing.

Pulsatile hormone secretion: A hormone is secreted in one or several pulses within a certain time period. It is also called episodic hormone secretion.

Rathke's pouch: The diverticulum of the embryonic fore-gut which forms the →adenophypophysis in most vertebrates.

Receptaculum seminis: Any structure used for sperm storage. In vertebrates mostly structures of the →oviduct (pockets, pits, folds or extended gland lumina).

Reichert's membrane: A thin homogenous acellular layer which is formed at the 7th day of →gestation in the mouse between the distal →ectoderm of the trophoblast and the →yolk sac and subsequently expands beneath the trophoblast. After the breakdown of the yolk sac tissue in the process of →yolk sac inversion, it makes close contact with the tissues of the →uterus.

Relaxin: A polypeptide hormone presumably produced in mammalian ovaries during →gestation.

Releasing hormone: A hypothalamic →neurohormone inducing the release of a →glandular hormone from the →adenohypophysis.

Residual body: The portion of cytoplasm the →spermatids lose in the process of →Spermiohistogenesis.

Restricted placenta: Anatomical type of →placenta: The placenta covers only a restricted area of the chorion.

Rete testis: A network of small canals within the →testis collecting sperm from the →semiferous ampullae or tubuli.

Rhipidion: An appendix of the →pterygopodium with unknown function.

Sclerotome: →Somites.

Scrotum: The cutaneous sac containing the →testes in mammals with →descensus testiculorum.

Secondary body cavity: →Coelom.

Secondary egg membrane: Any →egg membrane produced outside the ovary.

Secondary endoderm: A special →endoderm of the Sauropsida deriving from the →hypoblast. It possibly is an adaptation for quick →yolk mobilisation.

Secondary sex cords: Cords of tissue which derive from the →genital ridge epithelium ("cortex") of a vertebrate embryo containing primordial germ cells. They grow into mesenchyme towards the center of the →gonadal anlage and form the →granulosa cells in the course of the development of an ovary.

Secondary spermatocytes: The daughter cells of a →primary spermatocyte after the first →meiotic cell division. Also called second-order spermatocytes or prespermatids.

Secondary spermatogonium: The proliferation product of a →primary spermatogonium which is produced by →mitosis.

Secondary oocyte: The daughter cell for the →primary oocyte after the first →meiotic division. The latter as a rule is an inequal division in vertebrates which results in a sole functional secondary oocyte and a rudimentary one, the first →polar body. The secondary oocyte is the structure ovulated in most vertebrates.

Secondary urinary ducts: Ducts other than the →primary kidney duct or the →ureter leading urine from the kidney to the exterior (in some eleasmobranchs and Urodela).

Second order endocrine gland: A →endocrine gland which is influenced by the hormone of another (→first-order) endocrine gland to release its hormone.

Second order oocyte: →Secondary oocyte.

Second order spermatocyte: →Secondary spermatocyte.

Secretory phase: The time in a →menstrual cycle in which the uterine epithelium runs through its final differentation involving secretion of the uterine glands after →ovulation.

Sella turcica: A depression of the sphenoid bone containing the →hypophysis in many vertebrates.

Semiferous ampulla: Ampullar structure with the same characteristics as the →semiferous tubule within the →testis of Elasmobranchii and Urodela. The correct anatomical term is ampulla seminifera.

Semiferous tubule: Tubular structure within the →testis of most vertebrates which contains the developing and mature →spermatozoa. The only somatic cells within the semiferous tabule are the →Sertoli cells and the →Sertoli-cyst cells, resp. The correct anatomical term is tubulus seminiferus.

Semiplacenta: A type of →placenta in which the maternal blood vessels remain intact during birth.

Sertoli cells: The only somatic cell type within the amniote →semiferous tubules and ampullae. They have certain functions in the process of →spermatogenesis.

Sertoli-cyst cell: A cell forming alone or together with others the →spermatocysts in anamnia. It is probably homologous to the →Sertoli cell.

Sexual phase: The time in a reproductive cycle in which the behavioural activities are mainly directed towards sexuality culminating in spawning or copulation. It may consist of several subphases.

Sinus urogenitalis: A common unpaired terminal part of the genital and excretory duct system opening into other body structures or the exterior.

Sipho: Fluid-filled cavity within the musculature of the ventral body wall of Elasmobranchii which has functions in the course of →insemination.

Signal pheromone: →Pheromone.

Sinus terminalis: A ring vessel limiting the →mesoderm covering the →yolk sac in many mammalian embryos.

Somites: Segmentated →mesodermal structures lying dorsolaterally at each side of the →chorda dorsalis. In the course of development they each give rise to the →myotome (delivering a segment of trunk musculature), the sclerotome (delivering a part of the axial sceleton) and the dermatome (delivering non-epithelial parts of the skin). The somites are connected with the →lateral plates by the somite stalks which in later development become the excretory system so that the somites are detached from the lateral plates.

Somite stalk: →Somites and →nephrotome.

Spermatheca: Special type of →receptaculum seminis. Lacuna in the tissue of the →ovary in which sperm is stored (in some life-bearing teleosts, e.g.). →Receptaculum seminis.

Spermatids: The daugther cells of the →secondary spermatocyte after second →meiotic cell division. These cells are haploid and theoretically gamtes. To become functional gametes they have to run through the process of →spermiohistogenesis.

Spermatocytogenesis: The process of the formation of →primary spermatocytes from a primary →spermatogonium by →mitotic cell division.

Spermatogenesis: The four-phased process of the formation of functional →spermatozoa from a primary →spermatogonium. This term is also used for the entire process of the production of male →gametes in general.

Spermatogonium: Male →gonium. The primary spermatogonium derives from a →gonocyte at a certain developmental stage and then proliferates into secondary spermatogonia by →mitosis.

Spermatophore: Aggregation of →spermatozoa encapsuled by a "shell" which is transferred into a female's genital tract either by →copulation (e.g. some elasmobranchs) or by active uptake from the environment where it was deposited by the male (e.g. most urodelans).

Spermatozeugmatium: Aggregation of →spermatozoa without a differentiated capsule which is transferred into a female's genital tract.

Spermatozoon: The functional male →gamete of metazoans. There are two types of spermatozoa in vertebrates: the primitive type (most teleosts) and the modified (progressive) type (all other vertebrates.

Spermmiation: The release of mobile →spermatozoa out of a →spermatocyst or from the top of a →Sertoli cell into the lumen of a semiferous tubule or ampulla of the →testis.

Spermiduct: A duct leading only sperm out of the body.

Spermiogenesis: →Spermiohistogenesis.

Spermiohistogenesis: Last phase of →spermatogenesis. The transformation of the →spermatids into flagellated functional →spermatozoa. More correct term: spermioteleosis. Synonymous term often used in literature written in English: spermiogenesis.

Spermioteleosis: →spermiohistogenesis.

Spontanous ovulation: →Ovulation is regulated by the hormone system in the frame of an endogenous program (e.g. humans).

Subacrosomal material: Fibrious rod or threaed-like material between →acrosome and nucleus in the →spermatozoa of Petromyzontia, Elasmobranchii, Amphibia (excluding Ranidae, Reptilia and Aves other than passerine). It plays on important role in the →acrosome reaction.

Subgerminal cavity: The cavity below the →germinal disc in anamnia and below the →hypoblast in the germinal disc of most sauropsids.

Subplacenta: A structure within the →placenta of the guinea pig, e.g., which is responsible for the resorption of embryotroph.

Sulcus intercotyledonarius: The space surrounded by the →ectoplacental horseshoe which provides space for the development of an →amnion.

Superfetation: Several cohorts of embryos develop in the ovary at the same time but in different developmental stages. Despite the embryos present within the ovary, new oocytes ripen more quickly than the embryos develop. They are fertilised by spermatozoa present in the →spermathecae which derive from the same →insemiation as those who fertilised the first embryonic cohort.

Symplasma: A special form of degenerating tissue of the inner epithelium of the →uterus including that of uterine glands which is not a true syncytium.

Syncytiotrophoblast: A peripheral layer of the →trophoblast with greater or lesser thickness which becomes syncytial and which is characteristic for →invasive trophoblasts.

Syndesis: Pairing togehter of the homologus maternal and paternal chromosomes in the course of the reduction division.

Syngamy: The vertebrate type of →karyogamy. When the →male and →female pronuclei fuse the nucleus membrane is dissolved so that the →zygote nucleus stays without membrane. After the first →mitotic cell division of the embryonic development, the daughter nuclei are surrounded again by membranes.

Telolecithal egg: Egg with a →yolk gradient within its cytoplasm.

Territorial phase: The first subphase in the →sexual phase of many vertebrates in which a certain territory is occupied and defended. Territories, on the other hand, may also be established outside the reproductive cycle.

Testicond mammals: Mammals whose →testes lie at the dorsal wall of the body cavity as in all other vertebrates.

Testis: The male →gonad. Vertebrates as a rule have two testes. They are elongated or ovoid organs encapsuled by coelomic epithelium, and consist of somatic interstitial tissue and the compartments of →spergenesis, the elongated semiferous tubuli or spherical semiferous ampullae which are filled with germ cells and →Sertoli cells and Sertoli-cyst cells.

Testosterone: The main vertebrate →androgen.

Tissue hormones: Hormones produced within cells lying in a tissue of a different primary function and are released directly into the blood.

Theca: Exact term: theca folliculi. The "outer" layer of the →ovarian follicle. It consists of collagenous fibers, fibroblasts and endocrine cells. In most mammals it is divided in a fibrous theca externa and a cellular theca interna.

Total cleavage: All →blastomeres are separated totally in the course of →cleavage (e.g. Anura, Cyclostomata).

Trophoblast: The portion of the mammalian →blastocyst which does not form the embryo s.str., but becomes or is the peripheral layer which almost totally or partially contacts or invades the →mucosa of the →uterus for →placentation.

Trophoblastic syncytium: Syncytial parts of the tissue of the →trophoblast. →Syncytiotrophoblast.

Trophonemata: Secretory villi of the →uterus wall which extend through the spiracle or gill slit into the oesophagus of the embryos of →viviparous elasmobranchs where they release their secretion. →Embryotroph.

Trophotaenia: Specialised structures extending from an embryo which facilitate the uptake of nutrients from a →embryotroph.

Tuba uterina: A section of the differentiated →Müllerian duct lying between the →infundibulum and the →uterus.

Tunica albuginea: A thin layer of fibrous connective tissue surrounding the →gonads of most vertebrates just below the peritoneal epithelium.

Umbilical cord: The structure connecting the embryo or →foetus with the →placental region(s) by its blood vessels.

Ureter: The secondary urinary duct of the →metanephros.

Urethra: The outlet duct of the →vesica urinaria in mammals.

Urethral glands: Small glands within the wall of the →canalis urogenitalis in mammals.

Urogenital connection: The connection between the →testis and →mesonephric structures in the males of many vertebrate species which serves as a duct system for the transport of →spermatozoa from the →testis into a collecting duct. It may be direct or indirect. In the latter case a →lateral kidney canal is linked between the →ductuli efferentes and the →nephrons of the →mesonephros.

Urospermiduct: A duct leading urine as well as sperm out of the body.

Uterine milk: →Embryotroph.

Uteroglobin: A protein occurring within the fluid of the lumen of the →uterus in early →gestation in the rabbit. It is a →progesterone-binding protein and inhibits the activity of proteases, thus presumably opposing the histolytic activity of the →trophoblast.

Uterus: Caudal section of the differentiated →Müllerian duct which is well developed in →ovoviviparous and →viviparous species where it serves as a more or less highly specialised compartment for partial or total embryonic development. It is especially well developed in mammals, where several different types occur.

Uterus bicornis: A mammalian →Müllerian duct system in which two →uterus "horns" join to a more or less developed common end section which opens into the →vagina (e.g. in Insectivora).

Uterus bipartitus: A mammalian →Müllerian duct system with two →uteri which have a common opening into the →vagina (e.g. in Carnivora).

Uterus duplex: A mammalian →Müllerian duct system with two →uteri opening separately into the →vagina (e.g. in Rodentia).

Uterus simplex: A mammalian →Müllerian duct system with a single →uterus bearing two →tubae uterinae (e.g. in humans).

Vagina: The most caudal part of the →Müllerian duct connecting the uterus with the →cloaca or the →sinus urogenitalis. It is especially well developed in mammals.

Valitocin: A →neurohypophyseal peptide (→neurohormone) in Elasmobranchii.

Vasopressin: →antidiuretic hormone.

Vegetal pole: The part of an egg or the →blastomeres deriving from it which form the gastro-intestinal system and the gonads etc. In eggs containing →yolk as a rule it is the yolk-rich part. It lies opposite to the →animal pole.

Vesica urinaria: Urinary bladder.

Vestibulum vaginae: Sinus urogenitalis in female mammals. It forms the orifice of the vagina.

Villous placenta: →Placenta olliformis.

Vitelline membrane: →Primary egg membrane in amphibians, reptiles and birds comparable to the →zona pellucida in mammals.

Vitellogenin: A serum protein in vertebrates involved in →vitellogenesis which is synthesized in the liver and transported via bloodstream to the ovary, where it is taken up by the oocytes for →vitellus formation.

Vitellogenesis: The formation and deposition of reserve material (yolk, vitellus) in →primary oocytes.

Vitellus: A morphological term for certain reserve material in →oocytes, also called yolk. →Vitellogenesis.

Viviparity: The embryo develops after reduction of the egg membranes within the genital system of the mother, in vertebrates as a rule in a specialized area of the →oviduct, the →uterus, and is supplied by food substances and respiratory gases, for example. This type of internal parental care is mostly associated with a →placenta. (Some Elasmobranchii, possibly a few Teleostei, Urodela and Reptilia, most mammals). This definition of the term is used in this book; others are present in literature.

Wolffian duct: →Primary kidney duct.

Yolk: →Vitellus.

Yolk platelet: Flattened oval-shaped structure containing the proteins phosphitine lipovitelline in a special crystalline arrangement consisting of one or two "main bodies" and a superficial layer in the →oocytes of many vertebrates.

Yolk sac: A diverticulum of the →primitive gut containing the →yolk mass in vertebrate embryos with →meroblastic development.

Yolk sac inversion: The phenomen that the peripheral layer of →chorion covering the →yolk sac including the endodermal outer wall of the latter break down whereas the inner wall becomes concave by a "sinking down" of the embryo so that the yolk sac entoderm makes contact directly with the →endometrium. The →ductus vitello-intestinalis so opens directly into the lumen of the →uterus (e.g. in Lagomorpha and Rodentia).

Yolk sac placenta: →Omphaloplacenta.

Yolk sac stalk: The stalk connecting an vertebrate embryo with →meroblastic development with the →yolk sac. It consists of an →ectodermal and a →mesodermal layer covering two blood vessels (arteria and vena vitellina) and the yolk duct (ductus vitello-intestinalis).

Zona pellucida: →Primary egg membrane in mammals formed together by the →granulosa cells and the →oocyte itself. It is proposed to use this term for all comparable structures occurring in vertebrates.

Zona radiata: The light microscopice picture of a →primary egg membrane in vertebrates which perform vitellogenesis, which appears radially stripened by penetrating makrovilli.

Zygote: The product of →fertilisation. It represents the milten together →gametes including the processes of →cytogamy, →karyogamy and →syndesis.

Index of Animal and Plant Names

Asterisks refer to figures

Subject Index

Asterisks refer to figures